D1128645

TS
205
C45

Chaplin, Jack W.

Metal manufacturing
technology

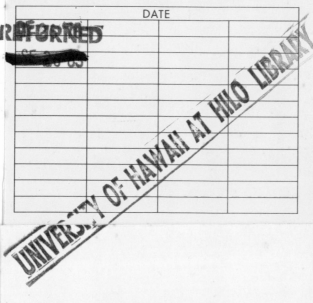

DATE

REFORMED

UNIVERSITY OF HAWAII AT HILO LIBRARY

LIBRARY
HAWAII COMMUNITY COLLEGE
1175 Manono Street
Hilo, Hawaii 96720

© THE BAKER & TAYLOR CO.

Metal
Manufacturing Technology

Metal
Manufacturing Technology

Dr. Jack W. Chaplin
Professor of Industrial Studies
Industrial Studies Department
San Jose State University

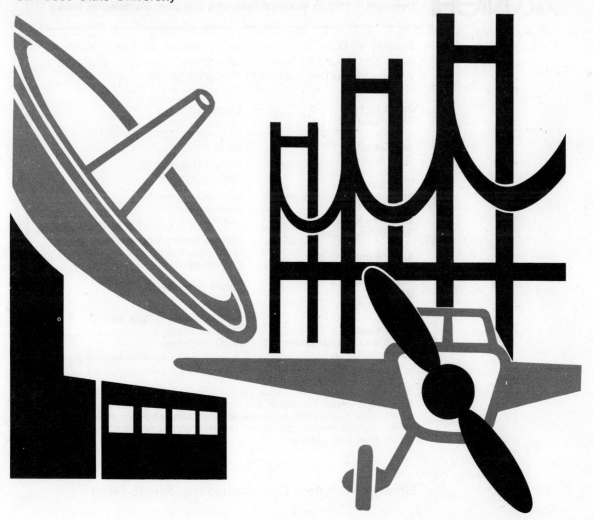

McKNIGHT PUBLISHING COMPANY
BLOOMINGTON, ILLINOIS 61701

FIRST EDITION

Lithographed in U.S.A.

Copyright © 1976 by McKnight Publishing Company, Bloomington, Illinois

Ronald E. Dale, Vice President - Editorial, wishes to acknowledge the skills and talents of the following people and organizations in the preparation of this publication.

Donna M. Faull
Production Editor

Ann Urban
Copy Editor

Elizabeth Purcell
Art Editor/Layout Artist

Willemina Knibbe
Production Assistant

Angie Burns
Judy Burress
Howard Davis
Bruce Nix
Technical Artists

Sue Whitsett
Proofreader and Researcher

Eldon Stromberg
University Graphics, Inc.
Carbondale, Illinois
Cover Design

William McKnight, III
Manufacturing

Gorman's Typesetting
Bradford, Illinois
Compositor

McLean County Graphics
Bloomington, Illinois
Preproduction

Illinois Graphics
Preproduction

R. R. Donnelley & Sons, Inc.
Chicago, Illinois
Printing/Binding

All rights reserved. No part of this book may be reproduced in any form or by any means, electronic or mechanical, including photocopying, recording, or by any information storage or retrieval system, without permission in writing from the publisher.

Library of Congress Card Catalog Number: 75-34620

SBN: 87345-132-5

iv

TS
205
C45

Foreword

Metal Manufacturing Technology introduces a number of new approaches to the study of metals. Emphasis is placed on the total technology of the metals industry, from locating and extracting deposits to automated manufacturing procedures. The book contains five major divisions:

1. **Planning for Manufacturing** — includes the study of how metals are located and extracted, the properties and characteristics of metals, and the planning and measurements which must be used in manufacturing.
2. **Metal Cutting and Forming** — presents the concept of metal separation and the numerous methods of forming and shaping metals.
3. **Combining and Conditioning Processes** — identifies the various methods used in bonding, including bonding by heat and pressure, with chemical heat, and with adhesives.
4. **Making Finished Products** — shows the principles and practices of assembling metal components into finished products.
5. **Automated Manufacturing** — directs the student's attention to the growing technology of automated manufacturing and assembling procedures.

The author has done an effective job of presenting **the study of manufacturing processes from a conceptual framework.** Instead of just speaking of the different methods used to separate metals, the author **presents the concept of metal separation.** The specific methods then flow from the basic understanding of the concept of metal separation. The conceptual approach is carried throughout the book, thereby placing emphasis on understanding and application of the major concepts involved in processing metals. This understanding helps the student identify and apply the correct process to different manufacturing processes.

Another important characteristic of this book is the **emphasis on careers in the metals industry.** Rather than just a chapter slipped in at the beginning or end of the book, career characteristics have been blended into the basic content of almost every chapter. As a result, students will learn about career requirements, entry level skills, and other career guidance information as they study and use the book. Through the use of this book, students will obtain insights into the multitude of occupations and professions available within the metals industry, as well as the knowledge and skills needed to obtain them.

The emphasis on occupations and professions keeps **the book centered on people and their roles in the metals industry.** This is a pleasant departure from the pure content basis of major technical publications. Future publications will see this approach expanded as society begins to demand more humanism and less materialism from both the public and private sectors of our society.

Cost effectiveness and the economic considerations which are a critical aspect of all industries are built into each section of the book. Cost effectiveness enters into every aspect of manufacturing — from design, assignment of tolerances, selection of manufacturing processes, to placing a finish on the product just before marketing. Students will gain an awareness of the important financial considerations which must be studied during each manufacturing step.

The book also emphasizes the study of metals as an industrial material. Special attention is given to the properties of metals and how these properties are changed by alloying, heat treatment, and working. Properties are also studied in relation to the manner metals are cut, formed and joined.

Metal Manufacturing Technology is planned as a basic metals textbook for secondary education. As such, it establishes new standards of quality for metals technology and will enable teachers to present current and appropriate instruction to their students.

Ralph C. Bohn
Dean of Continuing Education
San José State University
San José, California

Preface

Our lives are influenced daily by the use of metal products. Our food, transportation, and homes are heavily dependent upon metal. The metal manufacturing industry provides employment for millions of people who work at all levels of production and sales.

The metal manufacturing industry contributes heavily to the nation's economic well-being. These products provide a large part of the United States' trillion dollar economy of our gross national product.

Metal is studied so that you may acquire an understanding of its unique properties, its tremendous strength, its ability to carry electricity, and the ways it may be formed or shaped into industrial products to meet people's needs.

Metalworking industries make many career opportunities available. As you study about metals, you will become aware of many careers that may interest you. Choosing a career is an important decision for you in the next few years. The study of metal manufacturing will help you to understand how our economy functions, what career opportunities exist, what kinds of skills and abilities are needed, and what the trends in the metal industry are. In each chapter there is a list of careers which relate to the metal manufacturing processes studied in that chapter. The following is a general list of careers in metal manufacturing:

General Administration
Corporate executive
Lawyer
Accountant
Marketing research worker
Personnel worker
Public relations worker

Secretary
Stenographer
Industrial traffic manager
Purchasing agent
Bookkeeper
PBX operator
File clerk
Billing machine operator
Bookkeeping machine operator
Adding and calculating machine operator
Mail preparing and handling machine operator
Payroll clerk
Embossing machine operator
Duplicating machine operator
Typist
Transcribing machine operator
Data typist
Tape perforator operator
Keypunch operator
Industrial designer
Doctor, nurse, paramedic

Engineers
Aeronautical
Aerospace
Agricultural
Astronautical
Biomedical
Ceramic
Chemical
Civil
Electrical
Electronic
Industrial
Mechanical
Metallurgical
Mining
Geological
Design

Production
Craftsman
Machine operator
Assembler
Inspector
Warehouse worker
Shipping and receiving clerk
Stock clerk

Maintenance
Operating engineer
Machinist
Millwright
Bricklayer
Boilermaker
Carpenter
Plumber
Electrician
Painter
Oiler and greaser
Instrument repair worker
Maintenance mechanic
Hydraulic mechanic
Air conditioning and refrigeration mechanic
Janitor
Guard

These materials were analyzed through the Readability Testing Program developed by the Instructional Materials Center at the University of Texas, at Austin. This computerized program analyzes each reading in reference to average sentence length (words), average word length (syllables and letters), percentage of difficult words (not on Dale list), percentage of technical terms, and percentage of sentences with passive verbs. This material was rated as follows:

Readability Measure	Grade Level
Flesch (Reading Ease)	9.3
Dale-Chall	11.2
Farr-Jenkins-Paterson	10.2
Danielson-Bryan	8.6

The following multisyllable words were used in the reading material. These words influenced a higher reading score on the above tests. Proper teacher discussion of these words can help students with low reading ability comprehend the contents of this book.

Disassociated	Subcontractors
Characteristics	Massachusetts
Specifications	Manufacturing
Documentations	Understanding

This program is designed to build vocabulary by introducing new words. The author could have deleted these words to achieve a lower reading level. However, both the author and publisher agree that **increased student knowledge is the true goal of this program.**

Acknowledgments

The author wishes to thank the following for their generous cooperation and assistance for supplying photographs, illustrations, and technical information presented in this textbook.

Actron Industries, Inc.
Adhesives, Coatings & Sealers Div.
    of 3M Co.
Adjustable Clamp Co.
Airco Air Reduction, Inc.
Airco Welding Products Division
Ajax Electric Co.
Ajax Flexible Coupling Co., Inc.
Ajax Mfg. Co.
Allis-Chalmers
Aluminum Company of America
American Chain & Cable Co., Inc.
    Wilson Instrument Division
American Die Casting Institute
American Institute of Steel
    Construction, Inc.
American Iron & Steel Mfg. Co.
American Machine & Foundry Co.
American Telephone & Telegraph
American Tool Co.
American Wilhelmsburger, Inc.
Anaconda Company
A. O. Smith-Inland, Inc.
Armco Steel Corporation
A P Parts Corp., The
Atlas Welding Accessories, Inc.
Atomics International Div.
Babcock & Wilcox Co., The
Barber-Colman Company,
    Industrial Instrument Division
Barnesdrill
Bay Area Rapid Transit
Bay State Abrasive Products Co.
Beardsley & Piper Div.
Bell & Howell Company
Bendix Corporation

Bethlehem Steel Corp.
Black & Decker Mfg. Co., The
Blanchard Machine Co.
Bodine Corp., The
Boeing Co.
Bostitch Fastening Systems
Brescia, Carlota
Bridgeport Machines, Inc.
Brown & Sharpe Mfg. Co.
Bryington Steel Treating
Bucyrus-Erie Co.
Burgess-Norton Mfg. Co.

Carlton Machine Tool Company
Carmet
Carrier Air Conditioning Company
Cashco, Inc.
Caterpillar Tractor Co.
Cecil Equipment Company, Inc.
Cessna Aircraft Co.
CF & I Steel Corporation
Chemcut Corp.
Chemetron Corporation
Chevrolet Motor Div.
Chicago Bridge & Iron Co.
Chicago Wheel & Mfg. Co.
Cincinnati Milacron, Inc.
Commercial Shearing, Inc.
Communications for Industry, Inc.
    3M Company
Cook & Chick Co.
Copperweld Steel Co.
Corning Glass Works
Cross Company
Dana Parson's Photos
Department of Materials Science
    San José State University
DeVilbiss Co., The
DeVlieg Machine Co.
Dietert, Harry W., Co.
Digital Equipment Corp.
Dixon Automatic Tool, Inc.

DoAll Company
Dockson Corporation
Dow Corning Corp.
Eaton Corporation
  Engineered Fasteners Division
Eaton Manufacturing Company
Eaton Yale & Towne, Inc.
Ekstrom, Carlson & Co.
Electric Furnace Co., The
Entrekin Computers, Inc.
  A Cutler-Hammer Company
Ex-cell-o Corporation
F.A.G. Bearings Corporation
Falk Corp., The
Fansteel Metallurgical Corporation
Federal Products Corp.
Ferguson Machine Co.
Ferranti Electric, Inc.
FMC Corporation
Ford Motor Co.
Forging Industry Association
Fosdick Machine Tool Co.
Foxboro Company, The
Frigidaire
Frye, William
G. A. Gray Company, The
Gates Rubber Co.
General Dynamics Corp.
General Electric Company
General Motors Corporation
  Allison Division
  Central Foundry Div.
  Chevrolet Motor Div.
  Fisher Body Div.
  Oldsmobile Div.
Giddings & Lewis Machine Tool Company
Gorton Machine Corporation
Graphic Systems
Greenlee Bros. & Co.
Hamilton Standard
  Division of United Aircraft Corporation
Handy and Harmon
Hewlett Packard Co.
Hill-Rockford Company
Hi-Shear Corporation
Hoeganaes Corporation
Holo-Krome Co.
Hones, Charles A., Inc.
Houdaille Industries, Inc.
  Di Acro Division

Huntington Alloy Products Division
  The International Nickel Company, Inc.
IBM Data Processing Division
Illinois Bell Telephone Company
Ingersoll Milling Machine Co., The
Ingersoll-Rand Co.
International Harvester Co.
  Huntington Alloys and Solar Div.
International Nickel Company, Inc., The
International Silver Co.
International Telephone & Telegraph
  Corporation
Jacobsen Mfg. Co.
  Allegheny Ludlum Industries
Jones & Lamson, Div. of
  Waterbury Farrell, a Textron Co.
Jones & Laughlin Steel Corporation
J. Wiss & Sons Co.
Kaiser Aluminum Company
Kennametal Inc.
Keuffel & Esser Company
Keystone Carbon Co.
Klockner, Inc.
Koehring Co.
  HPM Division
Landis Tool Co.
Lebanon Steel Foundry
Le Blond Inc.
Leeds & Northrup Co.
Link-Belt Division of FMC Corporation
Littel, F. J., Machine Co.
Lockformer Co., The
Lockheed Missiles & Space Company, Inc.
L. S. Starrett Co., The
Malleable Founders Society
Mattison Machine Works
Matto, Allen
Max Manufacturing Co.
McMath, Jack
Mesta Machine Company
Milford Rivet & Machine Co., The
Miller Electric Mfg. Co.
Milwaukee Electric Tool Corp.
Minster Machine Co., The
Modicon Corporation
Moline Tool Co.
Monarch Machine Tool Co.
Morse Chain Company,
  Borg-Warner Corporation
NASA Lewis Research Center

National Acme Co., The
National Association of Metal Finishers
National Cash Register Co.
National Machinery Co.
National Radio Astronomy Observatory
New Jersey Zinc Company, The
Niagara Machine & Tool Works
Northrop Corp.
Norton Company
Numatics, Inc.

Oberg Manufacturing Co., Inc.
Ohio Gear, Div. of Towmotor Corp.
Omark-Winslow Aerospace Tool Company
Osborn Manufacturing Company, The
Pacific Press & Shear Co.
Peck, Stow & Wilcox Co., The
Peltzer & Ehlers Klockner, Inc.
Philadelphia Gear Corporation
Plasmadyne
    A Geotel Company
Pneumo Dynamics Corporation
Potter Brumfield Division of AMF
Pratt & Whitney, Inc.
Pravo Corporation

Racine Tool and Machinery Company
Rank Precision Industries
Ransome Company
Reed Rolled Thread-Die Co.
Republic Steel Corp.
    Steel and Tubes Division
Rex Chainbelt Inc.
Reynolds Metals Company
Rockford Machine Tool Co.
Rockwell International
    Power Tool Division
Rohr Corp.

Schramm Inc.
Schwinn Bicycle Co.
Sciaky Bros., Inc.
Shore Instrument & Mfg. Co., Inc., The
Simpson Electric Co.
Snap-on Tools Corp.
Snyder Corporation
South Bend Lathe
Standard Oil Company of California
Standard Pneumatic Motor Co.
Stanley Tools, Div. of The Stanley Works
Steel Founder's Society of America
Sunnen Products Co.

Sundstrand Machine Tool, Div. of
    Sundstrand Corp.
Superior Electric Co., The
Syntron Corporation

Tapmatic Corporation
Taylor-Winfield Corporation, The
Teledyne Precision Cincinnati
Teledyne Readco
Tempil Corporation
Thermal Dynamics Corporation
Timken Roller Bearing Co., The
Tinius Olsen Testing Machine Company
Toledo Stamping & Mfg. Co.
Torrington Company, The

Udylite Corporation
Uniform Tubes, Inc.
Union Carbide Corp.
    Coatings Service Dept.
    Linde Division
    Stellite Division
Unipunch Products Inc.
United Auto Workers
United States Steel Corporation
Universal Engineering Co.
U S I Clearing Div. of U. S. Industries, Inc.
USM Corporation
    Eyelet Division
    Fastener Group

Vacu-Blast Corporation
Varian Associates, Vacuum Div.
Victor Equipment Company
Virginia Electric and Power Company
V/R Wesson Company

Warner Electric Brake & Clutch Company
Warner & Swasey Company
Waterbury-Farrel, a Textron Company
Weldon Tool Co., The
Welsh Mfg. Co.
Western Electric Co.
Westinghouse Air Brake Co.
Westinghouse Electric Corporation
Wheelabrator-Frye, Inc.
Whitney, W. A. Corp.
Wirebound Box Manufacturers Association
Wire Reinforcement Institute
Woodings-Verona Tool Works
Wyle Laboratories, Inc.
Zinc Institute Inc.

Table of Content

List of Tables

Planning for Manufacturing

Section

1

Metal technology is the basis for much of the world's productivity of food, shelter, transportation, and communication. The products of metal technology are made possible because of the unique properties and characteristics of metal.

In manufacturing technology, these properties of metal are considered for the planning of a manufacturing system. Planning and organizing take place long before material is cut or assembled into a product. The product idea is obtained, finances to support the planning and production are supplied, the manufacturing processes and machines are acquired, the sources and supply of materials to be used in manufacturing are confirmed, and people are hired and trained to perform the work. People work at many levels in industry. All job levels are important to the success of the organization. Manufacturing technology requires many skills in organizing, planning, and production.

The application of many safety procedures must be a part of the manufacturing environment. Being conscious of safety rules while working with metals is of primary importance. Judgment is required for setting safe limits of operation for both human actions and mechanical devices. The correct use of protective equipment will help workers avoid common injuries and potential hazards in the metal area. Fire control and first aid are responsibilities of every person who comes in contact with a metal-working industry.

Manufacturing technology uses measuring and layout as a method of communication. It provides the information as to what is desired and what is actually produced. Accurate measurement makes the manufacture of interchangeable parts a reality. Basic measuring units of the International System of Units (SI) are used to determine length, mass, time, electrical current, amount of substance, luminous intensity, and temperature. The concepts of alignment, quantity, size, time, and content which are fundamental to quality control must also be measured. The techniques of measurement and marking, and the principles of geometry are applied in procedures for laying out. The reading of blueprints and techniques of layout for surface dimensions are needed both for making the workpieces and also for building the tooling for manufacturing.

Metal Technology

Chapter

1

Words You Should Know

Technology — The sum of the ways in which a culture provides itself with the material objects of its civilization.

Primary material — A basic material that serves all industry.

Value added — Change or work done on a material or product results in increased worth of the product.

Transfer machine — A machine that conveys or moves parts or products from one place to another within the machine while performing the work.

Durable goods — Products that are used for a number of years. They are not consumed or destroyed during use. Examples are appliances and machinery.

Toughness — The ability to resist rough treatment or shock.

Rigidity — The ability to resist deflection under stress or load.

Metal's loading — Tensile, tortional, and compressive.

Ductility — The ability of metal to be drawn out or rolled without cracking or breaking, to flow under pressure, to stretch, and to be deep-drawn.

Manufacturing properties — Machinability, formability, joinability, and castability.

Machinability — The relative difficulty by which metal can be separated by means of chip removal.

Tool life — The number of parts machined before the tool is worn out.

Cohesive bond — A bond or weld made by melting the metals. The metals intermingle and coalesce (freeze).

Adhesive bond — A bond is made when a different material diffuses into the other metals, making a new material which bonds the surfaces together.

Mechanical joining — Held together by a fastener such as a bolt, screw, rivet; or by clamping and crimping.

Castability — The relative ability of metals to be liquefied and shaped by pouring the metal into a mold to cool in a new shape.

Gray cast iron — A ferrous metal with 2% to 5% of carbon that appears as graphite flakes throughout the metal.

White cast iron — A ferrous metal made from gray cast iron by being rapidly cooled in the molted state. The carbon has not had time to form flakes.

Malleable cast iron — A ferrous metal made from white cast iron by annealing the castings (heat treating).

Ductile iron (nodular iron) — A ferrous metal made from gray cast iron by adding magnesium or cerium so that the graphite forms spheres in the metal.

Steel — A ferrous metal with 0.05% to 2.0% of carbon. A small change in the carbon content makes a great difference in the characteristics of the metal.

Rockwell Hardness — A test of a metal's hardness that reports relative hardness with a number.

Solid solution — An alloying element that dissolves in the base metal while the base metal is in the solid state.

Carbides — Alloying elements that precipatate out of a slowly cooling solid solution to form complex carbon compounds and create deep hardening and heat resistance.

Tensile strength — The ability of materials to resist being pulled apart. It is expressed in thousands of pounds per square inch (psi).

Hardness — Metal's ability to resist indentation, penetration, or scratching.

Corrosion — The wasting away of metals by chemical action.

Electrical chemical series — Chemical changes produced by electricity in which one metal replaces another.

Galvanic corrosion — The deterioration of metal by ionic action.

Cathodic — Having the properties of a cathode, which gives off electrons or negative ions.

Anodic — Having the properties of an anode, which gives off positive ions.

Conductivity — The ability to transmit heat or electrical current.

Thermal expansion — The increase of size due to an increase in heat.

Metal contraction — Shrinkage of metal, which occurs when metals cool from a liquid into a solid.

Strength-to-weight ratio — A metal's yield strength divided by its density (also referred to as specific strength).

Modulus of elasticity — A metal's ability to resist deflection (stiffness).

Stiffness-to-weight ratio — A metal's modulus of elasticity divided by its density (also referred to as specific stiffness).

Ladle — A vessel with a pouring tip or nozzle for conveying liquid metal from a furnace to another apparatus for further treatment or to a mold for casting.

Metal — a Primary Material

Metal is one of the most useful materials derived from nature. A civilization is frequently judged by its ability to develop a technology and the materials converted for its use. Improving the culture's standard of living and encouraging its creativity demand an understanding and use of special materials necessary to build a technology. One of the most important technologies today is **metal technology.**

Metal is a primary material that serves all industries. Metal can provide strength, rigidity, and toughness to manufactured goods and products. Metal is the material from which most tools and machines are manufactured to achieve today's standard of living.

Industrial Materials

Metal provides the industrial material for construction and manufacturing industries to produce their goods, Fig. 1-1. Industries may use industrial materials (metals) directly as

Fig. 1-1.   This six-story Centaur shroud protects the spacecraft. It is destined for a soft landing on Mars

cast metal, or they may convert the industrial material into standard stock — such as bars, sheets, and fasteners. Manufacturing also depends upon metal for constructing, fastening, and forming materials.

Manufactured Products

Metals provide the material for the manufacturing of many mass-produced objects. Metal can be formed, bent to shape, machined or cut to a dimension, and fastened together to make up manufactured products, Fig. 1-2.

Manufacturing is a system for making products. It may be as simple as bending and cutting wire to make a paper clip. Such products may be produced by the millions at a very high speed. Manufacturing may be as complicated as producing spacecrafts. Only a few very exact products are made in this case. With such complicated products, parts are formed and cut to size with extreme accuracy and cleanliness, Fig. 1-3.

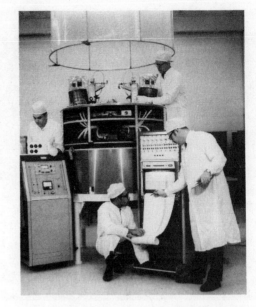

Fig. 1-3. Cleanliness, accuracy, exacting care, and craftsmanship make it possible for metal components to pass extreme environmental testing.

Fig. 1-2. A pulley is machined inside and out on a chucker-lathe.

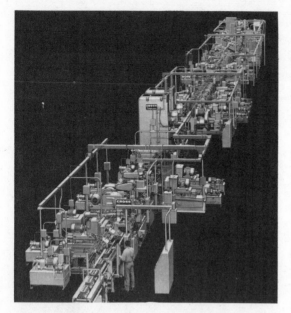

Fig. 1-4. Transfer machines that automatically machine automobile engine blocks.

Fig. 1-6. Automatic assembly transfer machines place valves and head into automobile engine.

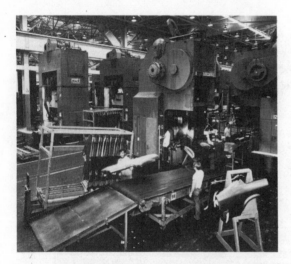

Fig. 1-5. Automobile bodies shaped in presses.

Testing of metals and components during manufacturing is carefully done to assure proper functioning of the craft.

Large volume manufacturing (mass production) is the most commonly known type of manufacturing. In large volume manufacturing, much of the bending, cutting, drilling, and joining of the metal is done automatically, Fig. 1-4. Large production machines which machine automobile engine blocks, stamp out parts for typewriters, or press auto bodies into shape are common, Fig. 1-5. Parts are assembled manually or automatically on conveyor assembly lines, Fig. 1-6. For large volume manufacturing, detailed planning and large financial investments in tools, machinery, and trained personnel are necessary. Products are carefully designed and redesigned for efficient manufacturing. The redesigning process is usually necessary to eliminate production parts or manufacturing processes from the product. Redesigning of a product often results in saving time and material so the manufacturing cost of the product can be reduced.

Metal functions well for products designed for automated manufacturing. Many manufactured objects are made from metal.

Uses of Metal

Metal is a primary material that contributes to all industries. Most of the articles

Fig. 1-7. Six plows in one prepare the field for planting.

A. Combine for harvesting grains.

B. Tomato harvester for mechanical picking of special varieties of fruit.

Fig. 1-8. Metal products are useful for farm machinery.

people use are made or processed by tools and equipment made of metals. The basic requirements of all people for food, shelter, transportation, and communication have been greatly improved because of the application of metal.

In the **food industry** metal has contributed both to the variety and amount of food available. The food growing and processing industries use large quantities of metals for their production equipment. The modern farming tools use metals for tractors, plows, cultivators, combines, and other farm machinery, Figs. 1-7 and 1-8. Farm products are transported to the market by trucks and trains which are made of metals. The food processing industries include equipment of all types such as slicers, sealers, cookers, or automatic packaging machines, Fig. 1-9. This food processing equipment is made primarily of steel, cast iron, and stainless steel. Other metals which are also used often are copper, brass, aluminum, and alloy steels.

In the **construction industry** industrial materials such as wood, concrete, stone, metal, and glass are used for housing. Metal is increasingly being used to help provide shelter. In large buildings the frames are nearly always made of steel. **Steel** provides

Fig. 1-9. Six-knife peach slicer with protective cover removed to show mechanism.

Fig. 1-10.   This 37-story building is constructed of 10,000 tons of steel.

the strength for the load-carrying capacity of the building, Fig. 1-10. Without the use of steel, very tall structures would not be possible. Metal is also used in the interior of buildings for stairways, door sashes, and decorative fixtures. **Aluminum,** a lightweight metal, is used extensively for exterior walls where weight, noncorrosion, and color are the desired requirements, Fig. 1-11.

Each decade the shelter people choose becomes increasingly dependent on the use of metals. Ecologists suggest that our shelter in the future will be more concentrated and planned as a self-contained city so that green space may be preserved for outdoor living area. Metals will be used to solve problems as our lifestyles change.

The **transportation industry** would not be available without the use of various metals. The strength and weight of steels and other metals have made ships, planes, trains, and automobiles possible, Fig. 1-12. Transportation is very dependent upon metals because the prime requirements of transportation are speed and strength, and metal

Fig. 1-11.   Steelwork has been brought to the exterior of this uniquely designed and engineered building.

Fig. 1-12.   Very large crude carrier in contrast with a regular-sized oil tanker.

provides both. The materials used in earlier periods of history would not withstand the loading or strain which high-speed planes, automobiles, trains, trucks, or ships require. Steel has become the most important material used in the world's mass transportation system, Fig. 1-13.

The **communication industry** is also dependent upon metal technology. Communication is essential to our way of living today. It may be person-to-person within a limited range or person-to-machine for large audience communication. The telephone is an example of two-way personal communication. Television and newspapers provide an example of one-way communication to large audiences, Fig. 1-14. In each case, vast amounts of metal are used to produce the circuits, microphones, receivers, and other devices necessary for communication equipment, Fig. 1-15. Die-cast zinc products are often plated with other metals to make them more attractive and to protect them from corrosion.

Fig. 1-14. Communication across the nation by television is made possible by microwave towers.

Fig. 1-13. BART (Bay Area Rapid Transit Co.) cars are light, strong, and safe.

Fig. 1-15. Television and telephone data are transmitted around the world by satellites constructed of light metals.

Fig. 1-16. Tensile strength is the ability of metal to resist being pulled apart.

FIG. 1-16A.
Tensile Strengths

Metal	PSI x 1000
Gray Cast Iron—Class 30	20—40
Ductile Cast Iron—Type 80-60-3	90—110
Malleable Cast Iron—Type 35018	83—60
Wrought Iron (Hot-Rolled)	48
Low-Carbon Steel (Hot-Rolled)	
C 1010	51
C 1018	69
C 1025	70
Medium-Carbon Steel (Hot-Rolled)	
C 1030	80
C 1040	91
C 1045	98
High-Carbon Steel (Hot-Rolled)	
C 1055	109
C 1070	128
C 1080	141
C 1095	142
Nickel Steel (Rolled) 2315	125
Chromium (Steel-Rolled) 5140	150

Metal Properties

The characteristics of metal discussed later — toughness and formability — are properties which make metal a good manufacturing material. Now you will study the properties of the metals themselves. This study of the technology of metals is known as **metallurgy.**

The ability of metal to withstand heavy loads without bending or breaking is its most important characteristic. This strength is relative to the metal's ability to resist deformation. A large load may be placed upon a metal product and it will not break, but wood, plastic, or concrete probably would.

Tensile Strength

One of the most important strengths that metal has is its ability to resist being pulled apart. **Metallurgists,** or metalworking people, call this **tensile strength,** Fig. 1-16. Metals are tested to determine their strength. The metal's strength is indicated by the number of pounds per square inch (psi) the metal can withstand without breaking.

The tensile test reveals a great amount of information about metals. Common metals have very high tensile strengths, Fig. 1-16A.

A piece of hot-rolled steel bar (c 1018) one inch square will lift a load of up to 34 tons before it will fail. This calculation is made from the preceding table. The tensil strength of the carbon steel is 69,000 pounds per square inch, and the number of square inches in the example is 1. There are 2000 pounds in a ton (T); therefore, 2000 divided into 69,000 equals approximately 34 T. If we take the same steel with a bar 2 inches on each side (4 sq. in.), the strength would be 4 × 69,000 = 276,000 lbs., or 138 T.

With heat treatment, strengths of the high-carbon steels may be increased. High-carbon steel C 1095 which has been water-quenched at 1450° F. and tempered to 800° F. will produce a strength of 200,000 pounds per square inch rather than the 142,000

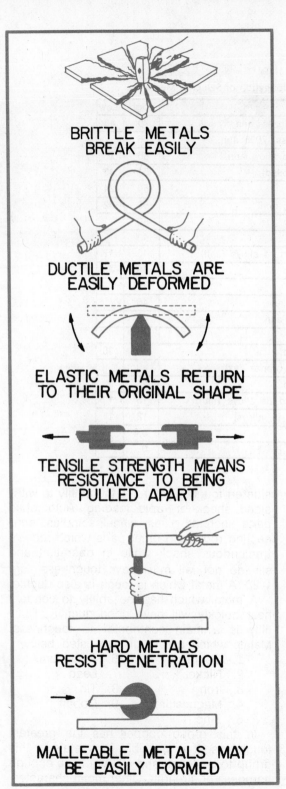

Fig. 1-17. Properties of metals.

listed for high-carbon steel, Figs. 1-17 and 1-18.

Higher strength materials are in research laboratories and are being developed by metallurgists. A new technology is emerging which involves **whisker-strengthened metals.** These are composites in which bundles of aligned aluminum oxide whiskers (thin crystals of metal of exceptional strength) have been infiltrated in the molten metals. At room temperature alumina whiskers have tensile strengths up to 4,000,000 psi. These ultra-high strengths are attributed to their high degree of crystalline perfection.

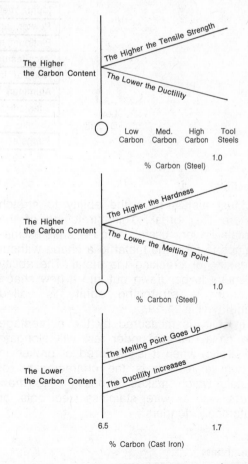

Fig. 1-18. Characteristics of steel and cast iron change as the carbon content changes.

FIG. 1-19.

Ductility or Percentage of Elongation of Selected Metals

	High	Low
Phosphor bronzes, ann.	70	48
Cartridge brass, 70% ann.	66	
Nickel and its alloy, ann.	60	25
Soft leads (rolled)	51	43
Monel, ann.	50	24
Nickel silvers, ann.	50	32
Pewter, C. R.	50	
Silver ann.	48	
Aluminum and its alloys (soft)	45	17
Copper, ann.	45	35
Gold, ann.	45	
Grade A tin, ann.	45	
Muntz metal, ann.	45	
Pewter, ann.	40	
Platinum, ann.	40	30
Architectural bronze (extr.)	30	
Nodular irons	25	2
Aluminum and its alloys (hard)	15	1.5
Titanium and its alloys, ht. tr.	12	1
Cobalt	0.4	
Tungsten, C. R.	0	

High Ductility ↑

Low Ductility ↓

Ductility

Most metals have the ability to stretch before they break. This stretching of the metals is very useful, since it enables machines to form the metal to a shape without breaking or cracking the metal. The ability of metal to be drawn out into a new shape through its **stretching ability** is called **ductility,** Fig. 1-19.

Ductility is measured by the percentage the metal sample under test will elongate or stretch when it is fractured or broken in a tensile test. It is the characteristic of metals which makes it possible to draw parts such as wire, stainless steel pots, or automobile fenders.

Toughness

Another characteristic of metals is **toughness.** "Difficult-to-break" metals are con-sidered tough. They have the ability to with-stand shock or rapid loading. Automobile parts such as axles, wheel spindles, con-necting rods, and other parts which receive tremendous shock while in operation and still do not fail must have toughness, Fig. 1-20. A metal which is tough is also ductile.

A metal which has the ability to conduct heat quickly will have good ductility. Duc-tility is a main contributor to toughness. Metals which are tough are listed below:

1. Copper
2. Nickel
3. Iron
4. Magnesium
5. Zinc
6. Aluminum
7. Lead
8. Tin
9. Cobalt

In this group copper has the greatest toughness. It is possible for metals in this group to be tough and still not have enough hardness or strength. Other metal character-istics must be added to make a strong,

Fig. 1-20. Metals which are difficult to break are tough. This front wheel axle must be tough in order to withstand impact loads in service.

FIG. 1-20A.
Moh's Scale of Mineral Hardness

1. Talc
2. Gypsum
3. Calcite
4. Fluorite
5. Apatite
6. Orthoclase
7. Quartz
8. Topaz
9. Corundum
10. Diamond

tough product. Some products which require great toughness are gears, chains, axles, spindles, and sprockets. Parts which are loaded and unloaded quickly in their function as a part of a machine must be tough.

Hardness

Hardness in metals is the ability of the metal to resist indentation, penetration, or scratching. Hardness results from the dislocation of the metal's atoms. Hardness is an indication of other metal characteristics such as strength, toughness, and brittleness. Hard metals are usually high in strength, low in ductility, and resistant to scratching or wear. Hard metals tend to have low impact strength or toughness, but they may be heat treated to improve toughness.

The ability to resist wear is an important characteristic of metals. Cutting tools, knives, chisels, lathe tools, springs, ball bearings,

gear teeth, music wire, grinders, crusher jaws, and high-strength steel depend on this characteristic.

Formerly, hardness was tested by comparing the hardness of one material against another. In the field of geology, one mineral was used to scratch another mineral. Minerals were rated on a scale, Fig. 1-20A. Hardness was tested in this manner: talc will not scratch gypsum, but gypsum will scratch talc. Therefore, gypsum is harder than talc.

By using the scale of hardness today, you can separate many degrees of hardness. The hardness difference between (9) corundum and (10) diamond has many degrees which need to be identified.

Hardness in metals is tested by these three methods:

1. Resistance to abrasion
2. Resistance to penetration
3. Elastic hardness

The **abrasion test,** which measures resistance to abrasion, is used to get a rough indication of hardness. The **file test** is used to see if the file will cut or bite into the metal. If the metal is hard, the file will slide over the metal and polish the contact spot. When the metal is soft, the file will cut the metal and leave a mark on it. This test may be applied after heat treatment, but it would only be used to subject the hardened batch of product to further tests or as a basis for rejecting the batch.

The **penetration test,** which measures resistance to penetration, is made by pressing a hard ball 10 millimetres in diameter into the metal with a standardized load. The **Brinell Hardness Tester** is a small hydraulic press or air press which pushes a 10-millimetre ball into the metal being tested. The load applied to the ball is 3000 kilograms to test ferrous metals and 500 kilograms to test nonferrous metals. The pressure is held on ferrous metals for at least 10 seconds, while 30 seconds is allowed for nonferrous metals. For the hardness reading, the diameter of the dent in the metal is measured by a microscope which contains an optical scale for measuring to a tenth of a millimetre, Fig. 1-21. The diameter of the measured dent is checked on a Brinell Hardness Chart, and the metal's hardness is found. The **Rockwell**

Hardness Tester is designed to measure the depth of penetration of a ball or a 120° conical diamond (called a **Brale**) under pressure. The depth of penetration under a specific load is measured. The most common Rockwell scales used are the B scale and the C scale. In the B scale, a 1/16″ ball and a 100 kilogram load are used to test soft materials. The C scale uses a diamond Brale and a 150 kilogram load to test harder metals.

The hardness numbers are on a dial and are read directly. A shallow indentation of the ball or Brale indicates a hard material and a high hardness number, Fig. 1-22. A deep penetration indicates a soft material and a low hardness number, Fig. 1-22A.

The **elastic hardness test** is referred to as **rebound hardness.** The Shore Scleroscope hardness tester measures the height of a diamond-pointed hammer. The harder the material being tested, the higher the rebound of the hammer. The rebound height is read on a graduated gauge which indicates the hardness number. The chief advantage of this type of hardness tester is that it is portable, and the mark left on the work is very small, Fig. 1-23.

A hardness test gives a quick reference to the uniformity of the metal, its tensile and other strengths in relation to its hardness, the exactness of temper in heat treatment, and the metal's homogeneity or consistency, Fig. 1-23A.

Geometric Form

Metals are made up of a number of types of strengths, each giving its unique contribution to the metal. Each metal may have distinctive strengths exclusive to the metal. However, the inherent qualities of the metal are important when considering strengths, as well as the form into which the metal is rolled. The form of the metal influences the compressive strength (crushing), the torsional strength (twisting), the shear strength (ripping or tearing), etc. If the cross-sectional shape is in the form of an I-beam, an H-beam, a channel, or a tube, the charac-

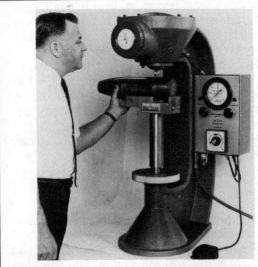

Fig. 1-21. The hardness reading is measured by the diameter of the dent in the metal. It is measured by a special microscope.

Fig. 1-22. Rockwell testers use a 1/16″ ball under load to test soft materials on the "B" scale and a conical diamond under load for hard materials on a "C" scale.

FIG. 1-22A.
Moh's Hardness Scale, Rockwell Hardness Scale

Moh's Hardness Scale	Rockwell Hardness Scale
1	
2	
3	74B
4	84B
5	39-45C
6	56C
7	61-64C
8	72C
9 Sapphire	75C
• Fused Alumina	77C
• Tungsten Carbide	80C
• Silicon Carbide	84C
• Boron Carbide (Norton)	85C
• Carbonado • Congo (Gray) • Congo (Yellow) } Diamond 10 Brazil Ballas	

FIG. 1-23A.
Hardness of Metals (Brinell Hardness)

Metal	Average Hardness
Low-Alloy Steels (13XX), H & T 800F	450
Low-Alloy Steel (46XX), H & T 800F	395
High-Carbon Steel H & T 700F	350
High-Carbon Steel (HR)	260
Gray Cast Iron	230
Low-Alloy Steel (43XX)	210
Medium-Carbon Steels CW	200
Aluminum Bronzes (sand-cast)	160
Yellow Brass (cast, high-strength)	150
Low-Carbon Steel CW	140
Standard Malleable Irons	130
Wrought Iron (HR)	101
Aluminum and Its Alloys T 72	95
Zinc Alloys (die-cast)	86
Yellow Brass (cast, leaded)	58
Gold C W	58
Aluminum and Its Alloys, Ann	49
Copper, Ann	40
Pewter, Ann	13
Soft Lead (chill-cast)	4.2

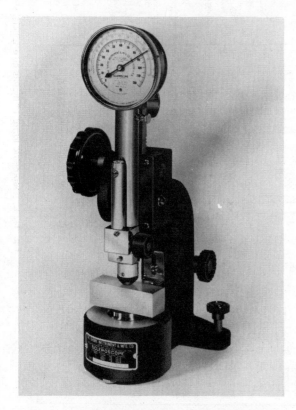

Fig. 1-23. Elastic hardness is measured by a rebounding hammer. The height of the rebounding hammer is graduated in a hardness scale.

STANDARD METAL SHAPES
GEOMETRIC FORMS

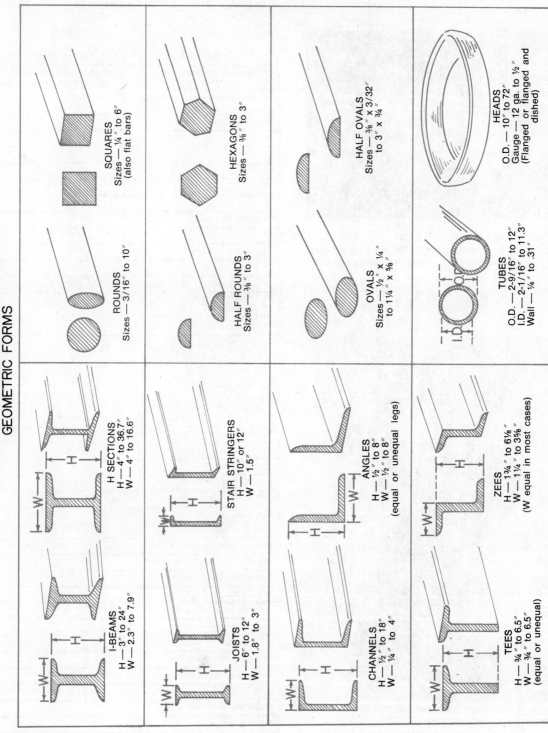

SQUARES
Sizes — ¼" to 6"
(also flat bars)

ROUNDS
Sizes — 3/16" to 10"

HEXAGONS
Sizes — ⅜" to 3"

HALF ROUNDS
Sizes — ⅜" to 3"

HALF OVALS
Sizes — ⅜" x 3/32"
to 3" x ¾"

OVALS
Sizes — ½" x ¼"
to 1¼" x ⅝"

HEADS
O.D. — 10" to 72"
Gauge — 12 ga. to ½"
(Flanged or flanged and dished)

TUBES
O.D. — 2-9/16" to 12"
I.D. — 2-1/16" to 11.3"
Wall — ¼" to .31"

H SECTIONS
H — 4" to 36.7"
W — 4" to 16.6"

I-BEAMS
H — 3" to 24"
W — 2.3" to 7.9"

STAIR STRINGERS
H — 10" or 12"
W — 1.5"

JOISTS
H — 6" to 12"
W — 1.8" to 3"

ANGLES
H — ½" to 8"
W — ½" to 8"
(equal or unequal legs)

CHANNELS
H — ½" to 18"
W — ¼" to 4"

ZEES
H — 1¾" to 6⅛"
W — 1¼" to 3⅝"
(W equal in most cases)

TEES
H — ¾" to 6.5"
W — ¾" to 6.5"
(equal or unequal)

Fig. 1-24. Geometric forms of metals give special strength characteristics to the metal.

16

teristics of the same metal differ. An H-beam form will stand a great crushing load, whereas a tube will stand crushing, but is primarily used to withstand a tortional load, Fig. 1-24. Manufacturing will take advantage of specific attributes in selecting the forms for products.

Corrosion Resistance

Corrosion resistance is a very important characteristic of metals. **Corrosion resistance** is the metal's ability to resist deterioration. This wasting away is caused by the action of the atmosphere, liquids, chemicals, gases, or galvanic action on the metal.

The most common corrosion is caused by rusting. Oxygen from the air forms oxides or rust on most steels. This is the reason ships and bridges must continually be repainted to protect them from corrosion.

These metals resist corrosion well:

lead	monel
tin	aluminum
gold	nickel
silver	chromium
stainless steel	

Corrosion frequently occurs where an electric current flows through or along metal and interacts with a nearby liquid. An electro-chemical reaction takes place and produces galvanic corrosion. Galvanic corrosion may be stopped if a metal higher on an electrochemical series is coated on the metal or is mounted close-by. The metal which is higher on the series will corrode more readily by giving up ions of metal to the solution, thereby protecting the original metal structure.

Metals arranged according to their relative resistance toward galvanic corrosion are shown below:

(High) magnesium
      zinc
      aluminum
      cadmium
      aluminum alloys
      iron or steel
      stainless steel (active)

      muntz metal
      yellow brass
      soft solders
      tin
      lead
      aluminum brass
      red brass
      copper
      aluminum bronzes
      nickel-copper alloys
      silver
      stainless steel (passive)
      monel
(Low) titanium

The above information is utilized to protect underground pipelines, ship parts, and boilers. A ship may be given cathodic protection by having pieces of zinc or magnesium fastened to the hull, stern, and rudder of the ship. The zinc or magnesium is sacrificed to protect the brass propeller and ship's parts from galvanic corrosion, Fig. 1-25.

Corrosion takes place in other ways than by seawater contact. Corrosion may be caused by stress concentration cells. Difference in the stress of the metal produces corrosion areas. Corrosion may also be produced in a crevice. Corrosion occurs when

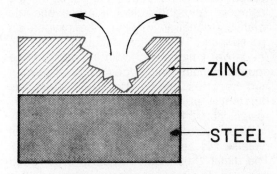

Fig. 1-25. Metal and alloys will corrode and offer protection to any metal which is lower in the galvanic corrosion series. The zinc is consumed, and the steel is protected for a period of time by galvanic action.

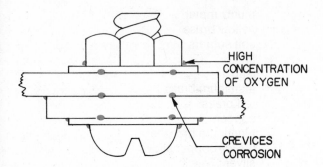

Fig. 1-26.   Stress corrosion is the result of ion flow, which is caused by chemical imbalance.

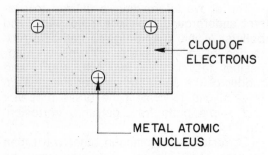

Fig. 1-27.   The highly mobile electron cloud is believed to be the reason that metals conduct an electrical current.

Fig. 1-28.   Copper and aluminum are excellent electrical conductors. This aluminum coaxial cable will carry 90,000 telephone calls simultaneously.

fasteners are overtightened. Oxygen is forced out from the contact points of the fasteners, and the adjacent areas become relatively richer in oxygen. The oxygen on the surface of the metal assumes a cathodic (negative) relationship with respect to the squeezed metal which does not have oxygen and becomes anodic (positive). When this differential of ion potential develops, corrosion starts, Fig. 1-26.

When corrosion takes place inside of metals, it is called **intergranular corrosion.** The metal "falls apart" when it is corroded this way. The most common example of this intergranular corrosion is dezincification in brasses. If brasses are made up of 12% to 15% zinc and do not have the minimal amounts of tin or antimony added to the brass, they corrode internally. If the brass is used in contact with water containing high percentages of carbon dioxide, the zinc is dissolved and removed from the metal, leaving the alloy weakened by corrosion.

Corrosion problems may be reduced either by using a high purity metal where the corrosion will not occur or by adding another corrosion-resistant metal to it. Metals such as silicon, aluminum, copper, nickel, tin, zinc, and lead are used to make corrosion-resistant alloys.

Electrical Conductivity

The ability to carry an electrical current is a notable property of metal. Metals are held together by a metallic bond in which a "cloud" of electrons is shared by many metal atoms. The cloud is made up of masses of negatively charged particles completely ringing the many nuclei of the metal. This highly mobile electron cloud is believed to be the reason that metals will

FIG. 1-28A.

Relative Conductivity of Metals

(Silver = 1.00)

Silver	1.00
Copper	1.10
Aluminum	1.78
Tungsten	3.53
Iron	6.29
Lead	13.80
Mercury	60.20
Nichrome	62.90

FIG. 1-28B.

Relative Thermal Conductivity of Materials

Silver	.97
Copper	.92
Aluminum	.50
Brass	.26
Iron	.16
Lead	.08
Mercury	.02
Glass	.0025
Water	.0014
Wood	.0005
Paper	.0003
Felt	.0001

conduct an electrical current, Fig. 1-27. Excellent conductivity of certain metals results because these clouds of electrons move electrical energy very rapidly, Fig. 1-28.

If the metal does not conduct electricity efficiently, the flow of electrons is slowed down. This may be thought of as "friction" which holds back the flow. When a metal has this property, it has resistance, or resistivity. If the resistance to flow of electrons is great, some of the electrical energy will be converted into heat energy. When the resistance of the metal is very high, heat is produced in large volume, as in an electric stove. When the resistance is low in a metal, the metal remains cool, Fig. 1-28A.

Thermal Conductivity

Another characteristic of metal is its excellent conductivity of heat. The rate that heat flows through a material is its **thermal conductivity.** Metals which heat quickly pass this energy to the molecules, which react by increasing their movement. Heat results from the energy of motion of the metal's molecules. The various metals have different rates of flow of heat.

Metal's thermal conductivity has many practical applications. Automobile radiators use this metal characteristic by spreading heat over a large area quickly and passing the heat into the air. Electric soldering irons, electric irons, grills, and cooking utensils are other products which are made possible because of this property, Fig. 1-28B.

Thermal Expansion

Thermal expansion is another characteristic of metals. **Thermal expansion** is the increase in size of the metal due to a change in its temperature. The molecules are very close together in a metal. When the metal is heated, it will expand. When a unit length of metal is measured and heated one degree, the change in length is called the **coefficient of linear expansion.** A steel railroad rail 30 feet long will increase its length about 1/2″ with a change in temperature from 32° F. to 212° F. It is easy to understand why rails must have spaces between their ends to keep the rails from buckling.

Metal expansion makes possible construction of many heat-sensitive instruments which employ bimetallic strips as the source of information.

Thermal expansion in metals may also lead to serious problems in metal fabrication. In welding, expansion of the metal may cause distortion in the metal parts being manufactured. Contraction of metal when it is cooling will cause stress and distortion of metal members if they are not welded together properly.

Metal contraction is a problem in the foundry when the molten metal cools and changes to a solid casting. This contraction in the metal is called **shrink** and may cause defects in castings unless the casting can

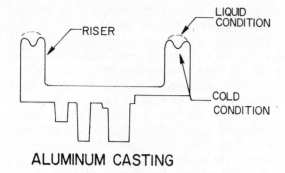

RISER
LIQUID CONDITION
COLD CONDITION

ALUMINUM CASTING

Fig. 1-29.   Use of a riser and sprue to feed metal to a casting as the metal shrinks on cooling.

be "fed" metal from a riser or another heavy section of metal to supply the needed metal as it shrinks.

Thermal expansion takes place on all the metal's dimensions: width, length, and thickness. All dimensions are taken when the metal is hot, and they are related to the metal's coefficient of expansion. When the volume of the metal is considered, a large amount of metal is being moved by expansion. The coefficient of cubical expansion is roughly three times the metal's coefficient of linear expansion. This cubical expansion, or contraction, is what is observed when an aluminum casting solidifies and the top of the riser changes its shape from convex to concave, Fig. 1-29. The shrinkage in castings is caused by the change in volume of the metal as it changes temperature.

Specific Strength

The point where a metal may be stressed or loaded without deforming the metal divided by the metal's density is known as its **strength-to-weight ratio.**

$$\frac{\text{Yield Strength (Stress in Pounds per Square Inch)}}{\text{Density (Pounds per Cubic Inch)}} = \begin{array}{l}\text{Strength-}\\\text{to-Weight}\\\text{Ratio}\end{array}$$

Design engineers are concerned with strength-to-weight ratio in the selection of

metals for the aerospace industry or any metal application where weight is a problem.

The metal characteristics that will determine the use or selection of a metal will relate to the metal's application. Most uses of metals will be concerned with the application of these qualities: (1) strength and rigidity, (2) space filling, and (3) quality and durability of surface finish. The characteristics that are most frequently sought in metals are strength and rigidity (strength-to-weight ratios or specific strengths). New alloys and material shapes that will produce greater loads with adequate safety factors are being perfected. Titanium alloys have a high strength-to-weight ratio of 1400, which is one of the major reasons titanium has become such an important metal to the space and aircraft industry. Such materials as reinforced plastic wound with glass filaments or titanium alloys (with possible combinations of plastics and metals) have very high values on a strength-to-weight table, Figs. 1-30 and 1-30A.

Stiffness

Structure form strength is a consideration that an engineer may make when designing lightweight structures. The configuration

Fig. 1-30.   Filament-wound, reinforced plastics produce tanks with a very high strength-to-weight ratio, higher than titanium alloys.

FIG. 1-30A.

Various Products with Strength-to-Weight Ratios or Specific Strength

Material	Ratio
Reinforced plastic, filament wound	4310
Reinforced plastic epoxy	1635
Titanium and its alloys	1400
Reinforced plastic polyester	1000
Martensitic stainless steels, H & T	982
Aluminum alloys	890
Ultra-high strength steels, H & T	882
Alloy steels H & T	821
Beryllium Ann.	821
Carbon steels H & T	502
Nodular iron	486
Carbon steel, H, R	297
Epoxies, molded	296
Nylon 11	250
Precious metals, CW	116
Copper, Ann.	31
Tin and its alloys, Ann.	4.9

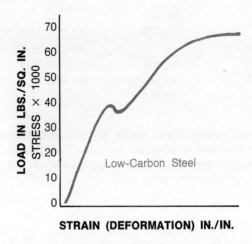

STRAIN (DEFORMATION) IN./IN.

Fig. 1-31.   The modulus of elasticity is the ratio of stress to strain on the straight part of the curve.

The **modulus of elasticity** is therefore the ratio of stress to the strain. The modulus of elasticity may be represented by the ratio number or the steepness of a line on a stress strain chart. Only the **straight part** of the line on the chart **represents stiffness.** The steeper the curve, the higher the modulus of elasticity. The **stiffer** the material, the **steeper** the **curve,** Fig. 1-32.

and the stiffness of the metal contribute to the rigidity of the structural member. The modulus of elasticity is a measure of stiffness and indicates the ability of a material to **resist deflection.**

The modulus of elasticity is obtained by dividing the material's **stress** in pounds per square inch (**load**) by the percentage of deformation (**strain**), the **elongation** per unit area, Fig. 1-31.

$$\text{Stress (Load/Unit Area)} = \frac{\text{Load in Pounds}}{\text{Cross Section Area in Square Inches}}$$

Stress is pounds per sq. in. (load).

$$\text{Strain (Deformation per Unit Area)} = \frac{\text{Elongation in Inches}}{\text{Original Length in Inches}}$$

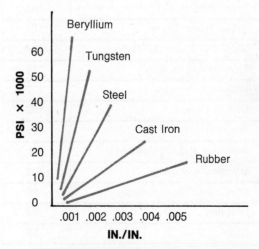

IN./IN.

Fig. 1-32.   Only the straight line portion of the curve reflects stiffness. The steeper the curve, the more the material resists deflection.

$$\frac{Stress}{Strain} = \frac{Steel\ 30,000\ psi}{0.001\ Inches/Inches} = 30,000,000\ psi\ (Modulus\ of\ Elasticity\ for\ Steel)$$

Stiffness-to-weight ratio is the **modulus of elasticity** divided by density. It forms the basis for selecting certain metals for certain job applications, Fig. 1-32A.

FIG. 1-32A.

Various Products with Stiffness-to-Weight Ratios or Specific Stiffness

Material	Ratio
Beryllium	657
Silicon Carbide	607
Boron Carbide	483
Alumina Ceramics	420
Mica, Natural	337
Titanium Carbide Cermet	248
Zircon	188
Tungsten Carbide Cermet	185
Boron Nitride	163
Glass, Fused Silica	129
Molybdenum and Its Alloys	127
Titanium and Its Alloys	111
Ultra-High-Strength Steels	107
Carbon Steels	106
Alloy Steels	106
Aluminum and Its Alloys	105
Magnesium Alloys	102
Reinforced Plastic Phenolic	96
Tungsten	84
Silver	29
Gold	17
Nylon 66	12

Information from the charts for strength-to-weight ratio and stiffness-to-weight ratio suggests that the strength of titanium and the stiffness of beryllium are reasons that these metals are important to the aerospace industry. As these materials find success in space applications, they will gradually be used more in the larger national manufacturing economy.

Factors Involved in Selecting Materials

Cost of materials is a prime concern to the manufacturing industry. Traditionally the cost of materials has been measured by **cost per ton** or **cost per pound.** This method is satisfactory when making cost analysis for structural steel and like materials. With the availability of high-strength materials, however, some different methods of analysis have come into view.

Most materials are chosen for use on the basis of their strength, rigidity, space-filling, or surface finish characteristics. The economical procedure is to select the material on the basis of the least cost and the highest number of desirable criterions possible. If you were building a dam, concrete would be the most probable material selection. In place concrete is one of our less expensive materials, especially when used on a massive scale. Strength, rigidity, and space-filling criterions would all be adequately met by the use of concrete.

For different applications, cost per ton may not offer the best solution. The **cost per cubic inch** may be a better method of analysis. This is different from cost per pound, because the factor of volume is being considered rather than weight, Fig. 1-32B.

When a cavity for a die-casting is filled by the plunger, the zinc casting alloy is injected into the mold. In this case, volume of **metal per piece** is the concern of the cost analyzer. When a carburetor is made, the engineering criterion must be considered. Is strength or rigidity the most important? Space filling? No, but the special shapes of the venturi and openings of the jets are

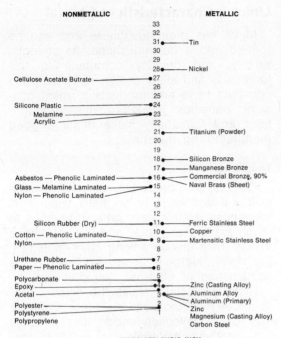

NONMETALLIC		METALLIC
	33	
	32	
	31 •	Tin
	30	
	29	
	28 •	Nickel
Cellulose Acetate Butrate —	• 27	
	26	
	25	
Silicone Plastic —	• 24	
Melamine —	• 23	
Acrylic —	22	
	21 •	Titanium (Powder)
	20	
	19	
	18 •	Silicon Bronze
	17 •	Manganese Bronze
Asbestos — Phenolic Laminated —	• 16 •	Commercial Bronze, 90%
Glass — Melamine Laminated —	• 15 •	Naval Brass (Sheet)
Nylon — Phenolic Laminated —	14	
	13	
	12	
Silicon Rubber (Dry) —	• 11 •	Ferric Stainless Steel
	10 •	Copper
Cotton — Phenolic Laminated —	• 9 •	Martensitic Stainless Steel
Nylon —	8	
Urethane Rubber —	• 7	
Paper — Phenolic Laminated —	• 6	
	5	
Polycarbonate —	•	
Epoxy —	• 4 •	Zinc (Casting Alloy)
Acetal —	• 3 •	Aluminum Alloy
		Aluminum (Primary)
Polyester —	• 2	Zinc
Polystyrene —	1	Magnesium (Casting Alloy)
Polypropylene —		Carbon Steel

CENTS PER CUBIC INCH

Fig. 1-32 B.

Fig. 1-33. Aluminum marine alloys become competitive for all-welded hull and superstructure when the cost/ton/tensile strength analysis concept is applied.

important, so the surface finish is probably the criterion which most needs to be considered. Zinc die-casting alloy rates 4 on the cents-per-cubic-inch scale, which is high. However, the unique requirements of the carburetor make zinc casting alloy the best choice. It could be made of cheaper steel, but the machining cost would make it uneconomical. It would be strong, rigid, and costly. The cost of the carburetor could be reduced, and the surface finish quality maintained if it were injection molded of polystyrene. This carburetor would work well until it got hot, and then it would lose its strength and rigidity.

In cost analysis where high strength and low weight are required for products such as air frames, ship superstructures, and some structural members, a different method of organizing data becomes helpful. The cost per ton per unit of strength (cost/ton/tensile strength) is used, Fig. 1-33. If this evaluation is used, the newer materials may cost more per ton than the traditional materials, but in some cases their cost per unit of strength becomes competitive.

Aluminum is in competition with steel for many products and structures. An outstanding example is the container industry. The aluminum soft drink container is light, strong, and easily formed, thus many processes of manufacture of the traditional can are eliminated. It has reduced the thickness of metal of the traditional can by 2/3. It resists corrosion, has eliminated the high cost of tin coating, and can be given various color finishes by anodizing or lithographing.

Aluminum is a "competitive metal" because of its cost-strength-weight ratio in ship building. Deck houses, superstructures, masts, lifeboats, and trim rails are made of aluminum. Smaller crafts may be entirely constructed of aluminum. In electrical transmission lines, aluminum is competing with copper. Its advantages of less cost and less weight, and its corrosion resistance overcome its disadvantage of being slightly less conductive than copper.

Other materials which will be competing on the cost per ton per tensile strength are titanium and its alloys, glass filament-wound, reinforced plastic, and glass-reinforced plastic epoxies. If it becomes possible to heatproof and fireproof the polymer materials, there will be a dramatic change in the manufacturing industries where strength and weight are primary requirements.

Fig. 1-34. The cone bits on this coal mining machine cut the coal from the earth.

Fig. 1-35. This turbine shaft does not deflect the 144″ swing lathe.

Unique Characteristics of Metal

Metal has great toughness and can be formed into a desired shape. Its strength, rigidity, and ductility contribute to this toughness and allow designers to plan and construct tools and products to meet consumer demands. These properties of toughness and formability make metal a good **manufacturing** material.

Toughness

One of the properties of metal which makes it adaptable to so many uses is its **toughness**. Metals have great resistance to rough loading or shock, and this resistance makes them unique among other materials, Fig. 1-34. Metal is tough because of its combination of physical characteristics.

Rigidity. Metals resist the forces produced by most tools and machines. Because they resist deflection under stress, metals are excellent materials for building machines and tools. The machines for metal part production must withstand the cutting pressure forces without changing their shapes. If a machine tool were to deflect or

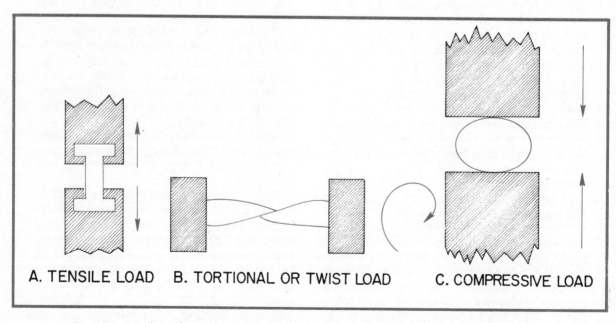

A. TENSILE LOAD   B. TORTIONAL OR TWIST LOAD   C. COMPRESSIVE LOAD

Fig. 1-36. Metals are designed to resist tensile loads, tortional or twisting loads, and compressive loads.

move, the various bearings which support shafts would be out of line and cause the shafts to bind. The part under manufacture would then be inaccurate. Cast iron and alloy steels are used to build most machines because they are rigid and will withstand great loads, Fig. 1-35.

The **loading** (weight or stress) that metals must resist are shown in Fig. 1-36 — tensile loading, tortional loading, and compressive loading.

Strength. Metal will withstand great force without yielding or breaking, Fig. 1-37. Metals vary in their strength, but when they are compared to other industrial materials, they are strong. Some of the strongest metals are alloy steels, stainless steels, and titanium and its alloys.

Ductility. Metal's resistance to impact or shock is another important property, particularly in the manufacturing processes. Metals which are capable of being drawn or rolled without breaking and cracking are ductile metals. Examples are copper, brass, aluminum, and steels.

Manufacturing Properties

The manufacturing properties of metals — machinability, formability, joinability, and castability — make it possible to produce parts from metal. These properties make the parts workable, give them shape, and give them strength.

Machinability. Machinability is an important property of metal. A metal's machinability refers to the ease or difficulty with which it can be separated or cut by means of chip removal. Machinability is measured by the length of time a cutting tool holds a sharp edge or by the rate of chip removal.

There are four major factors to consider in a machinability index:

1. The **characteristics** of the material.
2. The **cutting tool** material being used.
3. The **design** (geometry) of the cutting tool.
4. The **capacity** of the cutting tool.

The machinability of metals is a difficult concept to understand because of the interaction of the above four factors. Characteristics of the material being cut also affect machinability. These characteristics of a material include its hardness, its brittleness (which may cause the chip to break up), and its structure (uniform small grains or large spheroidal forms), Fig. 1-38. Treatment the material may have received from cold work-

Fig. 1-37. Nuclear submarine has great strength to resist the waves on the surface and the crushing pressure of the water at great depths.

Fig. 1-38. Machinability rating of metal is made up of factors of hardness, brittleness, and structure.

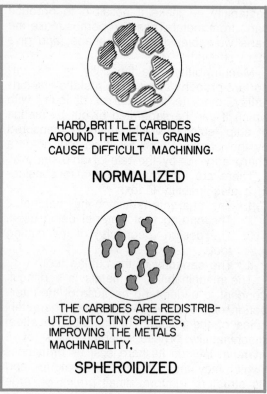

HARD, BRITTLE CARBIDES
AROUND THE METAL GRAINS
CAUSE DIFFICULT MACHINING.

NORMALIZED

THE CARBIDES ARE REDISTRIB-
UTED INTO TINY SPHERES,
IMPROVING THE METALS
MACHINABILITY.

SPHEROIDIZED

GRAY CAST IRON HAS BRITTLE
GRAPHITE FLAKES THAT BREAK
EASILY WHICH ALLOWS HIGH
MACHINABILITY.

COPPER HAS RANDOM GRAINS
THAT ARE SOFT AND DUCTILE.
A TOUGH CHIP LOWERS
MACHINABILITY.

Fig. 1-39.　Machinability may be improved by heat treating to change the metal's structure.

Fig. 1-40.　Brittleness, hardness, and ductility are factors of machinability.

ing, normalizing, or tempering also increases machinability. You will learn more about these processes in later chapters.

The machinability of some leaded steels may be improved by the addition of elements such as sulfur, phosphorus, or manganese. These steels are then called **free-machining** (or free-cutting) steels. The sulfur will cause a small decrease in the metal's impact strength, but the lead does not affect its mechanical properties.

Machinability may be improved by heat treatment of metals. Annealed tool steel with a carbon content of between 0.90% to 1.10% is difficult to machine because of the hard, brittle carbides around the metal grains. By heat treating (spheroidizing), the carbide is redistributed as tiny spherical particles, and the steel becomes easily machinable, Fig. 1-39.

Gray cast iron has a high machinability rating, because the chips break easily due to the flaky graphite structure of the particles, Fig. 1-40. **Copper** has a low machinability rating because the chips are flexible due to the softness and toughness of the metal.

Two properties greatly affect machinability of metal (1) hardness and (2) ductility. When the hardness of the metal workpiece increases, the tool has greater machining resistance, so the machinability decreases.

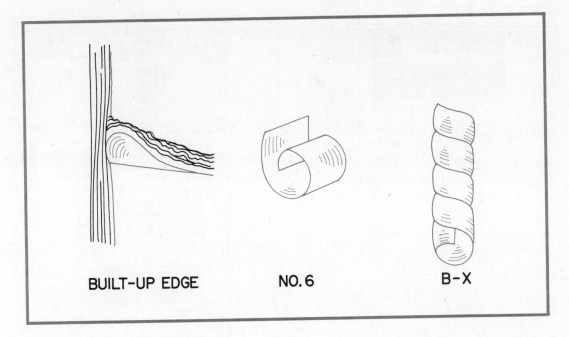

BUILT-UP EDGE   NO. 6   B-X

Fig. 1-41. Types of chips. Chips that look like a number 6 or a B-X electrical cable are the most efficient.

A highly ductile metal is a tough metal which produces a continuous chip when machined at high speeds, it generally has high machinability, Fig. 1-41.

Both the material and design (shape and geometry) of the cutting tools are carefully chosen. The most frequently used metals for cutting tools are **high-speed steels and carbides.** The shapes and angles vary with the characteristics of the metal of the workpiece and the material of the cutting tool itself.

The horsepower, the type of cutting fluid, the feed and depth of cut, the rigidity of the workpiece and the machine itself influence machining efficiency in actual production, Fig. 1-42. Measure of machinability can also be done by weighing the chips which have been removed per unit of time.

Formability. Another characteristic of metal is its ability to be formed. This is most important in manufacturing. Forming is achieved by **plastic flow,** or the redistribution of the metal into the shape desired,

FIG 1-42.
Machinability of Metals

Material	Machinability Index
Aluminum and Magnesium	300-1000
Brass and Cast Iron	120
Steel	60
Copper	40
Titanium Alloy	30
Inconel X	15
Hastelloy C	10
HS 21	6

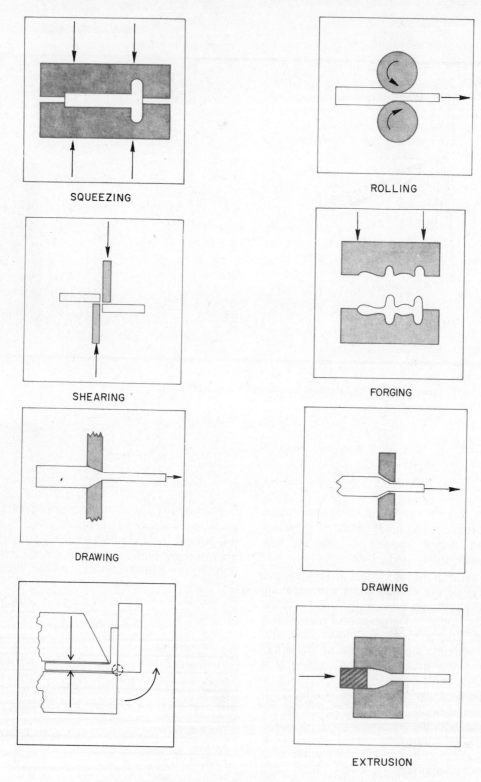

SQUEEZING

ROLLING

SHEARING

FORGING

DRAWING

DRAWING

EXTRUSION

Fig. 1-43. Processes for cold metal forming and hot metal forming.

Fig. 1-43. Since metals will flow under pressure, they can be twisted, bent, upset, stretched out, and fabricated into many forms and parts, Fig. 1-44. It is the fastest way to change the shape of metal and generally the most economical way to produce a high-strength part with the least amount of metal and labor invested.

Ductility is a necessary characteristic when metal is formed under pressure. A metal's ductility is dependent upon the **crystal structure,** the **alloying elements,** the **grain size**, and the metal's condition of **hardness.**

The primary **cold metal forming processes** are squeezing, shearing, drawing, and bending. The primary **hot metal forming processes** are rolling, extruding, drawing, and forging. Of these, forging and rolling are used to produce the largest amount of hot formed metal.

Joinability. Metals can be fastened together by several processes. In welding, metals are bonded with heat. The metals are first made fluid, then they run together, forming a **cohesive joint**, Fig. 1-45. They may be fastened by adding a different material within the joint. In **adhesive bonding,** the new material diffuses into the joined materials, Fig. 1-46. In soft soldering, the

Fig. 1-45.   Cohesive fasteners. When they are welded, the assemblies melt together into one unit.

Fig. 1-44.   Metal being squeezed to shape by a 35,000-ton United Press. Under this pressure the metal flows to the shape of the die.

Fig. 1-46.   The two metal plates connecting these chains were adhesively bonded and can support much weight.

A. Bolts, nuts, rivets, screws, clips, and clamps fasten parts together.

B. The titanium 747 aircraft landing gear beam is built up with mechanical fasteners.

Fig. 1-47. Mechanical joining with fasteners.

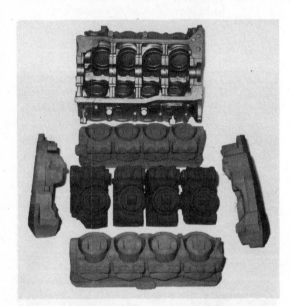

Fig. 1-48. Cast iron V-8 engine block poured with great accuracy and detail. The cores which produce the cavities and holes for water passages and cylinder bores are in the foreground.

solder behaves as a solvent and diffuses into the metal making a thin layer of **amalgam.** Amalgam is a **third material** which differs from the solder and the base metal.

This amalgam effects the bond. In adhesive bonding with organic materials such as epoxies, the wetting ability of the polymers (chemical compounds) allows it to diffuse into the surface of the metals and the catalyst causes the polymer to change its characteristics and become rigid. Thus it forms a strong joint.

One of the most frequently used methods of joining metal parts and components together is **mechanical joining,** Fig. 1-47. Parts may be joined together with a bolt and nut or fastened with a sheet metal screw. Metal parts may be joined with rivets or by crimping and clamps if they are to be fastened for long periods of time. For short period fastening, staples and banding may be used.

Castability. Castability is the property of metal which allows it to be liquefied and formed into parts or products by pouring it into a specially shaped mold, Fig. 1-48. Not all metals may be economically cast. Aluminum alloys, copper-based alloys, cast iron, and magnesium alloys have a high degree of castability and are most frequently used for commercial products. Metals such as silver and gold are highly castable, but it is not economically feasible to use them for large volume production.

Common Ferrous Manufacturing Metals

Although many metals are used in manufacturing, the bulk of products is manufactured from a few metals. The most commonly used ferrous metals today are:

- Cast iron
- Carbon steels
- Alloy steels

The ferrous manufacturing metals are predominantly composed of iron.

Cast Irons

Cast iron refers to a group of ferrous metals with a carbon content of 2% to 5%. The metal is mainly iron with carbon and small amounts of silicon, although it may contain some impurities. Materials such as manganese, phosphorus, and sulphur may be present because of the fuel and raw materials used in its manufacture.

By changing the carbon content and using heat treatment, cast irons with different properties are produced. Cast iron has low ductility and malleability, thus it cannot be forged or rolled into shape. It is a foundry metal. Cast iron melts easily and can be poured into almost any shape required. Because cast iron is very fluid when it is being poured, great detail and complicated shapes can be obtained by a rather inexpensive production process.

Cast iron has production advantages because the tools necessary to work it are not expensive, and all metalworking equipment will machine cast iron. Cast iron has good machinability and will produce a rigid, durable part. Cast iron has a vibration-damping (diminishing) capacity which makes it de-

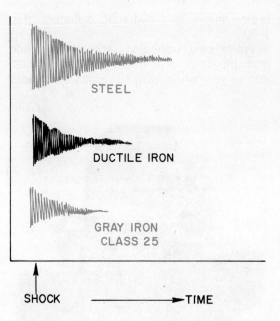

Fig. 1-49. Relative damping of shock and vibration. Steel, ductile iron, and gray iron are compared.

sirable for all types of machines and machine tool structural parts, Fig. 1-49.

The groups of cast iron are:

- Gray cast iron
- White cast iron
- Malleable iron
- Ductile (nodular) iron
- Alloy cast irons

Gray cast iron receives its name from the color of a fractured piece of cast iron. Graphite is present throughout the material. When you run your finger over a fresh break, it becomes dirty from the **graphite** rubbing off the surface. The quality of the cast metal is based upon its **tensile strength.** A **class 20 cast iron** has a tensile strength of 20,000 psi, and a **class 40 cast iron** possesses a tensile strength of 40,000 psi. Gray cast iron is the material used in automotive engine blocks, fly wheels, camshafts, machine tool bases, cast iron pipe and fittings, bathtubs, ornamental castings, gears, brake drums, and hydraulic cylinders, Fig. 1-50.

White cast iron also receives its name from the color of a fractured piece of cast iron. A very white fracture results from rapid cooling of the metal after the molten state. With the rapid chill, the iron and carbon cannot become disassociated, and graphite flakes are not formed.

White cast iron is used where extreme hardness is desired, because durability and abrasion resistance are its chief characteristics. Such products as rock crusher jaws, grinding rolls for processing clay and brick and sand, ball mills, drawing dies, and extrusion nozzles are made from it. White cast iron is used where hardness and wear resistance are important, but where its service does not require ductility. Brittleness is white cast iron's greatest disadvantage.

Malleable cast iron is made from white cast iron castings. These castings are annealed (heat treated) to make them much more ductile or malleable than white cast iron or gray cast iron. The material has its toughness greatly improved and is also softer than the other irons, Fig. 1-51. This added toughness makes these castings much more resistant to shock. The shock resistance makes possible many inexpensive tough parts for machinery and automobiles. Malleable cast iron is used to manufacture such parts as gear cases; brake supports; pipe fittings; anchors; rocker arms; gears;

Fig. 1-50. Gray cast iron has great detail, is easily machined, is rigid, and can dampen shock and vibration.

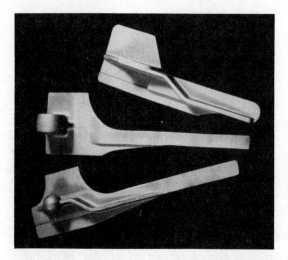

Fig. 1-51. Malleable cast iron is tougher and softer than other cast irons. It can withstand shock, so it is used extensively in agricultural machinery.

Fig. 1-52. Ductile iron is a tough cast iron. It has increased strength, because its ductility is increased. Thus, this crankshaft can withstand the shock of operation.

sprockets; tractor, truck, and automobile parts; and agricultural machinery.

Ductile iron (**nodular iron**) is made by increasing the strength and toughness of gray cast iron. Ductile iron is manufactured by adding a small amount of magnesium or cerium to the ladle just before the metal is poured into the mold. When the metal is solidifying in the mold, the graphite nodules or spheres form, producing a high-strength tough casting. These castings do not require heat treatment to give them their characteristics. Ductile iron makes inexpensive shapes which are tough and strong. Some applications of ductile iron are crank shafts, pistons, cylinder heads, drive pulleys, gears, cams, lead pots, track rollers, cam shafts, levers, and machine tool castings, Fig. 1-52.

Alloy cast irons are castings with other metals added. The other metal will usually be more than 3% of the new metal. High alloy irons may be made from gray cast, white cast, or ductile cast irons and will depend on the qualities desired. The alloy irons have three general use classifications: (1) wear resistance, (2) corrosion and heat resistance, and (3) special applications.

The alloying elements added to the cast iron change the crystalization time for the formation of graphite flakes. This crystallization time can influence the qualities of:
- wearability
- corrosion and heat resistance
- high-temperature strength

The addition of silicon, nickel, chromium, and copper **increases the corrosion resistance.** Silicon and chromium also **increase heat resistance.** Molybdenum and nickel **increase high-temperature strength.** If the alloying elements of molybdenum, chromium, copper, vanadium, and nickel are used, a **high-strength casting results.** When enough alloy is added to produce a martensitic (fine grain) structure, the casting becomes **wear-resistant and nonmachinable.**

Carbon Steel

Steel is a general term which includes many iron alloys. It is made of iron with a carbon content of 0.05% to 2.0%. Carbon steel accounts for 90% of the total steel production and is widely used in the manufacture of products. The primary characteristic of steel is its strength. Articles which must be very strong are made from steel. Examples are automobiles, appliances, or tools. Other articles may be made from another material, but steel is used as a reinforcing member. For example, in the construction industry, structural steel building frames add strength to the building.

Carbon is the principle hardening element in carbon steel. As the carbon content increases, the hardness and tensile strength also increase, but the ductility and weldability decrease. A very small change in carbon content makes a great difference in the characteristic of the steel, Fig. 1-53. The amount of carbon in the steel determines the classification of the steel:

- low-carbon steel.
- medium-carbon steel.
- high-carbon steel.

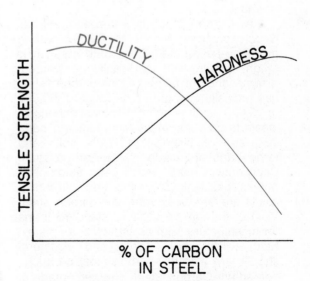

Fig. 1-53. Hardness and tensile strength are increased when carbon is added to steel. Ductility and weldability decrease.

FIG. 1-54.
Different Products with Varying Carbon Content in Steel

Percentage of Carbon	Products
0.05-0.10	Wire, nails, sheet steel
0.10-0.20	Screws, rivets, casehardened parts
0.20-0.35	Mild steel parts, structural steel plate, forgings
0.35-0.45	Machinery steel, axles, shafts
0.45-0.55	Large forgings, heavy-duty parts
0.60-0.70	High-carbon steel, dies, rails, tool steel
0.70-0.80	Cutting tool steel, cold chisels, band saws, hammers, shear blades
0.80-0.90	Drill rod, punching and blanking dies, rock drills
0.90-1.00	Small punches and dies, reamers, springs
1.00-1.10	Metal cutting tools, lathe shaper, planner, Small springs
1.10-1.20	Hand tools and cutlery small tops, dies, knives
1.20-1.30	Drawing dies, files, bearing races and balls, razors

The characteristics of each group largely determine the products which can be made from them, Fig. 1-54.

Low-Carbon Steel. Low-carbon steel has a percentage of 0.05% to 0.30% carbon and may be referred to as **mild steel.** It is tough, ductile, and can easily be welded, drawn, and machined. Low-carbon steel has a tensile strength in the range of 60,000 to 70,000 psi. It is one of the least expensive steels to produce. Many manufactured parts such as wire, chain, nails, rivets, bolts, and forge parts are made of low-carbon steel. Deep-drawn parts such as automobile bodies, gas tanks, oil pans, and differential cases are produced from low-carbon steel. Most of the mill products or standard forms for manufacture such as sheets, strip, plates, bars, wire, pipe, tubing, and structural shapes are also low-carbon steel products.

Medium-Carbon Steel. The percentage of carbon in medium-carbon steel ranges from 0.30% to 0.60%. This additional carbon makes the steel harder and stronger, but also more difficult to weld, machine, or forge. The steel's tensile strength varies with the amount of carbon, from 70,000 to 150,000 psi (with special heat treatment). Medium-carbon steel can be heat treated to provide the hardness, strength, and ductility required for products such as front wheel spindles or rear axles of automobiles. Medium-carbon steel can be forged to shape and machined to size to produce rails, axles, gears, connecting rods, automobile springs, and forgings, Fig. 1-55. Steels with a carbon content of 0.40% to 0.45% are called **axle steels.** They can withstand great twisting forces and are used for such articles as axles and high-stressed nuts and bolts.

High-Carbon Steel. The carbon content ranges from 0.60% to 1.50% in high-carbon steel. It is very strong and hard. Railroad equipment, automobile and truck parts, and farm machinery are made from high-carbon steel. Parts such as springs, grinding balls and bars, hammers, cables, axes, and

Fig. 1-55. Medium-carbon steel is forged and later heat treated to give the part great strength and toughness.

Fig. 1-56. High-carbon steel has great strength and hardness.

wrenches are also produced from high-carbon steel with a content of 0.60% to 0.95% carbon. This grade of steel has great hardness, toughness, and strength, Fig. 1-56.

When the carbon content is increased from 0.60% to 1.50% carbon, the strength and wear resistance increase, but the hardness does not continue to increase. These steels are referred to as **tool steels** and are used to manufacture cutting and forming tools such as dies, taps, drills, reamers, chisels, and forming and bending dies. The maximum hardness is obtained with 0.80% carbon in steel at Rockwell C-66. At 0.60% carbon, near maximum hardness is possible.

Alloy Steel

Alloy steels have had other metals added to the carbon steel to improve its properties. The properties that are often desired are increased strength and hardness, retention of the metal's properties at elevated temperatures, and improvement of the metal's resistance to corrosion.

Elements added to steel for alloying are chromium, nickel, manganese, vanadium, molybdenum, silicon, tungsten, phosphorus, copper, titanium, zirconium, cobalt, columbium, and aluminum.

These elements form two groups. The first group forms a solid solution in which the element dissolves in iron. Silicon, manganese, nickel, molybdenum, vanadium, tungsten and chromium increase the hardness and strength by strengthening the iron base of the alloy. The second group forms complex carbides when cooled slowly. These carbides precipitate out of a solid solution and are scattered throughout the metal. Chromium, manganese, vanadium, tungsten, titanium, columbium, and molybdenum are **carbides of deep hardening** which increase wear resistance and the ability to withstand heat. These carbides are possible only when the base metal has sufficient carbon in the metal to form alloy carbides.

Steel Classification

Steels are produced by a number of processes. The steels have been given prefixes which relate to the manufactured method and are designated as follows:

Prefixes	Steel Manufacturing Processes
B	Bessemer Steel
C	Open-Hearth Steel
D	Electric Furnace Steel

Steel is further classified by a four- or five-digit system which specifies the elements within the steel. The **Society of Automotive Engineers** (S.A.E.) and the **American Iron and Steel Institute** (A.I.S.I.) have set up a steel numbering system.

Carbon is a very important element in steel, so the last two digits of the numbering system represent the points of carbon. The first digit represents the element that is added to the alloy, and the second digit is the percentage of the element added, Fig. 1-56A.

FIG. 1-56A.
Steel Numbering System

S.A.E. or A.I.S.I. Number	Element Added
10xx	Plain Carbon
11xx	Lead-Free Machining Steel
13xx	Manganese
2xxx	Nickel
3xxx	Nickel and Chromium
4xxx	Molybdenum
5xxx	Chromium
6xxx	Chromium-Vanadium
7xxx	Tungsten
8xxx	Nickel-Chromium-Molybdenum
9xxx	Manganese-Silicon

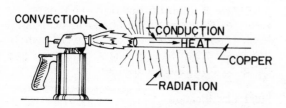

Fig. 1-57. Heat is energy. The movement of metal molecules spreads the heat to the surrounding atmosphere and other parts of the metal much like an automobile's radiator.

Thus C1040 is read as open-hearth steel with no alloy other than carbon with 40 points of carbon in the steel. In this system 100 points is equal to 1% of carbon in the steel. Therefore 40 points equals four tenths of one percent (0.4%) or C1040.

Common Nonferrous Manufacturing Metals

The nonferrous group of metals includes a large number of metals and alloys that contain no iron. These alloys have the following characteristics that are needed in manufacturing:

- Outstanding low weight
- Corrosion resistance
- Electrical conductivity
- Strength
- Formability.

The common nonferrous metals and alloys are based on copper, aluminum, and zinc.

Copper

Copper may be used as a pure metal or as a base metal for many alloys. Pure copper is usually used as some form of electrical conductor because of its electrical conductivity, resistance to corrosion, and its ease in forming and joining.

Copper also has the properties of excellent thermal conductivity. It is a nonmagnetic, tough metal of warm reddish color, Fig. 1-57. Copper is most frequently used as an alloy. Its properties may be shared with other metals to produce desired qualities. The strength of these alloys may be varied from the 30,000 psi of almost pure copper to the 200,000 psi of beryllium copper. Copper alloys are grouped into several general categories. These categories have a series of alloys under each group. The common categories are coppers, brasses, leaded brasses, phosphor bronzes, aluminum bronzes, silicon bronzes, beryllium coppers, cupro nickels, and nickel silvers, Fig. 1-57A.

FIG. 1-57A.

Atomic Weights

Name	Sym-bol	At. No.	International Atomic Weight 1959	Valence	Name	Sym-bol	At. No.	International Atomic Weight 1959	Valence
Actinium	Ac	89	(227)	....	Neodymium	Nd	60	144.27	3
Aluminum	Al	13	26.98	3	Neon	Ne	10	20.183	0
Americium	Am	95	(243)	3, 4, 5, 6	Neptunium	Np	93	(237)	4, 5, 6
Antimony,				•	Nickel	Ni	28	58.71	2, 3
stibium	Sb	51	121.76	3, 5	Niobium				
Argon	Ar	18	39.944	0	(columbium)	Nb	41	92.91	3, 5
Arsenic	As	33	74.92	3, 5	Nitrogen	N	7	14.008	3, 5
Astatine	At	85	(210)	1, 3, 5, 7	Nobelium	No	102	(254)	....
Barium	Ba	56	37.36	2	Osmium	Os	76	190.2	2, 3, 4, 8
Berkelium	Bk	97	(249)	3, 4	Oxygen	O	8	16.000	2
Beryllium	Be	4	9.013	2	Palladium	Pd	46	106.4	2, 4, 6
Bismuth	Bi	83	208.99	3, 5	Phosphorus	P	15	30.975	3, 5
Boron	B	5	10.82	3	Platinum	Pt	78	195.09	2, 4
Bromine	Br	35	79.916	1, 3, 5, 7	Plutonium	Pu	94	(242)	3, 4, 5, 6
Cadmium	Cd	48	112.41	2	Polonium	Po	84	(210)	....
Calcium	Ca	20	40.08	2	Potassium,				
Californium	Cf	98	(251)	....	kalium	K	19	39.100	1
Carbon	C	6	12.011	2, 4	Praseodymium	Pr	59	140.92	3
Cerium	Ce	58	140.13	3, 4	Promethium	Pm	61	(145)	3
Cesium	Cs	55	132.91	1	Protactinium	Pa	91	(231)	....
Chlorine	Cl	17	35.457	1, 3, 5, 7	Radium	Ra	88	(226)	2
Chromium	Cr	24	52.01	2, 3, 6	Radon	Rn	86	(222)	0
Cobalt	Co	27	58.94	2, 3	Rhenium	Re	75	186.22	
Columbium, see					Rhodium	Rh	45	102.91	3
Niobium					Rubidium	Rb	37	85.48	1
Copper	Cu	29	63.54	1, 2	Ruthenium	Ru	44	101.1	3, 4, 6, 8
Curium	Cm	96	(247)	3	Samarium	Sm	62	150.35	2, 3
Dysprosium	Dy	66	162.51	3	Scandium	Sc	21	44.96	3
Einsteinium	Es	99	(254)	....	Selenium	Se	34	78.96	2, 4, 6
Erbium	Er	68	167.27	3	Silicon	Si	14	28.09	4
Europium	Eu	63	152.0	2, 3	Silver, argentum	Ag	47	107.873	1
Fermium	Fm	100	(253)	....	Sodium, natrium	Na	11	22.991	1
Fluorine	F	9	19.00	1	Strontium	Sr	38	87.63	2
Francium	Fr	87	(223)	1	Sulfur	S	16	32.066*	2, 4, 6
Gadolinium	Gd	64	157.26	3	Tantalum	Ta	73	180.95	5
Gallium	Ga	31	69.72	2, 3	Technetium	Tc	43	(99)	6, 7
Germanium	Ge	32	72.60	4	Tellurium	Te	52	127.61	2, 4, 6
Gold, aurum	Au	79	197.0	1, 3	Terbium	Tb	65	158.93	3
Hafnium	Hf	72	178.50	4	Thallium	Tl	81	204.39	1, 3
Helium	He	2	4,003	0	Thorium	Th	90	(232)	4
Holmium	Ho	67	164.94	3	Thulium	Tm	69	168.94	3
Hydrogen	H	1	1.0080	1	Tin, stannum	Sn	50	118.70	2, 4
Indium	In	49	114.82	3	Titanium	Ti	22	47.90	3, 4
Iodine	I	53	126.91	1, 3, 5, 7	Tungsten				
Iridium	Ir	77	192.2	3, 4	(wolfram)	W	74	183.86	6
Iron, ferrum	Fe	26	55.85	2, 3	Uranium	U	92	238.07	4, 6
Krypton	Kr	36	83.80	0	Vanadium	V	23	50.95	3, 5
Lanthanum	La	57	138.92	3	Xenon	Xe	54	131.30	0
Lead, plumbum	Pb	82	207.21	2, 4	Ytterbium	Yb	70	173.04	2, 3
Lithium	Li	3	6.940	1	Yttrium	Y	39	88.91	3
Lutetium	Lu	71	174.99	3	Zinc	Zn	30	65.38	2
Magnesium	Mg	12	24.32	2	Zirconium	Zr	40	91.22	4
Manganese	Mn	25	54.94	2, 3, 4, 6, 7					
Mendelevium	Md	101	(256)	....					
Mercury,									
hydrargyrum	Hg	80	200.61	1, 2					
Molybdenum	Mo	42	95.95	3, 4, 6					

* Because of natural variations in the relative abundances of the isotopes of sulfur, the atomic weight of this element has a range of ±0.003.

** The 1959 atomic weights are based on 0 = 16,000.

Brasses. Brasses are alloys principally of copper and zinc. A 70-30 brass means that the alloy is 70% copper and 30% zinc. Other elements may be added in small amounts to improve the brass's machinability, strength, ductility, and corrosion resistance. The common names used for brasses are **red brass, cartridge brass** (yellow brass), and **architectural bronze.**

Red Brasses (85Cu - 15Zn) are low in the percentage of zinc in the alloy. A series of alloys results by changing the amount of zinc. Red brasses may be joined by welding, silver soldering, or soft soldering. They may be formed by bending, drawing, embossing, and cutting. Red brasses are used for electrical sockets, hardware, plumbing pipe, condenser and heat-exchanger tubes, radiator cores, weather stripping, flexible hose, jewelry, compacts, and nametags. These alloys work well cold, making drawing and stamping processes easily adaptable to production. The red brasses contain from 5% to 20% zinc.

Cartridge brass (70 Cu - 30 Zn) and **yellow brass** (65 Cu - 35 Zn) are alloys which lend themselves to the **deep-drawing processes** of manufacturing such as cartridge

Fig. 1-58.   These ship propellers are made of manganese bronze.

shells, automotive radiator cores, tanks, fasteners, springs, rivets, flashlight shells, lamp fixtures, and plumbing fixtures. Yellow brass contains between 20% to 36% zinc and is the most widely used of the brasses.

Architectural bronze (57 Cu - 40 Zn - 3 Pb) does not contain tin and thus is really a brass. This alloy has excellent hot-working characteristics, since it can be forged, extruded, or machined into many articles. It is used in industrial forgings, architectural extrusions, store fronts, trim, handrails, grills, hinges, and lock bodies.

Naval brass (**tobin bronze**) is very close to the composition of architectural bronze, except a small amount of tin has been added (60 Cu - 39.25 Zn - 0.75 Sn). The addition of tin increases the resistance to saltwater corrosion. This resistance is necessary for marine hardware. Tobin bronze is used for propeller shafts, piston rods, valve stems, nameplates, welding rods, turnbuckle bolts, and condenser plates.

Manganese bronze (58.5 Cu - 39 Zn - 1.4 Fe - 1 Sn - 0.1 Mn) has iron and manganese added in small amounts in addition to tin. When these elements are added to the basic bronze, high strength and excellent wear-resistance result. This alloy is widely used for ship propellers, which are subject to great wear, Fig. 1-58. Other applications are pump rods, valve stems, clutch disks, and welding rods. Manganese bronze has excellent hot workability, so it is widely forged and extruded.

Leaded brasses are used primarily to increase machinability and are used for screw machine products, Fig. 1-59.

Bronzes more accurately are alloys of copper and tin. As you will note above, some of the brass alloys are referred to as bronze. However, bronze generally is considered a better quality alloy than brass. Because of this sales appeal, some of the special brasses are referred to as bronze.

Commercial bronzes are made from copper and tin with the elements of aluminum, silicon, beryllium, and nickel added to produce the special characteristics needed from the metal.

Fig. 1-59. Leaded brasses increase machinability. Free machining is important for products machined on six-spindle automatic screw machines.

Aluminum bronzes are used where strength and corrosion resistance are required. These aluminum bronzes, which contain 10% aluminum, may be heat treated to increase their hardness and strength from 80,000 psi to 100,000 psi. Products such as gears, drawing and forming dies, propeller hubs, blades, bearings, bushings, decorative grills, paint pigments, and nonsparking tools are made from these alloys.

Silicon bronzes are applied when conductivity, weldability, strength, and corrosion resistance are required. These alloys have the strength of mild steel and the corrosion resistance equal to that of copper. This metal is used for hydraulic pressure lines, pressure vessels, boilers, pumps, shafting, electrical fittings, and marine hardware.

Beryllium bronzes are utilized when high strength, good electrical conductivity, and resistance to shock or repeated stresses are the desired characteristics. Beryllium coppers may be heat treated and made very hard so that they have strength up to 200,000 psi. Products using beryllium copper are electrical hair springs, instrument diaphragms, bellows, bourdon tubes, electrical fuse clips, and connectors.

Nickel silver (German silver) is an alloy of copper, nickel, and zinc. This alloy is noted for a silver-blue white color and its corrosion resistance to food. It is an excellent base metal for plating chromium, nickel, and silver. Products manufactured from nickel silver include table flatware, art metal, nameplates, costume jewelry, and zippers.

Monel metal (65 Ni - 28 Cu) is an alloy of nickel-copper ore found in Canada. It is often called a natural alloy. The monel series of alloys has high resistance to corrosion for such materials as salt solutions, food acids, strong alkalies, and corrosive gases (chlorine and ammonia). Monel is used in corrosive conditions which require strength, such as sinks, laundry equipment, valves, pump rods, and propeller shafts. Monel may be cast, welded, forged, and worked cold to produce the many parts and shapes which must resist corrosion.

Aluminum

Aluminum is a lightweight metal with a density which is one-third that of steel. Aluminum alloys are heat treatable and may develop strengths up to 88,000 psi for the 7178-T6 aluminum alloy. With this strength-to-weight ratio, aluminum has an advantage over steels in aerospace applications.

Aluminum also has excellent forming characteristics. It may be bent, extruded, joined, or machined with comparative ease. Aluminum is an excellent electricity and heat conductor, with the added advantage of having high corrosion resistance to our atmosphere.

A large number of alloys of aluminum have been developed by metallurgists for special applications and products. Aluminum alloys are these: aluminum copper, aluminum manganese, aluminum silicon, aluminum magnesium, aluminum silicon magnesium, and aluminum zinc.

Classification of Wrought Aluminum Alloys. The classification of wrought aluminum alloys was standardized by the Aluminum Association in October, 1954. The system developed was a four-digit number, with the first number indicating the alloy group, the second the impurity limit in the original alloy, and the last two digits identifying either the specific alloy or the purity of the aluminum, Fig. 1-59A.

FIG. 1-59A.

Classification of Wrought Aluminum Alloys

Aluminum Association Classification Number	Alloy Group
1xxx	Aluminum 99.00% Min.
2xxx	Copper
3xxx	Manganese
4xxx	Silicon
5xxx	Magnesium
6xxx	Magnesium and Silicon
7xxx	Zinc
8xxx	Other Elements
9xxx	Unused Series

FIG. 1-59B.

Uses of Wrought Aluminum Alloys

1060	Chemical equipment, railroad tank cars
3003	Ductwork, truck panels, architectural application, builders' hardware
3004	Hydraulic tubing for commercial vehicles, storage tanks, roofing
5052	Bus and truck bodies, aircraft tubing, kitchen cabinets, marine hardware
5454	Welded structures, saltwater applications
5456	Deck housing, heavy-duty structures, overhead cranes
5457	Auto and appliance trim
6061	Transportation equipment, heavy-duty structures, marine, pipe, furniture, bridge rails
7078	Structural aircraft parts

Designations of Aluminum Alloy Conditions

— F: As fabricated
— O: Annealed, recrystallized
— H: Strain hardened
— W: Solution treated, unstable temper
— T: Heat treated to stable temper
— T3: Solution treated and cold-worked
— T4: Solution treated
— T5: Artificially aged only
— T6: Solution treated and artificially aged
— T7: Solution treated and stabilized
— T8: Solution treated, cold-worked, and artificially aged
— T9: Solution treated, artificially aged, and cold-worked
— T10: Artificially aged and cold-worked

If an alloy has been strain-hardened or cold-worked, these designations may be used.

— H1: Strain-hardened only
— H2: Strain-hardened and partially annealed
— H3: Strain-hardened and stabilized

The alloy illustrated before was 7178-T6, which means that the metal which can withstand 88,000 pounds per square inch (psi) was solution-treated and artificially aged. The metal had a specific heat treatment to provide this exceptional strength, Fig. 1-59B.

An example of this numbering system would be an aluminum alloy 7178-T6. The first digit (7) indicates that the alloy is of the zinc group. The second digit (1) indicates the original alloy modifications and may use numbers 1 through 9 to show alloy modifications or impurity limits of the alloy group. The last two numbers (78) express the minimum aluminum percentage to the nearest 0.01 percent. The major alloying element is aluminum, thus the 7178 alloy has the purity of 99.78% of aluminum with a 1 level of control over alloy modification. Zinc is the alloy group being used.

A dash separates the numbers from the letter "T" and indicates the condition of the alloy.

Zinc

Zinc is a metal which has a low melting point. It can easily be cast at temperatures ranging from 750° to 800° F. Because zinc can be cast under great pressure at these low temperatures, it produces a very accurate and consistent product. It is cast in permanent steel molds (die cast). These molds can be used many times before they wear out, since there is no heat-cracking of the molds. Thin, detailed sections with

Fig. 1-60. Zinc die castings will yield detailed smooth castings of handles and parts. Formed parts of other metals may have the zinc cast around them.

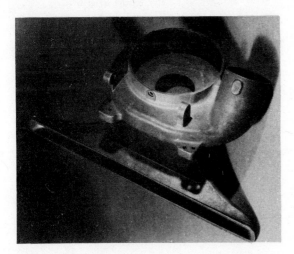

Fig. 1-61. Die casting of a vacuum cleaner housing.

smooth surfaces can be produced from the easy-flowing zinc, Fig. 1-60. Zinc is widely used in mass-produced articles, because it is a relatively inexpensive metal with moderate strength and toughness, and it can be economically cast. The appliance and electrical industries use zinc die casting for parts in washing machines, motor housings, vacuum cleaners, kitchen equipment, and utensils, Fig. 1-61. The automotive industry uses zinc to make carburetors, fuel pumps, door handles, and automobile grills.

Activities

1. Obtain a piece of cast iron welding rod and place the end in a vise. Place a rag over the cast iron and bend the cast iron rod. How would you describe the metal's characteristics? Repeat this activity with a piece of solder, then with steel.
2. Place a steel rod in a vise and cut the rod off with a cold chisel and hammer. Observe the cutting and list why steel is used for tools and industrial products.

Related Occupations

These occupations are related to metal technology.

Food processing technician
Building trade worker
Transportation worker
Communication worker

Grain farmer	Cable engineer
Cotton farmer	Cable splicer
Dairy farmer	Equipment worker
Fruit farmer	Installer — repair
Animal and livestock	worker
farmer	Pole framer
Truck farmer	Radio operator
Farm laborer	Radiotelegraph
	operator
Carpenter	Telephone repeater
Bricklayer	attendant
Cement mason	Television engineer
Floor covering	Television camera
installer	operator
Glazer	Industrial educator
Plasterer	
Painter	Aircraft pilot
Plumber	Aircraft flight engineer
Roofer	Aircraft flight mechanic
Sheet metal worker	Aircraft engine
Structural steel	mechanic
worker	Airport control operator

Cargo handler
Passenger agent
Bus driver
Bus dispatcher
Freight traffic agent
Taxi driver

Brake operator
Bridge operator
Bridge tender

Car-loading
  supervisor
Conductor
Depot master
Dispatcher clerk
Locomotive
  engineer
Porter
Railroad-car
  inspector

Telegrapher
Ticket agent
Wheel agent
Wheel inspector
Yard clerk

Barge captain
Barge operator
Cook, kitchen
  worker

Deck hand
Marine engineer
Port captain
Radio operator
Sailor
Seaman apprentice
Stevedore
Steward
Storage - wharfage
  clerk

Manufacturing Technology

Chapter 2

Words You Should Know

Manufacturing system — The application of energy, knowledge and the skills of people to produce a saleable product.

Planning — Identifying and organizing ideas into a system of production.

Manufacturing process — The methods used to produce the parts for a product. Examples are casting, forging, machining, etc.

Manufacturing materials — The raw and industrial materials, standard stock, and/or components used in the manufacturing of a product.

Industrial material — Industrial material is produced by converting raw materials such as ore into metals such as pig iron, steel, aluminum, copper, etc.

Standard stock — Industrial material is formed into bars, rods, tubes, rounds, structural shapes, rails, hot-roll strip and sheets, skelp, and plate to produce standard stock.

Ore dressing — Separating the grains of ore minerals from the waste rock.

Concentration — The enriching of a low-grade ore by removing the waste grains of rock.

Beneficiation — Concentrating the amount of iron in iron ore and preparing it for the blast furnace.

Classification — A process that separates ore from waste rock by specific gravity or other physical processes.

Sintering — Heating of fine concentrates until they bond into lumps or pellets.

Flotation — Mineral ores are separated from waste rock by adhering to an air bubble. The mineral is floated away.

Smelting — Melting ore or concentrates and combining them with chemicals so that metals are the result.

Leaching — The removal of metals from their ore by dissolving the metal chemically and later removing it from solution by precipitation or electrolysis.

Electrolysis — The passage of an electric current through a solution with the movement of ions to the electrodes.

Cryolite — A mineral used for the manufacture of sodium salts and as a flux in the electrolytic process of aluminum production.

Quality control — A system for verifying and maintaining a desired level of quality in a product or process by careful planning, continued inspection and corrective action where needed.

Administration — Managers of an industry are responsible for making long-range decisions for the organization.

Engineers — People who use scientific principles and knowledge to design and solve practical problems.

Time standard — A systematic analysis of the time required to perform a specific operation or to make a unit.

Millwright — A person who erects the machinery in an industrial plant.

Product research — A systematic investigation into a subject or device to obtain new information or to revise old concepts or devices.

Market research — The gathering and studying of data relating to consumer preferences and purchasing power.

Sales forecast — Used to make a prediction of sales before actual sales are made. It is based on data.

Industrial designer — Prepares the preliminary sketch or plans. A number of alternate solutions for the problem are worked out. The sketches are illustrated for a presentation and then completed. Frequently a three-dimensional model is built.

Prototype — The original full-scale working model of the product.

Field test — A product being used under actual operating conditions to evaluate its performance and the customer's response to it.

Product documentation — A system to supply evidence for legal or official specifications on which a contract may be based. May consist of engineering drawings, engineering specifications, bill of materials, and associated lists.

Tolerance — The permissible range of variation in the size of a dimensional object.

Manufacturing cost — The total expense of producing a product.

Scheduling — A production timetable; a number of events listed in sequence with a time when each will be completed.

Dispatcher — A person who sends out work and/or materials according to a schedule.

Quality control — A system for verifying and maintaining a desired level of quality in a product or process by planning, inspection and corrective action.

Bill of lading — A receipt which lists goods that have been shipped and also states the terms of the contract.

Fig. 2-1. Manufacturing is a system of organization for the purpose of production.

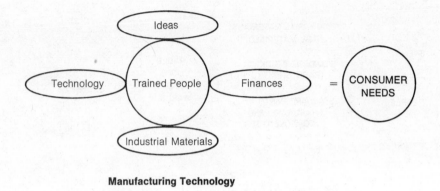

Manufacturing Technology

Fig. 2-2. Manufacturing technology depends on people to plan, make, and distribute products.

Metal Manufacturing

What is metal manufacturing? It is a system in which the various manufacturing processes are organized so that a metal product may be made economically, Fig. 2-1. It is a system that brings together the ideas of creative people, the technology that makes the production possible, and the finances that provide for the investment in plants, tools, materials, and labor. The most important ingredient of the organization is **people.** People do the planning, see that the product is made, and distribute the finished product, Fig. 2-2. In the past many jobs consisted of hard labor, and parts were produced by hand. Today energy is supplied to machines, which do the hard, dull work, Fig. 2-3. There are still many jobs available in industry, but they are primarily in the areas of planning, organizing, monitoring production, and delivering products.

A System of Production Processes

Metal manufacturing is a system which uses energy, knowledge, and skills of people to convert and process raw materials

Fig. 2-3. Precision-stamped panels are set into fixtures for welding.

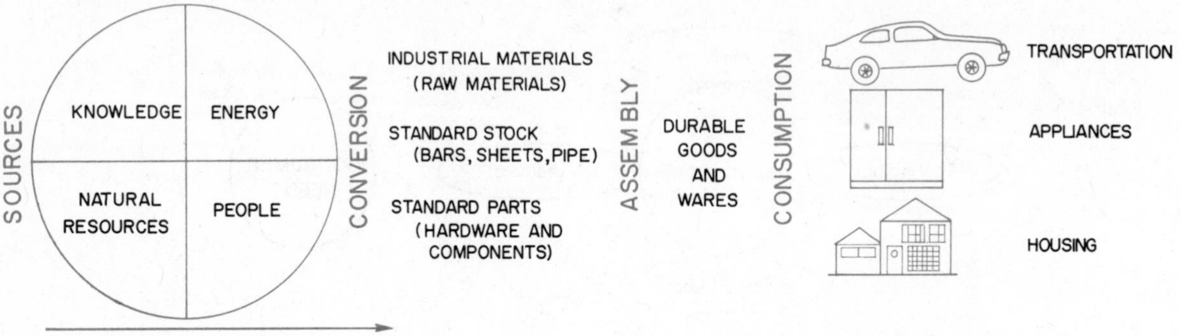

Fig. 2-4. Manufacturing: a system of production processes.

into industrial materials, Fig. 2-4. The industrial materials are converted to standard stock, which in turn becomes components and parts. The components and parts are assembled into the final product, which in turn are distributed to the consumers.

The Organization of Ideas — Planning. The organization of ideas (planning) is a most important function in a system of production. Planning is a part of the entire production. It includes these steps:

- Designing the product.
- Locating and designing the manufacturing plant.
- Selecting and organizing tools and machines at the plant.
- Specifying the sequence of operation of the production tools.
- Distributing the products.

Planning is deciding **what, how,** and **when** the manufacturing will be accomplished.

Manufacturing Processes. Manufacturers try to select the process which is both the most economical and which will produce the needed quality. Frequently a part is pro-

duced by forging, casting, machining, or stamping. The physical characteristics of the part and the plant tools and facilities available are analyzed to help in the decision of the process which will be used. Special part demands may result in the purchase of a new plant, system, or tool to produce the part.

Manufacturing Materials. After planning has been done and the manufacturing process has been chosen, the next step is taken. This includes analyzing the industrial materials needed to provide mechanical properties of the part. The physical characteristics of the material and the production process are interrelated and are studied together.

Industrial materials are the result of the conversion of the many raw materials used by industry. These industrial materials are in turn made into standard stock: metal bars, rods, pipes, tubes, sheet or structural shapes, Fig. 2-5.

Engineers spend large amounts of time and money organizing and selecting the processes and materials which will most

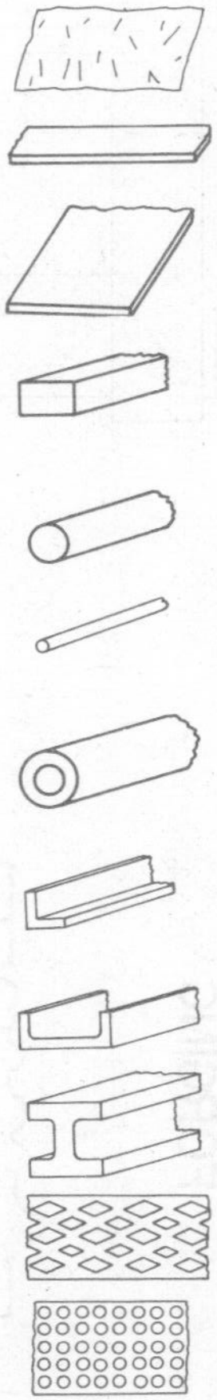

Metal Form	Metal	Thickness or Diameter	Size
Foil	Aluminum	0.0002" to 0.0055"	7" to 36", 10" x 48", Rolls 66" x 48" in dia.
Strip	Carbon Steel	0.250"	½" to 23 15/16" width
Sheet	Carbon Steel	0.0447" to 0.2299"	8" to 48" wide
Bar	Carbon Steel	¼" to 6" (square), ⅜" to 4 1/16" (hexagonal)	Lengths Rolled
Rod	Carbon Steel	7/32" to 4 7/16"	Coils
Wire	Carbon Steel	0.004" to 0.625"	Coils
Tube	Carbon Steel	3/16" to 10¾"	Wall thickness 0.028" to 0.250"
Angle	Carbon Steel	Equal leg and unequal leg ⅜" to 6"	Up to 40 ft.
Channel	Carbon Steel	Selection of web thickness and width	20 ft. to 40 ft. and special orders
I Beam	Carbon Steel	Same as above	Same as above
Expanded Sheet	Carbon Steel	30 gage to 14 gage	Up to 48" x 96"
Perforated Sheets (metal lace)	Carbon Steel	24 gage to 14 gage	Up to 48" x 120"

Fig. 2-5. Standards stock: mill forms.

economically accomplish the desired result, Fig. 2-6.

Finances. A manufacturing system needs capital to purchase machines and equipment, pay the wages of the company's employees, and acquire buildings and plant sites. Financial decisions must be made as the manufacturing system expands. Whether to borrow from a commercial bank, sell bonds or stock, or finance further expansion out of profits are all financial decisions.

People. Manufacturing production requires both **capital** and **labor,** Fig. 2-7. Industry has learned that better tools and equipment increase the productivity of the workers. Manufacturers may produce more goods and products with less time and labor by using automatic or automated machinery.

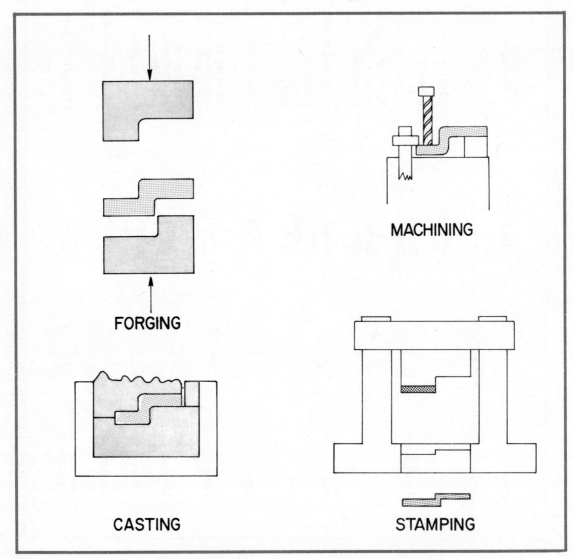

Fig. 2-6. Selecting the most economical manufacturing process.

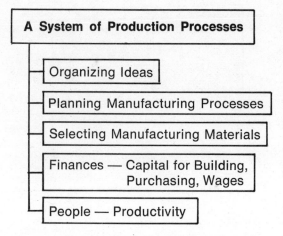

A System of Production Processes

- Organizing Ideas
- Planning Manufacturing Processes
- Selecting Manufacturing Materials
- Finances — Capital for Building, Purchasing, Wages
- People — Productivity

Fig. 2-7. Management's role in a system of production.

When simple machines are used, highly skilled workers are required to operate them. However, when more complex machines are added to production, the highly skilled workers are no longer needed to operate them. These workers can learn new skills to upgrade their positions. They are frequently retrained to install and operate new production equipment, or they may be transferred to help staff a new plant.

Fundamental Operations

The fundamental manufacturing operations carried out by a corporation are:
1. Obtaining the materials.
2. Handling and storing the materials and issuing the industrial materials.
3. Processing the industrial materials to convert them into standard stock and then into components.
4. Inspecting the product for quality control while it is being manufactured.
5. Assembling the components into the final product.
6. Testing, checking and adjusting the product after final assembly.
7. Packaging the product.

Obtaining Industrial Materials

Mining is the chief method used to initially obtain metals. Through mining, ores are extracted from the earth. **Ores** are rock which contain a high percentage of metal. They are used in the most economical way in the processing of raw metal, Fig. 2-8.

The Extraction Process. The ore minerals are removed from the earth by two primary methods: (1) open-pit mining and (2) underground mining.

Open-pit (or **opencut**) **mining** is usually used if the ore deposit is very large. The ore for copper, aluminum, and iron is frequently located in layers or pockets of ore hundreds of feet across. Nonproductive rock (**overburden**) often covers the deposit and must be removed by the most economical means possible. Large power shovels or earthmovers are used to remove the overburden and to open up the ore body. Iron ores and copper ores are obtained by digging down into the ore body and loading the minerals into railway cars or large trucks. The pits are quite deep, and a road or track is built in a spiral path down the benches (terraced shelves) to the bottom of the pit, Fig. 2-9. Power shovels and rail-

Fig. 2-9. Open-pit mine with track spiraling down the benches to the bottom of the pit.

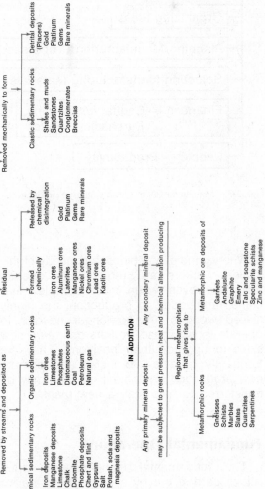

PRIMARY MINERAL DEPOSITS

Magmas within the earth are the original source of all known earth materials. On cooling and differentiation the average magma splits into

An acidic magma that on crystallization gives rise to
- Acidic igneous rocks
  - Granites
  - Syenites
  - Quartz-monzonites
  - Monzonites
  - Diorites

A basic magma that on crystallization gives rise to
- Syngenetic magmatic ore deposits of accessory minerals
  - Oxides
  - Native elements
  - Sulfides
- Basic igneous rocks
  - Gabbros
  - Peridotites
  - Dunites
  - Basalts

A mother liquor that is the source of numerous accessory minerals Forming Epigenetic magmatic ore deposits
- Contact metamorphic deposits
- Pegmatites
- Deep-seated vein deposits
- Intermediate vein deposits
- Shallow vein deposits
- Surface (magmatic spring) deposits

SECONDARY MINERAL DEPOSITS

Formed through Weathering

Chemical weathering Produces soluble products that are
- Removed by streams and deposited as
  - Chemical sedimentary rocks
    - Iron deposits
    - Manganese deposits
    - Limestone
    - Chalk
    - Dolomite
    - Phosphate deposits
    - Chert and flint
    - Gypsum
    - Salt
    - Potash, soda and magnesia deposits
  - Organic sedimentary rocks
    - Iron ores
    - Limestones
    - Phosphates
    - Diatomaceous earth
    - Coal
    - Petroleum
    - Natural gas
- Carried downward
  - Secondarily enriched ore deposits
  - Copper ores

Produces insoluble products that are
- Residual
  - Formed chemically
    - Iron ores
    - Aluminum ores
    - Laterites
    - Manganese ores
    - Nickel ores
    - Chromium ores
    - Lead ores
    - Kaolin ores
  - Released by chemical disintegration
    - Gold
    - Platinum
    - Gems
    - Rare minerals

Mechanical weathering produces fragments of minerals and rocks
- Removed mechanically to form
  - Clastic sedimentary rocks
    - Shales and muds
    - Sandstones
    - Quartzites
    - Conglomerates
    - Breccias
  - Detrital deposits (Placers)
    - Gold
    - Platinum
    - Gems
    - Rare minerals

IN ADDITION

Any primary mineral deposit or Any secondary mineral deposit may be subjected to great pressure, heat and chemical alteration producing
Regional metamorphism that gives rise to
- Metamorphic rocks
  - Gneisses
  - Schists
  - Marbles
  - Slates
  - Quartzites
  - Serpentines
- Metamorphic ore deposits of
  - Garnets
  - Andalusite
  - Graphite
  - Emery
  - Talc and soapstone
  - Specularite schists
  - Zinc and manganese

Fig. 2-8. Primary mineral deposit formation.

50

Fig. 2-11. Power shovel loading an iron ore car from the bank of a large open-pit mine.

Fig. 2-10. In open-pit mining, thousands of tons of rock and iron ore are loosened in a single delayed-action blast.

way cars are located on various benches of the pit to remove the ore from the pit.

Benches are widened as the ore is removed. Each spiral of the bench is kept about fifty feet wide. The height of the **bank** is the distance from one bench to the next. It is determined by the total thickness of the ore, the physical characteristics of the ore, the climatic conditions, and the method used for blasting the ore loose from the bank, Fig. 2-10.

The ore is loaded from the bank to the ore car and is taken up the sides of the pit to the mill. Another method used is to take the loaded ore cars down to the bottom of the pit and out a tunnel through the mountain to the mill, Fig. 2-11. In a large mine, the downhill method is preferred. Gravity helps haul the ore, and less energy is needed.

Underground mining is done where the amount of soil and waste rock is so great that removing them would be uneconomical. To mine, shafts (openings) are made in the rock to locate the ore. When the ore is discovered, it is removed from a stope (a step-like excavation) by caving the ore from the

roof of the stope, Fig. 2-12. The ore is loaded and hoisted up the shaft to the mill.

Ore dressing is the separation of the grains of ore minerals from the gangue minerals (waste rock). It is done by these processes:

1. Crushing and grinding
2. Beneficiation (washing and screening)
3. Milling
4. Flotation

The purpose of ore dressing is to concentrate the ore materials by removing the waste or tailings. To **concentrate** an ore means to enrich it by removing the waste material from it.

The two common processes of **crushing** and **grinding** produce the proper particle size for the freeing of the mineral. Crushing is usually used to break coarse material, and grinding to break fine material. **Concentration** separates the mixture of materials into mineral grains and waste grains. By ore dressing, a low-grade ore is concentrated and enriched, thus an economically profitable ore concentrate can be sent to the smelter. There, the metal is **refined**.

The **beneficiation** of iron ore is preparing it for smelting. Beneficiation is designed to prepare a low-grade ore by concentrating the amount of iron in the ore and to prepare

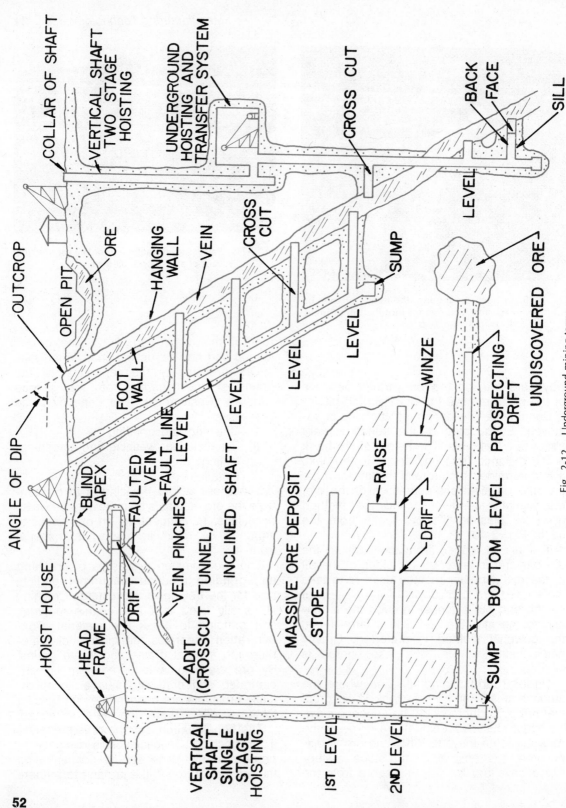

Fig. 2-12. Underground mining terms.

COLLAR OF SHAFT

VERTICAL SHAFT TWO STAGE HOISTING

UNDERGROUND HOISTING AND TRANSFER SYSTEM

CROSS CUT

BACK

FACE

SILL

LEVEL

OUTCROP

ORE

HANGING WALL

VEIN

CROSS CUT

SUMP

LEVEL

OPEN PIT

ANGLE OF DIP

FOOT WALL

FAULT LINE LEVEL

INCLINED SHAFT LEVEL

LEVEL

UNDISCOVERED ORE

PROSPECTING DRIFT

WINZE

RAISE

DRIFT

BOTTOM LEVEL

SUMP

HOIST HOUSE

BLIND APEX

FAULTED VEIN

VEIN PINCHES

DRIFT

ADIT (CROSSCUT TUNNEL)

MASSIVE ORE DEPOSIT

STOPE

HEAD FRAME

VERTICAL SHAFT SINGLE STAGE HOISTING

1ST LEVEL

2ND LEVEL

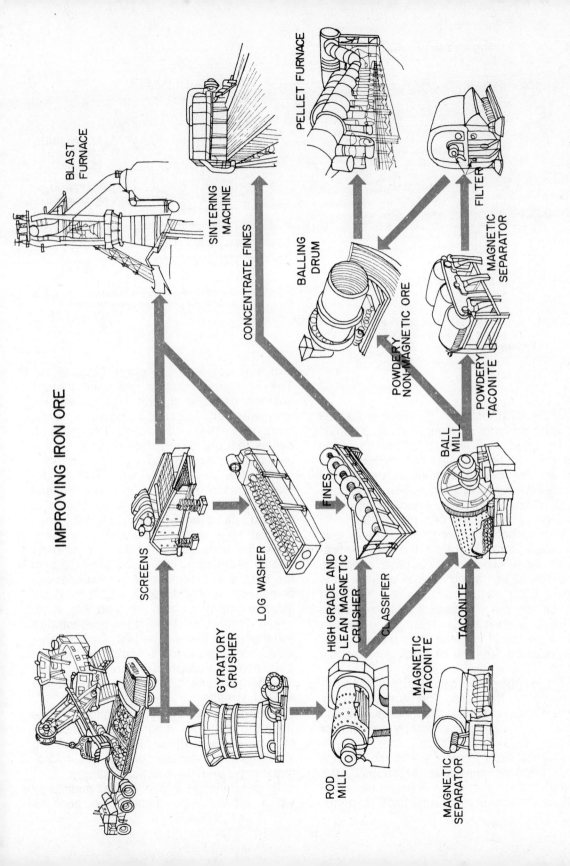

IMPROVING IRON ORE

BLAST FURNACE

SINTERING MACHINE

CONCENTRATE FINES

PELLET FURNACE

BALLING DRUM

FILTER

MAGNETIC SEPARATOR

POWDERY NON-MAGNETIC ORE

POWDERY TACONITE

SCREENS

GYRATORY CRUSHER

LOG WASHER

FINES

HIGH GRADE AND LEAN MAGNETIC CRUSHER

CLASSIFIER

BALL MILL

ROD MILL

MAGNETIC TACONITE

TACONITE

MAGNETIC SEPARATOR

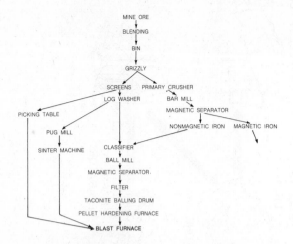

MINE ORE
BLENDING
BIN
GRIZZLY
SCREENS   PRIMARY CRUSHER
LOG WASHER   BAR MILL
PICKING TABLE   MAGNETIC SEPARATOR
NONMAGNETIC IRON   MAGNETIC IRON
PUG MILL
SINTER MACHINE   CLASSIFIER
BALL MILL
MAGNETIC SEPARATOR.
FILTER
TACONITE BALLING DRUM
PELLET HARDENING FURNACE
BLAST FURNACE

Fig. 2-13.   Beneficiation of iron ore improves the quality of the ore by removing silica sand, sulfur, and phosphorus.

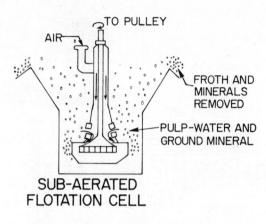

TO PULLEY
AIR
FROTH AND MINERALS REMOVED
PULP-WATER AND GROUND MINERAL

SUB-AERATED FLOTATION CELL

Fig. 2-14.   The flotation cell floats the ore on a froth to remove the ore from the waste material.

it for the blast furnace. The simplest method is **washing** and **screening.** Coarse lumps of iron oxide are washed by means of log washers to remove the fine material and sands. The coarse material is transported to the blast furnace. The fine iron oxide goes to a classifier, a machine which separates and concentrates the ore, Fig. 2-13.

Iron ore has a higher specific gravity than the waste rock. Therefore the ore is separated by a classifier. You have seen the principle of this along a creek or river edge where black sand appears where the sand is washed by a stream. The lighter silica sand is washed away, leaving the heavier black iron sand in place. This process concentrates the amount of black sand at that point.

The output of the classifier is a fine concentrate of iron ore. The concentrate is heated in a sintering machine to produce lumps or a cake of ore that is suitable for smelting in a blast furnace.

Fine magnetic iron goes from the rod mill to a magnetic separator. A rod mill uses metal rods for grinding. It is ground in a ball mill (which uses iron or steel balls for grind-

ing) with rock-hard, low-grade Taconite ore. The powdered taconite ore goes through another magnetic separator to recover the magnetic iron. Water is removed from the remainder. Then it is rolled into balls or pellets that are baked hard enough so that they may be placed into a blast furnace.

Flotation is a widely used method for the separation of nonferrous sulfide mineral ores. Flotation works on the principle that mineral particles will resist water when special hydrocarbons are added to them. These minerals adhere to an oil-coated air bubble and float to the top of the pulp with the bubble or froth. The waste material becomes wet and sinks to the bottom of the flotation cell, where it is removed, Fig. 2-14.

The froth containing the concentrated minerals is removed. The wet, fine-concentrated ore is dried and sintered (heated until it is nearly melted). The ore then sticks together in chunks and may be charged into a blast furnace.

Refining (**converting**). Refining means freeing metal from impurities or reducing it to a pure state. Three methods of refining are these: (1) pyrometallurgy, (2) hydrometallurgy, and (3) electrometallurgy.

Pyrometallurgy is a chemical metallurgy which depends upon heat action, such as

BLAST FURNACE IRONMAKING

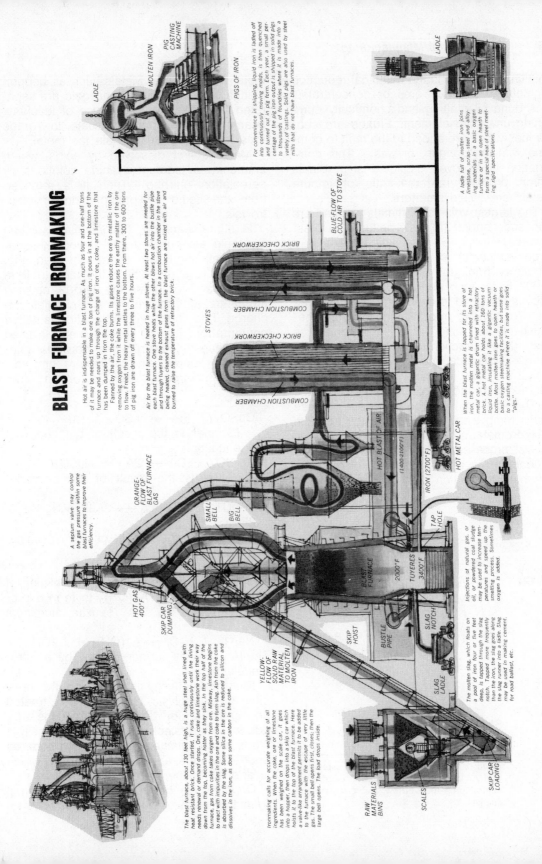

Hot air is indispensable in a blast furnace. As much as four and one-half tons of it may be needed to make one ton of pig iron. It pours in at the bottom of the furnace and roars up through the charge of iron ore, coke, and limestone that has been dumped in from the top.

Fanned by the air, the coke burns. Its gases reduce the ore to metallic iron by removing oxygen from it while the limestone causes the earthly matter of the ore to flow. Freed, the heavy metal settles to the bottom. From there, 300 to 600 tons of pig iron are drawn off every three to five hours.

Air for the blast furnace is heated in huge stoves. At least two stoves are needed for each blast furnace. One stove heats while the other blows hot air into the bustle pipe and through tuyeres to the bottom of the furnace. In a combustion chamber in the stove being heated, cleaned exhaust gases from the blast furnace are mixed with air and burned to raise the temperature of refractory brick.

For convenience in shipping, liquid iron is ladled off into continuously moving molds, is then quenched and turned out in pig form. Each year, a small percentage of the pig iron output is shipped in solid pigs to thousands of foundries where it is made into a variety of castings. Solid pigs are also used by steel mills that do not have blast furnaces.

A ladle full of molten iron joins limestone, scrap steel and alloying materials in a basic oxygen furnace or in an open hearth furnace to form a special heat of steel meeting rigid specifications.

When the blast furnace is tapped for its store of iron, the molten metal is channeled into a hot metal car, a gigantic drum lined with refractory brick. A hot metal car holds about 160 tons of liquid iron, insulating it like a gigantic vacuum bottle. Most molten iron goes to open hearth or basic oxygen steelmaking facilities, but some goes to a casting machine where it is made into solid "pigs."

A septum valve may control the gas pressure within some blast furnaces to improve their efficiency.

The blast furnace, about 130 feet high, is a huge steel shell lined with heat resistant brick. Once started, it runs continuously until the lining needs renewal or demand drops. Ore, coke and limestone work their way down from the top, becoming hotter as they sink. In the top half of the furnace, gas from coke takes oxygen from ore. Midway, limestone begins to react with impurities in the ore and coke to form slag. Ash from the coke is absorbed by the slag. Some silica in the ore is reduced to silicon and dissolves in the iron, as does some carbon in the coke.

Ironmaking calls for accurate weighing of all ingredients. When the coke, ore or limestone has been weighed on the scale car, it goes into a hopper, then drops into a skip car which hoists it to the top of the blast furnace. Here a valve-like arrangement permits it to be added to the furnace with the escape of very little gas. The small bell opens first, closes; then the large bell opens. The load drops inside.

The molten slag, which floats on a pool of iron four or five feet deep, is tapped through the slag notch. Tapped more frequently than the iron, the slag goes along the slag runner into a ladle. Slag may be used in making cement, for road ballast, etc.

Injections of natural gas, or oil, or powdered coal sludge may be used to increase temperatures and speed up the smelting process. Sometimes oxygen is added.

Fig. 2-15. The blast furnace smelts iron. Smelting is the chemical reduction of metals from their ore or concentrates.

55

smelting. Smelting is the chemical reduction of metals from their ores or concentrates. It is done by melting and chemically combining unwanted elements. **Fluxes** are substances which are added to the melt to combine with the gangue, or waste. This combination is removed as liquid slag. Plants which produce raw materials of impure metals are called **smelters.**

The largest volume of metals smelted are iron and copper. Iron is smelted in a **blast furnace,** Fig. 2-15. The blast furnace is a large vertical structure which is lined with refractory brick to withstand heat of over 3000° F. for long periods of time. The furnace is designed to chemically reduce oxide ores to metallic iron. This is possible because carbon (coke) at these temperatures has a great attraction for oxygen. After many complex chemical reactions, the net result is that the carbon removes the oxygen from the ore (rust) and leaves iron which contains carbon (pig iron) and burning gases. These gases remove the oxygen from the furnace.

Hydrometallurgy is the treatment of ores by wet processes, such as leaching.

Leaching is a chemical reaction process in which a solution dissolves most of the metal from the ore. Metals such as copper, manganese, zinc, silver, and gold are separated from the ore by the use of leaching. Leaching is used on large-scale, low-grade ores to inexpensively obtain the metal.

Metal is first dissolved by acids, water, ammonia, or sodium cyanide solutions by percolation (stirring). Then it is removed from the liquid by precipitation or electrolysis.

In **slime leaching,** the leaching solution and pulp are pumped into a wood or concrete tank lined with asphalt or lead, Fig. 2-16. Air is pumped into the tank to form an airlift and to agitate the pulp by causing the central pipe to overflow and circulate the charge. This chemical precipitation is called **cementation.** A less desired metal is used to displace the desired one from the solution. Iron, copper, zinc, aluminum, and carbon are commonly used to obtain other metals.

$CuSO_4 + Fe \longrightarrow FeSO_4 + Cu$
(Copper is the result of the chemical. equation.)
$Ag_2SO_4 + Cu \longrightarrow 2Ag + CuSO_4$
(Silver is the result of the chemical equation.)
$2AuCl_3 + 6FeSO_4 \longrightarrow 2Au2FeCl_3 + 2Fe_2(SO_4)_3$
(Gold is the result of the chemical equation.)

The preliminary processes in the refining of aluminum may use $Al(OH)_3$ to form a precipitate from a $Na AlO_2$ (bauxite, aluminum ore).

$NaAl_2 + H_2O \longrightarrow Al(OH)_3 + NaOH$
The $Al(OH)_3$ is further refined by electrometallurgy.

Electrometallurgy is based on the principle of electrolysis. Ore, previously treated concentrates, or leaching solutions are added to an electrolytic cell. The cell is made of a container, electrical bus bars, an anode, cathode, and electrical energy, which is applied to the system, Fig. 2-17.

Copper may be purified by extracting it from the leaching solution by electrolysis.

SLIME LEACHING

Fig. 2-16. Leaching solution and pulp are pumped into wood or concrete tanks lined with asphalt or lead. Air is pumped to form an airlift, causing the central pipe to overflow and circulate the charge.

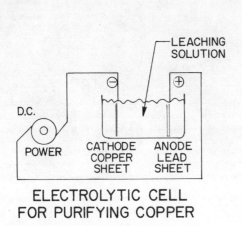

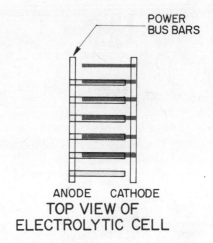

ELECTROLYTIC CELL
FOR PURIFYING COPPER

TOP VIEW OF
ELECTROLYTIC CELL

Fig. 2-17. An electrolytic cell is made up of a container, electrical bus bars, an anode, a cathode, and electrical energy.

$$2CuSO_4 + 2H_2O \longrightarrow 2Cu + 2H_2SO_4 + O_2$$

A thin sheet of cast lead that is insoluble in the leaching solution or electrolite is used to make the anode. A pure sheet of copper is used as the cathode. When the electrical power is applied to the cell, the copper is plated out of solution onto the copper plate. When the copper plate builds up to a few inches, it is removed from the solution and becomes an industrial material, Fig. 2-18.

Aluminum is obtained by dissolving alumina (Al_2O_3) in hot liquid cryolite or sodium aluminum fluoride ($Al F_3 \cdot 3Na F$) in an electrolytic furnace cell. A large carbon anode is placed in the solution. The walls and bottom of the cell are made of carbon with a cathode collector plate on the bottom of the cell. When electrical current is passed through the cell, heat is produced and the alumina is reduced to molten aluminum at the bottom of the cell, Fig. 2-19. The alumina reacts with the fluoride to replace the cryolite. The oxygen that is produced unites with the carbon anode and forms carbon dioxide, which is lost. Periodically the molten aluminum is drawn off and used in a mill or cast into aluminum pig molds.

Fig. 2-18. Lifting cathodes of 99.98% pure copper from electrolytic refining tank at Great Falls, Montana refinery.

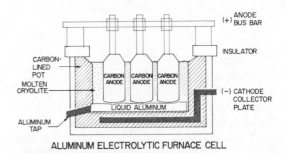

ALUMINUM ELECTROLYTIC FURNACE CELL

Fig. 2-19. Alumina is reduced to molten aluminum by the reaction of the alumina to the fluoride.

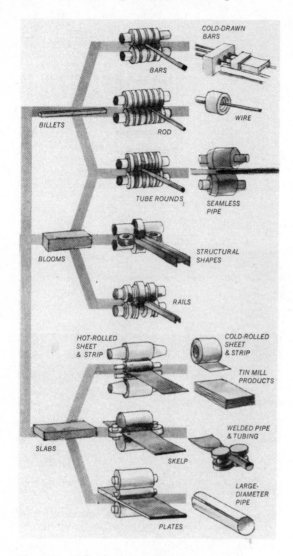

Fig. 2-20. A flowline on steelmaking.

Fig. 2-21.   Standard parts are used in the assembly of this bread-wrapping machine.

In a roughing mill steel ingots are reduced into blooms, billets, and slabs. Standard stock is produced from these. Bars, rods, tube rounds, structural shapes, rails, hot-rolled strips and sheets, skelp, and plates are standard stock, Fig. 2-20.

Standard parts or standard stock items are supplies which are made from standard stock and are used in the assembly of products. These parts include screws, bolts, rivets, springs, gears, shafts, or any of the many parts which may be assembled into components. They are made from long, continuous shapes such as wire, bar stock, pipe, sheet, and other standard stock. Standard parts are sometimes referred to as hardware. Standard parts may be assembled and used in innumerable ways to produce the components of production, Fig. 2-21.

Handling, Storing, and Issuing Standard Stock

Transportation Handling. Handling is the movement of raw material, industrial material, standard stock, or components from the place where they were produced to the place where they will be used. Handling varies because the bulkiness, the weight, and the expense of the product must be considered. Large and/or inexpensive products such as ore or finished products which must be transported long distances are moved by

Producing Standard Metal Stock

Steelmaking Processes. The raw materials for the making of steel are iron ore, coal, limestone, and air. The blast furnace takes these raw materials and combines them to form **pig iron.** Pig iron and scrap steel are added to the open-hearth furnace. Here the carbon content of pig iron is reduced, and steel ingots are produced.

Fig. 2-22. Ships are used to transport cargoes which are massive or bulky.

Fig. 2-24. Large, heavy work is handled by electric cranes that bridge the work area.

Fig. 2-23. Heavy, less bulky cargoes are transported by railroad.

Fig. 2-25. Industrial trucks move stock and components around the plant.

water in ships or barges, Fig. 2-22. Heavy, less bulky cargoes are transported by railroad cars and by trucks, Fig. 2-23. Small, light, expensive products are moved by delivery truck or air freight.

In-Plant Handling. Handling also includes the movement of products within a manufacturing plant. Large, heavy work is moved by electric cranes which are placed above the work area like a bridge and move on wheels and rails mounted high above the working floor, Fig. 2-24.

Industrial trucks are used to move smaller stock and components around the plant. These trucks are frequently called forklifts and are used in warehouses, Fig. 2-25.

Continuously moved parts within a plant are carried by monorails or conveyors suspended overhead, Fig. 2-26. If small parts and assemblies will not roll over or fall off, they are moved by conveyor belts and roller conveyors. To keep from falling, small parts may be placed in a tote tray and then conveyed.

Storage is the holding of materials and products in excellent condition until they are needed and used, Fig. 2-27. Storage and warehousing systems are responsible for unloading the various materials and products from their carriers. They are responsible for verification of the quantities received as recorded on the shipping invoice. Upon delivery the packages and materials are inspected for damage in shipment. If damage

such as freezing, wetting, or rough handling has occurred, a claim will be filed and the material reprocessed. The accepted materials are listed in the warehouse inventory and reported to inventory management.

To account for the vast number of items in a warehouse, punch card systems and computers are used to inventory, record, and locate the storage items. Large industries are automating their warehouses and building them very high in modules to make good use of space, Fig. 2-28.

Storage of products must provide protection from water, humidity, theft, and fire, so warehouses must be very carefully controlled and supervised. Materials are issued only when the proper identification is presented with a shipping order and bills of lading. The products are loaded upon carriers for delivery.

Maintaining Quality of Materials and Work

Industry must guarantee the customer that the product will have the quality and performance which is specified in the selling contract.

Quality Assurance. Seeing that reliability is built into the product is the duty of quality assurance, Fig. 2-29. To do this, the quality

Fig. 2-26. Continuously moved parts are carried by monorails within a plant.

Fig. 2-27. Storing and issuing materials from a warehouse.

Fig. 2-28. Automated warehouses are built to handle storage modules stacked very high.

(1) DESIGN DISCLOSURE
a. Specify quality req'mts
b. Define capabilities and specialities if unusual
c. Perform precontractual audit survey
d. Specify gov't source inspection req'mts

(2) SUPPLIER QUOTE
a. Verify quality req'mts are fully recognized
b. Proper placement of emphasis and priorities
c. Evidence of an acceptable quality plan

(3) NEGOTIATIONS
a. Verify capabilities exist
b. Assure quality system is acceptable
c. Ascertain quality management has organizational authority

(4) LET CONTRACT
a. Inject quality req'mts into work statement
b. Enforce quality interfaces
c. Define contract change mechanism
d. Specify milestone quality reporting

(5) ACTIVATE SUPPLIER SYSTEMS
a. Perform supplier system audits
b. Verify quality system is effective and integrates within the total system
c. Generate corrective action in deficient areas

(6) PRELIMINARY DESIGN REVIEW
a. Quality engineering participation:
1-Software
2-Design
3-Processes
4-Procurements (subs)
5-Procedures & systems
6-Methods
7-Contractual satisfaction
8-Specification satisfaction
9-State-of-the-art
10-Controls
11-Facilities (routine & special)
12-Certifications (personnel & equip.)
b. Report findings
c. Acquire resolutions to deficiencies
d. Transmit areas of concern to supplier
1-Items listed by LMSC as "not usable for new design"
2-Process techniques unacceptable (such as mechanical wirestrippers)
e. Review interface & ability to integrate with system

(7) ESTABLISH DESIGN BASELINE
a. Certify quality req'mts met
b. Configuration control established
c. Design change control enforced
d. Specify qualification req'mts
e. Identify process controls

(8) DEFINITIZE PROCESSES AND PROCUREMENTS
a. Certified processes
b. Certified operators
c. Calibration practiced and effective
d. Controls in effect & supported by objective evidence (trend data)

(9) FINAL DESIGN REVIEW
a. Review software for acceptability
b. Assure design satisfies quality concepts
c. Verify manufacturing and test programs are defined and effective
d. Assure test equipment, handling and features are defined
e. Assure processes are filed
f. Assure closure of all "open" samples from preliminary design review

(10) ENGINEERING RELEASE
a. Engineering meets specifications
b. Engineering well defined & clear
c. Change control in effect
d. Availability of related engineering
e. Approval of all top assembly drawings and interface documentation
f. Approval of all top level specifications

(11) DESIGN FREEZE
a. Identify hardware configuration
b. Identify test equipment
c. Identify processes
d. Freeze software
e. Identify suppliers (subtier & parts)
f. Specify packaging and handling
g. Stipulate environmental req'mts

(12) MANUFACTURING PHASE
a. Approved manufacturing plan
b. Acceptable discrepancy reporting system
1-Discrepancy detection
2-Failure reporting
3-Corrective action & feedback
4-Failure analysis
5-Preventive action
c. 1st and Nth article inspection
d. Controlled stores and impound areas
e. Controlled cann balization
f. Data accumulation and retention system
g. Management and technical reviews (periodic)

(13) SUPPLIER MANUFACTURING
a. Manufacturing plan issued
b. Log books issued
c. Adequate planning and work definition
d. Spar review authorized:
1-Processes
2-Data collected
3-Techniques, methods utilized
e. Collection of work authorized documents
f. Quality acceptance of all produced hardware
g. Equipment calibration and controlled tools

(17) ASSEMBLY
a. Assembly techniques approved
b. Assembly sequence specified
c. Mock-ups and prototypes
d. Traceability and lot-date code capability
e. Limited calendar life tracked
f. Quality inspection of all critical assembly operations
g. Discrepancy reporting
h. Controlled tooling

(18) MFG. TEST
a. Station proofing
b. Test procedure approval
c. Approved test plan
d. Environmental testing
e. Development testing
f. Qualification testing
g. Limited operating life tracked
h. Test history maintained
i. Failure reporting

(19) SYSTEMS TEST
a. Approved test procedures (TP's)
b. Review of qualification TP's
c. Safety limiters on test equipment
d. Test equipment validation
e. Test setup verification
f. Confidence testing
g. Operating life accumulation

(20) ACCEPT. TEST
a. Approved acceptance test procedure
b. Quality witnessed test data
c. Tested to TP defined test equipment
d. Tested within prescribed environments
e. Acceptable monitoring equipment (ovens, etc.)
f. Operating life within prescribed limits
g. Failure reporting:
1-Dispositioning by material review
2-Corrective actions
3-Failure analysis

(14) PROCUREMENT (RAW STOCK & PIECE PARTS)
a. Procurement specs defined
b. Procurements from acceptable suppliers
c. Certification of conformance available
d. Traceability and date code maintained
e. Purchase order review (quality)
f. Package and handling prescribed
g. Calendar life limits tracked

(16) RECEIVING
a. Data screening
b. Functional testing
c. Destructive physical analysis
d. Nondestructive testing (Pind, etc.)
e. Controlled logistics
f. Handling and storage req'mts specified
g. Quality acceptance of all accepted items
h. Proper identification & handling of defective parts

(21) DELIVERY
a. Data delivery req'mts
1-Failure history
2-Test history
3-Contractual compliance certificate
4-Configuration data
b. Final inspection
c. Consent to ship decision

(15) SUBTIER PURCHASES
a. Quality req'mts identified
b. Controls established to assure quality req'mts met
c. Subcontract review (quality)
d. Supplier spar activity
e. Pre-cap visual
f. Radiographic
g. Electron scanning microscope
h. Objective evidence of inspection acceptability

MODIFICATION (After Delivery)
a. Contractual authorization
b. Released-approved engineering
c. Approved test/rework plan
d. Approved retest procedure
e. Authorized parts/components substitutions
f. Approved rework techniques/processes
g. Quality witnessed rework/retest data

Fig. 2-29. The quality assurance group is actively engaged throughout the manufacturing process.

A. A quality control measuring center.

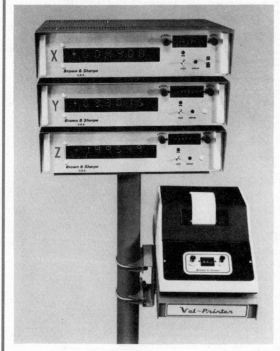

B. Simultaneous X and Y coordinates and diameter measurements to 50 millionths of an inch.

Fig. 2-30. Sensitive machines record data to ensure quality.

assurance manager and his or her group are actively engaged in the manufacturing processes by helping establish and write the material and operational quality requirements from the time of design disclosure to the delivery and acceptance of the product by the customer.

Quality assurance is different from the manufacturing line of authority. Decisions to make corrections may be made anywhere in the various levels of manufacturing, from getting the materials to inspecting the final product.

The quality control system is the manufacturing responsibility of the quality assurance program. Products in manufacture must meet the specifications released from the production engineering section. Quality control is a system of collecting data as the parts are manufactured. It includes monitoring, organizing, and reporting data to bring about correction of errors in the production system quickly, Fig. 2-30. If the errors are numerous, quality control shuts down production until the problem is solved.

Fig. 2-31. Inspection is the detailed checking of parts near the production machine. Approximately 64 critical dimensions are air-gaged simultaneously.

Inspection is the detailed checking of each part that is manufactured. The inspector is usually near the production machines and passes, rejects, or sends the parts back for rework, Fig. 2-31. The inspector provides the basic data for the quality control system, which in turn provides the data and information for the quality assurance system.

Assembling the Product

Assembly is the joining together of standard parts and components to build a finished product. It is also the bringing together of all the parts and organizing an efficient method of quickly and accurately combining components. You will study these types of assembly:

- Individual or single-station assembly
- Batch or lot assembly
- Continuous assembly

Individual or single-station assembly is being used when an individual assembles the whole product from parts, Fig. 2-32. The **batch or lot assembly** method is being used when the work is assembled in a group or lot size. This program depends upon the capacity of the equipment involved, such as the parts processed by a heat treating furnace or the hours in a working day. As production increases, **continuous assembly** becomes the most efficient method of assembly. The assembly procedure is arranged around a circle, dial, or a line with each worker attaching parts as the work moves past, Fig. 2-33. Assembly lines are the common methods of building large volume products. The work moves in a line, and all subassemblies come to the line and are attached to the product at the proper time. The trend in assembly is to perform more operations automatically, so assembly machines use part feeders, nutrunners, and clinching dies to fasten the positioned parts.

Fig. 2-33.   Continuous assembly in which each worker attaches parts as the work moves past.

Fig. 2-32.   An individual station assembler builds up the product from parts and subassemblies.

Fig. 2-34. In assembly by programmed industrial robots, automobile bodies are welded together.

Fig. 2-35. Testing and adjusting an engine while it is running under power on hot test stand.

Computers program and control automated machines such as industrial robots, Fig. 2-34.

Testing Performance and Adjusting the Product

Testing the manufactured product is a trial of the actual function of the product. It is different from inspection in that the product is used under actual operating conditions. Engines are tested, using regular fuel. They are tested while running under power. During this hot test, various adjustments are made so that the performance will meet specifications, Fig. 2-35.

Packaging the Product for Distribution

Packaging is the preparation of the product for shipment, sales, and customer use. The package is designed to protect the product and to provide an attractive sales

Fig. 2-36.   Packages for precision instruments are molded in Styrofoam cases.

display. Packaging for large objects such as engines, large valves, and machine tools involves mounting the product on skids or timbers and building a box or crate around the object. A shipping tag and a shipping list

are nailed onto the crate and an address is stenciled on it.

Smaller products such as instruments are packaged in specially molded, shock-resistant Styrofoam cases, Fig. 2-36. The cases are form-fitting to the instrument and give maximum protection from temperature change and rough handling.

People in Metal Manufacturing

Manufacturing corporations have developed efficient organizations that coordinate and perform the fundamental operations in manufacturing. **People** are needed to plan, design, organize, make, and distribute the nation's products — goods and services. Many outstanding career opportunities are available in metal manufacturing. Figure 2-37 shows where people work in the United States.

In a corporation, many people are required to perform work at various levels in

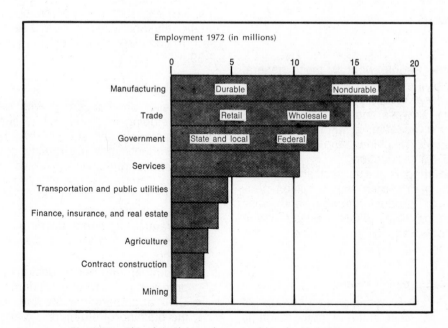

Fig. 2-37.   This chart shows where people in the United States work.

Fig. 2-38. Many career opportunities are available within each level of the manufacturing organization.

Fig. 2-40. Engineers are concerned with designing, planning, and developing methods to solve the practical problems of construction and manufacturing.

Fig. 2-39. Careers in business management include occupations of officers and managers of a corporation.

the industry. People work in these general employment areas: (1) general administration; (2) research; (3) product engineering; (4) process analysis and planning; (5) estimating, cost accounting, and pricing; (6) plant materials, machines, and tool procurement; and (7) production.

Manufacturing is the largest employer of major industries. About twenty million people work to produce durable and non-durable goods. Many different skills of many different occupations are needed in the organization, Fig. 2-38.

General Administration People

A small but very important group of workers in a manufacturing corporation is the group of professional and technical

workers. These people are responsible for making long-range decisions within the organization. Business administration includes the occupations of the officers and managers: the presidents, vice-presidents, secretaries, accountants, advertising workers, industrial traffic controllers, managers, market researchers, personnel workers, public relations workers, purchasing agents, and a complete range of engineers, Fig. 2-39.

Engineering People

Engineers are concerned with designing and developing methods for the conversion of raw materials and sources of power into useful products at a reasonable cost. They use scientific principles to solve practical problems. There are many kinds of engineers found in the industry, Fig. 2-40. Their jobs are described as:

- **Ceramic engineers** develop the products and processes of clay manufacturing.
- **Chemical engineers** design chemical producing plants and equipment.
- **Civil engineers** design and supervise the building of structures, roads, dams, tunnels, and industrial plants.

- **Electrical engineers** design and supervise the manufacture of electrical and electronic equipment and industrial plant electrical design. They also specialize in communication, computers, tracking systems, and related areas.
- **Mining engineers** design and lay out a system of extracting minerals from the earth and transporting the minerals to the processing plant. They design and supervise the construction of shafts, drifts, tunnels, and hauling procedures and are responsible for the efficient and safe operation of the mine.
- **Metallurgical engineers** are sometimes called material engineers or material scientists. These engineers study and invent ways of producing, improving, or using metals. Metallurgists design and develop new processes of metal production that will meet increasing industrial requirements such as space application of new metals and alloys. Metallurgists work in two general areas. (1) **extractive metallurgy,** which involves gaining metals from their ore, and (2) **physical metallurgy,** which involves investigating properties of metals and their alloys.
- **Mechanical engineers** are responsible for the best methods of using power. They design machines that use power and make manufactured products. They use mechanics, hydraulics, heat, and electricity for sources of power for the machines they design.
- **Industrial engineers** are used widely in manufacturing. They plan the most efficient methods to produce an object. They are concerned with the most efficient use of people, tools, and materials. Industrial engineers plan and develop management control systems that improve the efficiency of management, quality control, and production.

Research is conducted at various levels in the corporation. Frequently workers in marketing research provide business information about the development of sales for a new or present product. Research is carried out by engineers and scientists to develop and test new products and manufacturing processes. Research jobs require scientific or engineering training, Fig. 2-41.

Production engineering includes those positions which are concerned with the overall production of the product. Specifications are obtained and organized into the materials and processes needed to make the product. The people involved include engineers, designers, and a whole group of clerical and related workers, Fig. 2-42. Production engineering people in engineering testing verify the product's ability to comply to the set standards of reliability, quality,

Fig. 2-41. Research in a 10′ × 10′ supersonic wind tunnel to determine aerodynamic loads on the tail section and fuel tanks in flight.

Fig. 2-42. Designers work on new plans in the product engineering groups.

and performance, and its ability to be manufactured and then serviced. This work is supervised by an engineer, but much of the testing is carried out by engineering aides and laboratory technicians.

Process Analysis and Planning

The people who work in process analysis and planning carefully study the operations and processes required and the facilities and people needed to best produce an economical product. They measure the time to do the work, evaluate how well the work was done, and compute the average time to be allowed. When they have this information, they can establish a time standard as a basis for planning and scheduling work throughout the plant, Fig. 2-43.

The industrial engineering group plans the transportation and scheduling of materials throughout the plant. Operations and processes are studied to find ways to improve production or establish new processes and equipment. Responsibilities of this group include planning and maintaining machine tools, jigs, fixtures, dies, and patterns; controlling the inventory of all the tooling; and seeing that the tooling is in workable condition.

Estimating, Pricing, and Cost Accounting

Estimating the cost of a manufactured product is very complex. Usually it is done before the product is built to determine whether it can feasibly be built and where the materials and services will be purchased. Data from the product designer, product engineer, manufacturing engineer, and purchasing department will be carefully considered in the estimate, Fig. 2-44.

The estimator will conduct a cost accounting study in which each part is broken down and analyzed for the time of production, cost of materials needed, the cost of labor required to produce the part, and the overhead. Finally the estimator will include in costs the profit the company expects to make on the part. Cost estimating provides information for many management decisions and also provides the basic data for setting the price of the product, Fig. 2-45.

Plant, Materials, Machines, and Tool Procurement

Building and operating a factory uses the **skills** of many people. When an industrial plant is constructed, the people and skills

Fig. 2-43.   Planning and scheduling work through a plant is a responsibility of process analysis and planning.

Fig. 2-44.   The estimator will conduct a cost accounting study in which all costs to make the product are determined.

of the building industry are employed. Carpenters, structural metal-workers, plumbers, pipefitters, brick layers, cement masons, electricians, roofers, painters, and many other workers are needed, Figs. 2-46 and 2-47.

Fig. 2-45. A product has costs other than materials and labor.

Fig. 2-46. An industrial plant under construction.

Fig. 2-47. The building of an industrial plant requires all of the workers and skills of the building industry.

The procurement of materials and supplies for a manufacturing plant includes many jobs of a clerical nature. Planning, purchasing, and delivering the correct materials, tools, and labor to the place of manufacture at the right time involves detailed records. **Bookkeepers** record, file, and prepare bills for mailing. They keep records of accounts, prepare trial balances, and prepare income statements. **File clerks** make sure records are up to date and in their proper place. Much of their time is spent in retrieving information already stored in the files for the use of other members of the organization.

Shipping and receiving clerks account for goods that are coming into the plant and the products which are shipped out. They keep records of material that is transferred from one place to another within the plant. They direct the loading of trucks and railway cars.

Secretaries help office managers by doing the correspondence, arranging appointments, and supervising other office workers within the department, Fig. 2-48. **Typists** reproduce typed copies of printed, handwritten, office material. Typists prepare master copies for duplication by photography or electrostatic copy processes.

Machines and tool procurement means getting all equipment ready for production. **Tooling up** requires the efforts of a number of groups of people. The **manufacturing engineer** decides what tools, fixtures, jigs, dies, and machines must be purchased, built, or transferred to perform the manufacturing. The **industrial engineers'** studies of the best processes for making the product are carefully considered before decisions are made.

Millwrights are the skilled **craftsmen** (a level of skill identified by trade unions) who install the machine tools, automatic assembly equipment, and part conveyor systems in the plant, Fig. 2-49. They align and test the machines and equipment before they are signed over to the production group. They adjust, lubricate, and run the equipment during their testing.

Production Workers

Metal manufacturing provides jobs for a vast group of production workers. **Foundry**

Fig. 2-48. Office workers maintain the industry's records.

Fig. 2-49. Millwright checks newly installed crankshaft balancer for correct operation.

occupations supply the castings that are needed in manufacturing. **Patternmakers** are skilled craftsmen who build the patterns that are used to form the molds into which the molten metal is cast, Fig. 2-50. Patternmakers work from blueprints and make precise patterns of the product. These patterns allow for metal shrinkage and other pattern allowances. A patternmaker has a high degree of skill and knowledge about metals and metalworking.

A **machine molder** operates a machine which produces large quantities of identical molds. These molds later receive molten metal and become castings, Fig. 2-51. **Coremakers** prepare large, intricate sand cores that produce holes and hollow sections in castings when the metal flows into the mold. Cores are made by hand or by a machine and require workers with high manual dexterity. Coremaking frequently is a repetitive job because of the large number of identical cores required. An eighth-grade or high school education is the entry education level required.

Forging occupations provide parts which can deliver great loads. **Hammersmiths** lead the crew of workers in the operation of the open-die power forging hammer. They read blueprints and sketches and determine where the metal is to be struck. Hammersmiths determine when the metal is to be reheated and when the part has reached the correct shape and size. **Hammer operators** are skilled in using power hammers with dies. Drop forging calls for heaters and helpers, as well as the operator. The operator moves the metal under the blows of the hammer dies and determines when the metal needs heating or is complete.

Press operators control huge presses that squeeze the metal between open dies. They control the heating of the metal, the movement of the metal on the anvil, and the amount of pressure applied to the metal,

Fig. 2-50. A patternmaker builds patterns that produce molds into which molten metal is cast.

Fig. 2-51. A machine molder produces many identical molds.

Fig. 2-53. The all-around machinist is a highly skilled worker.

Fig. 2-52. Forging press that shapes automobile crankshafts with 6000 tons pressure per square inch.

Fig. 2-52. **Impression die-press operators** have die impressions that predetermine the shape of the forging. Identical forgings may be produced with a small crew.

Other occupational positions in forging are **heaters, inspectors, diesinkers, trimmers, grinders, sandblasters, shotblasters,** and **picklers.**

Machining Occupations

The machining workers are the largest group in the metalworking trades. There are about 1.2 million machining workers in the United States.

The **all-around machinist** is a highly skilled worker who can operate machine tools and has a broad knowledge of construction of machine tools, shop practice, properties of metals, and mechanical design, Fig. 2-53. Blueprint reading and precision measurement are required skills of the machinist. The machinist must have good eyesight and hand/eye coordination and must be able to stand for long periods of time. Machinists build the machine tools that other craftsmen use.

Machine tool operators can operate a few machine tools. Most of these workers are semiskilled machine tenders that place work in the machine and do the repetitive work of building hundreds of thousands of identical parts, Fig. 2-54. The machine is set up by a machinist or by a setup man who selects the speeds and feeds of the machine and the proper tooling. Machine operator jobs are drill press operator, lathe operator, boring mill operator, etc. The largest group of workers in the machining trades is the machine operators.

The **instrument maker** designs and makes scientific measuring devices and experimental machines. Instrument makers work from rough sketches, ideas, verbal instruction, and blueprints which are worked out by the engineer or scientist. They will frequently work for a long period of time fabricating parts from standard stock to build, test, and calibrate their instruments. Instru-

Fig. 2-54. Machine tool operators machine thousands of identical parts.

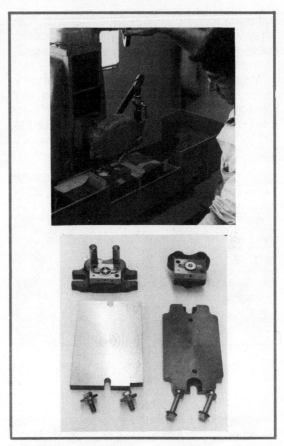

ment makers need knowledge and skill in mechanics, machining, electronics, and vacuum technology.

Tool and die makers are the workers who build the mass production machine parts that shear, bend, and form metals. These workers make the cutting, tools, dies, jigs, and fixtures used by industry to produce the parts required for production, Fig. 2-55. Tool and die makers need much experience, and their work is worth thousands of dollars. For this reason, their jobs are stable. Industry is very reluctant to lay off tool and die makers, because their experience with the equipment of the industry is so valuable. They can usually weather a slack period in production by working temporarily as instrument makers or machinists.

Machine tool setup technicians are employed in large machine shops and plants, Fig. 2-56. These workers set up the tools and fixtures and set the speeds and feeds for the machine operator. A setup man will usually be assigned a group of machines to set up, adjust, and control until the parts meet the required specifications. Then the machine operator takes over the operation of the machine and the setup man sets up another machine.

Manufacturing inspectors check to see if parts have been machined correctly and

Fig. 2-55. The tool and die maker makes the cutting tools, dies, jigs, and fixtures that are used by industry to produce parts.

Fig. 2-56. Machine tool setup worker sets up the tools, fixtures, speeds and feeds, and limit stops for the machine tool operator.

conform to the blueprint specifications. They may test and adjust the parts. The inspector uses gages, micrometers, and other measuring and testing equipment, Fig. 2-57.

Assemblers put together the parts to form components, subassemblies, and final assemblies. The work requires different skills, but the ability to position parts, bolts, wires, clips, and subassemblies quickly and accurately is needed. The parts are put into place, then fastened or sealed. Many men and women are employed as assemblers, Fig. 2-58.

Metal finishers prepare units for finishing by grinding, sanding, and cleaning the metal products for painting. **Metal sprayers** apply paint and other materials to the product to make it clean, beautiful, and weather-resistant, Fig. 2-59. Spray guns and other tools are used to apply the finish.

Polishers use portable tools and other power tools to buff metals and bring them to a high luster. This is a laborious job which is necessary for plating and final finishing work.

Maintenance mechanics repair machinery in factories when a breakdown occurs. They must quickly locate the trouble and repair it to get the machine back in operation as soon as possible. **Preventative maintenance** (fixing machines before they break) is an important part of the job. They inspect, grease, adjust, clean, and replace parts that are worn. Records of the performance and repair of the machines must be kept.

Maintenance electricians repair or replace wires, motors, switches, and other

Fig. 2-57. Manufacturing inspectors check parts to see if they conform to specifications.

Fig. 2-58. Assemblers put parts together to build up subassemblies and final assemblies.

Fig. 2-59. Metal finisher spraying paint to produce a clean, resistant finish on the product.

electrical equipment. A power failure or burned out motor will stop production immediately for many production workers.

There are a number of other maintenance jobs which vary with different manufacturing plants. **Maintenance welders, refrigeration mechanics, plumbers, carpenters, sheet metal workers,** and **painters** may all be employed in maintenance positions, Fig. 2-60. An efficient operating manufacturing plant requires a good maintenance department.

Packaging involves filling, sealing, adding instructions and warrantees, labeling, weighing, and shipping. The work may vary from filling small plastic containers to building large wooden crates around machine tools so they can be moved.

Product Design and Development

Obtaining Product Idea. A typical business will include a division such as product design and development. This division will receive, create, or develop new products for the corporation to manufacture. It is possible that an inventor may bring a prototype or a working model to the organization for production. The organization researches and tests the model, investigates the patent rights, and offers a contract to the inventor. However, most large organizations maintain

Fig. 2-60. Maintenance welders make repairs necessary to keep the plant operating and to make plant improvements.

a large staff of engineers to do research for products for the company or for products which are allied to their production speciality.

The research design and development division of an organization might also design and develop new products which can be produced with existing industrial facilities which are not being used to capacity and

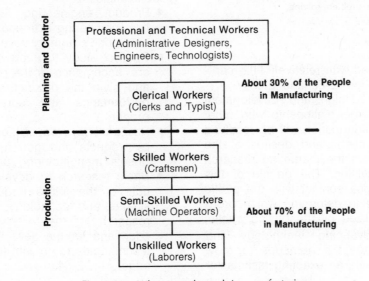

Fig. 2-61. Where people work in manufacturing.

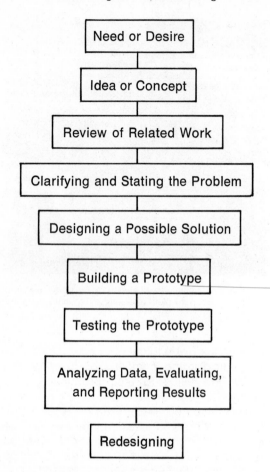

Need or Desire

Idea or Concept

Review of Related Work

Clarifying and Stating the Problem

Designing a Possible Solution

Building a Prototype

Testing the Prototype

Analyzing Data, Evaluating, and Reporting Results

Redesigning

Fig. 2-62.   Research-design and development; a scientific method of problem solving.

Fig. 2-63.   Prototype of a truck cab interior under design research.

could be improved with relatively little capital investment, Figs. 2-62 and 2-63.

The research design and development division has an interrelationship with other divisions. Much information will be received concerning the needs and desires of the buying public from the marketing research and planning division. The people of this division have close contact with the public and will be able to determine which products may be utilized by the buying public.

A rapidly advancing technology and society have made the demands for new and better products an exacting taskmaster for industry. Although the industry may already have a good product, it must create the idea that it is trying to improve the product even further. The improvement may be mechanical, functional, or stylistic. Industry must be alert to improve the product, or the product runs the risk of becoming obsolete or being relegated to a secondary position.

Product research design development in a large industrial concern will have these responsibilities:

- Product engineering, which involves the engineering or reengineering of the enterprise's primary developed product.
- Product study to improve the customers' acceptance of the product, design and style the product, and study the performance and durability of the product.
- Research to produce new products or by-products salvaged from the enterprise's manufacturing process.
- Basic research to develop new products of the future through organized science and technology. This is "pure" research for the extension of knowledge and for the development of new technologies to go with the new knowledge, Fig. 2-64.

Fig. 2-64. Basic research provides knowledge for new technologies.

Fig. 2-65. Government research centers develop new materials and technologies.

Today these basic responsibilities are generally carried out by a research team. This team may be within an industry and work full time improving that industry's products, or the research may be done through commercial laboratories. These commercial laboratories are relatively new. They are usually known as research institutes or industrial research centers.

The science of inventing and solving problems has become big business for the industrial researchers. The future will show a still closer relationship between pure science and industrial research. Government programs have become research centers for the development of new materials and processes and the application of new knowledge, Fig. 2-65. The space program with its many research centers has developed whole new technologies — such as integrated circuits, telemetering, special computer and radar analysis, and the many technologies related to life support systems. All the space research programs have "spin off" that will eventually make the knowledge and technology gained from these programs applicable in other industries, Fig. 2-66.

A business division which interacts closely with product research design and development is market surveying and planning.

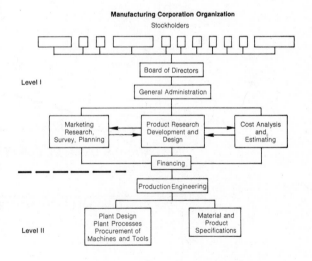

Fig. 2-66. Manufacturing corporation organization.

Market Research

Applied industrial research is concerned with the corporation's activities in these four areas: (1) marketing, (2) products, (3) materials, and (4) processes and equipment. These research processes perform in dif-

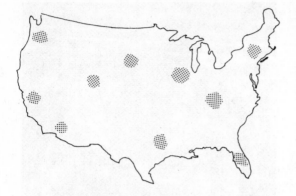

ferent places throughout the industrial organization. Each of these major divisions of industry has its own research function.

The research concern in the first level of general administration is **market research.** The primary question of market research is, "Will there be a demand for this product?" After this basic question is answered, precise information must be obtained to determine the potential demand for the product and the actual sales of the product, Fig. 2-67.

Fig. 2-67. Market research sample distribution map.

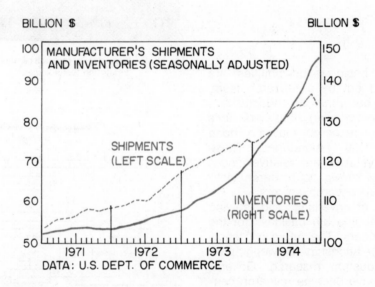

Fig. 2-68. Business trends are studied by the industrial forecaster in order to make projections.

The market research group must accurately answer these questions so that **production plans** may be formulated. This group must answer many questions. Will existing plant facilities and equipment be adequate to produce the product in the required volume? Will the raw materials be handled or stored, or can they be shipped to the place of manufacture? Is adequate transportation available to get the product to market on time? Will the needed and trained labor be available when it is needed for the manufacture? There are a great number of decisions which will be made on the basis of market research. These are only a few of the major ones.

Forecasting

The person responsible for providing this information for the corporation management is the **industrial forecaster.** The Federal Reserve Board publishes an index of industrial production which is adjusted for seasonal variations. This index graphically shows the trends of industrial production. It has subsections which give indices for several general types of production: consumers' perishable goods, consumers' semi-durable goods, consumers' durable goods, construction materials, capital goods, fuels, materials, and supplies, Fig. 2-68. Other agencies, public and private, compile data on economic trends. These studies include the following topics:

Industrial production
Volume of trade
Commodity prices
Wholesale prices
National income
Manufacturers' sales
Manufacturers' inventories
Employment
Car loadings
Exports
Imports
Building permits
Business failures
Consumer prices

Industrial forecasters are familiar with the national debt and can relate these trends to their corporations. Forecasters must supply and receive accurate information about their own corporations. They must have an understanding and insight to the corporation's past production and sales performances, as well as its past and current management and production policies.

The forecaster will gain current, specific information about the corporation's product sales before making a detailed analysis and recommendations to other general administrative divisions, Fig. 2-69. There are three major methods which are used to gain data for corporation forecasting: (1) opinion, (2) estimates by sales personnel, and (3) statistical sampling.

Opinion in Forecasting. Sales executives for corporations seldom depend strictly upon their own judgment when they make a decision. They consider the opinions of the **general administration** and compare them with **sales forecasts.** These sales forecasts are developed by market research methods. When the opinions of the general administration differ from the sales forecast, the two are studied carefully. This **opinion method** is useful in checking the corpo-

Fig. 2-69. The forecaster studies computer data.

ration's decision-making processes. It helps assure the accuracy of forecasting for a new product.

Sales representatives usually know the potential for sales in their own sales district. They know who will have an increased need for their company's product and which companies will no longer be using their product. Sales representatives know the quota of production allotted to a given company on a contract and the approximate number of units of their products which will be required for this contract. They can often estimate the growth percentage for sales in the forecast they transmit to the division office.

Division office managers tabulate the data by items and months. This information is totaled with additional information and recommendations supplied. These totals of the data represent the sales division's estimate of the sales forecast. The estimates are then passed to the regional or territorial sales manager. He or she takes this information, compiles it with other divisional reports and reviews the total situation. After discussing it with the staff, the manager makes revisions which are apparent at the regional level.

The sales executive receives this data and reviews the figures in light of various indices, and general economic conditions. The sales executive and the market research staff compile all the data and other modifying factors and draw up a final and complete forecast for the corporation's general administration to consider and settle.

Statistical Sampling in Forecasting. Statistical sampling employs the concepts of probability and applies them to a large market forecast. The forecaster will make an analysis to determine what group of customers is buying the majority of the corporation's sales. A ratio is established between the larger group and the smaller group of the total sales. The larger group or major purchasers is carefully studied and then a randomly selected sample is drawn from this group. The sample represents the true attributes of the larger group because each of the customers in the larger group had an equal chance of being in the sample. The sample group is small and therefore may be studied intensively. This group may be given questionnaires, interviews, or tentative orders for products under study, Fig. 2-70. These data are analyzed and a forecast is made on the basis of the sample. This is then projected to a large group. When the ratio between the larger and smaller group is known, a percentage rate of growth can be applied to the small group by using the growth percentage rate and applying it to the ratio. The result will be a statistical forecast of products needed for production planning.

An example illustrates this forecasting procedure:

Fig. 2-70. Taking opinions for forecasting.

1. A study of customers' purchases reveals that 70% of the corporation's sales are made by 1000 of the corporation's 100,000 customers.
2. The forecaster takes these 1000 customers and selects a random sample from them. A sample of 200 customers is very carefully studied by the use of interviews or tentative order blanks.
3. The data is compiled and multiplied by 5 to equal the 1000 customer needs. This forecast will be reasonably accurate for 70% of the corporation's sales.
4. The 30% of the corporation's sales may be adjusted using the growth rate from the sample intensively studied.

The results of the statistical forecast are presented to the sales executive, who presents the findings to the general administration for policy decisions to provide action for the increase or reduction of production.

Product Engineering

This division of a research, design, and development department is concerned with improving the corporation's products or creating new products for manufacture. Product engineers use ideas and information from any source — a need appearing in the marketing report; an industrial designer's ideas; basic research findings; or an idea from production, manufacturing, or an inventor. They are interested in new uses of materials, new materials, and new production procedures applied to new or old materials.

The number of new materials available to the industrial product designer is phenomenal. New materials or new combinations of materials are being made available almost daily. The developments in adhesive materials enable very unlikely products to be glued together. Textiles, insulation, and metals may be fastened with adhesives. Previously, products would have had to be bolted together, soldered, spot-welded, brazed, or fastened mechanically. Now, if the product remains in the low temperature range (below 150° F.), a whole group of products may be redesigned.

Many redesigning possibilities are also offered by the addition of plastics, powder metallurgy, cements, carbide cutters, super-alloys, ceramic magnets, exotic metals, etc. These applications of materials bring the industrial designer's contribution to industry to the front.

Redesigning becomes important when products are changed from conventional manufacturing to automation. The redesigning with new materials and new fasteners can often result in a vastly improved product which can be manufactured more economically.

Product Design Considerations

The industrial designer is concerned with taking a client's or a research division's generalized specifications and designing these specifications into a model of a possible manufactured product. As many fresh, functional, beautiful elements as possible are incorporated into the product to give it sales appeal.

The industrial designer needs to be very creative in developing models of potential products. He or she may develop any number of alternate designs for the solution of a problem. Sketches and illustrations are done before the models are actually made.

Three-Dimensional Models. Models may be made of paper, clay, plaster of paris, wood, or metal. They are made to scale and will accurately represent the form and color of the finished product, Fig. 2-71. The finished models and preliminary specifications will be presented to the general administrative management for approval. Before management's approval is given, surveys about the new product will be carried out by marketing and sales forecasting to determine the product's sales possibility.

Fig. 2-71.   Scale models of products are made for approval by management.

Cost Estimating Study

Cost analysis and estimating is also carried out to determine the cost of producing the article. A very detailed breakdown of production methods and costs will be studied so that manufacturing and selling prices might be estimated, Fig. 2-72. These are evaluated by the finance department, which considers the relationship of the cost of the new product to the total finances of the corporation.

Legal Patent Investigation

At this time an attorney assembles patent data as it relates to the product. The attorney investigates the patents which are necessary to protect the company from having its product made by other companies without agreements. He or she must also make sure that the product does not infringe on already existing patents held by other people.

When the attorney has conducted patent investigations, he or she starts applications for patents or negotiations for the corporation's use of patents already in existence.

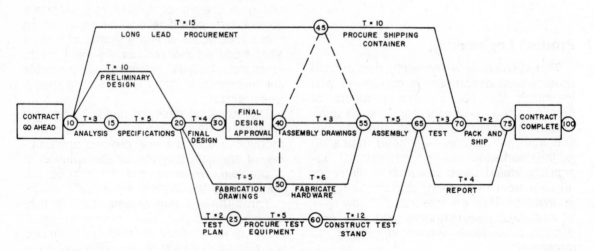

Fig. 2-72.   Manufacturing plan.

Patents may be issued on new products, materials, equipment, techniques, or on catalysts and additives to products. Patents may be issued on processes, portions of processes, or on any improvement in the items suggested. Most patents pass into public domain or general use after 17 years of service. However, the life of patents may be longer if improvements or other technical details are carried out. When the legal matters concerning the product are cleared, the plan must then be approved by the general administration.

Engineering Decisions

After the decision to produce the product has been made, the product engineer is responsible for carrying the project forward. Setting the quality requirements for the product and the specifications for manufacture are the next steps. The product engineer assigns the designer's model to the design engineering section so the engineering specifications may be drawn up, Fig. 2-73. The product engineer can do much to make the product economically feasible to produce by mass production means. However, the

Fig. 2-73. The design engineer is concerned with decisions, materials, manufacturing processes, structural characteristics, and production rates.

industrial designers have set the broad limits within which the production engineer must operate.

In the engineering of a product, many decisions must be made. How will the product be manufactured? What processes, machines, and materials should be used? Below are a few of the major decisions:

Engineering Design Considerations

1. **Choice of Materials.** The whole range is available — steel, plastic, aluminum, etc.
2. **Complexity of Part.** Holding, machining, and location of holes and punches are involved.
3. **Size.** Size helps determine which processes will be used.
4. **Forming Properties.** The characteristics of forging, bending, machining, etc. or the materials used will affect the process chosen.
5. **Precision Required.** Processes such as sand casting give tolerance of $\mp 1/16''$, while precision casting gives tolerance of .001 per 1″.
6. **Structural Characteristics.** Grain flow provides toughness as in forging.
7. **Smoothness Required.** Surface of a sand casting is poor, but that of a screw-machined part is excellent.
8. **Accuracy of Surface Detail.** Accuracy on sand casting is poor, but excellent on investment cast surfaces.
9. **Production Lead Time.** Wooden patterns take three to five days, but press forging dies require much time.
10. **Rate of Production.** Parts can be produced at spinning at 12-20 per hour, impact extruded parts can be produced up to 2000 per hour.

Prototype Development. Information and drawings from the design engineers must be carefully evaluated and tested to determine whether the new product is feasible and can be manufactured economically. Then a **prototype** is made. The prototype is a custom-built or hand-built, full-scale working model of the product. If standard parts exist, they are used. If not, parts are made.

During the actual building, it frequently becomes apparent that design changes are needed. It is then determined which tolerances and specifications are practical and which dimensions will need to be changed.

The product is refined until a working prototype meets the engineering group's requirements. The prototype is then ready for detailed engineering tests. This testing is carefully conducted. Subassemblies and completed products are given exhaustive tests under all types of climate and usage conditions. All the data gathered during these tests will be critically evaluated by design engineering, and changes are made where required.

Field testing will include customers' responses and reactions to the product. The prototype is sometimes given to a sample of customers so they will test it and give their reactions to it.

When the product has received complete approval, the manufacturing engineering group uses the prototype to help plan preproduction scheduling, tooling, manufacturing machines, layout, time, and methods needed.

Production Drawings and Specifications. The drawings, bill of materials, engineering specifications, and associated lists are called **product documentation.** The engineering design group is responsible for producing these documents. These materials become the binding documents that other corporations, the Department of Defense, and subcontractors use as a legal basis for bids and contracts.

Drawings are made of every part except those assemblies too complex to be easily portrayed. Detail drawings are made of these complex parts so that a bill of materials may be made from the drawings.

Assembly drawings are used as a reference when the product is put together. These drawings indicate part numbers, assembly procedure, field replacement parts, and breakdown assemblies of complicated assemblies.

A **bill of material** is a list of all parts and their numbers required to put together an assembly, Fig. 2-74. The bill of material contains the name of assembly part, the product or model number, the size of assembly drawing, engineering change number (if the part is being modified), and the date an engineering change is started. The body of the bill of material contains a parts description, a document code number of the part, the part number, the quantity of parts to be used, and the unit measurement used on the part.

Drawings may have a number of formats such as mechanical, electronic, hydraulic, vacuum, pneumatic, or others. The most common document is the **mechanical format.**

Detail drawing specifications and standards for documentation have been set down by a series of publications called **Military Standards.** These military specifications detail very carefully the drafting practices to be used by all contractors and subcontractors that do work for the U. S. Departments of Defense and Commerce, Fig. 2-75 (page 86). Industries that have any relationship with government contracts will have a drafting manual that is based upon military and other government standards. These specifications are mandatory for use by all departments and agencies of the Department of Defense.

Production Planning and Cost Analysis

After the engineering and the specifications for the product have been completed, production planning is begun. Production planning and manufacturing processes are

Bill Of Material						B/M No

Name	Product/Model	Type			E/C No
		☐ Factory			E/C Date
	Dwg Size	☐ Field			Sheet

Part Description	Doc Code	Part No.	Quan	Unit of Measure	Remarks

Fig. 2-74. Bill of material.

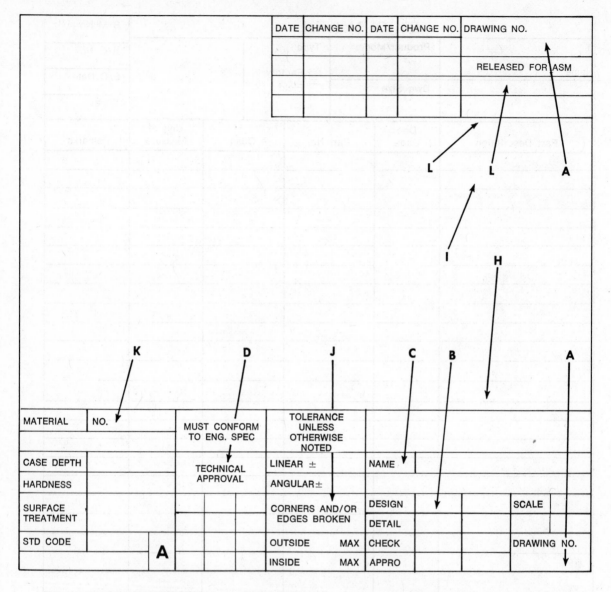

DATE	CHANGE NO.	DATE	CHANGE NO.	DRAWING NO.	
				RELEASED FOR ASM	

L   L   A

I   H

K   D   J   C   B   A

MATERIAL	NO.			TOLERANCE UNLESS OTHERWISE NOTED					
			MUST CONFORM TO ENG. SPEC						
CASE DEPTH				LINEAR ±	NAME				
HARDNESS			TECHNICAL APPROVAL	ANGULAR±					
SURFACE TREATMENT				CORNERS AND/OR EDGES BROKEN	DESIGN		SCALE		
					DETAIL				
STD CODE		A		OUTSIDE MAX	CHECK		DRAWING NO.		
				INSIDE MAX	APPRO				

A. The drawing number that is assigned at formal release is entered here.
B. Initials and dates are entered in these blocks for design, detail, check, and approval.
C. The descriptive name of a detail or assembly drawing is entered here.
D. The necessary technical approvals for release of the part are entered here.
E. Notes numbered 1, 2, 3, etc., should be grouped on sheet 1 of multiple-sheet drawings on the right side of all mechanical drawings.
F. Corner and edge conditions are entered here.
G. Material number, heat treatment, and surface treatment notations are entered here.
H. This calls out the assembly part number for which the part was originally released.
I. For multiple-sheet drawings, the sheet number and the total number of sheets (e.g. sheet 2 of 5) should be placed beneath the part number block.
J. Corner and edge conditions are entered here.
K. Material number, heat treatment, and surface notations are entered here.
L. This calls out the assembly part number for which the part was originally released.

Fig. 2-75. Generalized mechanical drawing format.

Fig. 2-76. This manufacturing engineer analyzes detailed information obtained from the drawings.

the purpose of compiling material specifications.

Material Costs

Determining Material Costs

1. Determine the number of parts in a component which may be cut or machined from standard bars, sheets, and tubing. A great number of parts can be produced from standard stock.
2. Determine the number of formed parts, forgings, and castings. These will usually be unique to the product being produced.
3. Determine the number of finished parts to be purchased. These parts may be washers, bolts, bearings, seals, wire, chains, sprockets, belts, pipe fittings, valves, etc.
4. Determine the number of purchased finished subassemblies. These parts may be gear reducers, motors, timers, electrical motors, switches, pumps, or voltage control equipment.

closely allied to the costs of manufacturing for the various production processes. The primary concern of the manufacturing engineer is to select the manufacturing processes which will enable the product to be produced with the least cost, Fig. 2-76. The product engineer examines the plant for tools which are not working to capacity or have a vacant period during the year. He or she procures any new equipment which is necessary for the manufacture of the new product. The materials for the product will have been specified by the design engineering department, but the tools and equipment will be analyzed in order to select the most economical process.

Costs considered in the selection of the manufacturing process will include these:
1. Raw material costs (standard stock)
2. Tooling costs
3. Economic production lot sizes
4. Direct labor costs
5. Tolerance required on parts
6. Finishing costs
7. Scrap loss cost

The process selected for the manufacture of the product will have a close relationship to the total cost of the production.

In the production analysis, the planning of the component parts may be classified for

This information about material specifications is useful in compiling the raw material costs and the purchasing of supplies. It indicates which processes are the most economical.

A study of the relationship among other component parts is necessary. The investigation of the other subassemblies may show that a number of component parts may have common type and material specifications, thus it is possible to combine stock and supply orders from other assemblies.

In addition to material, the study of component part relationships may indicate common machining or forming processes. Thus, the use of interchangeable parts routed through a common tool may make a much larger range of parts available at a more economical rate than if considered separately. A group of parts may have a similar general shape and later in their processing have different details. This type of component manufacture is done by using common mass production for the generalized shape and then a quick changeover of tools or a modified production line to finish the varying parts to the required details. When generalized shapes are used, manufacturing costs are cut, and standardization of parts is improved. Subassemblies and component parts, when examined critically, will be found to contain many elements which may be standardized, thus producing more saving in manufacturing costs.

Occasionally it may become more economical to purchase standard component parts from other manufacturers than to build these parts in the corporation's own plant. These standard parts may be purchased from a parts list which is used by many manufacturers. There are protections that the corporation must have. If the parts are to be purchased, the parts must meet quality specifications and delivery dates must be maintained, Fig. 2-77.

If the parts have high technical requirements, they are usually manufactured within the corporation's plant. This is especially true if there is production equipment within the corporation's plant which is not working at full capacity. By manufacturing within the plant, it may be possible to reduce the overhead cost of the outside parts and produce the part within the plant at a cost more equal to the material and direct labor cost. Thus, the component part is produced at a lower cost.

Tooling Costs

Tooling is one of the major expenses in setting up a new product for manufacture. Tooling consists of two major categories: machine tools and material handling.

Machine tools are tools which shape parts by removing chips, such as drilling, broaching, milling, sawing, grinding, turning, planing, and boring. Machine tools may also shape parts without the removal of chips. Bending, forging, casting, spinning, extruding, rolling, roll forming, squeezing, drawing, stretch forming, swaging, and metal powder forming are methods which shape without removal of chips. These tools all do work directly on the part to be produced. Their

Fig. 2-77. Subassemblies and components may frequently be purchased instead of being manufactured within the plant.

Fig. 2-78. Tooling costs include tools which hold, align, and gage parts, as well as cost for the machine, punches, dies, and cutting tools.

action is that which would be seen by an observer in a plant. There are other tools which contribute to tooling costs and are considered machine tool tooling. The tools which hold, align, or size the work are called jigs, fixtures, punches, dies, cutting tools, templates, molds, and gages, Fig. 2-78. These tools are generally expensive and wear out as the product is being produced.

Another type of tooling cost is **material handling.** The material or parts must be moved from one work station to another, thus such equipment as conveyors (chains, belts, monorails, transfer systems) are used. Hoists are also used to locate and move heavy or bulky parts (winches, cranes, fork lift trucks elevators, and overhead traveling cranes), Fig. 2-79. Positioners are used to locate or move the workpiece between working operations while work is being done on the part. Positioners are frequently used in welding shops. The part is turned so that the weld may nearly always be made in the welder's flat position. Other positioners will index, orient, or roll over the workpiece.

The loading of parts into a machine tool may be done manually or automatically. If the machine is to be fed by power, a loading machine will be needed and must be figured into the tooling cost.

Loading machines may be chutes, magazines, or transfer arms which must be timed with the machine to keep the parts moving steadily through the machine. Loading machines must be carefully adjusted to the production machine. They represent an additional tooling cost.

How Lot Size Affects Cost

Lot size is the number of units produced in each group or batch. The manufacturing lot size will be a reflection of the demand of the market. Basic information such as the size of the market is supplied by market research.

Not all products or parts are produced in a continuous manner. An estimate of how many units will be produced and the capac-

ity of the plant both affect the manufacturing lot size. The capacity of the machine tools will place limits on the lot size, because one section may be too slow to keep up with the production of the other machines.

The determination of the component lot size indicates the type and capacity of equipment to be purchased for production. The floor space which will be needed for the machines is determined by the lot size. This information is projected, and lot size costs and production time schedules may then be determined. The amount of storage which will be necessary to house raw materials, semifinished components, and the finished parts is also considered.

If the lot size is too large, an overinvestment in machine capacity will result. The investment cost in machinery will require too long a time to pay off, raising the cost of the unit product.

Fig. 2-79. Tooling costs also include material handling such as this monorail and hoisting equipment.

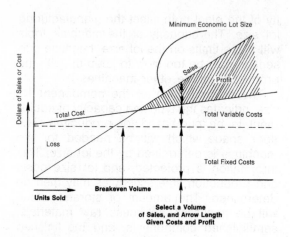

Fig. 2-80.   Breakeven chart for lot size.

A simple break-even analysis is helpful in understanding profits and costs at a given volume of production, Fig. 2-80.

Labor Costs

The cost of labor in the form of wages is a portion of the total cost of production. It is difficult to assign a dollar value to wages and the value of the wage earner's job to the total corporation.

This assigning a value to a job is difficult, because wage contracts are periodically changing, and the value of a worker's skill to the corporation may increase or decrease, depending upon the sale of the product.

In an effort to get a measure of what a job is worth to the corporation, a job may be classified in these two ways: (1) ranking methods and (2) point systems.

Ranking means the arrangement of jobs in order of their importance to the corporation. The most important job is at the top of the list, and the least important job at the bottom of the list. The ranking of the jobs within the top and bottom is made by committees who have knowledge of the job and skills and responsibilities as they relate to the job description. When this ranking has been completed, it is possible to assign a

pay level to the various job rankings. With these rankings, the number of jobs in each pay grade may be analyzed and the cost of labor determined for the time period of production.

The point system is the method industry most frequently uses to establish a job classification. Information may be added to include the various factors of skill, responsibility, working conditions, and effort. A job may be classified by the points value assigned to it by the judgment of experienced people. The job classification is made by comparison of job characteristics or point values against the factors the job requires. This job classification establishes a wage category. The wage category is divided into labor grades or pay grades and may be divided into any number of levels. A large industry may have 10 to 12 classifications.

Sample Wage Category	
Pay Grade I	100 to 120 points
Pay Grade II	121 to 140 points
Pay Grade III	141 to 160 points
Pay Grade IV	161 to 180 points
Pay Grade V	181 to 200 points
Pay Grade VI	201 to 220 points

Fig. 2-81.   Tool grinder: a position of the Grade V pay classification.

The points scale of factors and points is applied to each position in the corporation. For example, tool grinders may have 192 points assigned to their job. In this case, the tool grinders would be placed in Grade V, Fig. 2-81. Jobs with the comparable degree of skill required, responsibility, working conditions, and value to the corporation would be placed in the same labor grade by the number of points received per job.

Labor supply and demand is also a factor. If the corporation pays too low for a pay grade, the employees will take positions with other corporations. If the pay grade is too high as compared with the labor market, the corporation is increasing its labor costs more than is necessary. The product will then be less competitive with other products because of its increased cost.

To obtain a realistic appraisal of the pay grade in an area, a wage survey may be conducted. A sample of corporations who have similar manufacturing operations may be contracted to respond to key points on the corporation's pay grades. The survey will ask for information such as the high and low hourly rates for a job similar to the one described in the survey.

When the survey is completed, the various jobs can be tabulated from the various corporations and an average high or maximum and an average low or minimum figure established for each like pay grade. With this information, the corporation adds the information of labor supply and demand to the pay grade classifications by shifting the amount of money paid each pay grade.

The information is necessary to calculate the number of workers to operate the plant and the pay grade they will receive. The total labor cost can be calculated and an estimate of the labor costs achieved.

How Tolerance Affects Cost

Tolerance is the amount of error a machinist or machine can make on a part and still produce a part within the specifications and be accepted as a usable part. The tolerance which is set by the engineering depart-

ment has a considerable effect on the cost of the individual component manufacturing cost.

In order to keep machining and assembly costs to the most reasonable figure, the workpiece should have as large a tolerance as is compatible with the requirements necessary for the fitting of parts, the assembling of parts, and the adjustment of parts for alignment.

If size tolerance is small — such as ± 0.0005 tolerance — the cost is unjustly high. The cost is high, because a more exacting machine will be required to produce the ± 0.0005 tolerance. Instead of being finished on a milling machine, the part will need to be finished on a grinder. The cost has been raised, because an additional machine process has been added besides additional machine and labor time, Fig. 2-82. To carry out the close tolerances, a worker of great skill is required and thus the cost of a worker in the higher pay grades. The machine tool required to machine close tolerance is an expensive tool, so the unit cost per part will be raised. Finally, with a close tolerance required, there will be a greater percentage of parts which will have to be scrapped or reworked. Some parts are scrapped because upon gaging they are

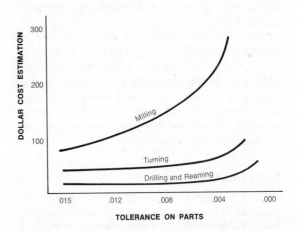

Fig. 2-82. The cost of parts has a relationship to the tolerance desired and the production process.

found to be beyond the limits of the close tolerance. When scrapping occurs, the cost of manufacture goes up very rapidly.

The tolerance on a component part is important during assembly also. If the tolerance is too coarse ($\pm$ 0.010), the part may be manufactured at a lower cost, but the assembly time (putting the component parts together) may increase. When the tolerance is close, parts may have to be put together by **selective assembly,** which means finding parts that are over or under size enough so that they will fit together. Selective assembly is slower, and assembly costs rise accordingly. If the tolerance is too close, time may be needed for hand-reworking the parts at the assembly area. This increases the cost per component part. When numerical control, jigs, or transfer machines are used, the cost of tolerance changes. The accuracy is built into the tool, and less dependence upon the craftsman is needed. With numerical control and transfer machines, very accurate parts may be produced without a large increase in cost for the additional accuracy of the parts.

Finishing Costs

After the components (the manufactured products) are made, they must go through a finishing process. The product may require deburring, cleaning, degreasing, etching, sandblasting, electroplating, painting, dip coating, or any other of a large number of finishing operations. The finishing operation will require floor space within the building: cleaning areas, process or coating areas, drying or curing areas, and short-term storage areas. These factors must be considered, as well as the labor and equipment involved in the finishing.

Scrap Loss Cost

Scrap losses come primarily from two sources: (1) residual material after forming and (2) material on parts damaged or lost because size specifications were not met. One of the most obvious examples of scrap

loss is the metal which is not used around a punch press blanking process. The parts are cut out to size, leaving the rest of the material which surrounds the part as waste. This material is classified as a scrap loss. However, it can be salvaged somewhat by baling the scrap and selling it at the going scrap metal price.

There is another scrap loss found in metal castings. The metal which has gone into risers, gates, runners, sprues, etc. cannot be used for parts. This metal is a scrap loss unless it can be remelted and still meet the metallurgical specifications of the product.

Castings which are porous or contain defects beneath the surface are usually discovered by nondestructive inspection. If the imperfections are not found until the machining operation, the cost is not only in scrap loss, but also in machining time.

Scrapped pieces may occur anywhere along the manufacturing line. Improper alignment of a forging die, deep scratches in a drawing operation, faulty welding, undersize grinding, below specification of surface finish, or warping and cracking of the piece during heat treatment all result in scrap loss.

Automation in manufacturing has reduced the scrap loss because of the close inspection of the workpiece between operations. The machines are corrected before the tolerance is great enough to cause scrap. Human error in manufacture is largely eliminated. Automation corrects the machines by feedback procedures so the machines are stopped if the parts cannot be held within tolerance.

The total amount of material which is not used or which is damaged during the manufacture is considered the scrap loss. The cost of the scrap loss is one of the overall manufacturing costs.

Inspection and Testing Costs

The purpose of inspection and testing is to maintain production quality control. In inspection the product will be examined for color, finish, required parts, fits, alignments, and overall operation. Testing is usually the final check and adjustment of the entire production unit. Both are extremely important, but whenever the testing and inspection are done, the cost must be added to the cost of manufacture.

Packaging Costs

Packaging is the final manufacturing cost. Boxes, cartons, and plastic foam molded containers are used to hold the product for transporting and carry identification of the manufactured product. The container becomes the sales display and shows the company's trademark or corporate image, Fig. 2-83. The package design and colors are carefully chosen because it is through the package that the customer receives the first contact with the product. Industrial designers carefully plan the package so that the product will have the maximum sales appeal and eye attraction.

Fig. 2-83. Labelling product with the corporation's identity.

Production Process Planning

Five basic steps are involved in production process planning. These are:
1. Analyzing specifications and drawings.
2. Studying basic operations.
3. Determining sequence of operation.
4. Combining basic operations.
5. Specifying tools and equipment.

Analyzing Specifications and Drawings

When the management of a corporation gives approval to proceed on a contract, the manufacturing process is set in motion. The production planning group will study carefully the specifications and prototype to find out exactly what is wanted. After they study the prototype, the process engineers may recommend a dimension or specification change that may aid in manufacturing without harm to the product. They may make notes on the print. If the suggestion is complicated, they may list the changes with comments in a table. These changes will be sent to the contractor for approval of these suggestions.

Studying Basic Operations

The basic operations needed to manufacture a product must be planned in order to determine the tools that will be needed and the sequence of manufacture. The prints are carefully studied to determine the manufacturing method to be used to produce the most effective component or part at the least cost. The engineering department will have specified the basic manufacturing process, but the manufacturing department will study the basic operations needed to manufacture the product.

The manufacturing methods group will determine whether the tools, personnel, and equipment are available in the plant to perform the manufacturing operations needed. In some cases they may suggest alternate methods of manufacturing which will make use of tools and machines already available in the plant. If no tools or equipment are available, tools and machines may be purchased to manufacture the part, or the part may be purchased from another company or subcontracted to another firm.

A list of operations needed for the manufacture of the parts is drawn up. Shearing, bending, machining, grinding, and drawing may be required. Additional required operations may be performed by spot welders, automatic lathes, bending machines, heat-treatment furnaces, or tooling for machines such as jigs, fixtures, and dies. After these operations are listed, details such as the number and capacity of the machines that will be needed to produce the parts are planned.

An example of a basic operation required would be the drilling of holes in a part. The operation is listed under the requirements of a machining operation. A drill press of the correct size will be needed to drill the hole. To drill a 7/8″ hole in a casting mounted in the drilling jig, a 20″ drill press with power feeds and holding tools, drills, and chucks would be required.

Determining the Sequence of Operation

The sequence in which manufacturing operations are performed frequently makes difficult jobs easier and more accurate, as well as increases the tool life. It is obvious that a hole must be drilled before it can be threaded. Other sequences in manufacturing may be very complicated and require experience and judgment as well as knowledge of the capacities of the machines that are available for use.

The planning for the optimal production requires sequences which make use of flat surfaces for locating the workpiece in jigs or holding fixtures. These large, flat surfaces may be used during the entire manufacturing operation, from the roughing cut to the total manufacture to inspecting and testing. For this reason the first operation in a sequence for a casting may be the planing or milling of a large or flat surface which becomes the reference surface for all later operations. Other operations following in the sequence will use that surface to locate the part in various jigs and fixtures, to locate holes, machine bosses and slots, and for the drilling and tapping of holes. When possible, the sequence will be designed so that all the work will be performed in a single holding fixture or pallet.

Combining Basic Operations

With the use of modern tools a number of basic operations may be combined, thus simplifying production. For example, a center drill bit for lathe work is designed to drill and countersink at the same time. Another example from woodworking is the "screw mate," a tool that drills a pilot hole, a shank-sized hole, and counter sinks the size of the top of the hole for a flat-headed wood screw. The best manufacturing example of combining basic operations may be found in the automated transfer machines which are used to mass-produce automobile parts. This principle of combining manufacturing operations is complicated, but it reduces the number of operations, and thus manufacturing costs.

Specifying Tools and Equipment

The manufacturing engineer and the manufacturing processes group are people who are thoroughly familiar with the manufacturing equipment and the production capacity of the equipment and machines. These people will receive all the above information and then will write a **process sheet, operation and tool record,** or a **route sheet.** The route sheet will vary considerably in detail from one company to another, but it will be used to provide the pertinent information for manufacturing, and it will supply the most economical route for the part to proceed through the manufacturing process, Fig. 2-84.

The route sheet will frequently contain the name of the part, the number of parts required, the operation number of the part, and the rough size of the stock. The production information will include the operation number, operation description, depart-

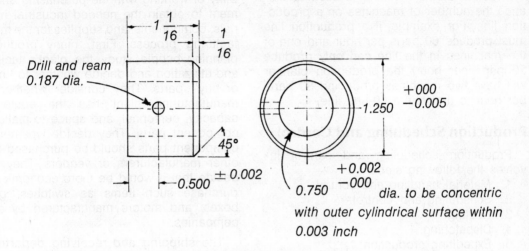

PART NUMBER
2-110

Drill and ream
0.187 dia.

45°

± 0.002

0.500

$+000$
-0.005 1.250

$+0.002$
-000

0.750 dia. to be concentric
with outer cylindrical surface within
0.003 inch

Material: AISI 1020 steel
Annealed or normalized

Note: Unless otherwise specified,
dimensional tolerances are
± 0.015 inch

PART NUMBER: 2-110
PART NAME: Collar
NAME OF COMPANY:
SCALE: Full size
DRAWN BY: JEC
CHECKED BY: HRD
DATE: 29 Jan, 1975

	Name of Part:					Part Number:			
	Estimated Quantity per Year:					Rough Size:			
Operation Number	Operation Description	Dept.	Machine	Cutting Tools	Jigs and Fixtures	Cutting Speed or RPM	Feed	Time	
								Setup Hours	Operating Pieces/Hour
5	Turn 1.250 dia., face end, chamfer, drill and ream 0.750 hole, cutoff	11	No. 2 turret lathe	See instruction sheet	—	150 rpm	0.008 inch/rev	2.5	34
10	Face unchamfered end	11	No. 2 automatic lathe	Facing tool	Fixture No. 4235	150 rpm	0.005 inch/rev	0.5	115
15	Drill and ream 0.187 hole	3	Drill press	11/64 drill 3/16 reamer	Drill jig No. 4236	Drill 2000 Ream 1000	0.004 inch/rev hand	0.4	64

Fig. 2-84. The operation or route sheet provides the information for manufacturing and the most economical route for the part to proceed through the manufacturing process.

ment doing the work, machine required, cutting tools required, jigs and fixtures required, cutting speed of the material, feed of the tool, setup time required, and the number of pieces per hour of production. The output of the machine is used to balance the number of machines on a production line. For example, if a production line must produce 60 parts per hour and one of the machines in the line can only produce 30 parts per hour, the production planner will have two machines producing 30 parts per hour to maintain a 60-unit line.

Production Scheduling and Control

Production scheduling and control involves the following steps:
1. Material handling and control.
2. Scheduling and control.
3. Routing.
4. Dispatching.
5. Expediting production.
6. Tool performance and control.
7. Quality assurance.
8. Value analysis.

Material Handling and Control

Material procurement has the responsibility of working with the purchasing department to obtain the needed industrial materials, components, and supplies for the manufacturing process. First, many production planning people study the plant's facilities and utilization and decide whether to "make or buy" parts. They consider whether the manufacturing plant has the equipment capacity, personnel, and space to make the component parts. They decide whether the component parts should be purchased from other manufacturers or vendors. They may decide that it would be more economical to purchase such items as switches, gearboxes, and motors manufactured by other companies.

The **shipping and receiving department** has the responsibility of checking to see if the proper material or component has arrived on time and if the specifications, number of parts, and quality are equal to the specifications set forth in the purchase order.

When the industrial materials and components are received, the supplies and materials become the responsibility of the **inventory or storekeeping** section of production control. Stores and store control have the responsibility of storing and issuing the materials, tools, and equipment needed for production. An inventory is kept for information about manufacturing materials that are on hand, where they are located, when and to whom they were issued, and how much material is left, Fig. 2-85. This inventory is often computerized and maintains accountability records of production and nonproduction materials, equipment, and plant facilities.

Tooling procurement is another responsibility of production planning. The tools, equipment, and facilities are analyzed to determine which will be needed in the manufacturing process. The manufacturing engineer works with this group to see if

Fig. 2-85. Inventory card used for storing and issuing tools and materials.

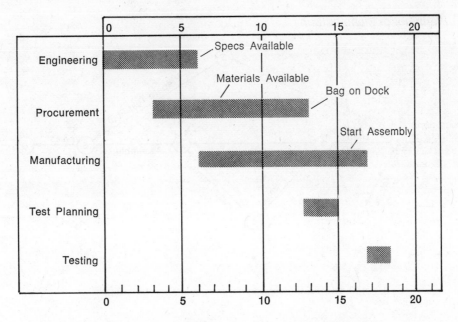

Fig. 2-86. A simplified Gantt bar chart of a back pack project.

existing plant machinery and groups may do the new manufacturing, or if new tools, machines, production lines, or a new factory may be needed. The tooling procurement department obtains the correct tools for manufacture, whether they are hand tools, jigs, fixtures, cutting tools, machine tools, or other necessary for gaging and inspection of the product.

Scheduling and Control

The major responsibility of production planning and control is to have the right worker, the right material, the right machine, and the right tools all at the right place at the right time, Fig. 2-86. This means that many materials, machines, tools, and people will have to be scheduled and controlled, Fig. 2-87.

Some of the scheduling functions include an operation schedule, dispatching, production expediting, and performance reporting.

Routing

Scheduling the movement of materials through the manufacturing process is a difficult, but very necessary part of manu-

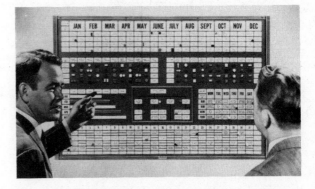

Fig. 2-87. Scheduling production so that workers, machines, and materials are available for production on time.

facturing. A schedule is made from a number of sources of information, but the chief source is the **route sheet**. The route sheet provides the schedule maker with the number of operations needed to make the part, the time for each operation, and the output for each machine in the operation. When a manufacturing schedule gets very complex, a **Program Evaluation Review Technique (P.E.R.T.)** network or critical path method

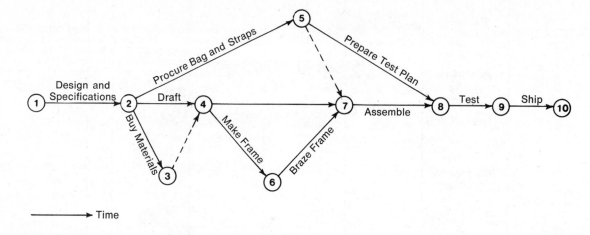

Fig. 2-88. P.E.R.T. network for the production of a back pack.

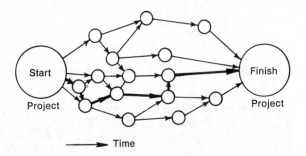

Fig. 2-89. P.E.R.T. network of a complicated product showing critical path. This evaluation sets a schedule so that all workers and machines will have a full work period. Slack periods may be identified and rescheduled.

may be established, Figs. 2-88 and 2-89. This network will control the manufacturing process and fix the responsibility dates for the various components to be designed, manufactured or purchased, assembled, and tested. In a complex manufacturing system where many vendors are supplying materials and component parts, a schedule may become so complex that it would be difficult for the mind to comprehend and keep all the facts available.

The schedule must provide for having the right workers and machines running eight hours per day, with all employees working at productive work. All of this must be accomplished with the least amount of inactivity of tools and materials.

Dispatching

The production schedule allows dispatchers proper time for releasing work orders to the shop, releasing materials from stores, releasing production tooling, and keeping records of the orders as they progress through the manufacturing processes. Dispatchers see that the following items are given to the workers to manufacture the product: the work order, route sheet, prints, material, cutting tools, fixtures, and gages. Dispatchers assign the equipment or machine to do tne work. They give the workers the materials, tools, and the authority to do the work. Dispatchers have the responsibility of following up to find out if the work that was scheduled was indeed done and on time. They report back on the movement of materials, operations completed on each part, the number of parts completed, and the time utilized. This information is generally sent back as data to scheduling, inventory control, cost accounting, and the payroll office.

Expediting Production

Expediting is part of a continuous follow-up plan that has been set down by the schedulers and dispatchers. Expeditors work in the plant with the production people to be certain that work is proceeding at the scheduled pace. They see that materials and tools are available for production on a continuing basis. Expeditors know the status of orders and will "trouble shoot" any problem that delays the production process. Expeditors are responsible for seeing that the work is done on time. They are the operational representatives of the production scheduling and control department.

Tool Performance and Control

The performance of tools and their control in manufacturing are recorded to provide information as to whom the tool has been assigned, which tools have failures, which have been lost, and which are worn out. The information from these performance records is analyzed to make decisions to improve and control tool usage.

The tool performance and reporting group is responsible for stocking and storing tools, dispersing tools to the workers, and providing a record of the tool's performance.

Quality Assurance

A quality assurance program in a manufacturing plant is an overall control program to coordinate these three things: (1) the acceptance of incoming material, (2) the inspection of the product during manufacturing, and (3) the maintenance and improvement of quality during manufacturing.

Quality assurance programs are **autonomous,** which means that the departments are independent of other departments in the plant. The programs are designed to provide manufacturing information directly to top management by fact finding and reporting. To obtain the information audits are made to determine the best product design, manufacturing design, design of inspection methods and procedures, and design of programs for the training and upgrading of personnel in order to meet the quality standards required during manufacturing. Another function of quality assurance programs is to examine the complaints of customers to establish what can be done to improve the product and the product sales.

The data that the quality assurance group assembles for decision-making comes from various sources: (1) reports from customers of poor service and of defects and failures of the product, (2) the auditing system carried on at the in-process inspection stations throughout the manufacturing process, (3) the comparison of competitive products on the market for the best dollar quality value.

Quality Control. Quality control has the responsibility of inspecting and checking the incoming materials and manufactured components as they arrive at the plant. They conduct tests and inspect the materials to determine whether these products meet the requirements specified in the purchase order. These materials will be accepted or rejected on the basis of the purchase specifications.

Quality control is responsible for inspection and in-process manufacturing to determine if the drawings and specifications of the engineering department have been reached during manufacturing. The inspection process includes testing and measuring of the products during manufacturing by the use of gages and laboratory tests.

An important part of in-process inspection is the auditing of the quality level of the parts being produced to see whether the product is in agreement with the manufacturing engineer's route sheet or process sheet.

A **realistic tolerance** is the part size that may be regularly obtained by the machine which is doing the work. Different machines have inherent characteristics as to the accuracy that can be expected under production conditions.

Close tolerances generally require specially built machines and tooling which increase costs of production. The production

FIG. 2-89A.
Realistic Tolerances for Machine Processes

Automatic Screw Machine	.001″ - .0005″*
Boring	.001″ - .0005″
Counterbore	.002″ - .001″
Drill	.002″ - .001″
Grind (Cylindrical)	.001″ - .0005″
Hone (Internal)	.00015″ - .00010″
Lap (Internal)	.00015″ - .00010″
Mill (Finish)	.001″ - .0005″
Plane	.002″ - .001″
Punch	.015″ - .005″
Ream	.001″ - .0005″
Saw	.015″ - .005″
Scrape	.001″ - .0005″
Turn (Smooth)	.001″ - .0005″

*Approximate values.

process group routes the work to machines that produce the required tolerances, not to machines which are capable of producing considerably closer tolerances than are actually required. Only parts which require ultra-precise tolerances should be scheduled into the highest quality machines, Fig. 2-89A.

Surface quality in the metal manufacturing industries pertains to the roughness of the surface. Different processes make different surface finishes, Fig. 2-90. The various finishes are measured by a **profilometer** and are assigned a microinch number which is the average roughness height in microinches (0.000001″) as a unit of measurement. A surface may easily be machined in

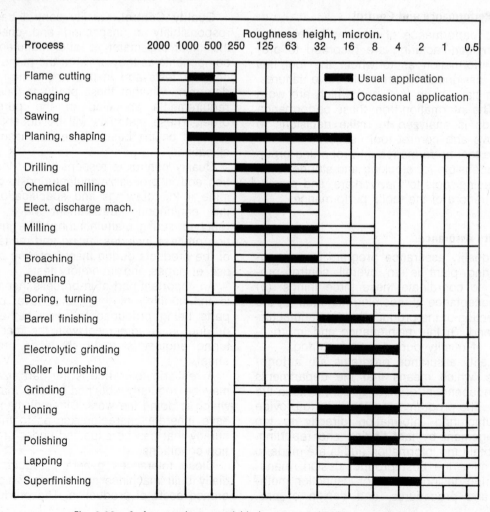

Fig. 2-90. Surface roughness available by common production methods.

a lathe to a 32 microinch finish at low cost. If an 8 microinch finish is required, grinding or honing will be the processes used.

A realistic finish is the coarsest finish that the .specifications or the customer will accept. Finishing will be done only on the surfaces requiring special finishes.

Value Analysis

A **value analysis** is the study of the manufacturing processes to determine whether the production cost may be reduced while keeping the value or sales appeal at the same or at a higher level. The concept is the **value of the product,** not the quality of the item. The quality is held at the recommended specifications, but the value of the product may be increased by redesigning, beautifying, or simplifying the units.

The study of the product will progress through this sequence of study:
  (1) gathering information about the article and its manufacturing process,
  (2) generating creative solutions to improve these,
  (3) choosing the most practical, economic, solutions which also have the most sales appeal, and
  (4) presenting a report with factual data for recommendation.

Frequently a value analysis study will show a way to simplify the manufacture of a product. A product may be completely redesigned so that an assembled unit is replaced by a cast unit or a fabricated or assembled unit is replaced by a pressed or stamped unit.

Cost reduction may be brought about by combining a number of parts or units and producing a number of parts as one part. The study may reveal that the same operational levels of the product may be obtained by using standard commercial hardware rather than special bolts, pins; hinges, belts, switches, etc.

Frequently a value analysis will suggest that a new or modern manufacturing process should be applied. An example of this technique may be the application of spot welding for riveting a fabrication of a sheet metal part or stamping or pressing a part rather than using a sand casting.

The production prints are examined for excessive specifications or places where customer surveys indicate that precision of certain parts is higher than necessary for the parts to operate well.

Materials are evaluated in order to check the product's requirements. It may be found that a more expensive or higher grade of material has been used than is really necessary to do the job. Plastics may be able to be substituted for many metal parts or aluminum for copper in some electrical circuits such as coils in electric welders.

The purpose of value analysis is to improve a product that may be economically manufactured and to increase its value to the customer so that sales will increase.

Activities

1. Interview a number of people who work in a manufacturing industry and develop a flowchart of the organization of the industry. Start with the stockholders and trace the responsibility of groups of people to the final completed product.
2. Examine newspapers' stock market quotations and list the industrial materials and standard stock that are mentioned.
3. Check the wanted advertisements in a large newspaper. What are the job descriptions and requirements? Organize the jobs into categories and make a file.
4. Design and plan the manufacture of a simple product.

Related Occupations

These occupations are related to manufacturing technology:

Corporate executive
Public relations worker
Market research worker
Industrial designer

Engineer and Technologist
  Aeronautical
  Aerospace
  Agricultural
  Astronautical
  Biomedical
  Ceramic
  Chemical
  Civil
  Electrical
  Electronic
  Industrial
  Mechanical
  Metallurigical
  Mining
  Geological
Drafting artist
Craftsmen
  Construction
  Mechanic and repair worker
Machine operator
Assembler
Inspector

Manufacturer's sales representative
Accountant
Purchasing agent
Industrial traffic manager
Computer console operator
Stenographer
Secretary
Lawyer
Bookkeeper
File clerk
Billing machine operator
Adding and calculating machine operator
Duplicating machine operator
Shipping and receiving clerk
Stock clerk
Waybill clerk
Typist
Transcribing machine operator
Data typist
Tape perforator operator
Personnel operator
Telephone operator
PBX operator

Safety in Metalwork

Chapter
3

Words You Should Know

Attitude concerning safety — A predisposition, manner, feeling, and behavior toward an object or thing. A concept of one's bodily welfare.

Mechanical advantage — A ratio of input force applied to a mechanism that supplies a larger output force.

Safety zones — A lane or walk area free from vehicle traffic. Lanes to tool cribs, offices, and restrooms.

Safety lines — Lines which indicate machine crane or truck overtravel. Railway cars project over an area wider than the track, so lines indicate the car's overtravel. Nothing should be stored in that area.

Safety area — An unsafe area such as in front of a shaper or planer, under a wire brush or incline with a high-speed belt or chain. A person should not stand or pass through this area.

Safety — a Learned Characteristic

People learn from their experiences. Attitudes of safe procedure can best be learned in the laboratory. People could have safety experiences by working with materials personally and by observing the results. However this could also be dangerous and inefficient. People can also learn about safety by **observing** an expert such as a skilled machinist while he or she is working or by watching a product made. During this observation, safe procedures are pointed out. Another way to learn safety is to study charts and books about the correct operation and safe use of tools, Fig. 3-1.

When people learn safety procedures, they develop an attitude concerning safety. The attitude they acquire when learning the safe operation of equipment will affect their behavior for the rest of their lives. A person who formulates a desirable attitude concerning safety will follow this procedure for operating a machine:

- Read the operation instructions.

Fig. 3-1. Learning safety instruction by using safe operating procedures, working with materials after demonstration, and studying safety and operation materials.

- Receive a demonstration of the operation of the machine, including an explanation of safety procedures.
- Under supervision, start operating the machine, following safety procedures.

The person who has the attitude that both the operation and the safety procedures of the equipment should be understood before the equipment is used will tend to be a safe worker.

Most accidents are caused. People who have accidents often do not have the proper knowledge, do not think, or they have an unsafe attitude — feeling that they can get by this time.

General safety characteristics that safe workers should learn include the following:

1. They should organize their work procedures and develop a plan of action. They should be **well organized**.
2. They should have a realistic view as to the tools and materials and what can be done in the time available. They should be **conservative.**
3. They will consider other people who are working nearby or with them and be **cooperative** and **courteous**.

The attitudes that students build about safety will be useful in all facets of their lives.

Planned Body Movements

Safety procedures require planned movements, or being organized. Safe workers have their work laid out in a neat and orderly manner; they know what tools they will need after studying their plan. The material, tools, and plan are laid out so that they may easily reach the tools they need without having to let go of the metal or part under construction, Fig. 3-2. When safe workers use a machine, they plan in advance the movements or sequence of operations they will be making. Then they lay out the tools and fasteners they will need.

Planned body movements are a part of safety. For instance, in operating a sheet metal squaring shear, great care must be taken. The machine will cut the edge of sheet metal square and straight in one cycle. Operators must plan movements to keep their bodies clear of all moving parts of the machine — their fingers away from the cutting blade and their toes out from under the foot treadle, Fig. 3-3. Safe workers also observe and caution any helper to stand clear of any springs and levers at the back and side of the machine. Safe workers plan their movements by understanding that the forces

Fig. 3-2. Planning ahead for work and organizing tools that will be needed to help create safer working conditions.

Fig. 3-3. Planned body movements ensure safe machine operation. All parts of the body are clear of the moving parts of the machine.

the machine generates may result in shifting work. This shifting work may result in damaged work or an injury.

No matter what power tool the workers use, they must do the thinking, planning, and setup of the tool before the tool is put into operation. The tool only supplies the work action after the energy has been applied. Safety is the result when the worker controls the tool and the material being processed. "Think before you act!" is the motto of planning safe body movements.

Power and Forces that Work for the Operator

Metal which is being cut or formed offers a considerable amount of resistance. Workers would be very inefficient if they used their bodies to overcome these forces

necessary to work metals. In order to provide the energy and forces necessary to shape metals, machines have been devised that increase operators' physical powers. The **lever** and the **screw** provide power and force to hold work being shaped, Fig. 3-4. The lever provides a mechanical advantage for operators by allowing them to clamp their workpieces to the table or fixture. As an example, safe operators will use the principles of levers by having the distance from the work to the fulcrum less than the distance from the fulcrum and clamp block. The holding power of a second class lever provides more force than could be applied by an operator, Fig. 3-5.

A bolt which holds work directly to a drill press table is also a device to overcome the counterforces which are developed in the cutting and shaping of metals. The common slip-joint pliers are like a strong pair of fingers. With the application of the lever and a pivotal point, the pliers provide the operator with a mechanical advantage that supplies power and force, Fig. 3-6. The pliers may be used to hold small pieces of sheet metal when a drill press is being used. They also provide powerful fingers and at

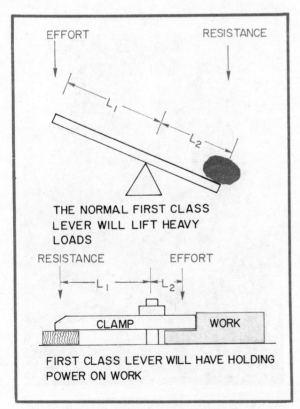

Fig. 3-4. The use of levers increases the worker's work-holding power.

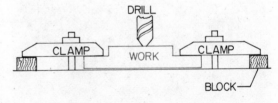

Fig. 3-5. Mechanical advantage increases the holding power on the work.

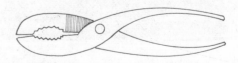

Fig. 3-6. Pliers increase holding power by acting as a lever.

Fig. 3-7. A barrier guard is placed in front of the cutting edge so fingers cannot get into the shearing blade.

the same time, keep the operator's hand away from the rotating drill.

The safe way to work is to take advantage of tools and make them do the work.

Mechanical Devices that Protect the Operator

Safety devices are often made up of simple machines that are mechanical, electrical, or pneumatic (operated by air). One of the simplest safety devices is the **guard**. The guard is a physical barrier that keeps fingers, hands, or other parts of the body away from dangerous areas, Fig. 3-7. Barrier guards are found in front of shears, over pulleys, around gears, and around grinding wheels.

Another type of guard is the **interlocking gate**, Fig. 3-8. When the machine is operated, a guard comes down and provides a barrier to the dangerous area. This may be

Fig. 3-8. A machine which has an interlocking gate guard will not operate unless the gate guard is down and in place.

Fig. 3-9. In the two-handed grip safety device, both hands must be on the control button at the same time or the machine will not operate.

found on a squaring shear, press brake, or punch press. These machines should not be operated unless the guard is in a protective position.

A **sweep guard** is a lever arm which sweeps across the danger zone. Operators must remove hands from the work area or the sweep will hit their wrist or hands and knock them out of the way before the work is cut or stamped.

The **two-handed grip** is possibly one of the most efficient devices used to protect workers. When this guard is used, both of the operator's hands must be on the buttons or switches before the machine will start and go through its metal-shaping cycles, Fig. 3-9. If only one hand is on the button, the machine will not move. The two-handed and multiple-control stop buttons are used on many modern machines, from punch presses to numerically controlled machines to automated production lines.

Safety Zones, Lines, and Areas

Safety zones, lines, and **areas** are floor markings which warn workers that hazards exist in, over, or around the areas, Fig. 3-10. A safety zone is a space where the workers should be free from moving vehicles such as fork lifts, cranes, trucks, or machinery. In a busy plant, safety zones are provided for people going to the tool cribs, rest rooms, or offices. It is the vehicle driver's responsibility to use extreme caution when crossing a safety zone.

Safety lines are painted on the floor. They indicate the space of overtravel a machine's table requires or the space an industrial truck with an overhanging load needs when it is transporting material. Safety lines may also mark the space railroad cars occupy in a warehouse so that people or products will not be within the line and cause an accident, Fig. 3-11. Safety lines may mark the travel of an overhead crane so that products being moved will not strike pedestrians or materials under the crane's path. When workers are inside or crossing between safety lines, they should be very alert.

Fig. 3-10. Bold black lines indicate a safety zone.

Fig. 3-11. Safety lines indicate the space industrial trucks and carts may move materials through a plant.

Fig. 3-12. Workers should not stand in or pass through a safety zone. This traveling spray booth with automatic painting machines paints 50-foot railway boxcars. Note the safety signs.

Fig. 3-13. Safety equipment for the eyes, ears, head, and lungs.

A **safety area** is a space where workers should not stand or pass through, Fig. 3-12. A safety area may be in front of a shaper (where the chips fly forward) or a planer (where the table moves out). It may also be the floor area under a wire brush. These safety areas are frequently painted with diagonal yellow and black stripes and indicate that no part of the body should be in these areas.

Safety Equipment

A worker is responsible for wearing the proper protective equipment. The body needs additional protection in hazardous work environment. Proper clothing and equipment can save lives and keep a worker on the job.

Eye and Face Protection

While operating machinery, workers protect their eyes and faces by wearing eye goggles, a face shield, or a helmet, Fig. 3-13. Workers avoid the danger of slight impact of particles by wearing plastic eye shields or glasses at all times in the shop area. Glasses are not sufficient protection for heavy grinding or machining operations. They may not give protection from particles which come in from the side. The face shield provides protection from such foreign objects and protects the whole face rather than just the eyes. These face shields are used when light grinding, buffing, or chipping are being done. They are worn in addition to glasses.

A helmet or hard hat is designed to protect the worker from falling articles such as nails, nuts, rivets, or small stones from areas above. Helmets will give protection from small tools which may fall or be accidentally swung. Helmets are also designed to protect the eyes and face. The arc welder's helmet performs these functions. The welder's eyes are protected by dark colored glass, which removes the harmful rays from the arc flash. The plastic and fiber shield protect the face and eyes from the ultraviolet rays and the spatter of welding flux and metal. Helmets give maximum protec-

Fig. 3-14. A foot guard of heavy steel over a foot control.

tion, whether the worker is a sandblaster, steelworker, or welder.

Finger and Foot Guards

Finger and foot guards are usually built into the machine. A finger guard is a barrier which is designed as part of the machine so that a finger cannot be put into a dangerous place. A steel barrier rail on the front of a squaring shear is an example of a finger guard. A worker should never try to work around or through a finger guard.

Foot guards frequently are bands of heavy steel that form an arch over a foot control, Fig. 3-14. The guard protects the foot from being crushed by falling steel or parts as the operator runs the machine with the foot control. Heavy hydraulic shears may have a movable foot control so that workers may position themselves for the most efficient cutting operation.

Industrial work shoes frequently have steel toe caps built into the shoe. This steel cap gives protection from falling objects. Two-by-fours, angle irons, wrenches, and pinch bars may fall on these shoes without toe damage.

Aprons, Gloves, and Leg Guards

Aprons, gloves and leg guards are used in many shop areas, but they are most

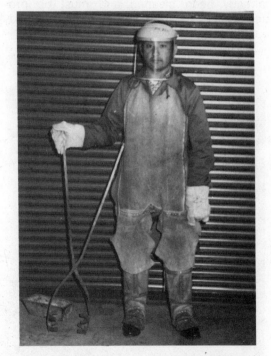

Fig. 3-15. Aprons, gloves, and leg guards are used when an operator works with molten metals.

frequently used in the foundry, Fig. 3-15. Aprons are made of canvas, leather, and asbestos. Canvas aprons are used for light duty and are used around work products and steel handling to protect clothing. Leather aprons are worn when a worker does heavy work such as unloading boxes, wood, steel, sacks of sand, or bricks. Leather aprons are used in foundries to protect workers from heat and the splatter of molten metal. Asbestos aprons are designed for hot work. They give the operator maximum protection from hot crucibles and molten metal. Asbestos aprons give good heat protection, but they do not give protection from any rubbing or abrasion.

Industrial gloves are of many types and materials. The common gloves are made of nylon, canvas, leather, and rubber. Nylon gloves are used in **clean rooms**, or where the product that is being processed has already been cleaned and must be protected from fingerprints, body oils, and body acids. Nylon gloves are used in the assembly of oxygen-free parts for electronic components. They allow for hand and finger movement and still leave the surface of the metal clean, Fig. 3-16.

Canvas gloves are used as general work gloves and give the worker general protection from dirt, cuts, and heat. Leather gloves are also used for general work. To make an inexpensive glove, canvas and leather are combined so that the worker's hand may be cool and yet receive maximum protection from dirt, cuts, heat, and pinches. Very flexible and lightweight gloves are used to protect the welder's hands from heat while TIG (tungsten inert gas) welding, Fig. 3-17. This type of welding requires accurate manipulation of the electrode so that aluminum welding may be successfully done. The flexibility of the glove is very important.

Leather will gradually get hard and stiff when it is exposed to heat and ultraviolet light, so the operator protects the gloves from heat as much as possible by using tongs or pliers to handle hot pieces of work.

Leg guards are used primarily in the foundry where heat and the splatter of molten metal may hit the lower legs and shoe tops. Leg guards are made of leather

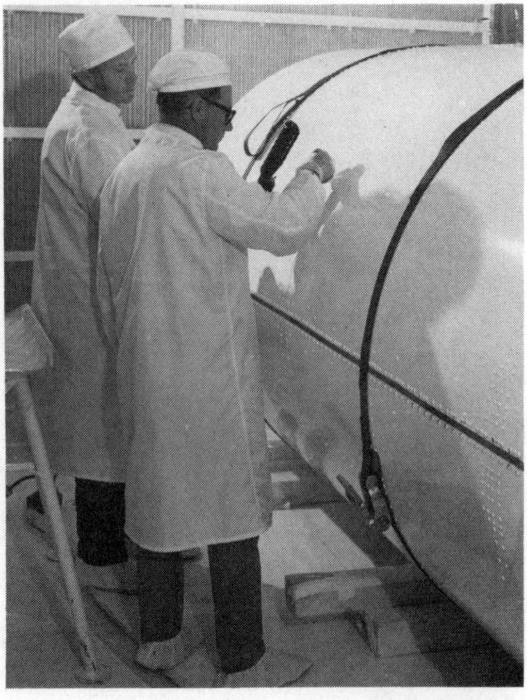

Fig. 3-16.   Nylon products are used in industrial clean rooms for protection from contamination from workers.

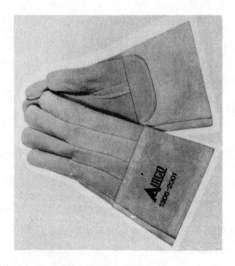

Fig. 3-17.   A welder's leather gloves.

or asbestos and will stop liquid metal from burning through clothing to the legs.

If a worker's clothes catch on fire, a fire blanket is used to extinguish the fire. A fire blanket is a chemically treated blanket which is wrapped quickly around the worker. Fire blanket containers are mounted on the wall of auto laboratories and welding and foundry areas.

Common Injuries

Safety in metalworking means knowing the ways to protect yourself from these common injuries and hazards:

- Burns
- Cuts
- Eye injuries
- Pinches
- Clothing caught in machinery
- Inadequate ventilation

Burns

Burns are one of the most common injuries which occur in a welding, oxyacetylene cutting, or foundry area. Most are first degree burns in which the skin is reddened and unbroken. These burns may be caused by picking up hot iron which does not show any color because of the heat. Castings and welded metal may be above a 500° F. heat for a considerable amount of time and are the source of a number of painful burns. The common practice for minimizing this type of burn is to mark the scrap or workpiece with the **date** and **time** and to mark "**hot**" on it, Fig. 3-18. The next person who comes into the area can judge whether the steel or casting has had time to cool by the date and time. The first aid treatment for this type of burn is to place the burned area in cold, clean water and then obtain further medical treatment.

Second- and third-degree burns are very serious. If a person receives such a burn, the skin will be blistered and broken, and the danger of infection is greatly increased. These burns may be the result of an oxyacetylene welding torch, a cutting torch, electric arc, or spilled molten metal. The first aid procedure for treating these burns is to place a sterile gauze bandage over the burn and then get a nurse or physician's aid. People with a second or third degree burn may develop severe shock, so they should be placed in a position where they cannot fall. They should be kept warm and quiet.

If clothing catches fire when a worker is welding or pouring metals, the flames should be patted out with a glove or apron and cooled with water. If a large area of clothing is on fire, it should be smothered with a coat, apron, or fire blanket. In any case, the person whose clothes are on fire should not run, because a severe burn could result.

Cuts

Cuts are painful and may be avoided if reasonable care is used. Sheet metalwork is an activity in which workers could easily cut themselves. If a file is drawn along the freshly cut sheet metal, the sharp or burr will be removed from the edge of the metal,

Fig. 3-18. Hot metalwork and scrap will have the date, the time, and "hot" written on the surface.

Fig. 3-19.   When sheet metal is sheared, it has a sharp edge. It is made safe by drawing a file along the sharp edge or burr.

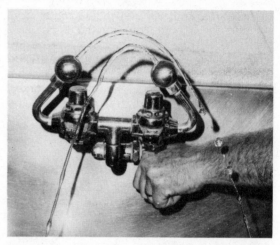

Fig. 3-21.   An eyewasher cleans with low-velocity water.

Fig. 3-20.   A push stick will prevent painful cuts.

and cuts will be minimal, Fig. 3-19. If pliers are used to remove notches in sheet metal, the worker is not as likely to get cut.

Cuts may result from the incorrect use of hand and machine tools. Gloves, a push stick, wrench, or other tools should be used to handle stock or cutting tools which are sharp, Fig. 3-20.

If a person is cut, the bleeding should be stopped with a sterile compress, cleaned, and dressed. If the cut is a major laceration the worker should be taken to a nurse or physician.

Eye Injuries

Most eye injuries can be avoided by wearing safety glasses in an eye hazard area. However, glasses do not give enough protection against eye injuries which are the result of the flash from an electric arc. The ultraviolet light from the arc can very quickly burn the retina of the eye. If the retina is severely burned, scar tissue will form, and the result will be impaired vision. When a person is arc welding, a helmet with a number 10 shade of colored lens should be worn to protect the eyes. The reflex action of the eyes will usually protect a person who is at least 10 feet away from the flash. At no time should a person look at the arc without eye protection. A person who receives a flash should receive immediate medical treatment. If chemicals are spattered into the eye, an eyewasher or low velocity water outlet similar to a drinking fountain should be used to wash out the eye, Fig. 3-21.

Fig. 3-22. Lowering heavy materials onto blocks to protect fingers and feet.

Fig. 3-23. Select the proper type wrench and use it by pulling on the handle.

Pinches

Pinches will cause pain as well as loss of time and efficiency. Pinches may be caused during the lifting, setting down, or sliding of heavy materials or objects. To protect feet and fingers, heavy objects should be set down on a block 1″ x 4″ until the fingers can be removed from under the object or material, Fig. 3-22. Then a pinch bar or pry bar should be used to raise the object enough to remove the block. The object is then set on the floor, and the pinch bar is removed.

When a worker is using wrenches, his or her fingers may become pinched if they get caught between a casting and the wrench that is tightening the nut. The easiest way to avoid wrench pinches and cuts is to first use the proper wrench. Workers can select from a box wrench, open-end wrench, or an adjustable wrench. The socket wrench is usually the first choice, and the adjustable wrench is the last choice. The second way to avoid wrench pinches is to always pull on the wrench, Fig. 3-23. If a bolt can be loosened only by pushing on a wrench, the hand should not be used. The foot or an extension should be used so a worker can change position and pull on the wrench.

Small tools such as pliers, hammers, and tool post wrenches may sometimes cause

Fig. 3-24. To avoid pinches, keep fingers from between moving parts.

pinches. The principle to keep in mind in avoiding pinches is to keep fingers from between the moving parts. When a lathe tool is removed from the tool holder, the tool post wrench should be put on in such a manner that the set screw may be loosened by squeezing the tool post wrench and tool holder in the palm of the hand, Fig. 3-24.

Fig. 3-25. Loose clothing can easily become caught in machinery such as this lead screw.

Clothing

Clothing caught in machinery can result in a very dangerous situation and should be guarded against by keeping all clothing tight and tucked in close to the body, Fig. 3-25. An apron or shop coat which keeps all loose clothing away from moving tools is important. Long hair must be held down and back by the use of a hat or headband so that hair will not be caught in moving machinery.

Rings and heavy wristbands must be carefully worn or removed because of the possibility of their being caught in lathe chucks or other rotating machinery. Metallic rings and wristbands are also hazards because of electrical shorts. If a person wears such jewelry while working around any battery terminal, he or she may become severely burned if the ring or band is shorted between the battery terminal and ground.

Ventilation

Ventilation is a safety concern in the welding area and in any area which deals with evaporating solvents. All types of welding give off gases, and if these gases are breathed over a long period of time, the welder may become ill. Ventilators are pro-

vided over welding booths to collect the carbon dioxide, iron oxides, cadmium oxide, lead oxide and zinc oxides and other materials such as dirt and grease, Fig. 3-26.

Spray booths, soldering areas, sandblasting, carburetor cleaning and engine test stands should have fans to remove undesirable airborne materials from the work area.

When gas-fired forges and furnaces are used in a laboratory, adequate ventilation must be provided, because forges and furnaces use a high percentage of oxygen from the room. Workers will experience sleepiness and headaches if ventilation is not provided.

Common Hazards

The operator must have an awareness of these common hazards and the correct procedures for avoiding them:

- Hot objects
- Intense light
- Flying projectiles
- Mechanical failures
- Electrical shock
- Moving parts of machinery
- Slippery floors
- Lifting heavy objects

Fig. 3-26. Ventilators collect fumes and airborne materials and remove them from the operator's area.

Hot Objects

Hot objects are another safety hazard. Steel may be 500° F. and look the same as room temperature steel. If forging tools or forging tongs are used and placed back into the rack while they are still hot, the next person who picks them up in a short time will receive a painful burn. These tongs or tools should be quenched in water before they are replaced on the rack so that such burns can be prevented, Fig. 3-27.

Pliers and clamps used around oxyacetylene welding and cutting tables are sometimes accidently heated and will burn when they are picked up.

Spot welding generates a large amount of heat in the sheets of metal welded. When small pieces of metal are welded, there is not a sufficient area for the heat to be dissipated. Therefore, small spot-welded pieces of metal should be held with pliers to protect the welder's hands and fingers.

Fig. 3-28.　Sharp wires may be broken and thrown away from the wire brush.

Intense Light

The greatest danger from intense light comes during the arc-welding processes. These light sources are rich in ultraviolet light which causes burns by radiation. Anytime a direct arc is exposed to the eyes, a colored lens is needed for protection. The welding techniques of shielded arc, the metal inert gas (MIG), and the tungsten inert gas (T.I.G.) all have **exposed arcs**. Ultraviolet rays radiate from these exposed arcs. The standard protection for these dangers is an arc-welding helmet.

Flying Projectiles

Flying projectiles are another safety hazard. Often wires which work-harden with use and fly free from a powered wire wheel are overlooked. These wires are sharp and may be thrown off the surface of the powered wire brush, Fig. 3-28. Although the wires do not have much weight, they are sharp and travel at a high speed. A face shield provides protection from these wires.

Flying projectiles may also come from a portable abrasive sander. When a worker is using either a portable grinding wheel or disc grinder, a shower of metal and abrasive particles may be thrown across the labora-

Fig. 3-27.　Tongs are used to prevent burns. They in turn must be cooled.

Fig. 3-29. A portable sanding disc will send particles flying across the room.

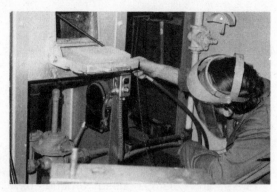

Fig. 3-30. Light a forge by standing to one side with a face shield and a long-sleeved shop coat.

tory, Fig. 3-29. If the worker is grinding an arc, sparks may fly tangent to the radius of the grinding point on the arc. The only protection from this hazard is for everyone to wear safety glasses or for the grinding to be done in an isolated area or in another room.

Unexpected projectiles are sometimes thrown when hot metal is being forged. Hot scale may be scattered during the heavy forging process. Safety glasses will usually give protection to the eyes from this hot scale. A shop coat will provide enough protection for body and arms.

When a gas-fired forge, crucible furnace, or heat-treating furnace is lit, pieces of scale and brick dust may be scattered. If the furnace is first blown out with air and then lit, the possibilities of an explosion are minimized. It is still possible to have a small blast when the flame of the burners start, scattering a small shower of brick dust and scale. For this reason, a worker should not stand in front of or look into the furnace door when the forge or furnace is being lit. A safety shield or glasses and a shop coat should be worn for further protection, Fig. 3-30.

In rare cases, a chuck key may be left in a lathe chuck or in a drill chuck. If either

tool is started with the key in the chuck, the key will come out of the chuck and fly across the room. This hazard can and must be stopped by having workers consider the chuck key "part of their hand." When the hand is removed from the chuck, the key comes with it.

The spot welding of galvanized steel will generate enough heat to liquefy the zinc coating on the steel. Often, as the welding circuit is closed, liquid metal spatters out in a plane parallel with the metal. Most of the time this molten metal will fall harmlessly to the side. However, if it hits a flange (a rim used for guiding or attaching other metal), the molten metal could be directed into the operator's face, Fig. 3-31. The safe solution to this problem is to watch carefully for flanges and deflectors and to wear a face shield.

Mechanical Failures

Mechanical failures of equipment may cause a dangerous condition for the machine operator. Do not leave a machine that has had a mechanical failure without first placing a note of warning on the control switch.

A general practice of safety against mechanical failures is to make sure your

Fig. 3-32. A person's body should not be in line with any high-speed device.

Fig. 3-31. During spot welding, a face shield should be worn, because molten metal may spatter from the surface of the metal.

body is not in line with machinery that is turning at a high speed or working at maximum load. For example, it is wise to always stand to one side of an abrasive cutoff wheel, chain, belt or pulley, or gear train, Fig. 3-32. If parts fail, they will generally fly in line with the force they were transmitting. Stand aside from heavily loaded machinery.

Electrical Shocks

Electrical shocks can produce a very dangerous safety hazard. Frequently, these shocks are caused by the breakdown of insulation. If machines are not properly grounded, a worker may receive fatal electric shocks. To be alert to such hazards,

workers should use grounded or three-wire portable tools.

The worker's body should be kept away from possible electrical grounds so the worker will not be in an electrical circuit. All plugs, cords, and switches should be examined regularly to make sure they are in operating condition. Extension cords and couplings should be laid out so that they will not pass through any water or puddle on a construction site. Natural weathering will cause small openings in the cord with time, and the cord will short-circuit.

If arc welders are well maintained, they are not an electrical shock hazard. All welders should be grounded so the power circuit will be broken if any breakdown of insulation or shorting occurs. When setting up for welding, it is necessary to make sure the welding ground wire is clamped to the work so a welding circuit may be established. In Tungsten Insert Gas Welding, it is possible to get a shock from the high frequency starting current if the welding ground clamp is forgotten.

Moving Parts of Machinery

Under working conditions, high-speed machinery is not always easy to see. For this reason, it is wise to keep hands at your sides. Do not point at something with your finger.

Fans, pulleys, belts, and gears should be guarded. If the machines are not guarded, fingers, rags, and loose clothing should be kept away from the machinery to avoid being entangled in the machinery. A common rotating shaft looks harmless, but it is very dangerous. A set screw, collar, or wrapping hitch will pull clothing or rags into the shaft. This type of hazard exists also on a propeller shaft of a pleasure boat or on other rotating shafts.

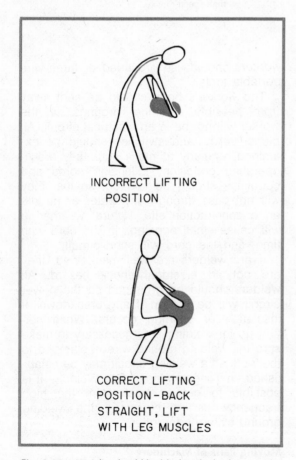

INCORRECT LIFTING
POSITION

CORRECT LIFTING
POSITION – BACK
STRAIGHT, LIFT
WITH LEG MUSCLES

Fig. 3-33. Weights should be lifted so the back is straight, and the load is on the leg muscles. The proper tool should be used for heavy work.

Slippery Floors

Spilled liquids must be wiped up immediately. Oil, water, solvent, cutting oil, and gasoline create a serious hazard. Rubber heels of shoes slide easily on concrete and oil, and the worker may have a serious fall.

Lifting Heavy Objects

Safety dictates that judgment be used when lifting heavy objects or pulling a load. The human body is a very versatile and efficient tool, but it is not designed to lift heavy loads or to hold sustained loads. The greatest single health hazard resulting in time off from work is **back injury.** Workers who are working too fast may try to lift too much weight, and they may receive a very painful back injury. Under regular working conditions, workers should not lift more than one half their own weight. When lifting, a worker should take great care to keep the back straight and do the lifting actually with the leg muscles, Fig. 3-33. Whenever you doubt that you can lift or hold a weight, always stop and select the proper tool, holding fixture, or hoist to perform the work. Tools may be replaced or repaired, but accidents to people are very costly in pain, expense, and future health.

Fig. 3-34. A prime hazard in lifting heavy loads is shifting, rolling, or falling loads.

Heavy loads may be lifted with block and tackles, differential chain hoists, hydraulic hoists, fork lifts or screw jacks, cribbing, and rollers. The prime hazards in lifting heavy loads are shifting, rolling, or falling loads. When heavy objects are lifted, the center of gravity of the load is a most important concern. Wooden blocks, braces, and cribbing must be placed under all necessary points to keep the load in control, Fig. 3-34.

Workers should keep alert to avoid breaking ropes, falling blocks, being crushed between timbers and wall or building supports, tripping on any number of tools, or dropping of tools, boards, machine parts, or rollers.

Work done around heavy loads should be done slowly. All cribbing, braces, and slings should be set well. The worker should stand back when the actual lift is being made and watch out for other workers.

Fire Control

To bring a fire under control, there are three factors to consider: heat, oxygen, and fuel. If one of these factors is removed from the fire, the fire is controlled. When the fuel is removed from a fire, the fire goes out. A campfire goes out at night because the wood has been consumed. If oxygen is removed from a fire, the fire goes out. Stepping on a burning match prevents the oxygen from reaching the flame. If a lid is clamped on a pan of burning grease, the fire will go out, because the oxygen cannot reach the flame. Heat is the third factor to consider. Each material has its own kindling temperature (point). If the temperature is below the material's kindling temperature, a flame will not start even when the fuel and oxygen are present. The range of kindling temperatures is broad and changes from time to time. Humidity has great influence on the kindling point of wood standing in the forest or products in storage.

Small Fires

Small fires which do not create a great amount of heat may be controlled quickly

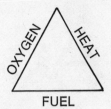

THE FIRE TRIANGLE

Fig. 3-35.   Three factors control fire: heat, oxygen, and fuel.

by spraying with water, thus removing the heat. Examples of these fires are trash can fires which are mostly papers burning. The water removes the heat by absorbing it and turning it to steam. The water finally overcomes the fire and wets the fuel enough to put the flame completely out. Water should not be used on oil or electrical fires.

Flammable Liquid Fires

Burning oil or gasoline are controlled by smothering the fire. A blanket of carbon dioxide gas or foam covers the burning liquid and keeps the oxygen from reaching the burning fluid. Fires in containers are quickly controlled by the smothering effect of carbon dioxide fire extinguishers.

Electrical Fires

Electrical fires are controlled by removing the oxygen from the area of the fire. If the motor control on the top of a motor generator arc welder unit caught on fire, the correct procedure would be to throw the circuit breaker on the main power panel to remove the electrical power and then use a carbon dioxide fire extinguisher. The nozzle should be pointed into the fire area and the trigger pulled. Carbon dioxide is a gas under pressure, and when it expands rapidly, it ab-

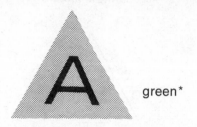

green*

FOR ORDINARY COMBUSTIBLES

Put out a Class A fire by **lowering its temperature** using a water or water-based extinguisher. Wet fire to cool and soak to stop smoldering; the burning combustibles with multipurpose dry chemical.

PRESSURIZED WATER — operates usually by squeezing handle or trigger — read instruction label (contains **water** or water with antifreeze chemical).

PUMP TANK — operates by pumping handle (contains water or water with antifreeze chemical).

SODA ACID — operates by turning extinguisher upside down. Has handle on bottom for inverting (contains water, soda mixture, acid — no antifreeze).

DRY CHEMICAL (MULTIPURPOSE) — operates by squeezing handle or trigger — read label (contains a powder commonly designated "A, B, C").

*New extinguisher label for type fire use (by letter, shape, and color).

Fig. 3-36. Class A fires are controlled by absorbing the heat by wetting the fire to cool.

red

FOR ORDINARY COMBUSTIBLES

Put out a Class B fire by **smothering** it. Use extinguisher giving a blanketing, flame-interrupting effect. Cover the whole flaming liquid surface.

CARBON DIOXIDE (CO_2) — operates usually by squeezing handle or trigger — see instruction label (discharges as a heavy gas that "smothers" fire).

DRY CHEMICAL — operates usually by squeezing handle or trigger — see instruction label (contains one of two general types of powder, not to be mixed. One is for Class B, C fires. One is for Class A, B, C fires). May be foam compatible.

FOAM — operates by turning extinguisher upside down (contains water and ingredient to make a smothering foam).

Fig. 3-37. Class B fires are controlled by sealing off oxygen with a carbon dioxide blanket or foam.

sorbs a vast amount of heat. When the heat is absorbed, the fuel no longer burns. Also, the whole area surrounding the burning wires is covered with carbon dioxide, and oxygen which is necessary for the fire to burn has been removed. Carbon dioxide thus fights fire two ways. It removes the heat, and it removes the oxygen.

Carbon dioxide is a good method to use in fighting oil, gasoline, and electrical equipment fires.

Intense Fires

Fires which generate a great amount of heat are called intense fires. An example of a very hot fire is burning magnesium. Mag-

blue

FOR ELECTRICAL EQUIPMENT

When live electrical equipment (Class C fire) is involved, always use a **nonconducting**

extinguishing agent to avoid receiving an electric shock. Shut off power as quickly as possible.

NONCONDUCTING
EXTINGUISHING AGENT
Such as **Carbon Dioxide (CO_2)**
　　　　Dry Chemical (B-C Type)
　　　　Dry Chemical (Multipurpose)
　　　　Vaporizing Liquid

Do not use soda acid, foam, or other water-type extinguishers until electric power has been shut off.

Fig. 3-38.　Class C fires (electrical fires) are controlled by reducing the heat and by removing oxygen by covering the area with carbon dioxide.

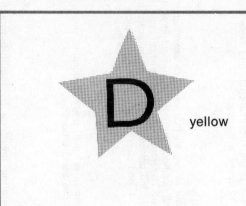

yellow

FOR COMBUSTIBLE METALS

Certain metals in finely divided forms (new Class D fire) require specially designed extinguishing agent to provide **smothering blanket or coating.**

SPECIAL DRY POWER EXTINGUISHANTS
Some are available in drums or pails and are applied by scoops or hand shovels and others in stored pressure or cartridge pressurized portable or wheeled extinguishers.

Fig. 3-39.　Class D (combustible metals) fires are controlled by covering the burning metal with a special dry powder.

nesium may accidentally get into the aluminum foundry remelt metal and cause a serious fire. When magnesium reaches its kindling point and oxygen is present, it burns with a bright blue-white flame and gives off intense heat. This fire will have to be controlled by methods other than water or carbon dioxide, because the heat is so great that it will disperse the carbon dioxide. The metal is generating more heat than the carbon dioxide can remove. If water is used,

the water will fall on the liquid metal and will instantly form superheated steam and explode, scattering the molten burning magnesium all over the laboratory. A new fire will be started wherever the metal scatters.

This fire may be controlled rather easily with a dry chemical fire extinguisher or by the use of a shovelful of foundry sand. Both of these materials will withstand high heat and will cover the burning metal. When the metal is covered, oxygen is unable to reach

121

SAVING A LIFE
BY ARTIFICIAL RESPIRATION

THE AMERICAN NATIONAL RED CROSS

If victim is not breathing, begin some form of artificial respiration at once. Wipe out quickly any foreign matter visible in the mouth, using your fingers or a cloth wrapped around your fingers.

Tilt victim's head back. (Fig. 1). Pull or push the jaw into a jutting-out position. (Fig. 2).

If victim is a small child, place your mouth tightly over his or her mouth and nose and blow gently into the lungs about 20 times a minute. If victim is an adult (see Fig. 3), cover the mouth with your mouth, pinch his or her nostrils shut, and blow vigorously about 12 times a minute.

If unable to get air into lungs of victim, and if head and jaw positions are correct, suspect foreign matter in throat. To remove it, suspend a small child momentarily by the ankles or place child in position shown in Fig. 4, and slap sharply between shoulder blades.

If the victim is adult, place in position shown in Fig. 5, and use same procedure.

Fig. 3-40. Restoring breathing by mouth-to-mouth resuscitation.

it, so the fire goes out. The metal slowly cools. When it has reached room temperature, it can be moved with a shovel to the slag pile and disposed of.

Another intense fire is a forest fire, in which the object is to remove the fuel by fire trails or backfires. **Fire trails** are barriers of cleared or plowed land intended to check a forest or grass fire. **Backfires** are fires which are purposely started in order to check an advancing fire by clearing the area. A burning building is an example of an intense fire. It is controlled by reducing the heat with a large volume of water.

Fire Reporting

All fires should be reported immediately by using the fire alarm system of the building. Quickly contained fires can be put out with facilities available in the building.

First Aid

First aid is the immediate treatment given to a person who is suffering as the result of an accident. There are four vital areas which should be checked immediately: breathing, bleeding, burns, and shock. Of these, breathing and bleeding must be given first priority.

Breathing

If a person has stopped breathing, he or she will not survive unless breathing is restored. Check to see what the cause may be — an electrical shock, choking, or inhaling carbon monoxide. If the breathing stoppage is due to an electrical shock, turn off the electric power and pull the contacting part away from the electrical source with a dry broom handle or folded dry shop coat while standing on a board, dry books, or newspapers. When there is no electrical danger to yourself, lay the injured person on his or her back, clear any obstructions from the person's mouth and start mouth-to-

mouth resuscitation. If choking is the cause, it is important to dislodge the obstruction and then give mouth-to-mouth resuscitation. If carbon monoxide has been inhaled, get an immediate supply of fresh air and then begin the artificial respiration. Continue the emergency treatment until professional help arrives. Never stop artificial respiration until further help arrives, even if the person does not respond.

Bleeding

If an accident occurs and blood is flowing in quick spurts, it means that an artery has been cut. Bleeding from an artery can usually be stopped by placing a compress directly on the wound and applying firm pressure with your finger at a point located about two inches above the wound. Bleeding from a vein produces a steady flow of blood and can usually be stopped by applying a sterile compress over the wound and pressing firmly with your hand. Elevating the wounded leg or arm will also help control the flow of blood. The injured person should sit on the floor with his or her back to the wall or lie down to avoid falls if he or she is faint or in shock. When the bleeding is controlled, the injured person should be taken to a doctor.

Burns

Using a fire blanket is the first action to take to extinguish any fire on a person's hair or body, Fig. 3-41. Burns can be very dangerous if a large surface of the body has been damaged. A smaller burned area should be plunged into icy water, or cold water should be poured over the area. Cover the burned area with a sterile gauze and keep the victim warm so he or she will not go into shock. If the hand is severely burned, the fingers should be separated by placing cloth between them. First-degree burns in which the skin is reddened may be cooled

Keep the Injured Person Lying Down

Do Not Give Liquids to the Unconscious

Control Bleeding

by Pressing on the Wound

Restart Breathing

with Mouth-to-Mouth Artificial Respiration

Dilute Swallowed

Poisons

Keep Broken Bones from Moving

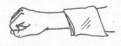

Cover Burns with Thick Layers of Cloth

Keep

Heart Attack Cases Quiet

Fainting: Keep

Head Lower than Heart

Cover Eye Injuries with

Gauze Pad

Always Call a Doctor

Fig. 3-42. General directions for giving first aid.

First Aid	
First Aid Kits	
Stretchers, Blankets, Oxygen	
Fire Protection	
Fire Hoses Hung Properly	
Extinguisher Charged/Proper Location	
Access to Fire Equipment	
Exit Lights/Doors/Signs	
Other	
Security	
Doors/Windows, Etc. Secured When Required	
Alarm Operation	
Dept. Shutdown Security	
Equip. Secured	
Unauthorized Personnel	
Other	
Machinery	
Unattended Machines Operating	
Emergency Stops Not Operational	
Platforms/Ladders/Catwalks	
Instructions to Operate/Stop Posted	
Maint. Being Performed on Machines in Operation	
Guards in Place	
Pinch Points	
Material Storage	
Hazardous and Flammable Material Not Stored Properly	
Improper Stacking/Loading/Securing	
Improper Lighting, Warning Signs, Ventilation	
General Area	
Condition of Floors	
Special Purpose Flooring	

Aisle, Clearance/Markings	
Floor Openings, Require Safeguards	
Railings, Stairs Temp./Perm.	
Dock Board (Bridges Plates)	
Piping (Water-Steam-Air)	
Wall Damage	
Ventilation	
Other	
Illumination — Wiring	
Unnecessary/Improper Use	
Lights on During Shutdown	
Frayed/Defective Wiring	
Overloading Circuits	
Machinery Not Grounded	
Hazardous Location	
Wall Outlets	
Other	
Housekeeping	
Floors	
Machines	
Break Area/Latrines	
Waste Disposal	
Vending Machines/Food Protection	
Rodent, Insect, Vermin Control	
Vehicles	
Unauthorized Use	
Operating Defective Vehicle	
Reckless/Speeding Operation	
Failure to Obey Traffic Rules	
Other	
Tools	
Power Tool Wiring	
Condition of Hand Tools	
Safe Storage	
Other	

Fig. 3-43. Monthly safety check.

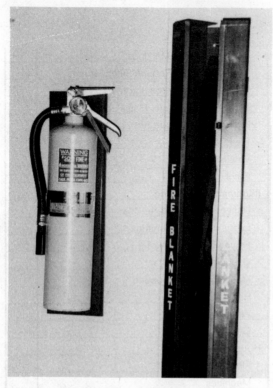

Fig. 3-41. A fire blanket is wrapped around burning clothing to smother the fire. First aid should follow immediately.

and covered with a burn ointment. All cuts or burns should be reported to the supervisor.

Shock

A person in shock may collapse or may merely seem dazed or stunned. The injured person may be pale, feel faint, act restless, and feel cool. Shock is a very serious condition and should be treated quickly. The victim should be isolated from the cause of the shock. For instance, he or she should not be allowed to view the scene of the accident. First, the injured person should lie down with his or her head level or slightly lower than the feet. He or she should be covered to stay warm. If conscious, the victim may be warmed with a hot drink. Vomitting may occur. A person suffering from shock must receive immediate health care.

Activity

1. Plan a program for leaving the building and safety procedures to follow in case of a fire, earthquake, or disaster from a blast.
2. Conduct a safety inspection of your laboratory. List items that have the potential of being a safety hazard.

Related Occupations

These occupations are related to safety.

National Occupational Health and Safety Administration Official
Safety engineer
Safety inspector
Safety investigator
Industrial health officer
Safety instructor
Environmental impact researcher
Radiation monitor
Doctor, nurse, paramedic
First aid instructor

Measuring
and Laying Out

Chapter
4

Words You Should Know

Interchangeability — Two parts can be used in place of each other.

Military Standards — A group of books that list the various government-approved industry specifications and standards that are used in the preparation of engineering drawings.

Reliability — The actual measured performance of a unit or part is equal to the planned performance. The part or units are always within tolerance.

Document — Any legal or official paper to claim evidence or data, such as engineering drawings, specifications, technical orders, policymaking papers, catalogs, sketches, etc.

Tolerance — The total permissible variation from design size, form, or location.

Light (candela) — Freezing platinum (2042K) 1/600,000 of a square metre radiates one candela of light.

Roundness — The circumference that a radius generates as it is rotated through an arc of 360°.

Brake finger allowance — The size of a work over the nominal size of the forming brake finger.

Sheet metal hem — The edge of sheet metal is folded over to strengthen the edge and to produce a round, safe edge.

Clean up a casting — A machined surface across the casting that does not leave any low spots on the cast surface.

Sine of a right triangle — The length of the side opposite divided by the length of the hypotenuse of the triangle.

Blueprint — Detailed drawings of white lines on a blue background which give shape, size and location of features of a product or part to be built.

Length — The unit of length is a metre and it is 1,650,763.73 wavelengths of Krypton — 86 long.

Mass — The resistance to change in motion of a body. Under the earth's gravitational pull, mass is measured as weight.

Time — The second was originally set as 1/86,400 of a rotation of a mean solar day. The metric standard is 9,192,631,770 oscillations, the radiation resonance frequence of Cesium 133.

Ampere — The current flowing through two wires one metre apart in a vacuum that produces 2×10^7 Newtons per metre against the wires.

Temperature — When a volume of gas is reduced to 0°, the temperature is absolute zero. Water freezes at 273.15° Kelvin; water boils at 373.15° Kelvin.

Mole — The number of particles of a material equal to Avogadro's number (6.023 $\times 10^{23}$).

Reasons for Measurement

Interchangeability of parts was only made possible when people learned to communicate measurements. When they learned to communicate length, weight, and time, they were then able to build instruments which could accurately determine whether these measurements were achieved. Measurement provides a means for determining whether parts are acceptable and will fit and function in a product, Fig. 4-1. The control for size is measured while the parts are being manufactured so the producer is assured that the parts will function correctly in a product.

Measuring to Make Things

Until the time of Eli Whitney all production was completed by hand fitting. Each part was filed and polished until it worked with its adjacent parts. This meant that a part taken from a machine, would not fit in a similar machine, Fig. 4-2. The part from the first machine would have to be fitted to the new machine. Custom work on one-of-a-kind products is done this way today, as well as a few products. Most custom work is done by craftsmen, and the work is usually a prototype, Fig. 4-3. A prototype is a full-scale model of the operating machine from which final measurements will be taken so similar machines may be built. The prototype will be tested and redesigned after

Fig. 4-1.   Measurement provides a means for deciding whether the product will function correctly.

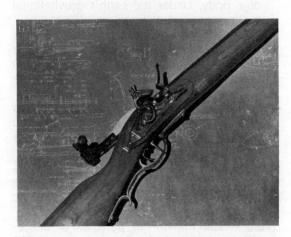

Fig. 4-2.   Handmade and hand-fit parts will not work in a like machine without being refitted.

Fig. 4-3.   Hand-fitting a prototype model.

production problems have been identified. When the performance of the product is satisfactory, accurate measurements will be made and all drawings corrected.

Because it is possible to measure to millionths of an inch if necessary, a drawing can reflect the size accuracy of the parts. These parts can be manufactured accurately according to the size required on the drawings. Every part produced will fit in any of the machines that require the part. This concept is called **interchangeability** of parts and is the basis of mass production. It is made possible by being able to measure accurately, Fig. 4-4, to control the size of parts.

Measuring to Control Manufacturing

Industry requires so many special skills, tools, and materials that it is no longer possible to build all the parts for a product in one factory. Electrical wire may be pro-

duced in Ohio, precision bearings in Massachusetts, castings in Michigan, solenoids in New York, pulley belts in New York, and electric motors in Missouri. Yet when all of these special materials and products are needed in a finished manufactured item, there needs to be a common language of control understood by both the buyer of parts and materials and the seller of materials. This language that is a short hand for detailed specification is called **specifications** or **standards**. The contracts are written and the drawings submitted to the vendor or subcontractor. These documents state specific standard tolerances on specific parts. A large part of the description has already been agreed upon by specifications set up by various government agencies. Military specifications, American Standards Associations, Society for Testing and Materials, and the Bureau of Standards have specifications that set the standards for performance, material, and tolerances of products. These specifications are a clear language that both manufacturers and their subcontractors can understand. The accurate measurements makes it possible to control that which the manufacturer buys from the subcontractor and gives the subcontractor the exact standards which must be met. Thus measurement is truly the controller of manufacturing, Fig. 4-5.

Product Reliability. The concept of **reliability** is based on the ability to measure accurately. When millions of units such as automobile engines are manufactured, the parts must be reliably accurate. One must be able to pick out any piston from the millions of engines being built during the year, measure it, and find that all the measurements are actually within the tolerances specified. If this is possible with any selected piston and the tolerances are always met, the product has **reliability**. For example, a piston has reliability when the **actual performance** in machining is the same as the **planned performance** for machining.

Product reliability is very important in our space flights. In order to make sure that the space mission will go as planned, an over-

Fig. 4-4. Fine, accurate measurements are the basis for interchangeability of parts.

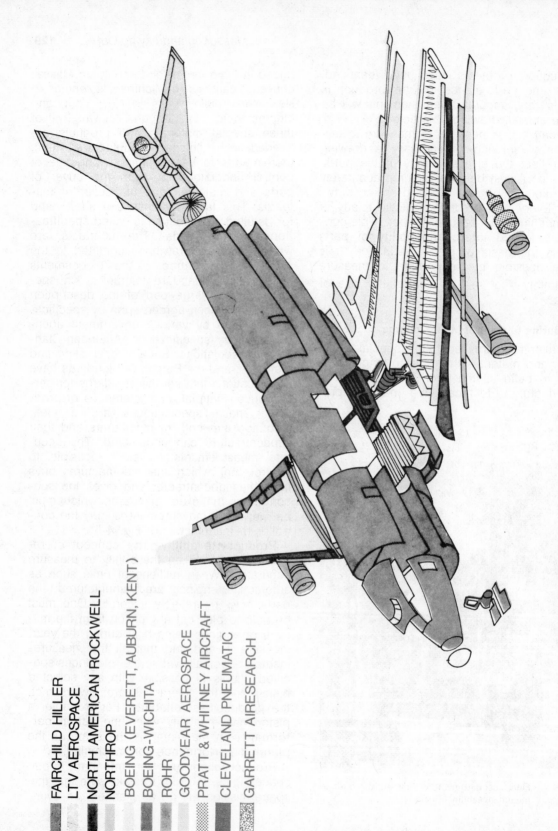

FAIRCHILD HILLER
LTV AEROSPACE
NORTH AMERICAN ROCKWELL
NORTHROP
BOEING (EVERETT, AUBURN, KENT)
BOEING-WICHITA
ROHR
GOODYEAR AEROSPACE
PRATT & WHITNEY AIRCRAFT
CLEVELAND PNEUMATIC
GARRETT AIRESEARCH

Fig. 4-5. Specifications and drafting standards data make it possible for subcontractors to have accurate measurements and data.

all space vehicle reliability is planned. This means that every part in the total rocket must have a reliability figure which is totaled so that the engineers are assured that everything will work at the planned time with the planned performance. It is very difficult and time-consuming work to establish all these reliabilities, but nothing is left to chance that may cause a mission to fail.

Control. Measurement is one of the important factors which make possible the control of manufacturing production. When a product is manufactured, its parts must be reliable or consistently within tolerances. The **direction** of **change** (whether it is in the upper or lower limits of the tolerance zone) of the parts being manufactured is also important. If the measurement is accurate enough to give this information, these actual sizes may be placed on a control chart and used for controlling the size of the part output, Fig. 4-6.

The performance of the machine producing the parts can now be controlled according to a plan in terms of the machine's actual part size performance or reliability. This requires a constant checking of both the parts produced and the performance of the machine. The need for tool sharpening or tool adjustments may then be reported to the supervisor in advance. As the part size approaches the upper or lower control limit, a correction is made on the machine or the cutting tools to maintain the part's reliability.

Measuring Units

Measuring manufactured parts to precise size, weight, and time requires accurate units of measure.

Weight. Weight is the pull of the earth on an object. It is a common measuring unit in manufacturing. Here are some common measures of weight:
- 16 ounces = 1 pound
- 2000 pounds = 1 ton
- 1 cubic foot of water = 62.48 pounds
- 1 U. S. gallon of water = 8.355 pounds

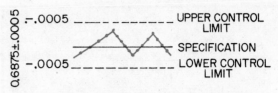

Fig. 4-6. The production of machines can be controlled according to a plan. Changes are made before the machine produces any out-of-tolerance parts.

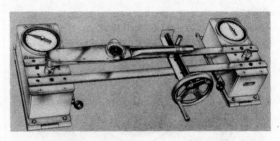

Fig. 4-7. Torque wrenches measure the force applied to nuts and bolts in assembly.

- 32 cubic feet of water = 1 ton (2000 lbs.)
- 240 U. S. gallons of water = 1 ton

Weight is used to specify industrial material. For example, 32 oz. copper means that one square foot of 32 oz. copper will weigh 32 oz. The thickness of the copper determines the metal's weight. Weight measurements are used to control countless industrial processes. The mixtures of industrial materials charging a blast furnace — iron ore, limestone, or coke — or the ingredients of many metals alloys are measured by weight. Weights are also used to calibrate (standardize) instruments and tools that are necessary to ensure that products will meet the desired requirements.

Measuring units of weight can be combined with units of length. The resulting units of measurement — inch pounds or foot pounds — give a new measure, a unit of work. Foot pounds and inch pounds are used to measure **torque**, the amount of twisting force used in tightening a bolt or nut during manufacturing, Fig. 4-7.

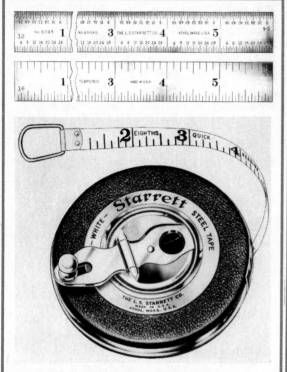

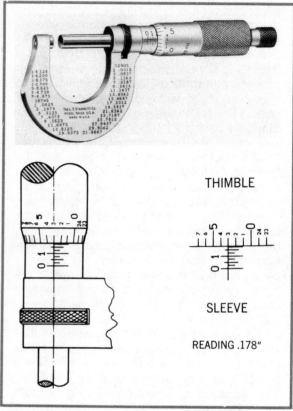

Fig. 4-8. Fractional measurements are transferred to tapes and rules for measurement.

Fig. 4-10. The micrometer utilizes the concept that one inch is divided into one thousand equal parts. The micrometer reads directly into thousandths of an inch.

Fig. 4-9. Decimal measurement was made by dividing the inch into 10 equal parts. Each one-tenth inch was divided into 5 parts to yield a 20-mil segment.

Length. Length is a most important unit of measurement for manufacturing industries. The common length measures, the fractional inch and the decimal inch, have long been used. Up through the Civil War, the common manufacturing dimension was 1/64″ for fine practical work. By World War I the practical standard had advanced to one-thousandth of an inch. In the 1960's measurements on machinery were commonly made to one ten-thousandth of an

inch. Measurements to one-millionth of an inch are not uncommon now.

The fractional inch units are made by dividing a foot into twelve equal parts to produce an inch. The inch is in turn broken down into a series by dividing the remaining fractions by two. These measurements have been transferred to rules and tapes for use, Fig. 4-8.

Decimal inch measurements are made by taking an inch and dividing it into ten equal parts. This generates a rule with a tenth of an inch scale, Fig. 4-9. Each one-tenth of an inch is equal to 100 mils. A tenth of an inch divided into five parts produces 20 mils segments.

The Ford Motor Company adopted the decimal system in 1930, and this procedure has been followed by more manufacturing industries each year.

- $1' = 12''$
- $1/10'' = 100$ mils
- $\dfrac{1/10''}{100} = 1$ mil
- 1 mil $= 1/1000''$

In the early 1900's the **micrometer** became the symbol of precision measurement. The micrometer utilizes the concept that one inch is divided into one thousand equal parts. Measurement was then made between the anvil and the spindle of the micrometer and read directly in thousands off the thimble and sleeve.

The micrometer has a screw thread with a pitch of 40 threads per inch, and a graduation on the sleeve of the micrometer which is twenty-five thousandths. The thimble of the micrometer is divided into twenty-five equal parts. Thus if you turn the thimble one graduation, the spindle will move from the anvil one-thousandth of an inch ($1/40''$ x $1/25 = 1/000''$).

The micrometer reads in decimal inch measurements and is easily related to the 1/10 and 1/100 rule, Fig. 4-10.

The Metric System. The metric system was devised in Europe after the French Revolution in 1789 in which the standards of measurements was an issue. The metre was adopted as the official standard in Paris in 1799. The words "metre" and "metric" have come from the Greek word metron meaning "measure," Fig. 4-11 (page 134). The French used the word metre, which they determined to be the distance one ten-millionth of a fourth of the earth's surface (one ten-millionth of the distance from the pole to the equator). The actual measurement was made from Dunkirk, France to Mont Jouy, Spain to establish the distance per degree of latitude.

The metric system uses multiples of ten rather than multiples of halves as the English system does. Calculations are easier in the metric system, because fractions are largely eliminated. When changing from one size unit to another, the number is either multiplied or divided by ten. Fewer errors in calculations result. Scientists have realized this for some time and have been doing their work in the metric system. The last decade has seen a rapid growth of the metric system in the United States because of the vast amount of research which has been done for the scientific and aerospace programs. This use of science in technology has brought the metric system into greater use in engineering also. With the increase of world trade, people in business have found that their customers are demanding materials, components, and parts which will interface with a metric system, Figs. 4-12 and 4-13.

One Millimetre		One Inch	
25.4 mm	= 1 inch	1. mm	= .03937 inches
2.54	= .1	.1	= .003937
.254	= .01	.01	= .0003937
.0254	= .001	.001	= .00003937
.00254	= .0001	.0001	= .000003937 (millionths)
.000254	= .00001	.00001	= .0000003937
.0000254	= .000001 millionth	.000001	= .00000003937

Fig. 4-12. Measurement comparisons.

International Standards of Measurement

Quantity	Name of Basic Unit	Symbol
Length	Metre	m
Mass	Kilogram	kg
Time	Second	s
Electric Current	Ampere	a
Temperature	Kelvin	k
Amount of Substance	Mole	mol
Luminous Intensity	Candela	cd

Fig. 4-13. The International System of Measurement: seven basic units.

THE MODERNIZED metric system

The International System of Units-SI

is a modernized version of the metric system established by international agreement. It provides a logical and interconnected framework for all measurements in science, industry, and commerce. Officially abbreviated SI, the system is built upon a foundation of seven base units, plus two supplementary units, which appear on this chart along with their definitions. All other SI units are derived from these units. Multiples and submultiples are expressed in a decimal system. Use of metric weights and measures was legalized in the United States in 1866, and since 1893 the yard and pound have been defined in terms of the meter and the kilogram. The base units for time, electric current, amount of substance, and luminous intensity are the same in both the customary and metric systems.

COMMON CONVERSIONS
Accurate to Six Significant Figures

Symbol	When You Know	Multiply by	To Find	Symbol
in	inches	[A]25.4	[B]millimeters	mm
ft	feet	[A]0.3048	meters	m
yd	yards	[A]0.9144	meters	m
mi	miles	1.609 34	kilometers	km
yd²	square yards	0.836 127	square meters	m²
	acres	0.404 686	[C]hectares	ha
yd³	cubic yards	0.764 555	cubic meters	m³
qt	quarts (lq)	0.946 353	[D]liters	l
oz	ounces (avdp)	28.349 5	grams	g
lb	pounds (avdp)	0.453 592	kilograms	kg
°F	Fahrenheit temperature	[A]5/9 (after subtracting 32)	Celsius temperature	°C
mm	millimeters	0.039 370 1	inches	in
m	meters	3.280 84	feet	ft
m	meters	1.093 61	yards	yd
km	kilometers	0.621 371	miles	mi
m²	square meters	1.195 99	square yards	yd²
ha	[C]hectares	2.471 05	acres	
m³	cubic meters	1.307 95	cubic yards	yd³
l	[D]liters	1.056 69	quarts (lq)	qt
g	grams	0.035 274 0	ounces (avdp)	oz
kg	kilograms	2.204 62	pounds (avdp)	lb
°C	Celsius temperature	[A]9/5 (then add 32)	Fahrenheit temperature	°F

[A]exact

[B]for example, 1 in = 25.4 mm, so 3 inches would be

$$(3 \text{ in}) \left(25.4 \frac{\text{mm}}{\text{in}}\right) = 76.2 \text{ mm}$$

[C]hectare is a common name for 10 000 square meters

[D]liter is a common name for fluid volume of 0.001 cubic meter

Note: Most symbols are written with lower case letters; exceptions are units named after persons for which the symbols are capitalized. Periods are not used with any symbols.

MULTIPLES AND PREFIXES
These Prefixes May Be Applied To All SI Units

Multiples and Submultiples		Prefixes	Symbols
1 000 000 000 000	$= 10^{12}$	tera (tĕr´à)	T
1 000 000 000	$= 10^{9}$	giga (jī´gà)	G
1 000 000	$= 10^{6}$	mega (mĕg´à)	M
1 000	$= 10^{3}$	kilo (kil´ō)	k
100	$= 10^{2}$	hecto (hĕk´tō)	h
10	$= 10^{1}$	deka (dĕk´à)	da
Base Unit 1	$= 10^{0}$		
0.1	$= 10^{-1}$	deci (dĕs´ĭ)	d
0.01	$= 10^{-2}$	centi (sĕn´tĭ)	c
0.001	$= 10^{-3}$	milli (mĭl´ĭ)	m
0.000 001	$= 10^{-6}$	micro (mī´krō)	μ
0.000 000 001	$= 10^{-9}$	nano (năn´ō)	n
0.000 000 000 001	$= 10^{-12}$	pico (pē´kō)	p
0.000 000 000 000 001	$= 10^{-15}$	femto (fĕm´tō)	f
0.000 000 000 000 000 001	$= 10^{-18}$	atto (ăt´tō)	a

National Bureau of Standards
Special Publication 304A (Revised October 1972)

For sale by the Superintendent of Documents, U.S. Government Printing Office, Washington, D.C. 20402 SD Catalog No. C13.10: 304A — Price 25 cents

REFERENCES

NBS Special Publication 330, 1972 Edition, International System of Units (SI), available by purchase from the Superintendent of Documents, Government Printing Office, Washington, D.C. 20402, order as C13.10:330/2; 30 cents a copy.

ASTM Metric Practice Guide E380-72, available by purchase from the American Society for Testing and Materials, 1916 Race Street, Philadelphia, Pa. 19103, $1.50 a copy, minimum order $3.00.

Rules for the Use of Units of the International System of Units, order as ISO Recommendation R1000; $1.25 a copy from the American National Standards Institute, 1430 Broadway, New York, N.Y. 10018.

Fig. 4-11. Basic units of the metric system.

SEVEN BASE UNITS

meter-m
LENGTH

The meter (common international spelling, metre) is defined as 1 650 763.73 wavelengths in vacuum of the orange-red line of the spectrum of krypton-86.

86Kr ATOM

An interferometer is used to measure length by means of light waves.

kilogram-kg
MASS

The standard for the unit of mass, the kilogram, is a cylinder of platinum-iridium alloy kept by the International Bureau of Weights and Measures at Paris. A duplicate in the custody of the National Bureau of Standards serves as the mass standard for the United States. This is the only base unit still defined by an artifact.

U.S. PROTOTYPE KILOGRAM NO. 20

second-s
TIME

The second is defined as the duration of 9 192 631 770 cycles of the radiation associated with a specified transition of the cesium-133 atom. It is realized by tuning an oscillator to the resonance frequency of cesium-133 atoms as they pass through a system of magnets and a resonant cavity into a detector.

Schematic diagram of an atomic beam spectrometer or "clock." Only those atoms whose magnetic momenta are "flipped" in the transition region reach the detector. When 9 192 631 770 oscillations have occurred, the clock indicates one second has passed.

CESIUM SOURCE

TRANSITION REGION (CAVITY/OSCILLATING FIELD)

DETECTOR

DEFLECTION MAGNET

DEFLECTION MAGNET

OSCILLATOR

ELECTRONIC/ATOMIC '86 SCALE SYSTEM

ampere-A
ELECTRIC CURRENT

The ampere is defined as that current which, if maintained in each of two long parallel wires separated by one meter in free space, would produce a force between the two wires (due to their magnetic fields) of 2×10^{-7} newton for each meter of length.

1A

FORCE = 2×10^{-7}N

1m

1A

1m

kelvin-K
TEMPERATURE

The kelvin is defined as the fraction 1/273.16 of the thermodynamic temperature of the triple point of water. The temperature 0 K is called "absolute zero".

On the commonly used Celsius temperature scale, water freezes at about 0°C and boils at about 100°C. The °C is defined as an interval of the Celsius temperature. 0°C is defined as 273.15 K.

1.8 Fahrenheit degrees are equal to 1.0°C or 1.0 K; the Fahrenheit scale uses 32°F as a temperature corresponding to 0°C.

TEMPERATURE MEASUREMENT SYSTEMS

°F		°C
212°	Water Boils	
	Body Temperature	
98.6°		
32°	Water Freezes	0
−40°		−40
FAHRENHEIT		CELSIUS

2045 Platinum Freezes

273.15

Absolute Zero

KELVIN

mole-mol
AMOUNT OF SUBSTANCE

The mole is the amount of substance of a system that contains as many elementary entities as there are atoms in 0.012 kilogram of carbon 12.

When the mole is used, the elementary entities must be specified and may be atoms, molecules, ions, electrons, other particles, or specified groups of such particles.

candela-cd
LUMINOUS INTENSITY

The candela is defined as the luminous intensity of 1/600 000 of a square meter of a blackbody at the temperature of freezing platinum (2045 K).

CAVITY

FREEZING PLATINUM

INSULATING MATERIAL

The SI unit of area is the square meter (m^2).

The SI unit of volume is the cubic meter (m^3). The liter (0.001 cubic meter), although not an SI unit, is commonly used to measure fluid volume.

The SI unit for pressure is the pascal (Pa).
$$1Pa = 1N/m^2$$

The SI unit for work and energy of any kind is the joule (J).
$$1J = 1Nm$$

The SI unit for power of any kind is the watt (W).
$$1W = 1J/s$$

1kg

ACCELERATION of 1m/s²

The SI unit of force is the newton (N). One newton is the force which, when applied to a 1 kilogram mass, will give the kilogram mass an acceleration of 1 (meter per second) per second.
$$1N = 1kgm/s^2$$

The number of periods or cycles per second is called frequency. The SI unit for frequency is the hertz (Hz). One hertz equals one cycle per second.

The SI unit for speed is the meter per second (m/s).

The SI unit for acceleration is the (meter per second) per second (m/s²).

Standard frequencies and correct time are broadcast from WWV, WWVB, and WWVH, and stations of the U.S. Navy. Many shortwave receivers pick up WWV and WWVH on frequencies of 2.5, 5, 10, 15, and 20 megahertz.

The SI unit of voltage is the volt (V).
$$1V = 1W/A$$

The SI unit of electric resistance is the ohm (Ω).
$$1Ω = 1V/A$$

The standard temperature at the triple point of water is provided by a special cell, an evacuated glass cylinder containing pure water. When a layer of ice is formed around the reentrant well, the temperature at the interface of solid, liquid, and vapor is 273.16 K. Thermometers to be calibrated are placed in the reentrant well.

THERMOMETER (ELECTRICAL RESISTANCE TYPE)

WATER VAPOR

WATER REENTRANT WELL

REFRIGERATING BATH

TRIPLE POINT CELL

The SI unit of concentration (of amount of substance) is the mole per cubic meter (mol/m³).

A 100-watt light bulb emits about 1700 lumens.

The SI unit of light flux is the lumen (lm). A source having an intensity of 1 candela in all directions radiates a light flux of 4π lumens.

TWO SUPPLEMENTARY UNITS

radian-rad
PLANE ANGLE

ONE RADIAN

The radian is the plane angle with its vertex at the center of a circle that is subtended by an arc equal in length to the radius.

steradian-sr
SOLID ANGLE

Area

ONE STERADIAN

The steradian is the solid angle with its vertex at the center of a sphere that is subtended by an area of the spherical surface equal to that of a square with sides equal in length to the radius.

Metric-English Conversion Table

mm	Inches	mm	Inches	mm	Inches	mm	Inches	mm	Inches
0.01	.00039	0.41	.01614	0.81	.03189	21	.82677	61	2.40157
0.02	.00079	0.42	.01654	0.82	.03228	22	.86614	62	2.44094
0.03	.00118	0.43	.01693	0.83	.03268	23	.90551	63	2.48031
0.04	.00157	0.44	.01732	0.84	.03307	24	.94488	64	2.51968
0.05	.00197	0.45	.01772	0.85	.03346	25	.98425	65	2.55905
0.06	.00236	0.46	.01811	0.86	.03386	26	1.02362	66	2.59842
0.07	.00276	0.47	.01850	0.87	.03425	27	1.06299	67	2.63779
0.08	.00315	0.48	.01890	0.88	.03465	28	1.10236	68	2.67716
0.09	.00354	0.49	.01929	0.89	.03504	29	1.14173	69	2.71653
0.10	.00394	0.50	.01969	0.90	.03543	30	1.18110	70	2.75590
0.11	.00433	0.51	.02008	0.91	.03583	31	1.22047	71	2.79527
0.12	.00472	0.52	.02047	0.92	.03622	32	1.25984	72	2.83464
0.13	.00512	0.53	.02087	0.93	.03661	33	1.29921	73	2.87401
0.14	.00551	0.54	.02126	0.94	.03701	34	1.33858	74	2.91338
0.15	.00591	0.55	.02165	0.95	.03740	35	1.37795	75	2.95275
0.16	.00630	0.56	.02205	0.96	.03780	36	1.41732	76	2.99212
0.17	.00669	0.57	.02244	0.97	.03819	37	1.45669	77	3.03149
0.18	.00709	0.58	.02283	0.98	.03858	38	1.49606	78	3.07086
0.19	.00748	0.59	.02323	0.99	.03898	39	1.53543	79	3.11023
0.20	.00787	0.60	.02362	1.00	.03937	40	1.57480	80	3.14960
0.21	.00827	0.61	.02402	1	.03937	41	1.61417	81	3.18897
0.22	.00866	0.62	.02441	2	.07874	42	1.65354	82	3.22834
0.23	.00906	0.63	.02480	3	.11811	43	1.69291	83	3.26771
0.24	.00945	0.64	.02520	4	.15748	44	1.73228	84	3.30708
0.25	.00984	0.65	.02559	5	.19685	45	1.77165	85	3.34645
0.26	.01024	0.66	.02598	6	.23622	46	1.81102	86	3.38582
0.27	.01063	0.67	.02638	7	.27559	47	1.85039	87	3.42519
0.28	.01102	0.68	.02677	8	.31496	48	1.88976	88	3.46456
0.29	.01142	0.69	.02717	9	.35433	49	1.92913	89	3.50393
0.30	.01181	0.70	.02756	10	.39370	50	1.96850	90	3.54330
0.31	.01220	0.71	.02795	11	.43307	51	2.00787	91	3.58267
0.32	.01260	0.72	.02835	12	.47244	52	2.04724	92	3.62204
0.33	.01299	0.73	.02874	13	.51181	53	2.08661	93	3.66141
0.34	.01339	0.74	.02913	14	.55118	54	2.12598	94	3.70078
0.35	.01378	0.75	.02953	15	.59055	55	2.16535	95	3.74015
0.36	.01417	0.76	.02992	16	.62992	56	2.20472	96	3.77952
0.37	.01457	0.77	.03032	17	.66929	57	2.24409	97	3.81889
0.38	.01496	0.78	.03071	18	.70866	58	2.28346	98	3.85826
0.39	.01535	0.79	.03110	19	.74803	59	2.32283	99	3.89763
0.40	.01575	0.80	.03150	20	.78740	60	2.36220	100	3.93700

For converting millimetres in "thousandths" move decimal point in both columns to left.

The L. S. Starrett Company, Athol, Mass., U.S.A.

Fig. 4-14.   The International System of Length Units. Metric to English conversion table.

The metric measuring unit of length is based upon the metre. The units of length for the metric system are these:

- 1 kilometre = 1000 metres
- 1 metre = standard of length
- 1 decimetre = 1/10 metre
- 1 centimetre = 1/100 metre
- 1 millimetre = 1/1000 metre

The International Standards of Measurement

The 11th General Conference on Weights and Measures in Paris in 1960 officially adopted a modernized metric system, Systeme International d'Unites (SI).

Measuring Units. The international standard of measurement is a coherent system of measurement based on seven basic units. They are all based on natural units of measurement — units which are constant, such as the resonance of atoms, the wavelength of light, the speed of light, the mass of objects, the freezing and boiling points of water, etc., Fig. 4-14.

The basic unit of length is the **metre.** It is technically defined as 1650763.73 wavelengths in vacuum of orange-red line of the spectrum of Krypton - 86. All other divisions of length — such as millimetres, centimetres, and kilometres — are obtained by multiplying or dividing this unit by 10's.

Mass is the resistance to change in motion. Under the earth's gravitational pull, mass is measured as weight. The basic unit of mass is the **kilogram.** Common units used are the **gram, kilogram,** and **tonnes.** A **newton** is the force applied to a 1 kilogram mass to accelerate the mass at 1 metre per second per second.

The basic unit of time is the **second.** This unit was originally set by the period of time involved in the rotation of the earth using the old English unit of two twelve-hour periods. One second of time is derived by calculating $1/60 \times 1/60 \times 1/24 = 1/86,400$ of a rotation (the mean solar day).

Industry requires very accurate clocks. The laws of physics have been applied to the period of oscillation of a vibrating slab of crystal. By applying electronics, an elec-

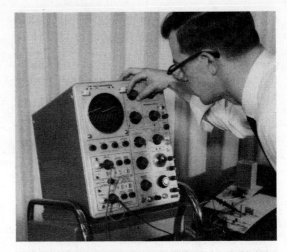

Fig. 4-15. Very accurate clocks (such as the oscilloscope) are required for industry.

tron crystal oscillator has been produced which is a very accurate short-time measuring clock. The oscilloscope makes use of this concept. Time is scaled on the horizontal axis and voltage on the vertical axis. This instrument is used for analysis of electronic circuits and research work, Fig. 4-15.

Atomic clocks which count vibrational frequency of atoms in a molecule are the most accurate clocks of all. Such a clock measures the half-life of Uranium 238 to lead. Since Uranium's half-life is 4.5 billion years, a very long-time measuring clock is possible.

The basic unit of the electrical current is the **ampere.**

Other electrical values are related to the ampere. The **volt** is defined as 1 watt divided by 1 ampere ($1V = \dfrac{1W}{1A}$). Resistance is measured by the Ohm, which is 1 volt divided by 1 ampere ($1\Omega = \dfrac{1V}{1A}$).

The **kelvin** is the unit of temperature. The kelvin scale has its point of origin, or zero, at absolute zero. The temperature of 273.15 kelvins is the freezing point of water. The boiling point of water is 373.15 K. The kelvin scale has a relation between the temperature and volume of a fixed amount of gas at constant pressure. As the temperature decreases, the volume decreases. Theoretically, when the kelvin scale reaches zero, the volume of the gas is also zero. As a practical problem, gases become liquids before this temperature is reached. The lowest temperature reached under strict experimental controls has been 0.0014 K.

The **mole** is the basic unit of the amount of substance. It contains as many elementary units as there are carbon atoms in 0.012 kilograms of Carbon 12. The elementary unit must be specified as to atom, molecule, ion, electron, photon, etc. One mole is the amount of material which contain Avogadro's number of particles (6.023×10^{23}).

The **candela** is the unit of luminous intensity. It is the luminous intensity of 1/600,000 of a square metre of freezing platinum (2042 K). A source of 1 candela radiates a light of 4 lumens. A common 100-watt light radiates around 1700 lumens.

For scientific and engineering purposes, there are other units of measurement which are derived from a combination of the basic units. These units are known as **derived units**. When a unit of length is combined with a unit of time, the measurement is length per minute. A new unit, a rate of motion or velocity, is developed. Metric units are easily converted to equivalent-based units so that nearly all the measuring units can be combined to give units of force, energy, and power, Figs. 4-16 and 4-16A.

FIG 4-16.
SI Units Derived from Basic Units

Quantity	Name of Derived Unit	Symbol
Frequency	hertz	Hz
Pressure and stress	pascal	Pa
Work, energy, heat	joule	J
Power	watt	W
Quantity of electricity	coulomb	C
Electromotive force	volt	V
Electric capacitance	farad	F
Electric resistance	Ohm	Ω
Electrical conductance	siemens	S
Magnetic flux	weber	Wb
Magnetic flux density	tesla	T
Electric inductance	henry	H
Luminous flux	lumen	lm
Illuminance	lux	lx

FIG. 4-16A.
Derived SI Units

Physical Quantity	Name of Unit	Symbol	Derived Units	Combination Base Units
Force	newton	N		$m \bullet kg \bullet s^{-2}$
Pressure and stress	pascal	Pa	N/m^2	$m^1 \bullet kg \bullet s^{-2}$
Energy	joule	J	$N \bullet m$	$m^2 \bullet kg \bullet s^{-2}$
Power	watt	W	J/s	$m^2 \bullet kg \bullet s^{-3}$
Conductance	siemens	S	A/V	$m^{-2} \bullet kg^{-1} \bullet s^3 \bullet A^2$
Dynamic-viscosity	pascal second	Pa$\bullet$s		$m^{-1} \bullet kg \bullet s^{-1}$
Surface tension	newton per meter	N/m		$kg \bullet s^{-2}$
Molar energy	joule per mole	J/mol		$m^2 \bullet kg \bullet s^{-2} \bullet mol^{-1}$

Measurement Concepts

The dimensions of length, roundness, flatness, angularity, surface conditions, level, and plumb must be accurately measured in industry. Many different types of instruments have been developed to provide these measurements.

Length. Length is a basic measurement in manufacturing because it may be applied to straight dimensions (Fig. 4-17), curved dimensions, or combinations of the two such as circumferences, radii, diameters, arcs, and chords, as well as width, breadth, length, and thickness. Length may be obtained by direct measurement (Fig. 4-18), transferred measurement, end standards, or comparison measurements, Fig. 4-19. Precision measurements can be made by the physics of light, Fig. 4-20.

Roundness. Roundness is the circumference that a radius generates as it is rotated through an arc of 360° When a diameter of a cylinder is measured with a micrometer, a measure of length is obtained. If a series of micrometer measurements is taken, you have a series of length measurements, not roundness measurement. A lobe or camlike surface may be produced by the machining process.

Fig. 4-17. A point has location. When it is extended, it has length.

Fig. 4-19. Comparison measuring instrument.

Fig. 4-18. Direct measurement instrument.

Fig. 4-20. Measuring instruments based on the wavelengths of light are obtained by the use of optical flats.

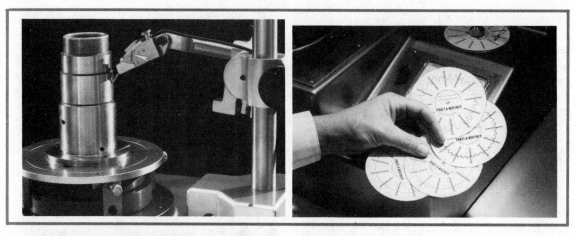

Fig. 4-21. Roundness is best visualized as the surface generated by a plane 90° to a rotating radius.

Fig. 4-22. A. A point may be extended into a line and a line into a plane. A plane may be inclined so that three points lie within the plane. Thus three points will define a plane.

B. A surface plate is the reference plane most frequently used to make measurements. Using reference planes for measurement.

A micrometer or ring-type gage will not detect this type of roundness. The measure will be made by using a lever arm to sweep through an arc the same as the part being tested. The arc path and the part surface may be compared.

Roundness may also be measured by using bench centers and rotating the workpiece on its machining centers, checking for a varying radius on the part. A V-block and precision indicator will also give indications of roundness.

Spherical roundness is best visualized by a surface that would be generated by a radius being rotated in one plane, while at the same time the center point of the radius is rotated continuously to 90° to the original radius plane of rotation. In this case true spherical roundness is produced, Fig. 4-21. The five axis numerical controlled milling machines are capable of producing these types of surfaces.

Roundness is most accurately manufactured by machines that generate the surface as in gear cutting or turning with a single pointed tool in a lathe.

Flatness. Flatness is best visualized as a plane. In this case, any three points in space may be connected so that together they establish a plane. This theoretical plane is used in a system of measurement as a reference plane. The reference plane enables

Fig. 4-23. Contact with a blued surface plate reveals a pattern of high areas on the work.

Fig. 4-24. The autocollinator detects angular displacement and checks surface plate flatness, machine slides, angular divisions, and straightness.

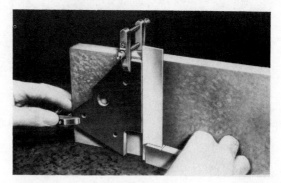

Fig. 4-25. The solid square, angle plate, and surface plate determine 90° angles.

one to measure a surface for flatness by using a group of instruments, Fig. 4-22.

A surface plate is the reference plane most frequently used to make flatness measurements. Flat surfaces may be checked by "blueing" the surface plate placing the surface to be checked in contact with the surface plate. A slight circular motion brings the surface in contact and the blueing will be transferred to the metal being checked revealing a pattern of high areas, Fig. 4-23.

Large areas are checked with the use of optical tooling and the application of an autocollimator which provides a reading of flatness, Fig. 4-24.

The laser beam is used to check flat surfaces also. The bright red beam travels over long distances and does not sag or bend. With the use of a laser, centering detector, and readout meter, any size area may be measured for flatness.

The principle that is used in measuring flatness is the comparison of the work to a reference plane.

Using a flat reference plane as a base, angles, length, and areas may be measured. The solid square is an angle measuring tool that is used to measure 90° angles and used at different levels of accuracy, Fig. 4-25.

Gage blocks and sine bars apply the functions of trigonometry to construct very accurate angles, Fig. 4-26.

Fig. 4-26. A sine bar or plate is positioned upon gage blocks to construct a very accurate angle between the surface plate and the top of the sine bar.

A dividing head is used to position work pieces so that angles may be accurately cut, Fig. 4-27.

Optical tooling with the application of a jig transit will measure angles accurately over long distances, Fig. 4-28.

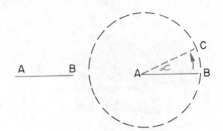

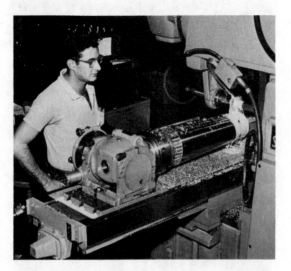

Fig. 4-27. The dividing head can locate work for cutting into divisions or accurately positioning angles for cutting.

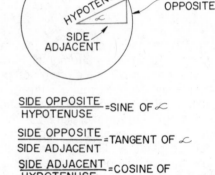

$$\frac{\text{SIDE OPPOSITE}}{\text{HYPOTENUSE}} = \text{SINE OF } \alpha$$

$$\frac{\text{SIDE OPPOSITE}}{\text{SIDE ADJACENT}} = \text{TANGENT OF } \alpha$$

$$\frac{\text{SIDE ADJACENT}}{\text{HYPOTENUSE}} = \text{COSINE OF } \alpha$$

Fig. 4-29. The triangle gives a system of measurement called trigonometry. Trigonometry makes possible the measurement of angles and lines and the location of a position in space.

Fig. 4-28. Measurement of angles over long distances is made possible by optical tooling with measuring tapes and jig transits.

Fig. 4-30. The laser is applied to very accurate angular measurements by the use of a digital readout.

Angularity. An angle is generated when a line is rotated around its end and stopped when the new direction is reached. Angles are defined with respect to a reference surface that may be the head of a protractor, index hand, base of a surface plate, or a base line, Fig. 4-29.

From the study of the triangle has come the system of measurement called trigonometry, which supplies the mathematical opportunity to measure angles, lengths of sides, and areas as they apply to metalwork.

The laser has added the accuracy of a digital readout from an interferometer to a system of angular measurement, Fig. 4-30.

Surface Condition. The measuring system for surface finish is a comparative measurement for roughness and other characteristics. A simple procedure is to look at the surface and compare it with other surfaces or to run your thumbnail across the surface, feel the roughness, and compare the feel with the standard roughness sample, Fig. 4-31.

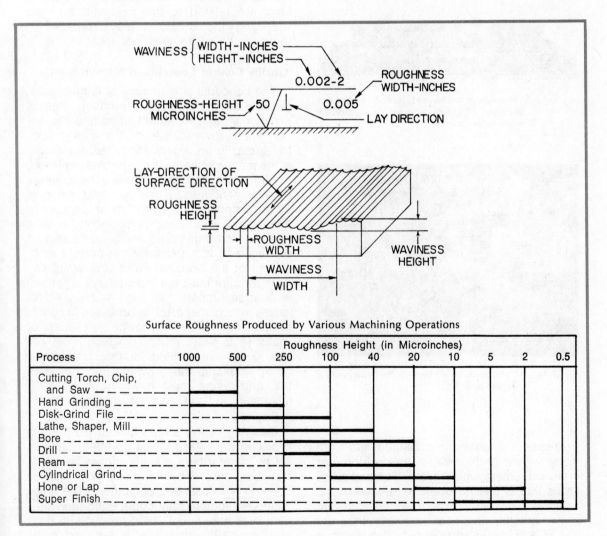

Surface Roughness Produced by Various Machining Operations

Process	\multicolumn{9}{c	}{Roughness Height (in Microinches)}
	1000	500	250	100	40	20	10	5	2	0.5
Cutting Torch, Chip, and Saw										
Hand Grinding										
Disk-Grind File										
Lathe, Shaper, Mill										
Bore										
Drill										
Ream										
Cylindrical Grind										
Hone or Lap										
Super Finish										

Fig. 4-31. Surface finish is measured by surface roughness. Various machine tools will produce a range of roughness.

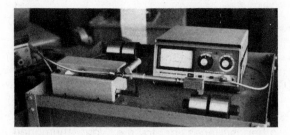

Fig. 4-32. The profilometer measures the surface roughness of this ground part at 22 microinches.

Fig. 4-33. The precision level is used for positioning so that the effect of gravity is equal throughout the machine or structure.

Fig. 4-34. An electronic level checking the bedways of a machine tool.

Precise roughness measurements are made with a profilometer. The **profilometer** is an instrument which has a diamond point stylus that is passed over the surface of the sample, Fig. 4-32. The movement of the stylus creates electrical impulses that are in turn amplified and read out on a meter in microinch units.

Level and Plumb. Levels may be thought of as "bubble instruments" because they employ a bubble floating on a nonfreezing spirit solution that gives its name to the instrument "spirit level." The precision of the level is gained by the radius and the finish grind of the glass tube. The level is a gravity-operated instrument. Therefore, the bubble moves in the tube until the bubble is directly over the center line of gravity for that point, Fig. 4-33 and 4-34. The tube or vial has calibration lines on either side of the level position. The units for reading a level are related to the calibrations of the vial, and the divisions are expressed in thousandths of an inch per foot.

Quality Control Concepts of Measurement

The concepts of alignment of components, quantity of parts, size of product, time of production, and material testing and inspection are responsibilities of the quality control group in an organization. Measurements is the means whereby the control is achieved.

Alignment of Components. The concept of alignment means that the centerlines of each component of a product are in correct adjustment with other components. It is a factor which enters into every manufactured product and is a basis for the correct functioning of the product. From very small objects to huge industrial machinery, alignment is of great importance, Fig. 4-35. A set of lenses which make up a subassembly for a photoelectric warning device, a series of shafts in a small machine which must run free to prevent binding, or the components of an airplane frame where the fuselage, tail, and wings must have the correct relationship to one another must all be in alignment.

Quantity of Parts. The quantity of parts in a box or package as well as their quality is set by standards or specifications. The parts may be counted by hand, or they may be counted by automated equipment. The automated equipment may count each individual part, or it may count some products in units of 10's or 12's, whichever is required in the specifications, Fig. 4-36. Large numbers of

Fig. 4-35. The assembly and alignment of components is essential for a smooth functioning of this five-axis, numerically controlled machine.

Fig. 4-36. Packaging an "item" might mean twelve cartons of eight telephone sets, each in a shipment.

Fig. 4-37. Incoming crankshafts are measured before they are installed in a huge diesel engine.

small items are "counted" by weighing the parts. Parts are added or subtracted until the correct weight is obtained.

Product Size. The actual size of the product is the primary concern in measurement. Quality control checks the sizes of industrial materials and components which come into the manufacturing plant as well as the product that is manufactured, Fig. 4-37. The size of the product is measured by those instruments which have been previously described.

Production Time. Most of the production employees of an industrial plant are paid by the hour. The number of hours that the workers are at a machine is the total number of possible hours of production that may be expected. This total number of hours includes hours other than machine running time, because setting up, tool changing, waiting for materials, waiting for finished parts to be moved, and time out for breaks are all factors involving time. Time studies of the utilization of machines are made to improve manufacturing productivity.

Material Content Testing. Material testing gives information about the physical characteristics of the material from which

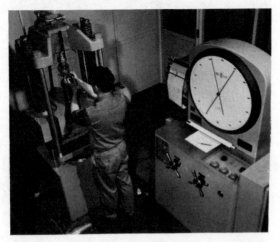

Fig. 4-38. Tensile-testing the strength of a steel sample. This is a destructive test.

the product is made. Testing may be done by destructive testing or by nondestructive testing. In destructive testing, the materials or parts are put under stress until they distort or fail. Common destructive tests to determine engineering properties of materials include tensile strength, compressive strength, torsional strength, shear, bend and impact, and hardness tests, Fig. 4-38. A variety of testing machines measure these properties. Destructive testing results in the bursting, crushing, or pulling apart of the product. The results give information about the product's maximum usefulness.

Nondestructive testing does not impair the function of the part or the material. These tests reveal imperfections or faults in the material or in the fabrication of the product. Such defects may occur during forging, drawing, welding, grinding, heat treating, rolling, and plating, Fig. 4-39. Nondestructive testing is also used to examine cracks

Fig. 4-39. The electron microscope will reveal flaws or imperfections in the micro-structure of materials.

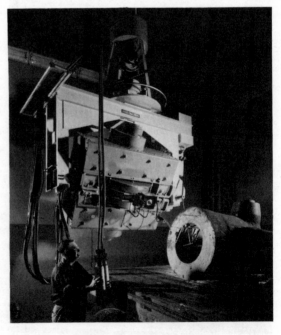

Fig. 4-40. Nondestructive tests do not damage the part. This 22-million-volt x-ray machine radiographs a casting.

that result from hard use or have developed from fatigue of the metal. Common non-destructive methods of testing are radiography (X rays or isotopes producing gamma rays), ultrasonic testing, magnetic testing (magnetic particles and eddy currents), penetrants (dye and fluorescent), and leak testing (Freon, Halide, mass spectrometer), Fig. 4-40.

Quality control is involved in the entire industrial cycle. Planning notices make certain that the vendor is instructed as to the requirements of the contract, drawing, and specifications. Purchase orders are screened to assure that the description of part number, of certification and inspection, and of test requirements is clear. Source inspection is carried out to make sure the vendor has the ability to produce the work and to perform inspection and tests which relate to the work.

Manufacturing process inspection, or product control, is carried out in the manufacturing plant. First piece or first article inspection is done, and all operations are checked thoroughly to assure the product specifications are met, Fig. 4-41.

Gages are used to control the product size. Gages are given checks and rechecks on a regular cycle of inspection to assure accuracy in the production, Fig. 4-42. Jigs and fixtures which control the location of holes and features on the work in manufacturing will also be inspected on a regular cycle.

In the final inspection, lots are sampled and tested. At this point the effectiveness of the control process can be determined, Fig. 4-43. If the product is not meeting specifica-

Fig. 4-42. Gage control is maintained by checking gages on a regular cycle, thus ensuring the accuracy of production work.

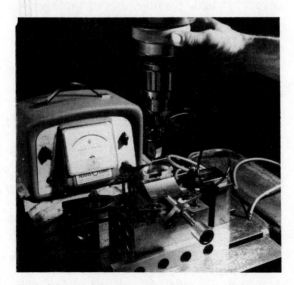

Fig. 4-41. The first piece is carefully checked in the manufacturing plant.

Fig. 4-43. Final inspection: parts are checked by an optical comparator for hole location.

Fig. 4-44. Laying out steel is the first manufacturing step in this inverted stern section of a river towboat.

Fig. 4-45. Large volume products will have the layout performed on the tooling that locates, punches, or bends the metal.

tions, then corrective action must be taken to find the weakness in the control process.

Laying Out Metal

Layout in manufacturing is done primarily for the manufacture of tools and tooling to be used in mass production and automation, Fig. 4-44. It is also done for custom work which is built to the buyer's specifications. This work may be very large and one of a kind, or it may be small and precise.

Large volume products such as refrigerators do not have a layout made on the metal of an individual refrigerator. Rather the layout is performed on the tooling which receives the metal, punches the holes, cuts the notches, and rolls or bends the metal, Fig. 4-45. The accuracy of the layout is built into the dies and tooling.

The layout presented here relates to custom work which a craftsman would need

in order to build a single product or to the manufacture of production tools.

Preparing Metal

Metal is laid out to make the outline of the metal part to be drilled, cut, bent, machined, forged, or welded visible to the craftsman. Layout should provide an accurate line, circle, or prick punch mark which will retain its definition and location during manufacture.

Metal is laid out in many areas of metalwork. Each type of layout work has its peculiar procedural techniques:
1. Sheet metal layout
2. Bench metal layout
3. Machine tool layout
4. Art metal layout
5. Forging layout
6. Foundry layout and
7. Welding and cutting layout

Layout is divided into three major classifications of work:
1. Linear dimensions
2. Angular dimensions
3. Surface dimensions

As they are received from the rolling mill, metals will have a grayish oxide, or in the case of cold-rolled steel, will be oiled to keep the steel from rusting. The gray color of the oxide on hot-rolled steel does not give a good contrast to line scribed on the metal. The bright finish on the cold-rolled steel is also nearly the same color as a scribed line. Therefore, a number of different materials may be added to the surface to provide contrast so that the layout lines may easily be seen.

One of the best surfaces for eye ease and accurate reading is a dark blue background and a white line. Construction and manufacturing industries have been using such a color combination successfully for years.

Layout Fluid. Cold-rolled steel or surface ground steel produces a silver colored line when a line is scribed on it. If a blue semi-transparent layout fluid is painted on the surface and scribed, the result is a clear, narrow, and well defined silver line on the blue background.

Oxide Color. Die steels may be laid out with a very durable surface that can withstand handling by heating the polished or ground steel to 550° F. and allowing it to cool slowly to room temperature. The surface of the steel acquires a dark blue oxide that may be scribed upon with accurate and durable lines.

Copper Sulfate. Copper sulfate or Blue Vitriol produces an excellent layout surface on steel. The copper sulfate is mixed in a water solution, and a few drops of acid are added to the solution. The surface ground or oxide scale-free steel may have the copper sulfate solution spread over its surface. Because of the electromotive series, a coating of copper is deposited upon the metal. When the metal has dried, a scriber will draw a silver line on the satin copper colored background. This layout material is durable, easily seen, and will not rub off easily.

Chalk. Materials like cast iron which have a rough exterior finish can be prepared for layout by painting the surface with a mixture of slaked lime to produce a flat white surface. If it is a small casting, chalk can be rubbed into the surface. When the surface is ready for layout, a scriber will produce a line that resembles a pencil line on a white background.

Soapstone. Hot-rolled steel and rough surfaced material can be laid out with the use of soapstone (talc). This line is rather broad and is used for rough work such as oxyacetylene cutting. Steel and other materials are also marked and roughly laid out with yellow crayon. Crayon gives the same order of accuracy as soapstone and is used to produce layouts, marks, or notes.

Whitening. Art metal or artistic work requires a different or "freer" layout procedure. Poster paint or Chinese White may be painted over the metal, brass, copper, or silver and allowed to dry. The artwork will be drawn on tracing paper or on a hard paper. Carbon paper will be placed over the metal with the paint on it and the artwork over the carbon paper. The edges of the paper will be taped down to the back so that the carbon paper, metal, and art work will not move in relation to one another. A hard pencil is used to trace over the art work and thus transfers the design to the metal. When the papers have been removed, a sharp black line is visible on a white background.

Cementing Layout. If the artwork is large, the layout may be made by drawing the project full-scale on butcher paper and adhering the paper to the metal with rubber cement. This layout procedure has the advantage of being fast in that transfer from a carbon is eliminated, and the paper also helps protect the metal surface when highly polished metals are being worked. Scratches, nicks, and tool marks are kept to a minimum when the layout paper is cemented to the metal.

Scribing Metals. Metals like galvanized steel and aluminum may be laid out directly on the metal with a scriber. The scriber will cut through the zinc or oxide coating and reveal a narrow, accurate silver line. Scriber lines should not be left on galvanized steel if they will be exposed to a corrosive condition. The lines should be regalvanized, painted, or soldered.

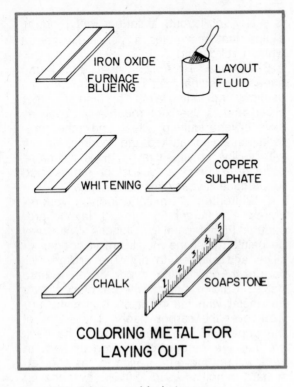

Fig. 4-46. Coloring metal for laying out.

Colored Pencil. Aluminum is sometimes laid out with a graphite pencil or a colored grease pencil. These markings are good because they will not scratch the metal, but they cannot be relied upon for close tolerances, Fig. 4-46.

Layout Lines and Tools

A layout line is a **single** line. If the line is scribed more than one time, the accurate line could not be defined. A line should be sharp and narrow so that it can be read accurately. The line should not be deep because it will be necessary to remove the layout lines before the work is complete, Fig. 4-47. The lines are usually removed during the shaping processes, Fig. 4-48.

Layout tools are designed for accurate work. By using hand tools such as the scale,

TOO WIDE   TWO LINES   CORRECT

Fig. 4-47. A layout line should be narrow, of a single stroke, sharp, and clear.

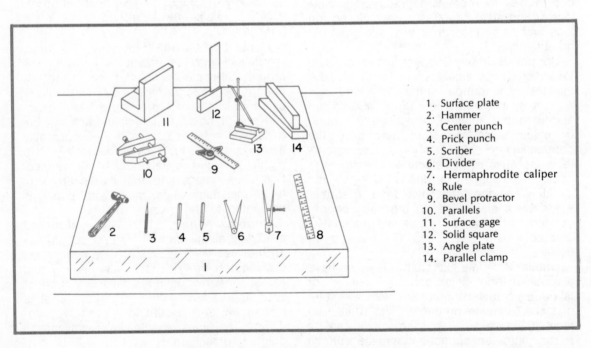

1. Surface plate
2. Hammer
3. Center punch
4. Prick punch
5. Scriber
6. Divider
7. Hermaphrodite caliper
8. Rule
9. Bevel protractor
10. Parallels
11. Surface gage
12. Solid square
13. Angle plate
14. Parallel clamp

Fig. 4-48. Common tools used in laying out.

dividers, hermaphrodite caliper, square, prick punch, or combination sets, the craftsman should be able to maintain accuracies of less than ±0.010″, Fig. 4-49. With the use of tools such as the surface plate, vernier height gage (Fig. 4-50), and angle plate,

layout may be accurate to ±0.001″. A gage block, height gage holder, and scriber produce lines with accuracy within a few millionths of an inch.

In the layout process, scribers, dividers, and prick punches are most frequently used, Figs. 4-51 through 4-58 (page 152). The prick punch is a simple, but very accurate tool. The prick punch makes a small mark that provides a bearing for a divider to swing an arc or circle and also offers a permanent location for lines and holes. Common procedure is to prick-punch all line intersections, base lines or reference lines, centers of holes, and arcs. Irregular curves are prick-punched approximately every ¼″ so that the curves remain if the coloring is worn off.

Fig. 4-49. Combination sets contain a square head, center head and reversible protractor head. These tools are frequently used for regular layout and measuring.

Laying Out Linear Dimensions

A linear dimension is measurement along a line, but the line may have other characteristics. Five examples of lines are shown below, Fig. 4-59. These basic lines may be combined to draw figures or to design manufactured products.

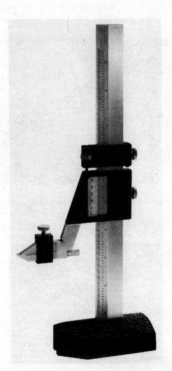

Fig. 4-50. Accuracies of one-thousandth of an inch are made possible by vernier height gages, angle plates, and surface plates.

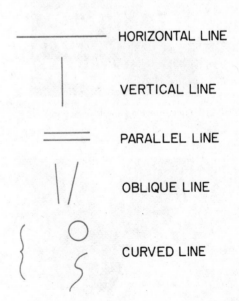

HORIZONTAL LINE

VERTICAL LINE

PARALLEL LINE

OBLIQUE LINE

CURVED LINE

Fig. 4-59. The characteristics of lines.

Fig. 4-51.   Preparing for layout.

Fig. 4-52.   Scribing a line with a solid square.

Fig. 4-53.   Prick-punching line intersections.

Fig. 4-54.   Marking line intersections with an automatic center punch. A hammer is not used.

Fig. 4-55.   Circles are drawn around a prick-punch mark with dividers.

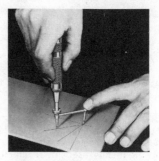

Fig. 4-56.   The space attachment for automatic center punches spaces distances accurately from a central point.

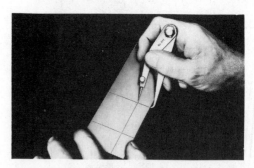

Fig. 4-57.   The hermaphrodite caliper will scribe lines an equal distance from an edge. The line may be curved as well as straight.

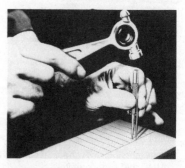

Fig. 4-58.   The combination center punch and spacing tool will punch holes which are an equal distance from a reference line.

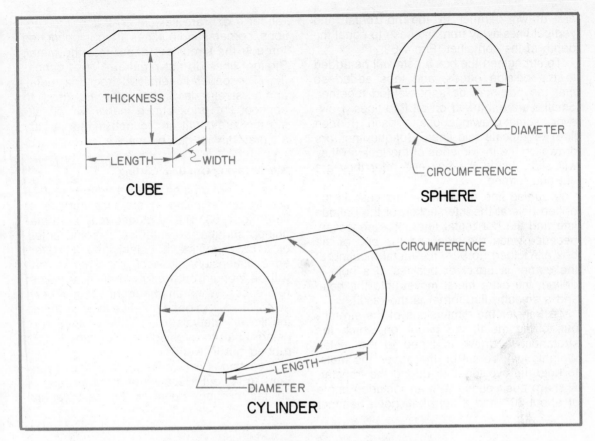

Fig. 4-60. Basic forms establish dimensional measurements.

Linear dimension is the measurement along the length of the line. Accurate measurements of figures can be described because of the linear dimension measurement. These shapes establish the fundamental dimensional measurements: cube (length, width, thickness dimensions), cylinder (diameter, length, circumference dimensions), and sphere (radius, circumference dimensions), Fig. 4-60.

Linear dimensions are used in sheet metal layout. Sheet metal articles that are designed from rectangular solids are numerous. These rectangles make application of basic lines, Fig. 4-61. The basic rectangle is drawn out full-scale on paper, making the vertical and horizontal lines slightly longer than an even inch such as 6⅛″ x 12⅛″ or 3³⁄₁₆″ x 5¾″.

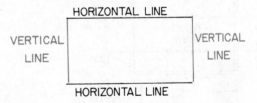

Fig. 4-61. The basic rectangle is drawn from vertical and horizontal lines.

The reason for the odd number is that most of the fingers of a box and pan brake (which is used to form rectangles) are in even inches.

The horizontal and vertical lines that make up the rectangle make up the size of a box. The box has to have depth, so parallel lines are drawn parallel to the horizontal and vertical lines away from the base to equal the depth of the container, Fig. 4-62.

To strengthen the box a hem will be added to the outside edges, and tabs added so that the box may be spot welded together. Small x's are marked on all fold lines. However, some drawings use a dash (hidden line) to indicate fold lines. Oblique lines are drawn on the inside ends of the tabs so they will slide inside the box end after they are cut and formed.

A curved line or circle in this case is the drilled hole at the intersection of the oblique line and the horizontal lines. It is necessary because when the side and the end of the box are folded up 90°, the metal meeting at the corner is upset or buckles. If a hole is drilled, the extra metal moves into the hole, and a smooth, flat corner is the result.

To transfer the dimensions of the stretchout to the metal, the paper on which the stretchout is drawn is taped to the metal. Weights may be set on the paper and metal to hold the two together during the transfer. A sharp prick punch with an included angle of about 30° and a light ball peen hammer

(about 4 oz.) are selected. All line intersections, corners, and outlines are prick punched through the stretchout paper into the metal. Straight lines do not require a lot of punching. The paper is removed from the metal, and a steel rule and scriber are used to connect the prick punch marks. By checking the stretchout and punch marks, a very accurate layout can be made.

Linear Laying Out on a Casting

Laying out on a casting is begun by establishing or assuming a reference surface or line, Fig. 4-63. The reference may be a machined surface, a centerline of some holes, or a baseline that is established for reference dimensions. In each case dimensions will be worked out from one of these layout references. When an important line has been established, it is good practice to prick punch the line at intervals to offer the protection of being able to reestablish the line if it is rubbed out in work.

A casting is one of the more difficult problems to lay out. Because of the draft (angle) on the casting, none of the cast vertical

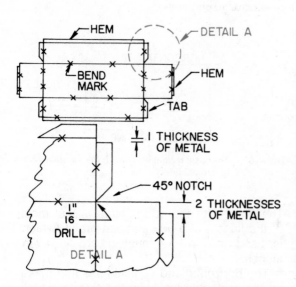

Fig. 4-62.   Characteristics of lines applied to a rectangular sheet metal box.

Fig. 4-63.   Laying out a casting. Note the jacks to level the casting.

sides will be square. There may also be cored holes through the casting. For layout, it is customary to plug the cored holes with soft wood and place a small piece of copper or steel on the wood in the center of the hole. The sharp corners of the pieces of sheet metal are bent down 90° and the metal driven into the wood. The center distance between two or more holes may be accurately laid out on the casting by scribing the correct lines on the sheet metal. When the centers are located, a slight prick punch mark may be made in the metal and a divider leg point may be placed at the intersection and a bolt circle or other holes may be laid out.

Another consideration that must be made when laying out castings is whether or not the casting will have enough material so that the cut will continue across the entire casting surface. In the foundry a core may shift or a surface may shrink or warp so a preliminary layout should be made to assure that all machined surfaces will have sufficient metal to "clean up" across the total face of the cut. If this is not the case, the layout may be shifted enough so that the casting will clean up and be useful, Fig. 4-64.

Laying Out Angular Dimensions

Angular dimension is the measurement of the relationship between two straight lines. When two straight lines meet, an angle is formed. Angles are described as right, acute, or obtuse, Fig. 4-65. Figures with

various combinations of angles are used to design manufactured products.

Angular dimensions are measured from a reference. The reference in this case is one leg of the angle being measured. A line that is being rotated about one end will generate angles when referred to the starting position. If the line moves ⅛ of a rotation, the angle generated will be 45°. If rotated ¼ of a rotation the angle will be 90°, Fig. 4-66. When one complete revolution is made, the angle is 360°, and a circle will have been generated. The 90° angle (square, or right angle) is the most common angle used. This is called a **reference angle.**

Squareness

The concept of squareness has come down from ancient times. The early Egyptians established squareness or a method of

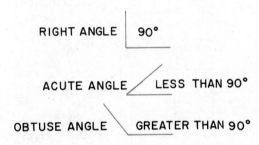

Fig. 4-65. An angle is described as a right angle, acute angle, or an obtuse angle.

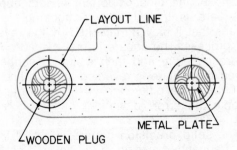

Fig. 4-64. A cored casting with wooden plugs and sheet metal layout plates.

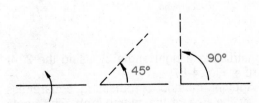

Fig. 4-66. A line that is rotated about an end point will generate an angle with reference to the original position.

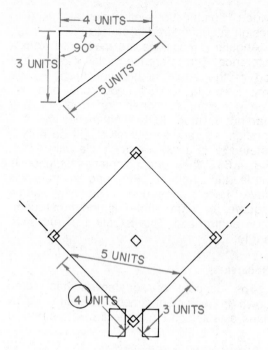

Fig. 4-67. Laying out a square corner by using the 3-4-5 rule. At 90° the hypotenuse will be 5 units long.

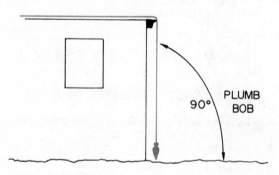

Fig. 4-68. A plumb bob will construct a 90° angle with level ground.

constructing a right angle by using the 3, 4, 5 rule, Fig. 4-67.

Another method of obtaining squareness is by the use of the **plumb bob**. The angle between the plumb line and the level ground, horizontal line, or reference point is 90°, Fig. 4-68.

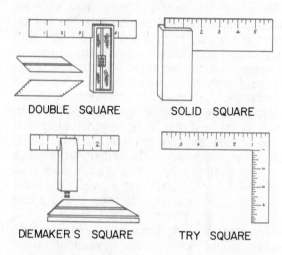

Fig. 4-69. Squareness may be checked by a variety of squares.

These concepts of squareness are used today in laying out large structures. The **builder's transit** is also used to make these measurements.

In manufacturing, the dimensions of squareness are usually small. These angles require more precision, so different instruments have been developed. Some of these instruments (such as the dial indicator) do not appear as squares but involve the geometry of squareness.

The **square** is the most efficient tool for constructing or checking a 90° angle. Squares for different uses will be of different sizes and accuracies, Figs. 4-69, 4-70, and 4-71.

One of the most important factors about a square is that it is **self-checking**. This means that it may be checked by the reversal process. The square is first placed on a reference edge and then a line is scribed. When the square is reversed and the line is rescribed, the difference in the two lines is double the error (deviation) of the square, Fig. 4-72.

The **solid precision square** is a tool that will frequently be used to lay out and check manufactured parts for squareness. The **solid square** is often used with a surface

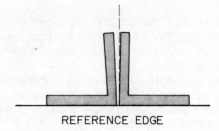

REFERENCE EDGE

Fig. 4-72. Check a square by striking a line and then reversing the square. This example illustrates the square is untrue.

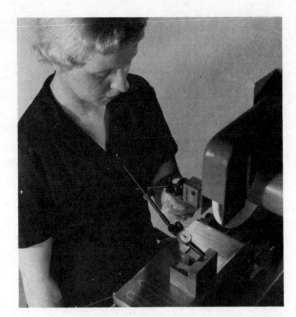

Fig. 4-70. A dial indicator is used to check the work for its squareness with the machine in two directions.

Fig. 4-73. The work error is indicated by a wedge-shaped line of light along the contact line.

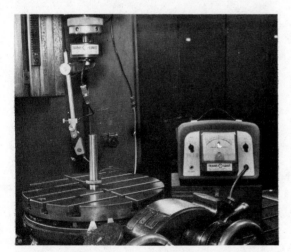

Fig. 4-71. The shaft is checked for squareness with the machine by an electronic comparator.

plate to check squareness or to lay out angles 90° to an edge or surface.

The **cylindrical square** is one of the most interesting squares because it does not appear to be a square at all. The cylindrical square is a cylinder of heavy walled steel that is carefully ground. The ends are ground and lapped so that they are square with the cylinder's axis, Fig. 4-73. The ends have small notches cut in their surface so that when the square is rotated on a surface plate, any dirt particles will be swept into the notches and not be under the square. With this tool a 90° angle has previously been constructed with the reference surface at as many points around the circumference as desired. One of the chief advantages of this type of square is that it will only contact the workpiece at the point of tangency of the square. This makes it possible to accurately read the crack of light between the square and the work. When no light passes between the square and the work, the workpiece is square.

The concepts of layout of angles came from geometry and many angles are laid out by construction using arcs, radii, and lines, Figs. 4-74 through 4-76. Angles may be constructed to lay out two inclined machined surfaces, the holes on a bolt circle, or to check for squareness by constructing a perpendicular line.

Laying Out a Circle

The bolt circle on a pipe flange may be laid out with the use of a steel rule, dividers, and a prick punch. The center of the pipe is crossed with a piece of wood wedged into the hole. A small piece of copper is tacked, or its corners are bent and then it is driven into the wood.

Locating the Center. A center head is used on the blade of a combination set ro-

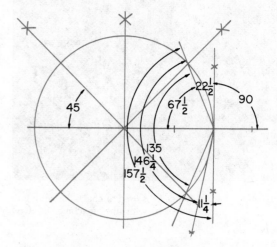

Fig. 4-75. An octagon will yield 11¼°, 22½°, 45°, 67½°, 90°, 135°, 146¼°, and 157½° angles.

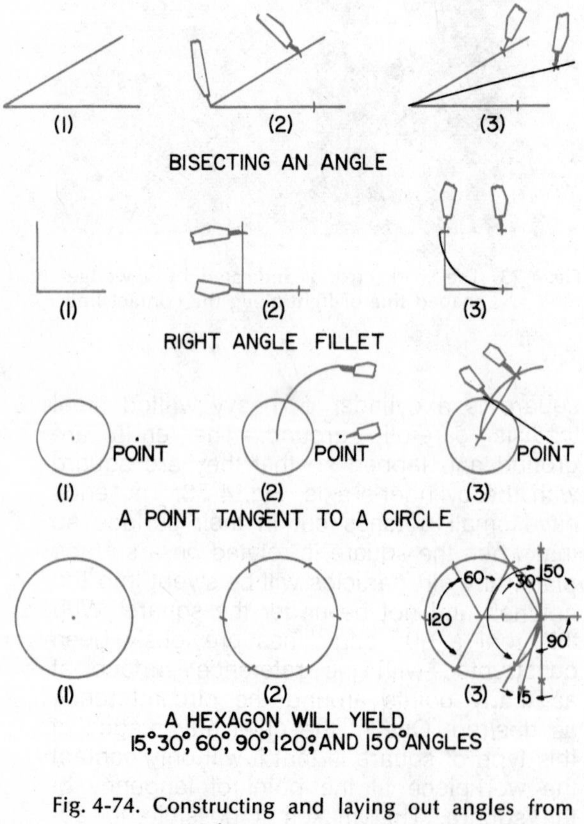

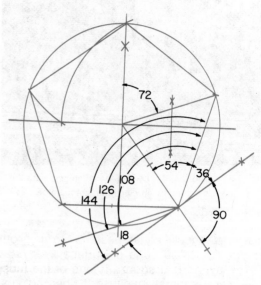

Fig. 4-76. A pentagon will yield 9°, 18°, 36°, 54°, 72°, 90°, 108°, 126°, and 144° angles.

(I)   (2)   (3)

BISECTING AN ANGLE

(I)   (2)   (3)

RIGHT ANGLE FILLET

POINT   POINT   POINT

(I)   (2)   (3)

A POINT TANGENT TO A CIRCLE

(I)   (2)   (3)

A HEXAGON WILL YIELD
15°, 30°, 60°, 90°, 120°, AND 150° ANGLES

Fig. 4-74. Constructing and laying out angles from geometry.

tated around the outside edge of the flange, and lines are scribed through the center point of the copper square. This is done at least three times to locate the center of the flange. At the intersection of the lines, a small prick-punch mark is made. This mark becomes the center of the layout. Layout fluid is painted where the center line and bolt holes will be drawn on the flange. The dividers are set to the radius of the bolt circle by placing the point of the leg into the graduation of the steel rule, Figs. 4-77 and 4-78.

The other divider point leg is moved until it fits into the other graduation and the length of the radius desired. The divider, if carefully set, should be within 0.003″ of the actual dimension. One leg of the divider is placed in the prick-punched hole in the copper and a center line is carefully drawn on the flange. This line is now the diameter of the bolt circle.

Marking Hole Positions. When several holes align with each other and mating flanges, the position (orientation) of the first hole should be noted before the hole layout is begun. If the hole orientation has been established, a center head is used to mark the first hole on the center line. A small prick punch mark is made at the intersection of the centerline and the scribed line. A six bolt hole circle will be made with the dimension of the radius of the circle set on the divider. One leg of the divider is set in the prick punch mark and the second arc scribed across the centerline. This process is continued until all bolt holes are located.

The size of bolt that will be used in these holes should be determined and then 1/64″ added to the diameter of the hole to allow for clearance of the bolts. If the bolt size is ⅝″, the divider points are set so that one leg is in the graduation of the steel rule and the other leg 5/16″ away from the first but on the large side of the graduation. The bolt holes are drawn with one divider leg point in the prick-punch hole on the flange and the other leg scribing the circle, Fig. 4-79.

Prick Punching. With the completion of all bolt hole circles, the line intersections are prick-punched at the center line and arcs around the bolt hole. This is done to

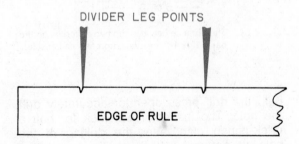

Fig. 4-78. Divider points are adjusted until they fit into the correct graduation for accurate measurement.

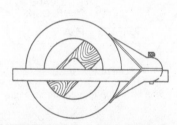

CENTERHEAD LOCATING THE CENTER OF THE FLANGE

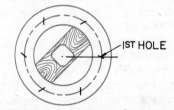

BOLT CIRCLE LAYOUT (6 HOLES)

Fig. 4-77. Bolt circle layout.

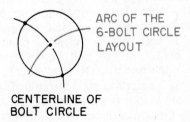

CENTERLINE OF
BOLT CIRCLE

Fig. 4-79. Layout of bolt hole.

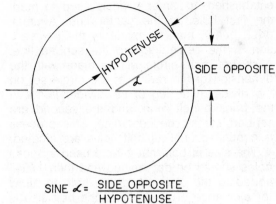

$$\text{SINE } \alpha = \frac{\text{SIDE OPPOSITE}}{\text{HYPOTENUSE}}$$

Fig. 4-80.   When instruments are substituted for the elements of a triangle, precision angles are constructed.

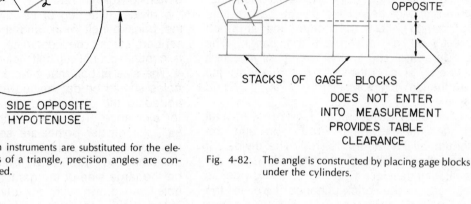

Fig. 4-82.   The angle is constructed by placing gage blocks under the cylinders.

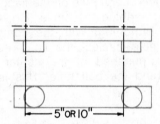

Fig. 4-81.   The distance between the two cylinders on the bar is 5″ or 10″, the two common sine bar sizes.

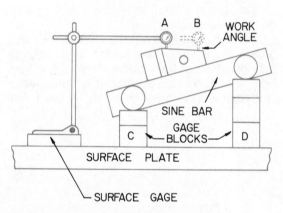

Fig. 4-83.   Sine bar setup to measure an angle.

help the drill press operator accurately drill the hole. The inspector will look for half a prick-punch mark after the drilling of the hole to determine if the hole was drilled where the layout was. If the drill chips wear the coloring off the metal, the inspector will still be able to see one half of the four prick punches if the work has been done properly.

The last layout operation is to take a **center punch** with an included angle of 60° to 90° and enlarge the prick punch mark in the center of the bolt hole. Changing the 30° prick-punch hole to a 60° to 90° hole makes it easier to start the drill. The process of hole layout is repeated for all the holes.

By the use of geometric construction, these six holes are 60° apart on a bolt circle.

Precision Angular Layout

For the very precise angular measurement or layout of angles, a sine bar is used. The angles used are produced by using the functions of a triangle. The **sine** of any angle is its opposite side divided by its hypotenuse, Fig. 4-80.

When instruments are substituted for the elements of the triangle, precision angles may be constructed.

The **sine** bar is constructed so that the distance between two cylinder centers is

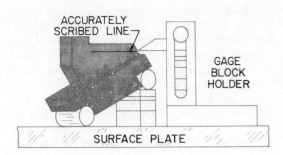

Fig. 4-84. Gage blocks assembled to scribe very accurate lines and angles.

Fig. 4-85. The work may be fastened to sine plates and blocks, which have large surfaces.

5″. This distance may be used as the hypotenuse of a triangle, Fig. 4-81.

When the hypotenuse is held constant, the angle may be constructed by adding gage blocks under the cylinder, Fig. 4-82.

Sine of 30° = 2.50″ with a 5″ sine bar, Fig. 4-83. When 2.5″ of gage blocks are placed in the opposite side, the angle made by the top of the sine bar will be 30°. The number 2.5 is obtained from a table of sines for a 5″ sine bar. Angles from 0°0′ to 59°60′ will appear in a common table of sines.

If the dial indicator reads the same number while traveling from point A to point B, the surface is parallel to the surface plate. The angle of the sine bar and the angle on the workpiece are the same, but **opposite** in direction. Thus the dial indicator shows no change. The height of the gage blocks (C) may be subtracted from height D and the length of the opposite side will be found.

Height of gage blocks (D)		4.723500″
Height of gage blocks (C)	−	1.000000″
3.723500″ = 48°8′		3.723500″
= the sine of 48°8′		

Gage blocks are assembled as end standards and scribe very accurate lines; very accurate angles are scribed when the gage blocks are combined with the sine block, Fig. 4-84 and 4-85.

Trigonometric Function for Special Measurements. For special measurements other trigonometric functions may be used to solve angular problems. The diagram

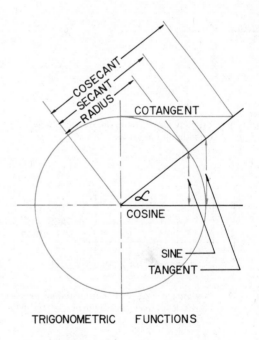

Fig. 4-86. Angular problems are solved by trigonometric functions.

does give a visual presentation of what the trigonometric functions are, Fig. 4-86.

Laying Out Surface Dimensions

A most convenient method for both laying out and measuring is from a surface plate,

Fig. 4-87. Surface plates are the reference plane for accurate layout and measurements.

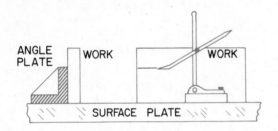

Fig. 4-88. A surface gage can transfer measurements from a standard to the work.

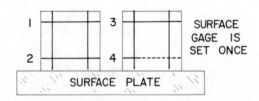

Fig. 4-89. The surface gage will transfer dimensions to many sides of workpieces with one setting of the surface gage.

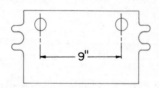

Fig. 4-90. The die shoes is coated with layout fluid and mounted to an angle plate.

Fig. 4-87. Surface plates come in many sizes, from 6″-8″ to plates 8′-10′. In special cases, they may be much larger. Surface plates are made from steel, cast iron, and granite. They form a reference plane and many tools make measurements from their surface.

The surface gage is an instrument used to transfer measurements from a standard to various parts of an object being laid out. It is also used for drawing parallel lines, Fig. 4-88.

The surface gage is especially helpful in laying out if the work has edges that are 90° to each other and the work can be rotated, Fig. 4-89.

Rough Layout of a Die Shoe

A **die shoe** is used to hold and locate the upper and lower dies mounted in a punch press. The punch press is one of the most efficient metal shaping tools.

A **surface gage** is used to make a rough layout of a die shoe. This stock is 10″ x 12″ and 1¼″ thick. The stock has been squared and surface ground. Two holes are to be drilled on 9″ centers, Fig. 4-90.

To make a rough layout of this die shoe for a punch press die set, the surface-ground surface is coated with dykem or layout dye to provide color contrast. The top edge surface is placed down on the surface plate and the back clamped against an angle plate resting on the surface plate. A combination square with the blade resting on the surface plate is used to adjust the scriber, with the aid of a magnifying glass, until the scriber fits into the graduation of the steel rule, Fig. 4-91.

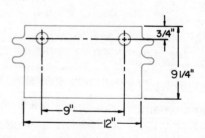

Fig. 4-91. Read the blueprint.

The surface gage is slid across the surface plate and across the length of the die shoe. The die shoe is removed from the angle plate, and the length of the line measured. A line is scribed at the midpoint of the line. From the midpoint, a measurement is made 4½" to the left. Again with the aid of a magnifying glass, a short temporary line is scribed there. The back of the die shoe is placed against the angle plate in a vertical position on the surface plate, while a combination square is used to square the top edge vertically from the surface plate. Parallel clamps hold it in place. By using the surface gage and a magnifying glass, the setting of the temporary line is picked up and scribed across the first scribed line. The surface gage setting is then set back to the steel rule. The rule reading where the scriber fits into the graduation is observed and a note of this dimension is made. This next setting of the surface gage is the most important measurement, because it determines the center-to-center distance of the layout and holes. Nine inches are added to the reading of the surface gage setting, then the scriber point is fitted into the rule graduation. The line for the second hole is scribed across the first scribed line, Figs. 4-92 and 4-93.

The die shoe is removed from the angle plate and placed on a bench plate for the hole layout. The line intersections of the two holes are carefully prick-punched with the aid of a magnifying glass. These punch marks become the center point for layout of the holes. A 4" divider is set for 7/16" between the points. This is done by placing the point of one of the divider legs in the 1" graduation and adjusting the other graduation until it fits into the 7/16" graduation. One leg of the divider is placed in the prick-punched hole of the die shoe and a ⅞" diameter circle is carefully drawn. The second circle is drawn in the same manner for the other hole.

Where the circle crosses the straight layout lines, light prick-punch marks are made with the aid of a magnifying glass, Fig. 4-94. The last rough layout operation is to open

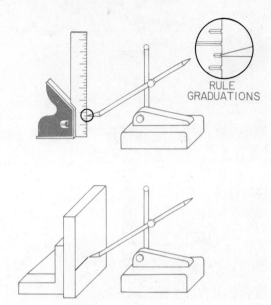

Fig. 4-92.   Measuring the dimensions and setting the surface gage.

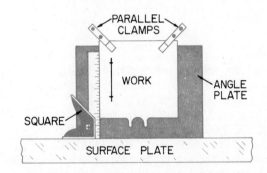

Fig. 4-93.   Scribing lines and rotating work to locate holes.

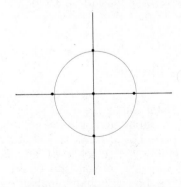

Fig. 4-94.   Where lines cross, they are prick-punched with the aid of a magnifying glass.

Fig. 4-95. Toolmakers' buttons are accurately checked by the use of a micrometer.

Fig. 4-96. The jig bore machine is a measuring and boring machine which will very accurately locate, space, and size holes and surfaces.

out the center prick-punch mark with a center punch so that the craftsman may start the drill correctly.

If the layout has been done carefully, the points may be within 0.003″ of the correct setting. In most work this tolerance is acceptable for the layout. When the tolerance needs to be closer, the dimensions may be improved in the machining by the use of tool makers buttons, Fig. 4-95, or by boring machines, Fig. 4-96.

Reading Blueprints

Blueprint reading is a skill in which the craftsman translates the information of the engineering department into the actual manufactured object. The drawing will include all the information, specifications, and instructions needed to complete the product. It is a precise method of communication between engineer, craftsman, and buyer.

Blueprint Organization

Actually, a blueprint is made up of a series of blueprints which are rolled together and bound on one end. The cover sheet has the job title, file number, date, and description of the job, such as where it fits into the total product.

The **drawing format** has additional information when the drawings deal with government projects. This information will be found around the margins. It relates to classification of title, accountability, release date, security classification, group marking, proprietary notice, and others.

The **title block** contains information concerning the drawing title, scale, weight, signatures of those who have done the drawing, checking, approving for stress, approving of the system, and approving of the project. Other information includes the contract number, tolerances, material, number of drawing sheets, drawing size, and drawing numbers.

Orientation of Blueprint Views

A blueprint contains accurate information so that the product may be built, inspected,

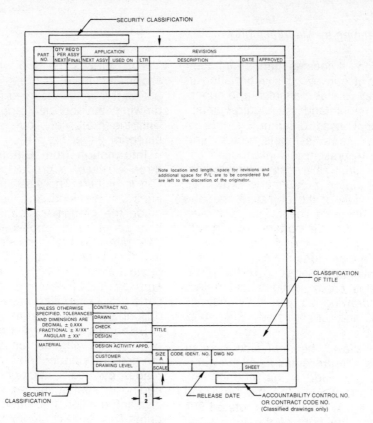

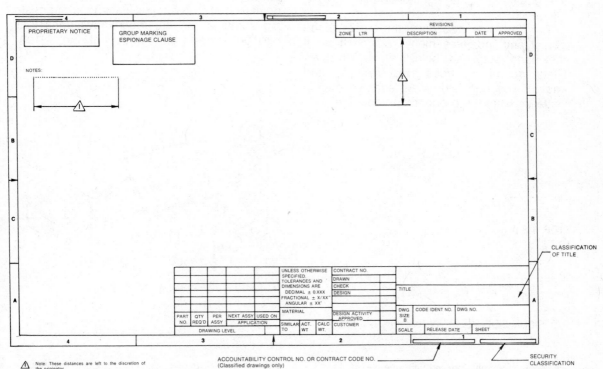

Fig. 4-97. The drawing sheet format supplies information in addition to the object drawing.

and assembled. It locates part information and numbers. It conveys this information in a clear manner so that craftsmen and supervisors easily understand the requirements.

The blueprint uses different weights of lines, shapes, views, symbols, notes, and specifications to present this information.

Front View. Just as a map is drawn with North at the top, a blueprint is drawn so that the front view of the object is near the center of the blueprint sheet. The front view furnishes most of the important surface with the characteristic shape and dimensions.

The front view provides the craftsman with the general overall dimensions of the product and locates cross sections and details of the total drawing so that they may be easily found and coordinated, Fig. 4-97 (page 165).

Right Side View. The right side view of the product is a representation of the information of the right side of the product. The front view orients the part on the page and the right side of the front view is rotated 90° and appears to the right of the front view, Fig. 4-98.

The right side view gives the profile of the object and indicates many relationships of the parts such as holes, slots, grooves, fillets, rounds, surface relationships, and angles. The right side view enables one to visualize what the product will look like.

Top View. The top view presents the shape and location of features on the product as if it were observed from the top. On a drawing, the top view appears in alignment with the front view, but is above it on the blueprint page, Fig. 4-99.

Information from Blueprints. With these three views of a product, detailed information is gained. The size and location of features on the work are very important and guide the craftsman in the layout. Informaiton about arcs, circles, and cylinders as they relate to the parts under manufacture; the dimensions of holes and angles as they relate to other holes and surfaces; and tolerances and finishes of the manufactured parts are all supplied as well as notes which describe special information.

Blueprints are very carefully drawn and can be read with ease once the symbols and the notes are understood. A skilled blueprint reader can gain additional information by adding or subtracting lengths and angles to give information not directly shown on the print. This skill in blueprint interpretation is pointed out by a simple illustration suggesting some of the common knowledges that may be gained from a blueprint, Fig. 4-100.

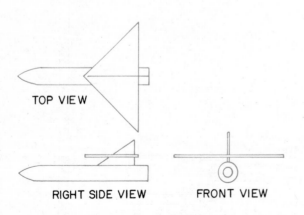

Fig. 4-98.   The front, top, and right side views.

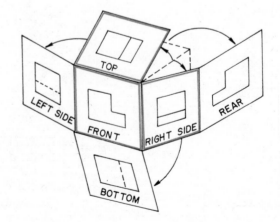

Fig. 4-99.   Six-view projection of a product.

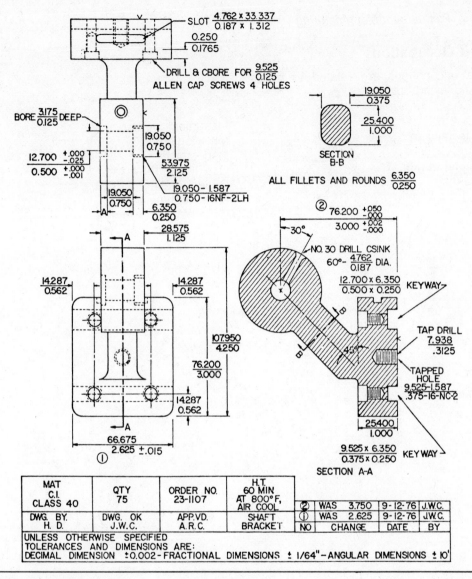

SLOT $\frac{4.762 \times 33.337}{0.187 \times 1.312}$

$\frac{0.250}{0.1765}$

DRILL & CBORE FOR $\frac{9.525}{0.125}$ ALLEN CAP SCREWS 4 HOLES

BORE $\frac{3.175}{0.125}$ DEEP

$\frac{19.050}{0.750}$

$12.700 \begin{smallmatrix}+.000\\-.025\end{smallmatrix}$
$0.500 \begin{smallmatrix}+.000\\-.001\end{smallmatrix}$

$\frac{53.975}{2.125}$

$\frac{19.050}{0.750}$

19.050 - 1.587
0.750 -16NF-2LH

$\frac{6.350}{0.250}$

$\frac{28.575}{1.125}$

$\frac{14.287}{0.562}$

$\frac{14.287}{0.562}$

$\frac{107.950}{4.250}$

$\frac{76.200}{3.000}$

$\frac{14.287}{0.562}$

$\frac{66.675}{2.625} \pm .015$

SECTION B-B

$\frac{19.050}{0.375}$

$\frac{25.400}{1.000}$

ALL FILLETS AND ROUNDS $\frac{6.350}{0.250}$

② $76.200 \begin{smallmatrix}+.050\\-.000\end{smallmatrix}$
$3.000 \begin{smallmatrix}+.002\\-.000\end{smallmatrix}$

30°

NO. 30 DRILL CSINK
60° - $\frac{4.762}{0.187}$ DIA.

$\frac{12.700 \times 6.350}{0.500 \times 0.250}$ KEYWAY

45°

TAP DRILL
$\frac{7.938}{.3125}$

TAPPED HOLE
9.525-1.587
.375-16-NC-2

$\frac{25400}{1.000}$

$\frac{9.525 \times 6.350}{0.375 \times 0.250}$ KEYWAY

SECTION A-A

MAT C.I. CLASS 40	QTY 75	ORDER NO. 23-1107	H.T. 60 MIN AT 800°F, AIR COOL				
				②	WAS 3.750	9-12-76	J.W.C.
				①	WAS 2.625	9-12-76	J.W.C.
DWG. BY. H. D.	DWG. OK J.W.C.	APP.VD. A.R.C.	SHAFT BRACKET	NO	CHANGE	DATE	BY

UNLESS OTHERWISE SPECIFIED TOLERANCES AND DIMENSIONS ARE:
DECIMAL DIMENSION ±0.002 - FRACTIONAL DIMENSIONS ± 1/64" — ANGULAR DIMENSIONS ± 10'

1. The specifications for the material.
2. Stress relieving heat treatment is required after rough machining.
3. A 19.050-1.587 (.750-16NF-2LH) thread is required on the part.
4. The overall length of the part is 107.950 mm (4.250").
5. The angle of chamfer of the starting end of each thread is 30°.
6. The angle of the countersink is 60°.
7. The computed dimension of A is 3.175 mm (0.125").
8. The specifications for the flat keyways are 12.700 mm x 6.350 mm (0.500" x .250").

9. Changes were made from the original drawing:
   1. was 2.625 9/12/76 by J.W.C.
   2. was 3.750 9/12/76 by J.W.C.
10. The bored hole is 3.175 mm (0.125") deep.
11. The center-to-center distance of the elongated slot is 28.575 mm (1.125").
12. The specifications of the counterbored holes are to drill and counterbore for 9.525 mm (0.375") Allen cap screws, 4 holes.
13. The counterbored holes are 9.525 mm (.375") deep.
14. All fillets and rounds are 6.350 mm (.250").

15. Finished machined surfaces are indicated by a V.
16. The upper limit dimension for the 45° angle is 45° 10'.
17. The upper and lower limits of the reamed hole are:
    upper — 12.700 mm (0.500")
    lower — 12.675 mm (0.499")
18. The specification for the tapped hole is 7.938 mm Drill, Tap 9.525-1.587 (.312" Drill, Tap .375-16NC-2).
19. Drawings dimensioned in English and metrics are called dual dimensioned drawings.
20. The part number is 23-1107, and 75 parts are required made of Class 40 cast iron.

Fig. 4-100. Sample blueprint which shows both metric and standard dimensions.

Related Occupations

These occupations are related to measuring and laying out:

Engineer
Layout technician
Machinist
Sheet metal worker
Boilermaker
Tool-and-die worker

Instrument worker
Metrologist
Inspector
Quality control worker
Optical tooling worker
Laser technician

Metal Cutting and Forming

Section

2

Material is separated by being sliced, wedged, or scraped to produce the shape of the part. Metal cutting edges for separating work fall into three large groups: (1) pointed edges, (2) straight edges, and (3) shaped edges. Many different cutting tools are produced with these various edges. Separation of heavy metal by cutting is achieved by sawing, surface cutting, filing, disc and belt grinding, shaping, planing, milling, grinding, and turning. Electrical chemical separation is done by electrical discharge machining, electrochemical machining, and chemical machining. Sheet metal shearing operations include square shearing, cutting off, parting, blanking, punching, notching, slitting, lancing, trimming, and nibbling performed by power tools. Basic separation operations may be done by hand tools.

Nearly all manufactured products require holes upon which to mount subassemblies or parts to complete the product. The drilling of holes is done by portable drilling tools or by drill presses. Drilling holes is the primary concern of drill press operations, but the processes of countersinking, counterboring, boring, reaming, and spot facing are other important operations done by drilling machines. Different types of drill presses are needed to handle the scope of drilling work, ranging from sensitive drill press work to way drilling. Drilling procedures necessary to set up, locate, and properly drill holes require the understanding of the drill size systems, drilling speeds and feeds, and the use of drill jigs and fixtures. Recently hole size and location by numerically controlled drilling has brought about a rethinking of hole production in that a precision hole may be produced by numerical control at the same cost as regular tolerance holes by production tooling.

Practices of metal forming fall into two major groups: cold working of metal and hot working of metal. Cold forming is done by bending, stretching, and compressing. Hot metal forming brings materials to shape by compression. Forging is the primary squeezing process of hot metal compression forming. The metal is heated so it can be deformed when it is worked. Many heavy shapes needed by industry are formed by forging.

Forming metal by casting begins with an understanding of how metals melt and freeze to produce sound castings. Different types of patterns can be made to produce the complexities of castings. Cores are required to make holes, slots, and recesses in the castings. Sand characteristics such as mold strength, resistance to flowing metal, ability to vent mold gases, and ability to withstand high temperatures are considered when the sand is chosen to be used as the molding sands. Sands are compounded to use their most desired qualities. Green-sand molding is most frequently used in the foundry industry. Other methods of casting are permanent mold casting, centrifugal casting, die casting, plaster mold casting, shell molding, and investment casting.

Principles of Metal Separation

Chapter 5

Words You Should Know

Velocity — The rate of motion or operation of a body or tool.

Discontinuous chip — A broken, segmented, or intermittent metal chip.

Chip breaker — A device which further stresses a flowing chip so that it breaks into short segments.

Vitrified — Fused into a glass-like substance.

Form cutter — A device which has a cutting edge produced by two surfaces intersecting to form a curved line, which is the cutting edge.

Parting — Separating metal into two sections or parts.

Kerf — The width of a cut made by a cutting tool.

Contour cutting — A band saw that cuts straight, radius, or irregularly shaped pieces.

Gullet — The space between two cutting edges.

Feeding — The movement of the work into a cutting tool or vice versa to produce a cutting action. Feed is expressed as inches per minute of metal/cutter engagement.

Parallels — Hardened steel bars ground as a pair. They are used to raise the height of the work in a vise or on a machine table.

String planing — Machining of a number of small identical parts at the same time.

Parts are grouped together on the machine table.

Arbor — A rotating shaft that holds rotating cutting tools such as milling cutters.

Conventional milling (up) — The direction of rotation of the milling cutter is opposite the movement of the table feed.

Climb milling (down) — The direction of the rotation of the milling cutter is the same as the movement of the table feed.

Dressed — To true a surface or make it straight or flat. Grinding wheels may be dressed flat or dressed to a special shape.

Plunge grinding — The workpiece is rapidly moved directly onto the surface of the grinding wheel until a specific size is reached.

Side clearance — The metal is removed from under the cutting edge of the tool so that the flank of the tool will not rub as the tool is fed into the workpiece.

Side rake — The angle on the top of the tool produced by grinding the metal away from behind the cutting edge. Differing angles are required for various metals for efficient cutting action.

Foreshortening — To reduce an angle or line by intersecting with another compound angle.

Tailstock — The movable or sliding support for the dead center on a lathe or grinder.

Servo-controlled — An electronic control system which activates a hydraulic-, mechanical-, or pneumatic-powered system.

Electrical discharge machining — Removing metal with the use of an electric spark to erode the work to the shape of a carbon tool.

Electrochemical machining — Removal of metal with an electrical current to deplate the work to the shape of the electrode tool.

Chemical machining — Removal of metal by dissolving it with an acid or base. The metal is etched, and the shape is controlled by a chemical-resistant mask.

Theory of Cutting

Metal can be separated into parts by using a tool and some source of energy to cut, Fig. 5-1. The energy for cutting is supplied by the machine, and the tool utilizes one or a combination of three tools: (1) a slicing tool, (2) a wedging tool, or (3) a scraping tool. These three cutting tools differ in the amount of pressure which is applied to the cutting edge, the direction that the pressure is applied to the material, and the velocity that the cutting edge is moved in relation to the material.

Slicing

A knife blade cuts with a slicing action. Light pressure is applied while cutting, but the velocity of movement of the cutting edge is relatively high. Because light pressure is used, a relatively fragile tool may be used. The edge is made by the intersection of two planes at a slight angle to one another. The knife edge is narrow without a great deal of strength or rigidity supporting the cutting edge, Fig. 5-2. The knife slices if it has velocity and direction. The cutting edge must have side clearance to make it cut, Fig. 5-3.

Fig. 5-2. Knife edge.

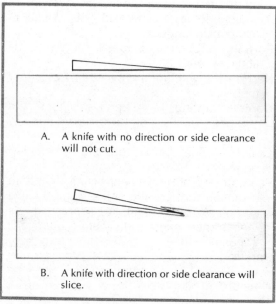

A. A knife with no direction or side clearance will not cut.

B. A knife with direction or side clearance will slice.

Fig. 5-3. A knife must have velocity and direction in order to slice.

Fig. 5-1. Metal separation is obtained by applying energy to a specially shaped tool.

Soft, ductile materials are frequently separated by slicing, Fig. 5-4. The greatest application of slicing is to organic materials. In wood turning, a skew cutting tool is used. The wood fibers are sliced off the top of work and a smooth, plane-like surface is obtained.

Wedging

A cold chisel cuts with a wedging action, Fig. 5-5. The pressures applied are high, so additional strength is needed. The cold chisel has wide angles to provide support. The concentration pressure in a very narrow zone in the metal causes the concentration of stress under the cutting edge. This pressure to the metal causes first crystal elongation, then plastic deformation, and finally a shear zone where rupture or separation occurs along the cutting tool's edge.

Tool wear is the result of this shear zone. Pressure and heat are necessary to bring the metal to a high enough stress so that the metallic bonds within the metal will be overcome and will compress, deform, and slip. Multitudes of these slip plane movements will rupture and make possible the removal of chips from the metal's surface.

Applications of cutting by wedge action are found chiefly in chiseling, punching, and blanking operations.

Scraping

Materials which require a great amount of pressure or stress upon the metal to remove a chip apply all three cutting actions — pressure, direction, and velocity.

A metal-cutting lathe tool cuts under extreme pressure and requires rigidity and support for the cutting edge. A large wedge angle is needed to provide this support and to dissipate the heat. The planes forming the cutting edge meet at nearly a right angle to provide maximum support.

Scraping is applied to most metals which have great strength and require heavy pressure to remove material. All three of the cutting actions are applied: pressure, direction, and velocity.

In a lathe, **velocity** is provided for the cutting action of the tool by the spindle speed of the machine, **pressure** by the horsepower that is applied to the spindle,

Fig. 5-4. Soft materials and organic materials are frequently separated by slicing.

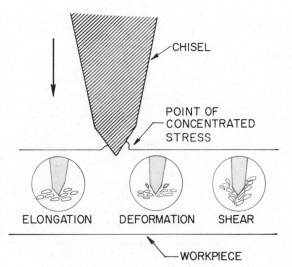

Fig. 5-5. Wedge action applies great energy along a narrow zone.

Fig. 5-6.   The cutting action of the lathe is made up of velocity, pressure, and direction.

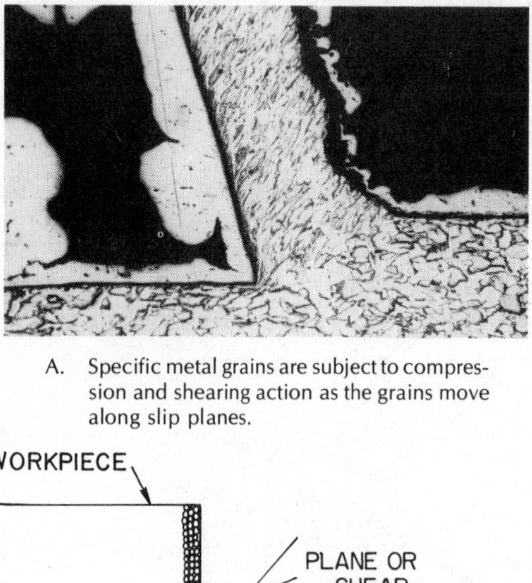

A.   Specific metal grains are subject to compression and shearing action as the grains move along slip planes.

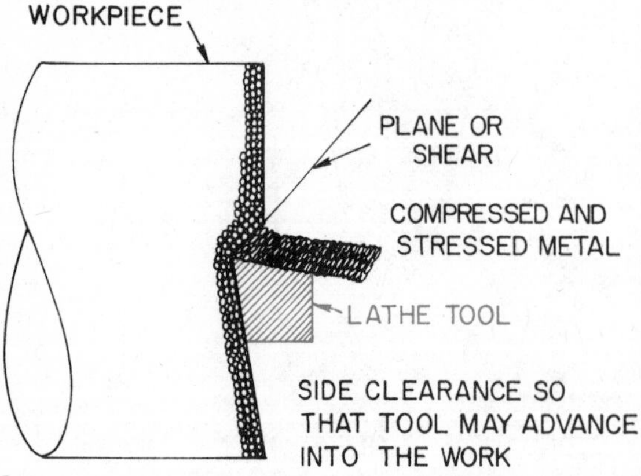

WORKPIECE

PLANE OR SHEAR

COMPRESSED AND STRESSED METAL

LATHE TOOL

SIDE CLEARANCE SO THAT TOOL MAY ADVANCE INTO THE WORK

B.   The metal grains are compressed, sheared from base metal, and carried away as a thickened chip.

Fig. 5-7.   Metal is composed of layers of grains.

and **direction** by the feed of the machine, Fig. 5-6.

These three cutting actions combine with the correct tool geometry to place metal deforming forces into such a position that a chip is formed. Scraping forms chips by deforming the metal by developing a shear plane that compresses and separates the chip.

Metal is made up of innumerable grains. Pressure applied to one grain or a layer of grains is passed on continuously to other individual grains. Pressure created by the wedging action of the cutting tool passes from grain to grain of the metal. Because pressure is not applied equally to all surfaces, specific grains are subjected to compression and shearing action, Fig. 5-7. Grains move along their slip planes, and the metal is compressed, sheared, and carried away in the form of a thickened chip.

Cutting tool chips have been classified into three types of chips: **Type I**, a discontinuous or segmented chip; **Type II**, a continuous chip; and **Type III**, a continuous chip with a built-up edge, Fig. 5-8.

Type I — Discontinuous Chips. A discontinuous chip is segmented and highly stressed, Fig. 5-9. It is the chip produced by brittle metals such as cast iron and hard bronzes. Brittle materials fracture ahead of the cutting edge and break into short chips. When ductile materials such as a mild steel produce a discontinuous chip, this indicates that the cutting speed is too low, the feed is too large, or the rake angles are too small for the machining condition. The results of segmented chips on ductile materials are poor surface finishes, high tool wear, and high cutting temperatures, all making for less efficiency in cutting.

Type II — Continuous Chips. Continuous chips flow from the cutting edge because of continuous metal deformation taking place ahead of the cutting edge, Fig. 5-10. A ribbon of chips is produced because a great number of slip planes are being utilized in the deformation. Continuous chips are formed in ductile materials at high cutting speeds. Efficient cutting is associated

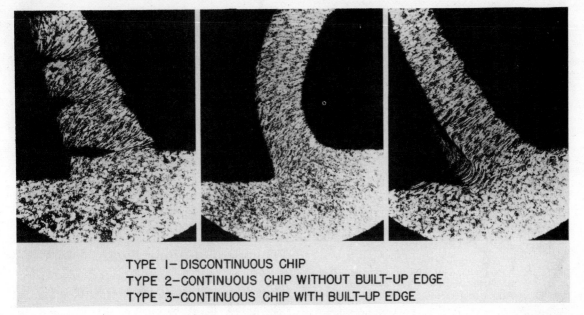

TYPE 1— DISCONTINUOUS CHIP
TYPE 2— CONTINUOUS CHIP WITHOUT BUILT–UP EDGE
TYPE 3— CONTINUOUS CHIP WITH BUILT–UP EDGE

Fig. 5-8.   Classification of cutting tool chips.

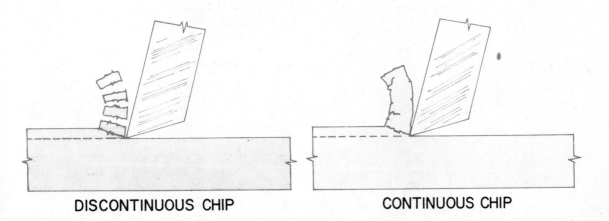

DISCONTINUOUS CHIP

CONTINUOUS CHIP

Fig. 5-9.   Discontinuous chips are not efficient in ductile materials.

Fig. 5-10.   Continuous chips are efficient, but they may produce a dangerous snarl.

with this type of chip generation. The result is low friction between the chip and the tool, a good surface finish, and low heat generation which extends tool life.

A continuous chip makes a snarl and thus makes it difficult to remove the chips from the machine, therefore it is frequent practice to add a chip breaker to the cutting tool to further stress and break the chips into short

lengths. The continuous chip is considered the most efficient cutting action.

Type III — Continuous Chip with a Built-Up Edge. When tools made of high-speed steel machine mild steel dry, frequently there is a metal build-up on and behind the cutting edge. This is caused by the chip material flowing across the face of the tool and partially sticking to the cutting edge and

175

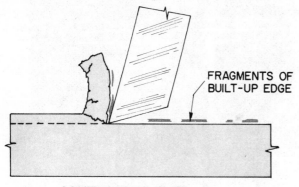

CONTINUOUS CHIP
WITH BUILT-UP EDGE

Fig. 5-11. A continuous chip with a built-up edge is constantly breaking down, leaving material on the machined surface.

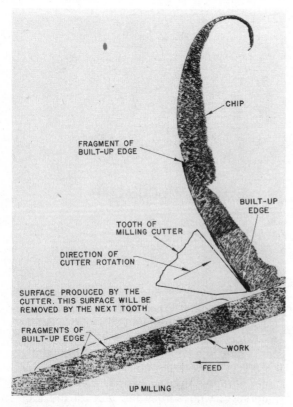

Fig. 5-12. A milling cutter with a built-up edge.

surface. For the given material, if the speeds, feeds, and rake angles are not ideal, build-up occurs, Fig. 5-11. The built-up edge is constantly welding metal to the cutting edge and breaking down, causing excessive friction, heat, and a rough surface finish. A built-up edge on a cutting tool indicates that the tool is not cutting as efficiently as possible, Fig. 5-12. The cutting may be improved by speeding up the machine, producing thinner chips, increasing the rake angle, and applying a cutting fluid, Fig. 5-13.

Theory of Cutting Edges

The cutting edge performs the work. Therefore, the most efficient form must be designed to provide the greatest cutting efficiency. The most fundamental forms are these:

- Pointed edge
- Straight edge
- Shaped edge

Pointed Edges

Pointed edges, Fig. 5-14, are used on such tools as scribers, handsaws, circular saws, rasps, and grinding wheels. When hard or tough materials are cut, the pointed edge does not stand up well unless the cutting process is very slow.

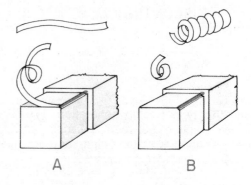

A   B

Fig. 5-13. Continuous chips will produce a snarl (A). Continuous chips with a chip breaker stress the chip and a BX cable chip or a No. 6 shaper chip is formed.

A cutting edge must have a mass of metal behind it to provide the metal with rigidity and to conduct the heat of cutting away from the cutting surface. When a pointed cutting surface is used, the cutting edge must be able to withstand a large amount of heat without becoming dull.

A piece of grit that is vitrified (made glassy by burning) into a grinding wheel has a multitude of single-pointed cutting edges presented to the workpiece, Fig. 5-15. Each working piece of grit cuts a chip from the product being shaped. When grit becomes dull from wear, heat, or both, the pressure and heat are increased on the grit. Finally, the pressure becomes so great that the glass posts holding the grit in the wheel are broken and the grit falls away. When the dull grit point falls out of the wheel, a new grit is uncovered, and the point cutting action starts anew. Pointed cutting edges are not used on hard materials unless they are self-renewing, easily replaced, or reconditioned.

Straight Edges

Straight cutting edges are used on milling cutters, files, chisels, drills, countersinks, and some tapered reamers, Fig. 5-16. The straight cutting edge is formed when two flat surfaces intersect. The intersection forms the cutting edge, as with a cold chisel. This cutting edge may be designed with sufficient metal to support the cutting edge and thus becomes an efficient cutting tool.

The largest group of cutting tools used falls into the straight cutting edge classification. Cutting edges of many complicated cutters are simply multiple applications of the straight-edged cutter.

The straight-edged cutter is the easiest type of cutter to resharpen and to keep in condition. Regrinding does not change the shape of the cutting tool, and straight-edged cutters may be strengthened by the addition of primary and secondary relief, Fig. 5-17.

Fig. 5-15.   A grinding wheel is made up of a multitude of single-pointed cutting edges.

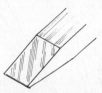

Fig. 5-16.   A straight cutting edge is the easiest cutting edge to make and to keep sharp. Thus, it is widely used on all types of cutters.

Fig. 5-17.   Regrinding does not change the shape of the cutter. It provides clearance for the cutter with a primary and secondary relief ground surface.

Fig. 5-14.   Pointed edges are generally used for cutting soft material such as wood, plastic, and leather.

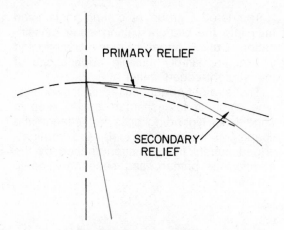

Fig. 5-18. A narrow primary relief keeps the bulk of the metal close to the cutting edge to give it support and to provide heat dissipation.

Fig. 5-20. Form cutters cut many curved shapes (such as gear teeth), as well as irregular curved shapes.

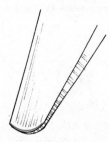

Fig. 5-19. When two formed surfaces intersect, they form a curved line. This makes a curved or shaped cutting edge.

The milling cutter or other tools may be designed so the bulk of the metal may be kept at the cutting edge by the primary relief, thus keeping the cutting edge cool, strong, and rigid, Fig. 5-18.

Shaped Edges

Shaped cutting edges are used on such tools as convex and concave milling cutters, spiral milling cutters, various shaped lathes, planer tools, and various chisels. The shaped cutting edge is formed when two surfaces intersect to form a curved line, Fig. 5-19.

Special cutting tools, called form cutters, have many special shapes. Two examples are radius tools and gear cutters. These tools are used to form the many shapes which are made up of curves, arcs, angles, and straight lines, Fig. 5-20. The cutters used for making the curved faces of a gear tooth are examples of **shaped-edge tools.**

Cutting by Hacksawing

In **hand hacksawing** metal is parted by a blade that cuts with a series of individual cutting teeth. The tool is very useful because of its flexibility and ease of operation. It is light and can be used to cut in almost any position. The hand hacksaw has a hardened steel blade which will cut all metals except hardened steels. It has a frame to hold the blade. The frame is frequently adjustable so that 8", 10", or 12" blades may be fitted into the frame. The saw has a handle that provides a comfortable position

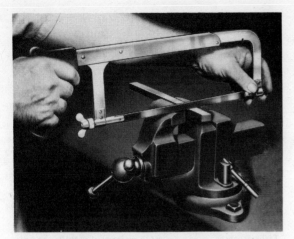

Fig. 5-21. The hand hacksaw may be guided to produce an accurate cut in all metals except hardened metals.

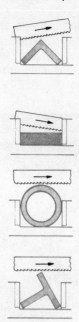

Fig. 5-22. Hand hacksawing: Thin materials require light pressures and a greater number of teeth in contact with the work. Heavy materials require a fewer number of teeth per inch.

for the hand so that the saw may be powered and guided in an accurate cut, Fig. 5-21. A wing nut and screw are attached to the frame and blade so that the blade may be tightened in the frame.

The blade for the hacksaw is selected in relation to the thickness of the metal to be cut. If the metal to be cut has a thin cross section, the number of teeth in the selected saw blade will be high, such as 32 teeth per inch. When the metal to be sawed is thick, the number of teeth in the selected saw blade will be low, such as 18 teeth per inch. Correct blade selection is important, because if too few teeth on a saw engage the material to be cut, the teeth on the saw will straddle the metal, and the teeth will be torn from the blade.

Use of the Hand Hacksaw

The hacksaw blade should be selected so that at least 2½ to 3 teeth of the saw blade will be cutting at all times, Fig. 5-22. The work to be sawed should be held rigidly in a vise. The cut should be made close to the vise to prevent vibration of the workpiece. For accurate work a file is used to

make a starting notch at the layout line. The saw is started in the notch so that the blade will not slip and mar the surrounding work.

The cutting is done when the saw is pushed away from the operator. Therefore, the cut is started with short strokes near the front of the saw until the cut is established. When the **kerf,** or cut, is located, the cut is continued by using long but slow strokes. The full length of the blade should be used. The speed of sawing varies with the hardness of the metal cut, but it should be in a range of 40 to 60 strokes per minute. The harder the metal, the fewer strokes there will be per minute. A slight pressure should be applied to the frame and blade on the cutting or forward stroke, and the pressure should be lifted upon the return stroke of the saw.

The height of the vise and the position of the operator's body should be such that the operator is comfortable and will not twist or cramp the blade. When the operator is tired or is talking to another person, it is wise to remove the saw from the cut and

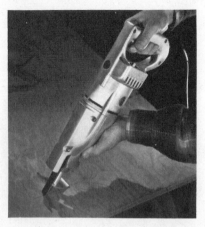

Fig. 5-23. The bayonet saw provides power and flexibility for short cuts, irregular cuts, out-of-position cuts, and cuts that are difficult to reach.

Fig. 5-24. Aluminum extrusions, sheet steel, pipe, bar, and round steel may be cut with a bayonet saw.

rest. This will save breakage of the blade and time lost in restarting a cut.

Bayonet Sawing

The bayonet saw is used to cut wood, plastics, or metal. Therefore, it is really a general purpose saw, Fig. 5-23. The bayonet saw is a hand tool that has been redesigned. Electrical power has been added to it to provide the sawing force.

Aluminum extrusions, galvanized sheet metal, pipe slitting, flat or bar steel, and round steel may be cut with a bayonet saw, Fig. 5-24. A metal cutting blade is required. This blade is made of hacksaw quality stock with metal cutting teeth that vary from 10 to 24 teeth per inch. The 24-tooth metal cutting blade is used for cutting all types of tubing and extrusions. Sheet metal is often hard to cut, because it bends very easily and is likely to tear. The bending and tearing may be controlled by placing hardboard or plywood scraps on both sides of the work and cutting the wood and metal together as a "sandwich."

Sawing heavy metals such as solid bar stock, rod, or thick sheet is made easier by applying some stick wax to the blade before starting the cut. Additional wax may be applied to the blade for lubrication when it is needed.

Soft metals may clog in the teeth of the saw. If this occurs, it is necessary to change to a blade with coarser teeth or a more open tooth blade. Chips may be removed from the saw blade with a stiff brush.

When a bayonet saw is used, the work should be clamped to a table top or sawhorse or put in a vise. Check to see that the saw will not extend down and cut other material such as the holding device or electrical cord. Hold the saw solidly against the surface and gently push the saw into the work. Do not force the blade. The saw will do the work. When finishing a cut with the saw, support the saw and waste material while the final completion of the cut is made. Allow the saw to stop running completely before the saw is placed in its holder.

Cutting By Power Sawing

The cutting of metal can also be performed by power tools. Correct operating procedures should be learned before such saws are used.

Reciprocating Power Hacksaws

A power hacksaw is a machine designed to cut standard stock into a workable size quickly and accurately. A saw found frequently in metal manufacturing is the reciprocating power-driven hacksaw. The blade is power-driven and is moved back and forth through the workpiece. The machine removes the pressure on the return stroke. This action is generally provided by a cam, gravity, or a spring and ratchet mechanism. Some saws have a hydraulic drive which provides great control of pressure for fragile workpieces, Fig. 5-25. Frequently, small diameter stock is cut in multiple-banded pieces, Fig. 5-26.

The power hacksaw must be completely stopped when tension adjustments are made, when the blade is changed, or when work to be cut is placed in the vise. The work should be held by the full width of the vise jaw. Work that is too short to be held this way should be cut by another method. The vise is tightened on the work securely so that the work will not shift or roll in the vise during cutting, Fig. 5-27.

Fig. 5-26. An overhead clamp is used to cut off bar stock in bundles.

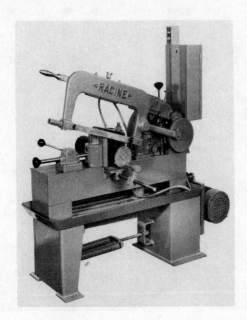

Fig. 5-25. A power saw cuts standard stock into workable lengths. The saw is fast and provides length control for rough machining.

Fig. 5-27. A single, short piece must be held tightly in a vise. The piece must be at least as long as the vise jaw.

Long or heavy work that is to be cut needs to be supported on both sides of the saw. When heavy or long stock is cut, it may shift, roll, or fall when the cut is completed, causing an accident or injury. A rag or other means is used to stop the cutting fluid from running down the inside of long pipe or the surface of flat stock onto the floor. The cutting fluid should be channeled back into the sump of the saw rather than letting it run onto the floor.

If long stock projects into a walkway or aisle, a red flag should be placed on the stock for a warning. This is especially necessary in a sawing area because of the intermittent use of the aisle when long stock is sawed.

Once a power hacksaw is started, no adjustments or attempts to oil the machine should be made, because the operator might injure his or her hands and arms. The removal of the metal from the saw must be done carefully to avoid the sharp burr at the saw cut. The burr should be filed as soon as possible in order to remove the hazard.

Circular Sawing

Manufacturing frequently requires accuracy and excellent surface finish on power sawing of heavy stock. To meet these requirements, a **cold saw** is used. A cold saw is a cut-off saw which uses the circular saw designed like a milling cutter. It is designed for cutting cold metal.

The cold saw provides a fast, continuous, smooth, and accurate means of cutting surfaces, Fig. 5-28. It is capable of holding a length tolerance of 0.003″ and a milling machine finish with no burrs on the end of the stock. The cut stock will be square and the proper length for subsequent machining operations.

Fig. 5-29. Band cutoff saws are capable of accurate, thin cuts on heavy stock. These disks are 8″ in diameter and are .010″ thick.

Fig. 5-28. Cold sawing produces a cut with quality surface and length.

Band Sawing

In cutoff band sawing, a continuous band welded into an endless cutting tool is used. The blade runs around a set of wheels which provide the support and power to the blade. The wheels are set at an angle to the work so that the returning side of the band will clear the work. Guides twist the blade and hold the band in alignment usually 90° to the stock so that an accurate cut is produced.

The **band saw** is thin, so it does not waste large amounts of stock when many cuts are required, Fig. 5-29. The speed of the saw may be controlled to suit the hardness and type of material being cut. The cutting action is continuous. Each tooth cuts only for a short period of time and then cools as it moves around the band track. For this reason, greater cutting speeds may be realized.

The feed pressure of the saw is frequently hydraulic, and this gives accurate control to the cutting action. With the application of hydraulic accessories, the movement and location of the stock and cutting may be done automatically.

Angular and bundled stock cutting may be done on production cutoff band saws so that many workpieces may be cut at once.

The **contour band saw** is a tool with a continuous band of metal cutting teeth along one edge. It is a faster cutting saw than the reciprocating saws, because it does not have to waste time as the saw moves back to the starting position; it cuts continuously. The saw is not restricted to a straight line, it has little limitation as to angle, direction, or length of cut. The saw is capable of cutting accurate straight lines, radius cuttings, and irregular shaped pieces, Fig. 5-30. It can do contour cutting, angular cutting, notching, slotting, slitting, and three-dimensional shaping.

Contour band sawing cuts out shapes with only one piece of waste material. Previously the shape would have been laboriously machined to shape with end milling or other machining processes.

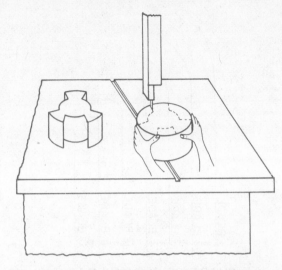

Fig. 5-30. The contour-cutting band saw has little limitation as to the angle, contour, or length of cut.

The operation of the metal cutting band saw requires an understanding of basic procedures such as selection of the correct blade width, the blade speed, setting of the post, and the holding and feeding of the work.

Selecting the Blade. The saw width that is selected for a job is controlled by the smallest radius or arc that the saw will have to cut. A narrow blade will cut a short radius. A ¼″ saw will cut a ⅝″ radius of an arc, but no smaller. If the saw is forced around a shorter radius, the saw will bind in the cut and produce wear on the saw and saw guides and will produce poor work. The correct saw blade for a given radius may usually be selected from a chart supplied with the contour saw.

If too narrow a saw is used for straight-line cutting, the saw may wander or weave about the cutting layout line. For most efficient cutting, it is wise to select a saw width that is as wide as possible and still designed to cut the radius that is on the workpiece, Fig. 5-31.

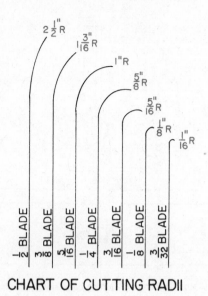

CHART OF CUTTING RADII

Fig. 5-31. Select the widest blade that is designed to cut the radius size. .

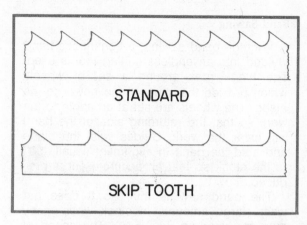

Fig. 5-32. Standard tooth shape is used when fine or accurate cuts are made in steel. Skiptooth saw blades are used on soft materials that require large chip clearance.

The selection of the types of tooth construction of the band saw is related to the finish and accuracy of the cut and the material. The two most common saw blade constructions are precision (standard tooth) and buttress (skip tooth), Fig. 5-32.

The standard saw blade is used where fine or accurate cuts are to be made. In many cases this cut may be the workpiece's finished cut. The skip tooth saw blade has a tooth missing between the saw teeth. This blade has a large chip clearance and thus has an advantage in cutting soft metals and/or thick workpieces. The skip tooth saw produces a rougher cut, but it is accurate for further machining or welding.

The Blade Speed. Band saw blades are made of carbon steel, high-speed steel, and bands with carbide inserts. The most common band saw blade used is the carbon steel blade.

The carbon steel blade has a maximum speed. If the steel were to exceed that maximum, the heat produced would destroy the blade. Heat draws the hardness from the blade, thus the blade would become soft

and would soon dull. For this reason the speed of the blade, given in feet per minute, is a controlling factor.

Materials of different thicknesses and compositions have different cutting blade speeds, Fig. 5-33. The saw velocity must be within the range of the work, or the saw blade will quickly heat and destroy the cutting edge. The feet per minute required for the saw velocity for the various thicknesses and materials is found on a chart supplied with the machine tool.

The Setting of the Post. The post is the support for the upper saw guide and needs to be adjusted to provide clearance for the work and the saw guide. This guide provides support for the back edge of the saw and should be adjusted so that there is not more than 1/8" clearance between the workpiece and the upper saw guide. If too little clearance is given, the work will jam and create a hazardous condition. If the clearance is too great, inaccurate work will be the result.

The contour band saw is designed to cut with the blade traveling down, thus placing the cutting load directly on the machine table. Complicated holding fixtures are not needed to hold the work, so greater cut-

Cutting Speeds for Band Sawing
for a ½" Blade (Dry Cutting)

Metal	Speed
High-Carbon Steel (C1090)	75 ft./min.
Medium-Carbon Steel (C1040)	100 ft./min.
Cast Iron	125 ft./min.
Alloy Steel (C4140)	125 ft./min.
Low-Carbon Steel (C120)	175 ft./min.
Brass and Bronze (hard)	200 ft./min.
Copper	250 ft./min.
Aluminum	250 ft./min.
Brass (soft)	500 ft./min.

Fig. 5-33. Cutting speeds for band sawing for a ½" blade (dry cutting).

Fig. 5-34. A push stick keeps fingers away from the saw blade when cuts are being finished.

ting freedom is gained. A great deal of work may be hand-held on the table and guided to accurately follow a curved or straight layout line. Work that is not flat requires some holding or feeding fixture, such as a hand screw, vise, or specially designed tool.

At all times parts of the body should be kept away from the saw blade. Push sticks should be used to protect the fingers, especially when finishing cuts, Fig. 5-34. The wooden push stick may be easily replaced if it is cut. A fixture may hold the workpiece on a contour saw to ensure safe operation, Fig. 5-35.

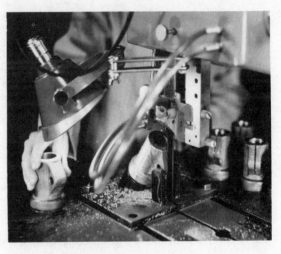

Fig. 5-35. The work is held at an angle to ensure safe cutting action.

Surface Cutting

Surface cutting of metal can be performed in these ways:
1. Filing
2. Grinding
3. Shaping
4. Planing
5. Milling
6. Turning

Filing

The production of an accurate square surface by filing requires the development of considerable hand skill. Hand filing is useful for removing burrs on machined edges, for fitting small parts, and for some prototype part development.

Hand files come in a large assortment of shapes, sizes, and cuts. They are designed to cut the workpiece in order to produce the needed part. The file has a cutting action equivalent to the band saw tooth, milling cutter tooth, and other metal cutting edges. However, the file has a very small amount of chip clearance, Fig. 5-36.

The file must be given special care to prevent **pinning**, which is the lodging of metal chips in the gullet of the file. To prevent this, the proper cut and file size for a given metal should be selected and the face of the file should be covered with chalk. The chalk absorbs any oil on the surface of the file and also fills the rough tool marks in the gullet. This action makes it easier for the chips to fall out of the gullets of the file tooth when the file is rapped on the bench to clear the chips from the file.

There are two primary cuts of files: (1) **single-cut,** which are used on smooth work and finishing work and (2) **double-cut,**

Fig. 5-36. A file has a cutting action similar to other cutting tools, but it has a limited amount of chip clearance.

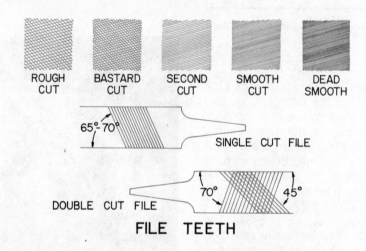

ROUGH CUT   BASTARD CUT   SECOND CUT   SMOOTH CUT   DEAD SMOOTH

65°-70°   SINGLE CUT FILE

DOUBLE CUT FILE   70°   45°

FILE TEETH

Fig. 5-37. The single-cut file is for smooth finish work, while the double-cut file will cut rapidly and will rough-cut work.

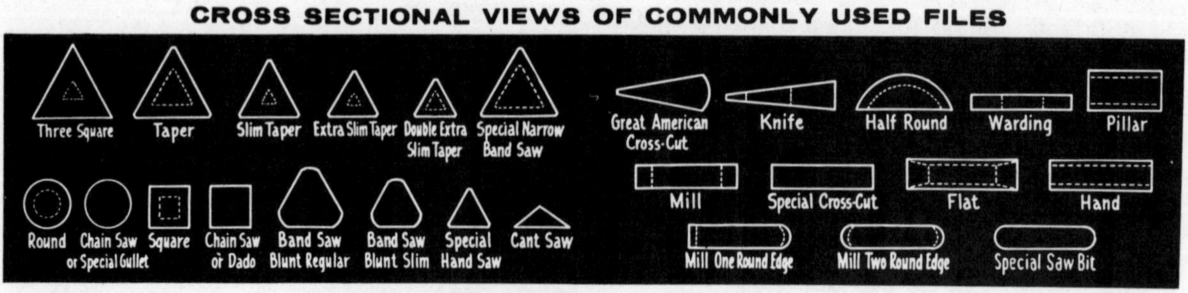

CROSS SECTIONAL VIEWS OF COMMONLY USED FILES

Three Square   Taper   Slim Taper   Extra Slim Taper   Double Extra Slim Taper   Special Narrow Band Saw   Great American Cross-Cut   Knife   Half Round   Warding   Pillar

Round or Special Gullet   Chain Saw   Square   Chain Saw or Dado   Band Saw Blunt Regular   Band Saw Blunt Slim   Special Hand Saw   Cant Saw   Mill   Special Cross-Cut   Flat   Hand

Mill One Round Edge   Mill Two Round Edge   Special Saw Bit

Fig. 5-38. File shapes are designed to fit the kind of job such as a slot, curve, or flat surface. Files are designed for filing in lathes as well as for bench work.

which are used on roughing and fast metal removal, Fig. 5-37. The double-cut file is used to bring the workpiece to rough size.

Files have different tooth spacing and are classified in these ways:

- Rough
- Coarse
- Bastard
- Second-cut
- Smooth
- Dead smooth

The rough file has the widest tooth spacing, and the dead smooth has the closest tooth spacing.

The length of the file is related to the tooth spacing. The tooth spacing also varies with the length of the file. The common lengths of files are 6″, 8″, 10″, 12″, and 14″. For example, a 10″ bastard-cut file has a tooth spacing that is close to the tooth spacing 12″ on a second-cut file. The combination of file length and the degree of tooth coarseness makes it possible to select an exact file for a specific job.

Files come in a number of shapes to fit the filing job requirement, Fig. 5-38.

Hand Filing. Cross filing is used to remove material rapidly and to square the ends of small stock. It is done by filing across the metal, Fig. 5-39. The metal part is held in a vise and filed. The file is held at a 60° angle and moved parallel to the edge while traveling across the surface. The work is removed and checked for square and/or level, and the high spots are marked with a lead pencil. The work is placed back in the vise in the same position. The filing position is rotated so that the file cutting marks are across the previous filing marks. The cutting is concentrated upon the pencil-marked high spots. The process is repeated until the surface is square and/or level.

Draw filing is a filing operation used for finishing. It is done by using the long-tooth shearing angle which is produced by the file when it is held in a different position, Fig. 5-40. The file is pushed and pulled over the surface to be finished, and a fine long chip is removed. This long-angle contact produces a fine surface finish on the metal.

Disc and Belt Grinding

Disc grinding is a method for cutting a surface on metal. Disc grinders utilize an abrasive disc mounted on a flat rotating surface. The work is held flat upon a table and moved across against the rotating abrasive disc, and a flat surface is cut. The disc grinder speeds up many filing and sur-

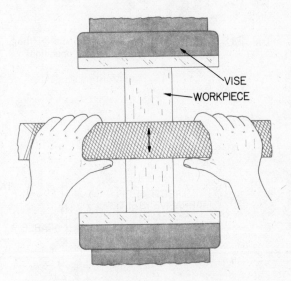

Fig. 5-40.   Draw filing will produce a fine finish on a flat surface. The cutting action parallels the elongation of the grains of metal.

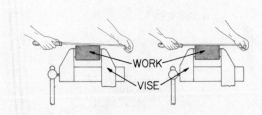

Fig. 5-39.   Cross filing. The right hand of the filer is adjusted up or down until the file cuts the penciled high marks on the surface being squared.

face cleaning operations, as well as the finishing operation, Fig. 5-41.

The disc grinder should be used only on the half of the disc that is moving down toward the table. The grinding forces will then be transferred directly down upon the table. Goggles should be worn by the operator when the disc grinder is used.

Belt grinding and polishing machines use continuous abrasive belts which cut the metal. Belt grinders fall into two general classes: (1) vertical belt design and (2) horizontal belt design, Fig. 5-42. These endless belt grinding machines may be built with a platen for flat and edge work.

The **contact wheel** makes possible grinding round and flat work and work with a large arc surface, Fig. 5-43.

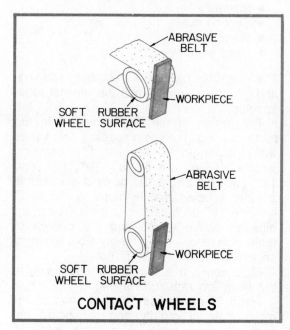

CONTACT WHEELS

Fig. 5-41. The disc grinder speeds up many filing, surface-cleaning, and finishing operations.

Fig. 5-43. The contact wheel will grind or cut away imperfections from the surface of round, flat, or curved surfaces.

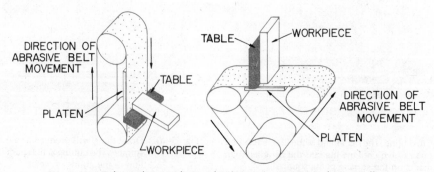

Fig. 5-42. Belt grinders are designed to be used vertically or horizontally.

The **formed wheel grinder** is a modification of the contact wheel grinder and has additional formed rolls which shape the abrasive belt to the finished shape of the surface, Fig. 5-44.

The **flexible belt** is a grinder adapted to curved and irregular surface grinding, Fig. 5-45. It is used for grinding radii and arcs on small work.

Shaping and Planing

Shaping and planing are metal separating processes that are intended primarily for the development of flat surfaces. Most manufactured products, machine tools, or manufacturing machine tools have accurate flat surfaces. The shaper and planer are methods of obtaining these flat and angular surfaces, Fig. 5-46.

The **shaper** is smaller than the planer and has a different cutting movement. The shaper tool moves over the workpiece, which is supported by the ram. The ram cuts on the forward stroke and the table and workpiece move over (feed) on the return stroke so that another chip may be removed, Fig. 5-47. The surfaces shaped are horizontal, vertical, angular, and curved.

The work done by the shaper is controlled by the position of the work against the solid

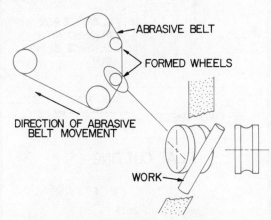

DETAIL OF FORMED WHEEL

Fig. 5-44. A formed wheel belt grinder may be designed to finish specific radii and rounded products being manufactured.

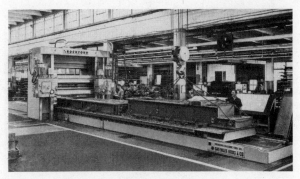

Fig. 5-46. Planers will cut large, flat surfaces such as these die casting machine bases.

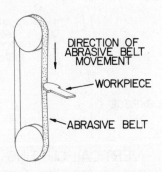

Fig. 5-45. A flexible belt is used to finish small radii and irregularly curved work.

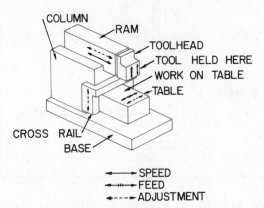

Fig. 5-47. The shaper is used on prototypes and small work.

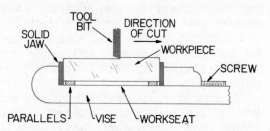

Fig. 5-48. The work done in a shaper vise is positioned against the solid jaw and the work seat. The work is raised to the top of the vise by the use of parallels.

vise jaw and work seat of the shaper vise and the plane which is cut by the tool bit, Fig. 5-48.

The work seat must be parallel with the plane generated by the tool bit. Also, the solid jaw of the vise must be perpendicular

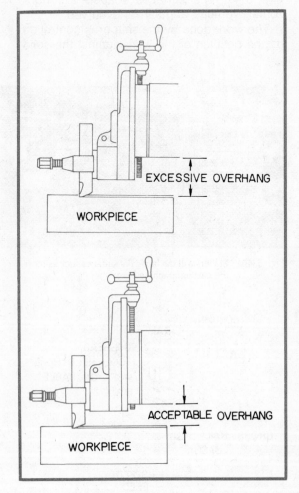

Fig. 5-49. The tool slide is retracted until the bottom of the head is even. The work and table are raised to produce the proper height for machining.

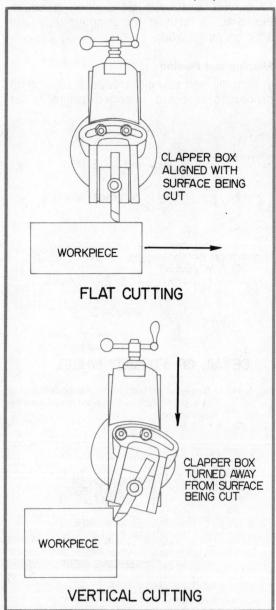

FLAT CUTTING

VERTICAL CUTTING

Fig. 5-50. The correct angle of the clapper box allows the tool to clear the workpiece on the return stroke.

to the tool bit plane in order to do square and accurate work. The surfaces of the vise are aligned and squared with the tool bit before the work starts.

The work is set up on **parallels,** which rest on the vise work seat. The vise is tightened with a square or assumed square surface against the solid jaw. When the work is a rough casting, a piece of soft aluminum sheet metal is placed between the solid jaw and the casting so that the accuracy of the jaw will be preserved. The high points of the casting will press into the aluminum and provide greater holding power. The casting is tightened in the vise and rapped down upon the parallels with a lead hammer. The workpiece is aligned with the work seat of the vise and the solid jaw.

The **tool slide** should be retracted until the bottom of the head is even. The tool bit and tool holder are mounted in the tool post, and the work and table are raised until they have minimum clearance, Fig. 5-49.

For flat cutting, the head should be set perpendicular to the surface of the cut and the clapper box parallel to the tool slide. When angular cuts are made, the head will be swiveled to the desired angle. The clapper box will be turned **away** from the surface being cut, Fig. 5-50. The length of the

stroke for the shaper is adjusted so that the tool bit goes beyond the workpiece ¼" on the return stroke and ½" to ¾" on the cutting stroke, Fig. 5-51.

The speed of the ram, or cutting speed, is adjusted by selecting the number of strokes per minute. The number of strokes per minute may be calculated by this formula:

$$\text{Strokes per Minute} = \frac{\text{Cutting Speed of the Material} \times 7}{\text{Length of the Stroke (in inches)}}$$

The cutting speed of the material will vary, Fig. 5-51A.

Fig. 5-51A.
Shaping and Planing
(Cutting Speed of Material
in Surface Feet per Minute)

Material	High-Speed Tool	Carbide Tool
Cast Iron	60	100
Machine Steel	80	150
Tool Steel	50	150
Brass	160	300
Aluminum	200	400

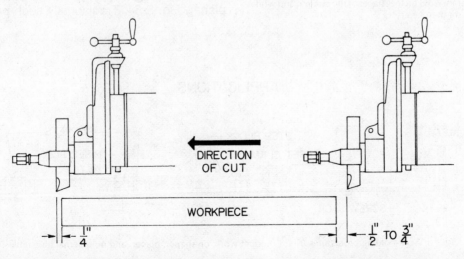

DIRECTION OF CUT

WORKPIECE

¼"   ½" TO ¾"

Fig. 5-51.   The length of stroke should be adjusted so that the operator does not lose machining time while waiting for the tool to contact the work.

Shaper and Planer Safety

CAUTION: When work is set up on the shaper or planer table, the cross rail or tool heads should be adjusted so that the vise or work does not strike the ram slides, cross rail, column, or work on the first return stroke of the shaper or planer. If the machine is set for angular cuts and the tool slide becomes overextended, a hazardous situation is created, Fig. 5-52. The operator or other persons must not stand in front of the shaper when it is running. When the cutting tool finishes its cut, the chip is shot forward with great force. The operator should work from the **side** of the machine. Before the machine is started, the vise and all clamping bolts must be securely fastened, Fig. 5-53. Wrenches should be returned to their racks so they won't fall into the machine or work when the machine is in operation. Chips and scales should be removed from the machine with a brush only when the machine has stopped. Burrs and sharp edges should be carefully removed from the workpiece with a file so that painful cuts will be avoided. Eye protection is always necessary around machine tools, and a face shield offers added protection from flying chips.

The **planer,** a massive machine with a heavy table, is capable of holding very large castings. The planer is a production tool that is used primarily to build other very large machinery. Some planers can accommodate workpieces up to 40 feet long.

The cutting movement of the planer is produced when the table moves back and forth under a tool held stationary by cross rail and tool head. Planers generally use single-pointed tools. However, as many as three or four tool stations may be cutting at one time on the planer as the table (with the work bolted to it) travels by the cutting tools. Most large, heavy castings receive their first roughing and truing cuts on a planer, Fig. 5-54. Traditionally such parts as

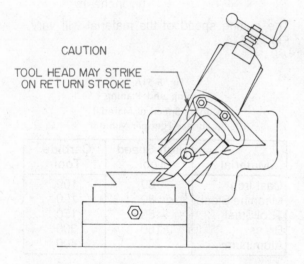

CAUTION

TOOL HEAD MAY STRIKE
ON RETURN STROKE

Fig. 5-52. The tool head on the shaper or planer should be adjusted so that it will not strike the end of the ram ways on the shaper or the column and work on the planer.

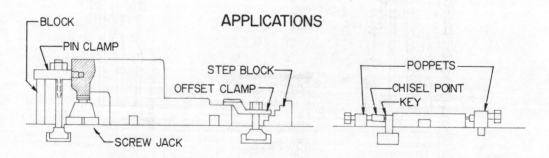

APPLICATIONS

Fig. 5-53. Work-holding procedures used to mount work on shaper, planer, and milling machine tables.

machine tables, beds, columns, and rams are machined on the planer. Surfaces which are flat or angular over long distances are also machined on it, Fig. 5-55.

The practice of setting many duplicate pieces in rows on the planer table and machining them all in the same stroke makes it possible to economically machine smaller parts on a planer. This type of planer work is called **string planing.**

Planers have been specialized for different kinds of work and classified according to their main structure: open-sided, double-housing, plate, and pit planers.

The **open-sided planer** has a housing on only one side of the bed, Fig. 5-56. This allows extra wide work to extend over the side of the bed. Because one side of the planer is open, it is somewhat easier and faster to load and set up the work.

The **double-housing planer** is of massive construction with two heavy columns and cross rail so that a number of heavy cuts may be made simultaneously. The double-housing planer is sometimes redesigned into the planer-type milling machine. In this case milling heads are mounted on the columns and the cross rail. Heavy cuts are made as the planer table moves very slowly, Fig. 5-57. The milling cut is made in one slow stroke.

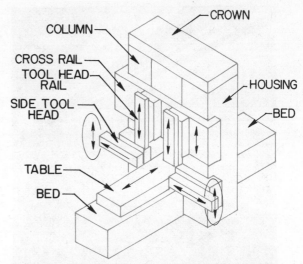

Fig. 5-55. Flat, angular parts are machined on the planer.

Fig. 5-56. The open-sided planer allows wide work to extend beyond the table for special setups.

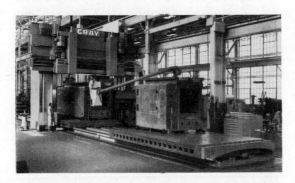

Fig. 5-54. The planer is a massive machine with a table designed to hold large, heavy castings. It is frequently used to machine the bases of machine tools.

Fig. 5-57. A planer-type milling machine will take heavy cuts as the planer table moves very slowly.

Fig. 5-58. A plate planer prepares the edge of a plate for welding or boiler construction.

Fig. 5-59. The milling machine employs a multitooth rotary cutter to rapidly bring metal to shape by the removal of chips.

The **plate planer** is a lighter specialized planer. It is designed for squaring or beveling heavy plate for boiler work and for preparing the plates for welding, Fig. 5-58.

A **pit planer** is heavier in construction than the double-housing planer. The pit planer is used when the workpiece is too heavy or bulky for the standard machines. The table of the planer is recessed into the floor as a stationary table. The columns carrying the cross rail and tool heads ride along on **ways** (guiding surfaces) on each side of the table. The clapper boxes holding the tools are designed so that they will allow planing in both directions.

When heavy work is to be moved, a hoist or crane will be needed so that an accident will not occur.

Planers may perform the same basic operations as shapers and milling machines except that planers will work large, heavy, bulky castings and weldments. The work is supported by a very strong table which makes it possible to take very heavy roughing cuts. The planer offers more machining capacity at the same cost as a milling machine. The cost of tool sharpening is less for the same amount of production in nearly the same time as the mill.

Milling

The **milling machine** is designed to use a multitooth cutter which has the workpiece fed against it to produce the desired cut, Fig. 5-59. With the application of a multitooth rotary cutter, these additional machining operations are made possible: slab milling, face milling, gear cutting, keyway cutting, end milling, side milling, and form milling. The heavy-duty cutters frequently make use of inserted tooth cutters, Fig. 5-60.

Most of these milling cuts may be produced by more than one method, depending on the kind of milling machine available, the cutter used, the shape of the work, and the position of the cut.

The work is mounted on the table of the milling machine and is moved into the cutter by a power feed. The cutter is mounted on necessary around machine tools, and a face an arbor (a shaft which holds the cutter), which is turned by the spindle. The milling machine will usually have power feeds on three axes: longitudinal, vertical, and transverse table feeds, Fig. 5-61.

A number of types of milling machines are built to meet the special needs of manufac-turing. There are two basic types of milling machines: (1) the **bed type** and (2) the **knee-and-column** type. The bed type is designed for mass production, Fig. 5-62. This machine will manufacture small parts such as typewriter, sewing machine, and automotive parts. The table cannot be raised or lowered and has only a longitudinal feed,

Fig. 5-60.   Carbide-inserted tooth cutters of the throw-away inserted type.

Fig. 5-61.   Dial-type universal milling machine.

Fig. 5-62.   Bed-type milling machine with hydraulically operated tooling and clamping.

Fig. 5-63. The fixed bed-type milling machine is designed for general work, but the rigid bed allows for heavy production cuts.

Figs. 5-63 and 5-64. The spindle is located in an adjustable head, making it possible for the depth of the cut to be set. The only transverse adjustment is by the position of

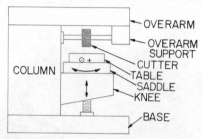

OVERARM
OVERARM SUPPORT
CUTTER
COLUMN
TABLE
SADDLE
KNEE
BASE

KNEE-AND-COLUMN HORIZONTAL UNIVERSAL MACHINE

Fig. 5-65. The horizontal knee-and-column milling machine has great flexibility in machining. However, because of the special setups required, a skilled machinist is needed to operate the machine.

Fig. 5-64. The bed-type duplex milling machine has little vertical adjustment on the table of the machine. Small adjustments may be made by moving the head up or down. This type of machine is rigid and will mass-produce small parts.

Fig. 5-66. The horizontal milling machine has hand and power feeds. Movement of the machine's table is on three axes or in the direction of movement.

the cutter on the arbor, location of the work on the table, or slight adjustment of the spindle.

The knee-and-column type of milling machine is designed with a base, a column to mount the spindle and cutters, and a knee sliding on the front of the column to hold the table. The knee-and-column has great flexibility as to the kind of work it will manufacture, Fig. 5-65. A skilled machinist must operate the machine to carry out its full capability because of the attachments available to machine special setups.

The knee-and-column milling machines are further subdivided into horizontal and vertical milling machines. The **horizontal milling machine** may be a plain or a universal milling machine, Fig. 5-66. In the horizontal milling machine there is movement of the table in three axes, with power or by hand. The work may be adjusted under the cutter by hand and the cut performed under power feed. The cutter is mounted on an arbor. The spindle of this arbor is horizontal to the table, and the arbor is supported on the outboard end by an overarm which extends from the top of the column. Cast iron braces are available to connect the arbor support and overarm to the knee for rigidity when heavy cuts are made.

The **universal machine** can swivel the table in the saddle on the knee. This makes it possible to machine a helix (spiral) on the workpiece. The chief manufacturing advantage of the universal milling machine is the variety and broad capacity of work possible that this machine can produce with common attachments.

The **vertical knee-and-column milling machine** is similar to the universal milling machine except that the spindle is vertical to the table. This machine lends itself to end-mill cutting and heavy facing operations. Vertical milling machines are frequently used for hole-making operations, Fig. 5-67. Hole location, size, and roundness may be accurately machined with this type of machine. The vertical mills are refined into precision measuring machines

that have the ability to bore very accurate holes. This machine tool is called the **jig borer,** Fig. 5-68.

In light manufacturing (such as the electronic industries), light and medium vertical machines have become very popular, Fig.

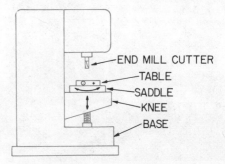

KNEE-AND-COLUMN VERTICAL
UNIVERSAL MACHINE

Fig. 5-67. The vertical milling machine has the spindle and tools set 90° to the table.

Fig. 5-68. The jig borer is a specialized machine designed to precisely locate and size holes. Jig borers are frequently used to build tooling such as jigs, fixtures, and dies.

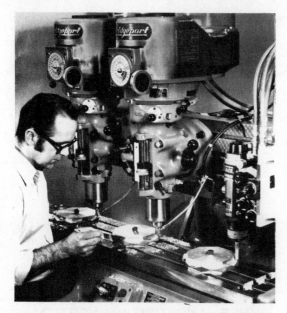

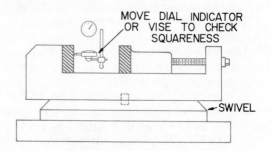

Fig. 5-70. Checking the solid jaw for squareness by using the dial indicator.

Fig. 5-69. A vertical milling machine with a hydraulic synchro-trace unit. Irregular parts can be reproduced from templates or a master part. Production can be increased by adding multiple heads to the machine.

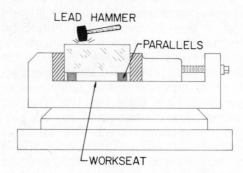

Fig. 5-71. After the vise has been tightened, the work is driven down onto the parallels and work seat.

5-69. They are used in tool rooms, prototype development, and school shops.

The operation of the horizontal and vertical milling machines has many common principles. The horizontal mill has more basic operations and is described for a setup procedure.

When machine tools are being prepared for a setup, it is advisable that the operator examine all clutches to see whether or not the power feed has been left engaged. When clutches are in the disengaged position, the table feed handwheel should be turned to be sure that the table is free and unlocked. The table may be moved out toward the operator so that the table surface may be easily reached. The table is cleaned and checked for burrs. The mill vise is cleaned, and the base is wiped to be sure that no chips or dirt are lodged between the vise and the table. The vise is pushed toward the spindle so that the vise keys will

both contact the same side of the T-slot in the milling machine table. The vise is then bolted down. The vise swivel should be checked to see if the index mark on the vise reads either 0° or 90°, depending on the position of the vise needed for the milling operation. A dial indicator is attached to the column and the solid jaw of the vise indicated to test if the vise jaw is parallel or 90° to the table movement, Fig. 5-70.

Parallels are placed in the vise. The work is placed on the parallels and the vise tightened, Fig. 5-71. The work is driven down with a lead hammer to ensure that the work is down on the parallels and that the parallels are down on the work seat of the vise.

The table with the work is moved to the approximate running position, and the arbor

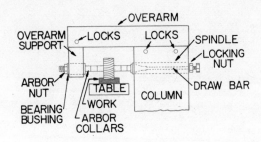

Fig. 5-72.   Cutter and arbor mounting in a horizontal milling machine. When the machine is set up, the arbor nut is the last item tightened. When the machine is taken down, the arbor nut is the first item loosened.

and cutter are selected from the rack. With the working position, it is easy to judge where the cutter should be placed on the arbor. The cutter should be keyed to the arbor, because the collar pressure will not drive the cutter under heavy loads. The arbor collars are put in position and the arbor bearing located. When assembled, the arbor support and bearing will clear the vise and the work. Each of the parts is cleaned and assembled.

The arbor nut is tightened only finger tight at this time because pressure without the support could damage the arbor.

The arbor is placed in the milling machine spindle and the draw bolt tightened just enough to hold the arbor in place, Fig. 5-72. The overarm is extended and the correct arbor support selected and locked to the end of the overarm. The arbor support and overarm are slid carefully back until the bearing on the arbor slips into the bearing sleeve of the arbor support, and all locks are then tightened. The nut on the draw bolt is tightened and finally the nut on the end of the arbor is tightened. The operator's rule for the milling machine is, ''Tighten the arbor nut last when setting up the machine, and loosen the arbor nut first when taking the machine down.''

Milling cutters may be damaged by the use of incorrect speed or rotation, so a simple calculation will give the proper revolutions per minute of the cutter, Fig. 5-72A.

$$\frac{4}{\text{Diameter of the Cutter}} \times \begin{array}{c}\text{Cutting} \\ \text{Speed of} \\ \text{the Material}\end{array} = \text{R.P.M.}$$

Fig. 5-72A.
Milling
(Cutting Speed of Material
in Surface Feet per Minute)

Material	High-Speed Cutter	Carbide Cutter
Cast Iron	70-100	150-300
Machine Steel	80-120	300-700
Tool Steel	60-100	120-300
Brass	140-250	200-600
Aluminum	200-800	250-1400

Milling cutters are mounted on arbors or shanks, or directly on the spindle nose. The cutters frequently found on a dial-type universal milling machine are mounted on the arbor. Cutters are classified according to the type of cuts they make and their shape, Figs. 5-73 and 5-74.

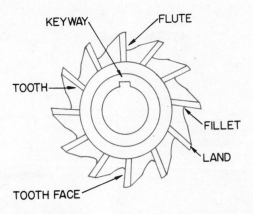

Fig. 5-73.   The basic construction of an arbor-mounted cutter.

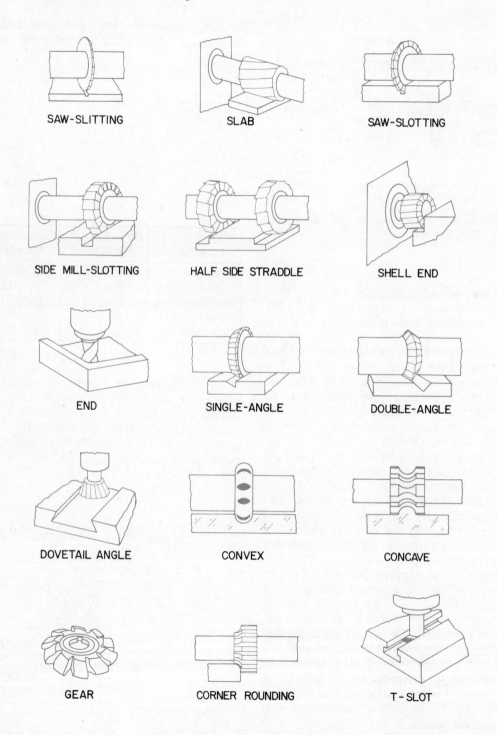

SAW-SLITTING

SLAB

SAW-SLOTTING

SIDE MILL-SLOTTING

HALF SIDE STRADDLE

SHELL END

END

SINGLE-ANGLE

DOUBLE-ANGLE

DOVETAIL ANGLE

CONVEX

CONCAVE

GEAR

CORNER ROUNDING

T-SLOT

Fig. 5-74. Milling cutters come in many forms and shapes. The standard milling cutters are shown here.

Using the Milling Cutter

1. The **saw** or metal-slitting cutter is very thin. It is used to part work or make narrow slots. The cutters range in width from 3/16″ to a few thousands of an inch in thickness.

2. The **plain-milling cutters** have cutting teeth only on the outer edges. Cutters wider than 5/8″ and up to 4″ have helical teeth and notches to break up chips and to control chatter on heavy slab cuts.

3. The **side-milling cutters** have the same construction as the plain-milling cutter except that they have cutting edges on the sides of the tooth. This allows cutters to straddle mill work and machine four surfaces at one time. These cutters may have straight, helical, staggered, or interlocking cutting teeth.

4. The **angle cutters** are delicate cutters. They cut single angles or double angles for dovetails and V-grooves.

5. **Form cutters** cut irregular shapes. They can cut flanks for gear teeth and milling threads. They round off corners and mill convex and concave cuts. These cutters are used for making drills, reamers, and other tools.

6. A **shell-end mill** is mounted on a short arbor that goes into the spindle. These cutters are rugged and they remove a large volume of material from the work efficiently. Large diameter cutters are called **face-milling cutters**, Fig. 5-75.

7. **End mills** come in many diameters and lengths. These cutters have cutting teeth on the sides and end and are used to form slots and to rout out small detail areas. The numerical control machines use this cutter with its wide selection of sizes most frequently.

8. **T-slot cutters** make a kerf in metal that looks like an inverted T. Tables and accessories for all machine tools with slots for standard size T-bolts are cut by T-slot cutters.

9. **Inserted tooth cutters** are designed to receive a carbide cutting tool edge and are built into almost any shape and size. The inserted tooth cutters are very rigid and can remove the highest volume of material per hour of all cutters, Fig. 5-76.

Fig. 5-75. Face-milling a steel block.

Fig. 5-76. Inserted teeth are capable of high cutting speeds and heavy loads. Inserted tooth cutters are rugged and perform well under heavy production cutting.

The edges of milling cutters engage the metal in two directions, depending upon the direction of the table feed and the rotation of the cutter.

Up milling, or conventional milling, is performed when the teeth of the cutter and the direction of the table feed move in opposite directions, Fig. 5-77. Up, or conventional, milling was used in the past before the development of rigid machines which have little back lash between the table feed nut and the lead screw. The up-milling process kept all machine slack out of the system by keeping the pressure in a constant direction on the lead screw, Fig. 5-78. This milling direction left an irregular surface and a less efficient cutting action, Fig. 5-79.

Down milling or climb milling, is performed by the teeth of the cutter and the table feed moving in the same direction, Fig. 5-80. The higher cutting force is on the start of the cut when the chip is thick. Most

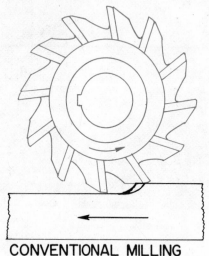

CONVENTIONAL MILLING

Fig. 5-77.   Up or conventional milling is used on older machines that do not have a lead screw backlash adjustment.

Fig. 5-79.   The surface condition of the various parts of an up-milled or conventional workpiece.

Fig. 5-78.   In up or conventional milling, the cut is started with a thin chip and increases in thickness as the cut proceeds. This passes any deflection that may occur in the arbor and/or lead screw to the surface of the work.

of the resulting arbor or machine deflection is relieved when the thin section of the chip is being cut, Fig. 5-81. This cutting action pushes the work firmly into the holding fixture to produce a self-clamping action. When down milling is done, the chips are removed from the cut and deposited behind the work zone. This reduces the possibility that they will be carried around the cutter and damage the surface finish.

Down milling is the most efficient milling procedure, and it produces the best finish, Fig. 5-82. However, up milling is best for producing thin cuts, long key ways, and interrupted cuts such as crossing slots.

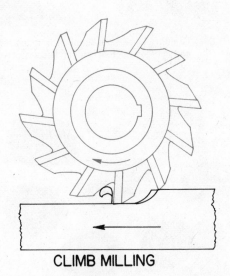

CLIMB MILLING

Fig. 5-80. Down or climb milling is performed when the cutter and feed are moving in the same direction. The most efficient cutting action results from this procedure.

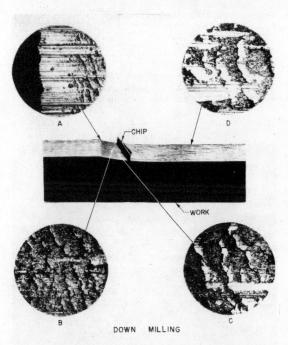

DOWN MILLING

Fig. 5-82. The surface condition of the various parts of a down- or climb-milled workpiece. Cutting efficiency and surface finish are superior.

Fig. 5-81. In down or climb milling, the cut is started with a maximum thickness chip and decreases as the cut proceeds. This cutting action requires a rigid machine with a backlash eliminator on the lead screw. Down milling produces the most efficient cutting method.

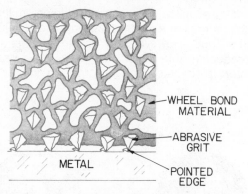

Fig. 5-83. When an abrasive is applied to a wheel, it supplies a hard, pointed cutting edge of grit to cut a small chip away from the metal.

Fig. 5-85. Grinding wheels are supplied in many shapes and sizes for different kinds and quality of work.

Fig. 5-84. Abrasives are mounted on belts, discs, points, and cylinders. These forms are designed for cleaning and smoothing metal surfaces and edges.

Fig. 5-86. Different sizes of aluminum oxide grit.

Grinding

Grinding is applying an abrasive material to a workpiece to cut away small pieces of metal, Fig. 5-83. A pointed edge of abrasive grit reduces the product to desired size, shape, and finish.

The cutting tool is the abrasive and must be put into a shape where it will cut the metal. The forms in which the abrasives are most frequently used are wheels, belts, and free or loose abrasive, Fig. 5-84. There are other forms used by industry for specialized uses, Fig. 5-85.

Nature has provided abrasives. Emery, rouge, tin oxide, flint, garnet, rotten stone, and diamonds are natural abrasives. Manufacturing industries use primarily man-made abrasives such as aluminum oxide and silicon carbides, because they are more uniform in hardness and readily available, Fig. 5-86.

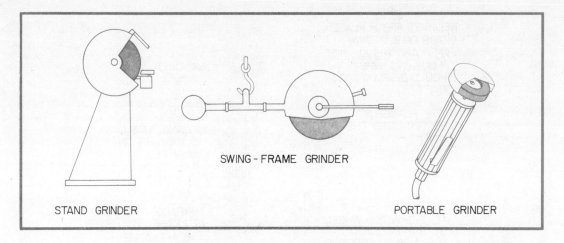

Fig. 5-87.   Three grinders used for rough grinding.

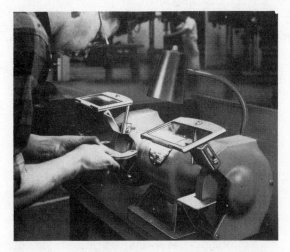

Fig. 5-88.   The bench grinder is widely used for grinding small parts and for tool sharpening.

Fig. 5-89.   A clear ring in the wheel is heard when the wheel is tapped with a nonmetallic tool. This sound is an indication that the wheel is not cracked.

Grinding processes fall into two large groups: (1) **Rough grinding** and (2) **precision grinding.**

Rough grinding. In rough grinding an operator guides the grinder or the work. Standard floor and bench grinders, swing frame grinders, and portable grinders are used for rough grinding, Fig. 5-87.

The standard floor or **bench grinder** is used essentially for sharpening tools and shaping small parts, Fig. 5-88. Certain pro-

cedures should be followed if grinding is to be done successfully.

Vitrified grinding wheels are designed to rotate with a surface speed of 6500 ft. per minute, or a surface speed of a little over 60 miles per hour. A wheel traveling at this speed must not be cracked or damaged or it will explode and cause an accident, so one must test the wheel for cracks, Fig. 5-89. For this reason these procedures are

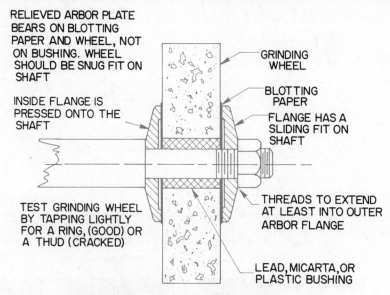

RELIEVED ARBOR PLATE BEARS ON BLOTTING PAPER AND WHEEL, NOT ON BUSHING. WHEEL SHOULD BE SNUG FIT ON SHAFT

INSIDE FLANGE IS PRESSED ONTO THE SHAFT

TEST GRINDING WHEEL BY TAPPING LIGHTLY FOR A RING, (GOOD) OR A THUD (CRACKED)

GRINDING WHEEL

BLOTTING PAPER

FLANGE HAS A SLIDING FIT ON SHAFT

THREADS TO EXTEND AT LEAST INTO OUTER ARBOR FLANGE

LEAD, MICARTA, OR PLASTIC BUSHING

MOUNTING A GRINDING WHEEL

Fig. 5-90. Mounting a grinding wheel on a grinding arbor requires cleaning flanges, fitting the bushings, applying blotting paper, and carefully tightening nuts. Note that nuts must not be overtightened.

given to protect the operator and the equipment, Fig. 5-90.

Safety Procedures

1. With the power off, turn the grinding wheel slowly and look for a powdery white line or a gouge across the face of the wheel. If none is present, the wheel has not been jammed by a workpiece and may be safely used.
2. While turning the wheel slowly, examine the flanges and nuts to see if they are snug against the grinding wheel blotter. If they are tight, the wheel will be free of all parts of the machine.
3. Examine the tool rests to see if they clear the grinding wheel by 1/16" to 1/8" so that it is impossible to get work between the wheel and the rest.
4. Examine the face of the wheel to see if it is flat without high sides or grooves. If the surface is not serviceable, it will need to be **dressed**, Fig. 5-91.
5. On a newly mounted wheel or on a grinding wheel that is suspected of being misused, tighten all guards, stand to one side and start the grinder. Let the wheel run for one full minute as a test before attempting to use the grinder.
6. When using the grinder stand to one side of the wheel. The body should not be in line with the rotating wheel.
7. Do not use rags to hold work when grinding. Rags are very easily and quickly caught in a high-speed wheel, thus causing an accident.
8. Hold work in your fingers when grinding. The nerves of the fingers are more sensitive to the movement and heating

Fig. 5-91. The correct method of truing and grinding the grinding wheel on a bench grinder.

of the metal than pliers would be. Cool the metal frequently so that it will not be dropped because it is too hot.

9. Heavy work is ground by moving the work across the full wheel face so that the grinding wheel wear will be even and keep the face of the grinding wheel flat.

10. Grinding should be done so that the grind marks run across the full plane of the work piece being ground. This makes a quality craftsmanlike finish.

11. The grinding of soft metals such as copper, aluminum, brass, and lead requires special silicon carbide grinding wheels rather than the usual aluminum oxide wheels. These soft metals quickly fill the chip clearance within the wheel and load the surface of the wheel, drastically reducing the cutting action of the grit in the wheel of an aluminum oxide wheel.

12. While grinding, clothing, hands, and hair should be kept away from the rotating grinding wheel. Caution prevents many painful injuries.

13. Goggles or a face shield are always worn when a person is in the vicinity of a grinding machine.

Fig. 5-92. Two types of portable grinders. The 90° grinder has a cupped wheel which is frequently used for surfacing. The straight grinding wheel is used on weld beads and for general purpose grinding.

Another class of rough grinding is the **swinging frame grinder.** This type of grinder is frequently used to grind castings after they are removed from the foundry and cleaned. This grinding method is fast for large castings. It is simpler to move the grinder over the casting than to move the casting, thus the name swing grinder.

Snag grinding is another class of work. It removes fins, bosses, and small pieces of gates and risers that remain on the casting. It is heavy, dirty work, but the casting is smoothed up and projections blend into the surface of the casting rapidly.

Portable grinders are used throughout manufacturing plants to perform many grinding tasks, Fig. 5-92. They remove burrs and

Fig. 5-93. Electric portable grinder used to finish a weld bead before it is painted.

sharp corners and edges from products. They grind weld beads and remove fins and bulges from castings, Fig. 5-93. Metal surfaces are cleaned and smoothed by air grinders and discs.

Portable grinders are frequently driven by compressed air so they develop high speed and power. The rough work which they do requires a very tough grinding wheel, Fig. 5-94. The wheel is made with a resinoid bond that withstands shock at 9500 surface feet per minute speeds.

Precision Grinding. Precision grinders are grinders that are capable of great accuracy and high surface finishes, Fig. 5-95. Specialized grinders are capable of grinding to dimensions of 5 millionths (0.000005″) accuracy. The range of accuracy in these machines makes possible the production of very precise parts needed in our space, science, and industry technologies.

Precision grinding falls into three general groups of operations: (1) **cylindrical grinding,** (2) **surface grinding,** and (3) **internal grinding,** Fig. 5-96. These three categories have subgroups which produce the many special grinding processes used on modern products of industry.

Fig. 5-94. Portable air grinders remove irregularities from casting before they are machined.

Fig. 5-95. High precision and surface quality are obtained by the grinding processes.

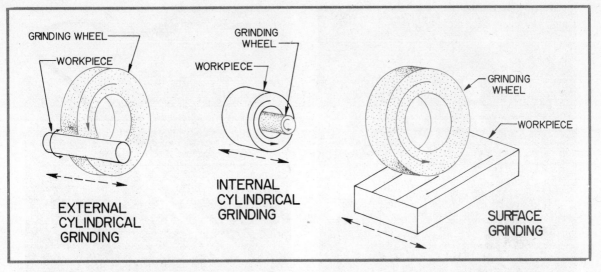

Fig. 5-96. Precision grinding falls into three groups of operations: cylindrical grinding, surface grinding, and internal grinding.

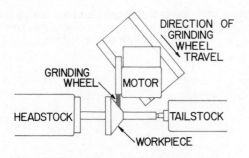

Fig. 5-97. Cylindrical grinding finishes the surface of a straight cylinder or a conical shape.

Fig. 5-98. Cylindrical grinding is frequently done with the work mounted between two centers. The work moves past the grinding wheel. The work is ground parallel to the center line of the work.

The classification of **cylindrical grinding** is given to work that is ground on the outside of a cylindrical form. The work may be shaped as a portion of a cylinder, cone, or a formed shape, Fig. 5-97. The work is most frequently mounted between centers and moved back and forth past the grinding wheel. The grinding surface is usually parallel to the center line of the grinding wheel. The parts most frequently produced on the cylindrical grinder contain straight or tapered round sections, fillets, shoulders, and faces, Fig. 5-98. These include such parts as machine rolls, shafts, spindles, and columns.

With attachments the machine will grind crank shafts and cams.

The work done by the cylindrical grinder is the same type of work that is done by the lathe. However, the grinder may be controlled more accurately. The grinder will bring heat-treated (hardened) parts to size with excellent finish and accuracy. Often to

Fig. 5-99. Plunge grinding to a precise size. An indicator is measuring the size of the work as it is ground down. When the correct size is reached, the machine moves to the load position.

Fig. 5-101. The centerless grinding of ball bearing races. The races are ground to size as they continuously pass through the machine.

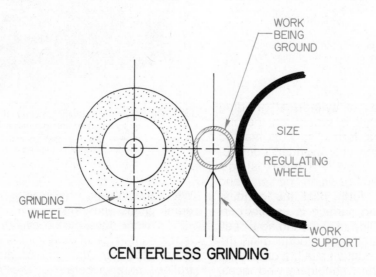

CENTERLESS GRINDING

Fig. 5-100. Centerless grinding is applied to many round and cylindrical parts (such as roller bearings) that must be mass-produced to precise dimensions.

save grinding time and cost, manufactured parts are brought to rough size in a lathe and then heat treated and finished to a very high quality workpiece in a grinder.

Usually the work traverses in front of the grinding wheel of the cylindrical grinder. In **plunge grinding** a different action is applied. The work is done between centers. The width of the wheel is about 2″ across its face, Fig. 5-99. The wheel face is dressed to shape and the wide grinding wheel is fed directly into the workpiece. The wheel may be dressed to a contour and the work ground to irregular shapes, or the grinding may be done on straight surfaces.

Centerless grinding is an economical process for finishing the surface of cylindrical products such as bearing races, piston pins, drill rod, and standard stock, Fig. 5-100. The work is not held between centers, but is held against the grinding wheel by a regulating wheel. The work is supported on a knife edge and is free to rotate as the grinding takes place. The grinding wheel is large and maintains the relationship between the size regulation wheel and the work support. The work is caused to move past the grinding wheel by tilting the regulation wheel a few degrees away from a true horizontal position. Centerless grinding lends itself to automation, because continuous grinding can be done with an automatic loader, Fig. 5-101.

Internal grinding is an application of the cylindrical grinder, but it is designed to finish the inside surfaces of holes and bores, Fig. 5-102. Surfaces such as the internal diameter of bearings, sleeves, gages, gears, cutters, or bushings are thus finished. Because the grinding wheel works on the inside of the hole, the grinding wheels are quite small. For this reason the internal grinding wheels must revolve at very high speeds in order to do the grinding.

Internal grinding machines fall into these three groups, which are based upon the machine design:

1. The work rotating type with the work mounted on a face plate, chuck or

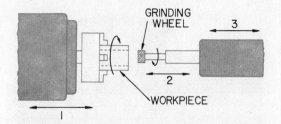

Fig. 5-102. Internal grinding is used to precisely finish the insides of holes or bores.

Fig. 5-103. Internal grinding will produce a straight, round, and accurately sized surface in hardened steels.

holding fixture turning slowly and traversing back and forth. The wheel rotates at high speed in a stationary position.

2. The work rotating type in which the work rotates slowly in a stationary position while the wheel rotates and is moved back and forth through the length of the hole, Fig. 5-103. This type is often used in production.

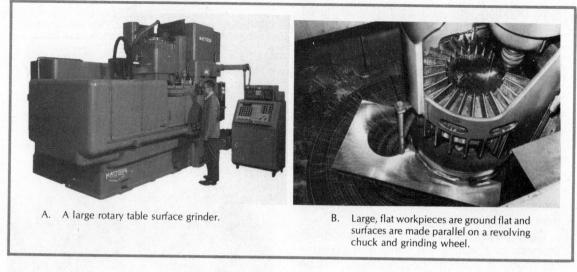

A. A large rotary table surface grinder.

B. Large, flat workpieces are ground flat and surfaces are made parallel on a revolving chuck and grinding wheel.

Fig. 5-104. Two types of surface grinders.

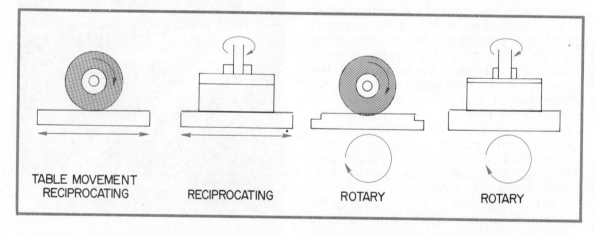

TABLE MOVEMENT
RECIPROCATING

RECIPROCATING

ROTARY

ROTARY

Fig. 5-105. Wheel and table movements that make possible four surface grinding machine designs.

3. The planetary action type has the work stationary, but the rotating wheel spindle is moved through an eccentric motion to make the hole. This grinding is used on large work that is difficult to rotate.

Surface grinding does the grinding in one plane. The grinding wheels may be mounted on spindles that are horizontal to the grinding table or vertical to the table.

The table movements of a surface grinder may utilize either a reciprocating table grinder or a rotary table surface grinder, Fig. 5-104. These combinations of units make possible four machines, Fig. 5-105.

The most commonly used surface grinder is the **horizontal spindle reciprocating table machine.** The horizontal spindle grinder is designed so that the grinding wheel cuts on the face of a straight wheel. The workpiece

Fig. 5-106. A large rotary table surface grinder showing stock and magnetic chuck.

Fig. 5-107. Large pieces of tool steel are surface-ground on a machine with a rotary table. Notice the circular pattern on the metal's surface.

is usually held down on the table by a magnetic chuck. Ferrous materials may be held in place on a rotary or reciprocating table for grinding by the use of magnets, either with chucks made with permanent magnets or with electro-magnets, Fig. 5-106. Nonferrous materials may be fastened to the table by T-bolts which fit into the table's T-slots for clamping. Machines may be used to grind dry, but more frequently a cutting fluid is used for rapid grinding.

Long flat surfaces or vertical surfaces can be ground on a large rotary table, Fig. 5-107. Angular surfaces may be ground with the use of a sine chuck, thus producing very accurate ground angles. Ninety-degree angles are ground on surfaces by using angle blocks or plates, Fig. 5-108.

The surface finish imparted to the metal will depend on the attributes of the grinding wheel and the wheel's speed. The surface speed of grinding wheels used for this type of work falls in the range of from 4,500 to 6,000 feet per minute. The movement of the work under the grinding wheel is the result of the table movement. The surface grinders are driven with a hand wheel or with a hydraulic drive, Fig. 5-109. The hydraulic controlled table movements up to 100 inches per minute are available for table feeds. The hydraulic crossfeed of the table may be

Fig. 5-108. Surface-grinding long workpieces on a reciprocating table. The work is clamped to angle blocks, and the total assembly is held by a magnetic chuck.

Fig. 5-109. Surface grinder feeds which are controlled by hand wheels and by hydraulic power.

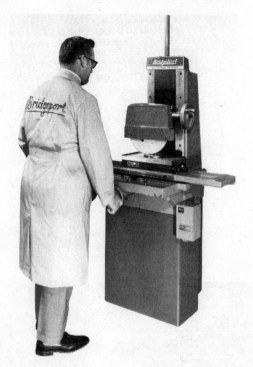

Fig. 5-110. A hand-operated dry surface grinder.

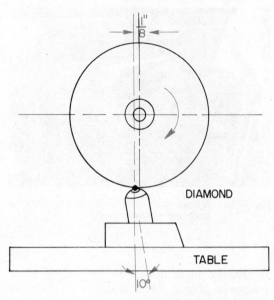

Fig. 5-111. A grinding wheel is dressed by having the diamond contact the wheel ⅛" beyond the center when the wheel is turning clockwise.

varied from a few thousands of an inch up to the width of the grinding wheel. These movements are very accurate and smooth. Smaller surface grinders may operate dry, Fig. 5-110.

Setting Up for Surface Grinding

1. Inspect the grinding wheel for chips and gouges on the face of the wheel. Check the wheel, flanges, and nut which mount the wheel on the spindle to see if they are tight. Put on safety glasses.
2. Check the oil level of the spindle. If it is low, refill with spindle oil.
3. Replace all guards on the wheel and the machine. Be certain that the wheel clears all parts of the machine or work.

Jog the spindle motor 5 or 6 times so that the spindle will come up to speed slowly. This allows the spindle oil to lubricate a cool spindle before it reaches maximum speed. Always stand out of line of the speeding grinding wheel.

4. Be sure to keep fingers and hands away from the grinding wheel. Allow the coolant to wash down the magnetic chuck to remove any dust or previous grinding residue if the machine is designed for coolant.
5. Stop the machine. When the wheel has stopped, use a rubber squeegee and wipe off the top of the magnetic chuck and rock the diamond wheel dresser on to the chuck. Be sure to turn on the chuck which clamps the diamond down to the table.

6. Check to see if the diamond point clears the wheel. Set the diamond holder so that the 10° inclination of the diamond holder points in the direction of the wheel rotation and the point is offset 1/8″ to 1/4″ from the centerline of the wheel, Fig. 5-111.

7. Carefully locate the highest point on the face of the wheel. (This will usually be the center if the wheel is fed into the work from both edges.) Lower the grinding head .002″ and by using the crossfeed handwheel, move the ·diamond across the face of the wheel. If the grinding is to be done dry, dress the wheel dry. If the grinding is to be done wet with coolant, the wheel should be dressed wet.

8. Continue dressing the wheel in stages until the face of the wheel is flat and the complete circumference has been dressed. The dressing of the wheel is finished by lowering the grinding head .0005″ and making several lively movements across the face of the wheel.

9. Turn off the grinder when the wheel has stopped. Turn off the magnetic chuck and rock the diamond onto one edge of the holder and remove. Raise the head so that there will be clearance to clean the chuck and to load the work.

10. Clean both the face of the chuck and the bottom of the workpiece. Remove any burrs before placing the work on the chuck.

11. Align the work with the wheel travel and turn on the magnetic chuck. Test to see if the work is held tightly by pulling it with your hands.

12. If the table has hydraulic controls, adjust the table reverse dogs so that the wheel travels about 1″ beyond both ends of the work, and then tighten the dogs to the table.

13. Set the crossfeed control for a roughing cut .025″ to .050″, depending on the workpiece. The grind wheel is moved over the highest point of the work and the hand lowered to 1/8″ over the workpiece.

14. Start the grinding wheel and lower the wheel until light contact is made. A few sparks will indicate the position. Raise the wheel .005″ above the work and engage the table and crossfeed levers. Allow the machine to travel over the entire workpiece while you watch for high spots on the work. If the wheel was originally sitting on the highest spot of the work, air will be cut on the first pass. If there is a higher spot on the work, it will be found and a wheel-damaging cut will have been avoided.

15. The second cut will cut metal. Lower the wheel head .005″ to .010″ for a rough cut and engage the feeds. Continue to lower the head between surface cuts until the surface has ground clean or reached its near-dimension.

16. Make the finish cut by lowering the grinding wheel .001″ to .002″ and changing the crossfeed to .005″ to .020″. A finish cut is ground and then a final cut is made without changing any machine settings. This last operation is called **sparking out.**

17. Stop the machine and raise the grinding wheel away from the work. Be certain the wheel is not turning.

18. Turn off the magnetic chuck, and rock the work up on one edge and remove. Be careful, because surface-ground work will have very sharp edges. The work should be blown off and deburred immediately to prevent cuts from handling.

19. Wipe off the magnetic chuck and center the machine so that the weight will be equally distributed over the castings while standing. Turn off the spindle oil if it is of a drip oiler type.

Toolroom grinding is a precision grinding that is applied to the production and maintenance of cutting edges on milling cutters, hobs, taps, reamers, and single-pointed cutting lathe and planer tools. Tool-room grinding is also the finishing shape and size given to jigs, fixtures, dies, and gages. Toolroom grinding supplies the geometry for cutting or aligning surfaces on the tools. The work is performed on a special grinding machine which is designed to

Fig. 5-112. A shell end mill being sharpened on a universal tool and cutter grinder. The finger positions each tooth as its cutting edge is ground.

Fig. 5-113. A form cutter cutting gear teeth is sharpened to keep the edges cutting to the highest efficiency. The tooth is sharpened on the flat gullet surface.

make light cylindrical and/or internal grinding cuts, Fig. 5-112.

In manufacturing a large amount of toolroom grinding is necessary to keep the cutting tools in proper condition. These tools will be resharpened before they become completely dull in order to reduce the power required to cut, maintain the proper surface finish and accuracy, and reduce the heat produced, Figs. 5-113 and 5-114. Sharp, uniform, and accurate cutting angles are required to maintain cutting efficiency.

Turning Metal

Turning metal is removing material from the outside surface of a rotating workpiece by the use of a cutting tool, Fig. 5-115. The surface may be straight (a continuous size), tapered (changing in size, stepped (having a series of straight surfaces), contoured (having a shaped surface), or bored (enlarging a hole by cutting away material on the inside of the hole).

The cutting tool is the important agent in shaping the materials to a finished shape and diameter.

Single-Point Tool Grinding. A single-pointed cutting tool must be ground so that one edge and one edge only is presented

Fig. 5-114. Grinding a carbide face mill with a flaring cup grinding wheel.

to the workpiece, Fig. 5-116. If more than one edge were presented, the tool would not cut efficiently because of the possibility of another surface rubbing the workpiece. Clearance and relief angles prevent secondary surfaces from rubbing the work and therefore are very important to tool cutting geometry. The tool angles ground by the cutting tool vary with the material being cut and the material from which the cutting tool is made, Fig. 5-117 (page 218).

Front clearance is ground so that only the cutting edge will contact the work. The grinding and the position of the tool control the front clearance.

Side clearance is ground so that when the tool is moving into the workpiece, the lower portion of the tool bit will not rub against the metal that has just been cut. Too little side clearance or too much feed will cause the tool to rub the flank of the tool.

Relief angle is the clearance between the newly machined surface and the front of the tool. This angle is also controlled by the tool grinding and the position of the tool in the lathe.

Lead angle is not a clearance angle, but is important to the cutting of metal. The lead angle forms the cutting edge and when it has a positive direction, the sum of the forces acting on the tool pushes the tool and tool post back away from the work, removing any backlash in the controlling screws of the lathe. This force against mat-

ing surfaces produces the accurate movement by the turning of the crossfeed dial.

Side rake is not a clearance angle, but it is designed for efficient removal of chips away from the cutting edge. The side rake is very important because it determines the wedge angle of the tool and thereby the type of cutting action and chips produced. The side rake angle varies with the material being cut, Fig. 5-117A.

FIG. 5-117A.
Side Rake Angles for High-Speed
Steel Lathe Tools

Cast iron, brass, bronze	0°- 3°
Tool steel, alloy, or hardened steels	8°-10°
Mild steel	14°
Stainless steel	18°
Aluminum and magnesium alloys	20°

Back rake, like the side rake is not a clearance angle. It aids the side rake in the type of chip formation. The back rake may be ground into the tool bit, but more com-

Fig. 5-115.   World's largest generator shaft undergoing rough machining.

Fig. 5-116.   A carbide cutting tool presents a single cutting edge to the workpiece at high speed.

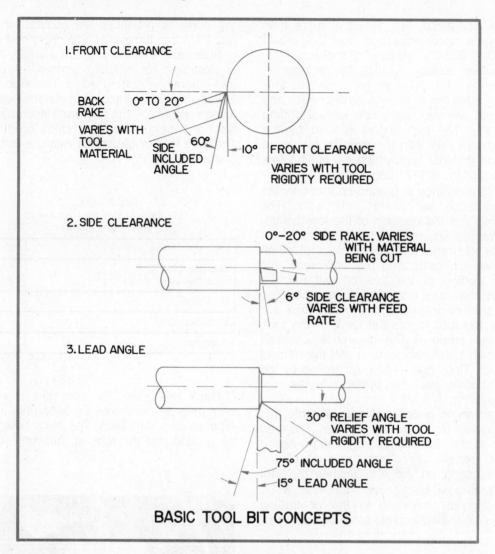

1. FRONT CLEARANCE

BACK RAKE   0° TO 20°
VARIES WITH TOOL MATERIAL

SIDE INCLUDED ANGLE   60°

10°   FRONT CLEARANCE
VARIES WITH TOOL RIGIDITY REQUIRED

2. SIDE CLEARANCE

0°–20° SIDE RAKE. VARIES WITH MATERIAL BEING CUT

6° SIDE CLEARANCE VARIES WITH FEED RATE

3. LEAD ANGLE

30° RELIEF ANGLE VARIES WITH TOOL RIGIDITY REQUIRED

75° INCLUDED ANGLE

15° LEAD ANGLE

BASIC TOOL BIT CONCEPTS

Fig. 5-117.   Concept of rake angles, clearance angles, relief angles, and lead angles of single-pointed cutting tools.

monly it is imparted to the tool by the broached hole in the tool holder or a slot in the tool block. The position of the tool in the holder and its relationship to the work also affects the back rake, Fig. 5-118.

The turning tool will have different shapes, depending on whether it is designed to cut toward the left or to the right. The term

right-hand turning tool means that the cutting starts on the operator's right side and proceeds toward the left as it cuts toward the lathe's head stock, moving toward the left. When the cut starts at the operator's left by the head stock and cuts toward the tail stock (moving toward the right), the tool is a **left-hand cutting tool,** Fig. 5-119.

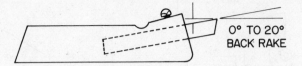

Fig. 5-118. The tool holder positions the tool and affects the back rake of the tool.

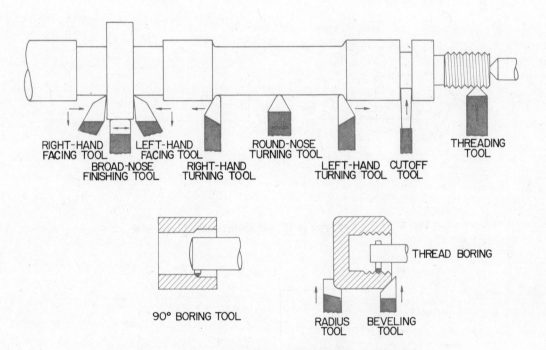

Fig. 5-119. Single-pointed tool shapes used in turning operations.

Grinding of a High-Speed Steel General-Purpose Lathe Tool Bit

When it is received, the tool bit blank will have a bevel on both ends, but the bevel is not great enough for the front clearance.
1. The tool bit blank is held between the thumb and fingers and ground on the face of the grinding wheel. The tool is moved back and forth so that a groove will not be worn in the surface of the wheel. The angle is 60° and the grind marks run across the full face of the tool blank. The tool can be checked with a 60° center gage, Fig. 5-120.
2. The lead angle and side clearance are ground by placing the tool against the face of the grinding wheel at 15° and rolling the tool under to 6°, Fig. 5-121.

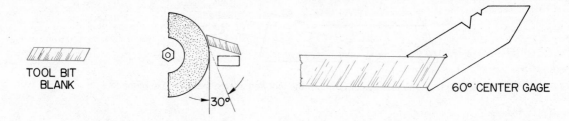

TOOL BIT
BLANK

30°

60° CENTER GAGE

Fig. 5-120.   Front clearance is ground on the tool blank by grinding a 60° included angle on the end of the tool blank.

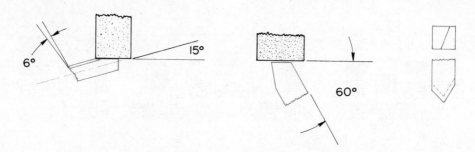

6°

15°

60°

Fig. 5-121.   A lead angle of 15° and a side clearance angle of 6°.

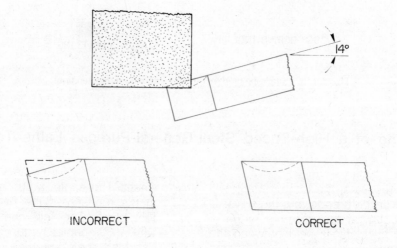

14°

INCORRECT

CORRECT

Fig. 5-122.   Side rake is ground at 14° for mild steel. The rear of the tool bit must be moved towards the center of the spindle equal to the lead angle. The side rake becomes another compound angle to grind.

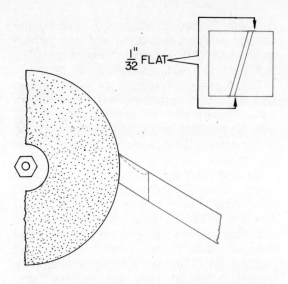

Fig. 5-123. A 1/32″ flat is ground on the front of the tool. Surfaces ground on a wheel will be hollow-ground or lower in the center of the grind.

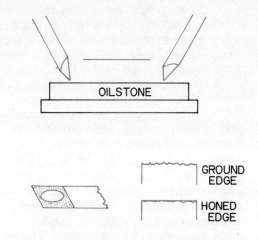

Fig. 5-124. The flat on the front of the tool is honed into a radius on an oilstone.

3. The relief angle is ground on the face of the wheel so its surface is 60° from the centerline of the tool. The back of the tool is lowered so that the original 60° front clearance is retained.

4. The side rake is ground by placing the top of the tool against the wheel, rolling the tool under 14°, dropping the back end of the tool 15° and moving the back end toward the grinder spindle and grinding on the edge of the wheel. When the grind is completed, the side rake plane should meet the side clearance plane without **foreshortening** the back rake, Fig. 5-122.

5. Grind the small flat down the front of the 60° angle of the tool. This flat will be made into a radius, Fig. 5-123.

6. Hone all angles on an oilstone to remove grind marks. The surfaces will be hollow ground and will be honed only around the edges where the work is done.

7. Hone all hollow-ground planes so that finished cutting edges are produced, Fig. 5-124.

8. When the tool is honed, the surface will last longer, because the sharp points left from grinding are removed. The isolated points will break down and cause progressive dulling; honing prevents this from occurring.

Cutting Tool Materials. Carbon tool steels formerly were one of the commonest of lathe tool materials. Today they are largely obsolete except for specially formed or low production rate applications. The material is inexpensive; it may be forged, machined, and heat treated to produce a specialized tool. The tool must be used slowly, because the hardness in the tool will start to draw out and soften at 400° F.

High-speed steel is an alloy of tungsten and other metals. High-speed tools commonly contain 18% of tungsten, 4% of chromium, 1% of vanadium, and carbon steel. Other metals are sometimes added, such as cobalt to increase red hardness and molybdenum to replace most of the tungsten.

High-speed steel may be easily ground to the finished shape of the tools. The steel is strong and will withstand heat to 1100° F. Above this temperature the tool steel will dull and fail quickly. A large number of drills, milling cutters, and turning tools are made of high-speed tool steels.

Nonferrous alloy tool steel is designed to cut very abrasive materials such as the surface of sand castings of cast iron and bronze. The surface of these materials is hard because of chilling, and sand may also be on the surface of the metal. "Stellite" is a commercial tool material which contains 45% cobalt, 18% tungsten, 35% chromium, and 2% carbon. This alloy holds its hardness at a red heat, and the cobalt makes it tougher at elevated heats. This material must be cast and ground to shape, since it is not machinable.

Cemented carbides are cutting materials made of the carbides of tungsten, titanium, tantalum, cobalt, and boron. These mate-rials are formed to shape by pressing the carbide powders into a die and later heated or sintered and ground to final shape. The tool materials will operate up to 2200° F.

Throw-away tool edges or silver brazed carbide inserts are mounted into medium carbon steel tool bit holders, Fig. 5-125 and 5-126. The tough steel backs up the brittle carbide and produces a very efficient cutting tool.

Ceramic cutting materials are made from metal oxides. The primary material is aluminum oxide. Other powders may be added as binders. The sintered oxide materials are manufactured by a powder metallurgical process similar to the sintered carbides and finished on diamond grinding wheels. The ceramic tools are very wear resistant and thus are excellent for long finish cuts. Frequently the workpiece is roughed out with a carbide tool and the work is completed with a ceramic tool finishing cut.

The ceramic tool will retain its hardness and strength to 2000° F. and surface speeds up to 3000 f.p.m. For efficient operation, the tool requires a rigid, high-speed machine with ample horsepower to produce a high quality finish and precisely sized parts.

Fig. 5-125. Carbide throw-away inserts are rotated so that the cutting edge on each side may be used. In some cases the throw-away may be turned over and the additional edges used.

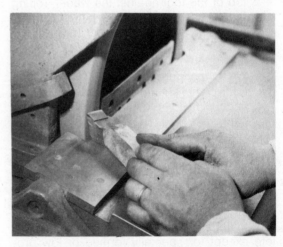

Fig. 5-126. A silver-brazed carbide insert is sharpened with a silicon carbide or diamond abrasive wheel. A tilting table provides accurate clearance angles.

Diamonds may be used on nonferrous materials such as plastics, hard rubber, brass, and aluminum, Fig. 5-127. They are used on these abrasive materials to give a fine finish and an accurate size. Diamonds are used for finish cuts at very high speeds, up to 5000 surface feet per minute. Vibration or chatter will damage a diamond, so very rigid machines are required.

Turning in a Lathe. The lathe is a machine tool that produces its work by revolving the workpiece. A cutting tool is moved against the work, and a chip is removed, Fig. 5-128.

Lathes are used for many sizes of work and special operations, Fig. 5-129. They perform operations that fall into two general classifications: **straight turning** and **chuck work**.

Straight turning is performed with the workpiece between centers rotating between conical points, or with one end mounted in a chuck or collet and the other end on a dead center. A cut may be taken the length of the work so that the work size may be reduced, or a number of different dimensions may be cut into the workpiece.

Fig. 5-127. A diamond-tipped tool was used to turn the cylinder of extremely hard Grade A lava 1-1, an abrasive that is machining graphite.

Fig. 5-129. Straight turning is done with the work between centers, with one end in a chuck, or with one end in a collet.

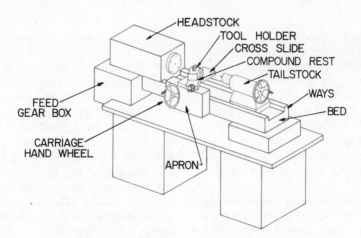

Fig. 5-128. The engine lathe causes the workpiece to revolve, and a cutting tool is moved against the ways to produce a straight cut on the work.

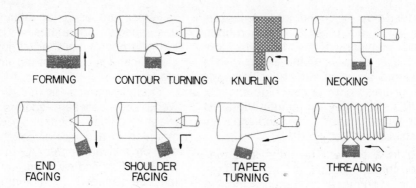

FORMING   CONTOUR TURNING   KNURLING   NECKING

END FACING   SHOULDER FACING   TAPER TURNING   THREADING

Fig. 5-130.   Turning operations between centers.

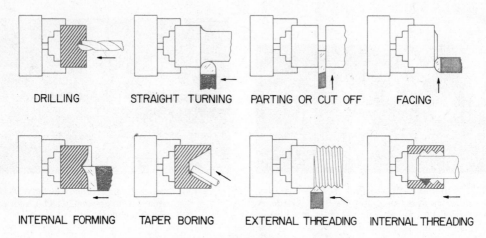

DRILLING   STRAIGHT TURNING   PARTING OR CUT OFF   FACING

INTERNAL FORMING   TAPER BORING   EXTERNAL THREADING   INTERNAL THREADING

Fig. 5-131.   Chucking operations in turning.

Other turning operations may be performed by modifying the tool or its movement. Facing, threading, taper turning, contour turning, knurling, and other operations may then be done, Fig. 5-130.

Chuck work is carried out by supporting the workpiece only by the chuck. This allows the operations of straight turning. In addition, drilling, boring, internal and external threading, and internal forming may be performed.

Engine Lathe. An **engine lathe** is the general purpose lathe found in schools, tool rooms, and small production shops, Fig. 5-132. The name originated at the turn of the century when this tool was powered

from an engine in one corner of the shop. These early tools were belt-driven and thus were limited as to power and speed. Modern engine lathes have gear head stocks and motors capable of delivering power as well as speed to their spindles. The engine lathe has considerable flexibility with accessories and can do a broad range of work. Repair and light manufacturing are frequently done with engine lathes. The engine lathe may be identified from other lathes by the fact that it always has a tailstock.

The engine lathe is either mounted on a bench or a cabinet, or it is a floor model lathe with its own legs. The common size of these tools varies greatly, from a 9″ swing to a 50″ swing, and with bed lengths from 3 ft. to 16 ft. in length.

Toolroom Lathe. The toolroom lathe is a redesigned engine lathe built to higher standards with greater range of spindle speeds and carriage feeds, Fig. 5-133. The lathe often has a gear-driven headstock with many attachments to give it the accuracy and the flexibility required of the toolmaker. The lathe may have a series of collets to supplement the 3 and 4 jaw chucks for holding the work. Other accessories may include a center steady rest, taper attachment, relieving attachment, and coolant pump, as well as quick change gearbox, lead screw, and a feed rod.

The toolroom lathe is used to manufacture small tools and instruments, tooling jigs and fixtures, and testing gages. It is capable of performing intricate operations on small parts with a high degree of precision.

Turret Lathe. The turret lathe is classified by the movement and arrangement of the tool slides. The two types are (1) the ram and (2) the saddle, Fig. 5-134.

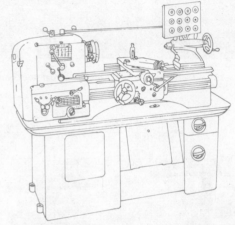

Fig. 5-133. Toolroom lathe is designed to manufacture tools and parts for tools to a high degree of precision. The toolroom lathe has many attachments; an important one is a set of collets.

Fig. 5-132. Engine lathes have gear head stocks capable of delivering power and different speeds to the spindle. They also have attachments to increase their work capability.

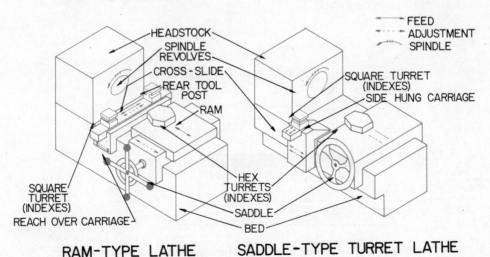

Fig. 5-134. Turret lathes are classified according to the slide the turret is mounted upon and the class of work it performs.

Fig. 5-135. Ram-type turret lathe finishing a bar stock part.

The **ram machine** has the major part of its tools on a tailstock turret which slides forward and back on a tailstock ram mounted to the lathe ways. This type of machine is very adaptable to bar stock and light work, Fig. 5-135. The weight of the ram style turret is less than the saddle style, thus it is faster for the operator to complete tool and operation change.

The **saddle type** is a more rigid machine, because the turret is mounted on the saddle of the lathe, which moves back and forth, Fig. 5-136. The turrets may be designed for hexagonal or octagonal turrets. This type of machine is adaptable to large deep bores,

Fig. 5-136. A saddle-type turret lathe is a rigid machine designed to machine larger casting, forging, or ring stock. Boring, facing, and turning are quickly performed in this machine.

Fig. 5-137. The vertical lathe is used to turn and bore large or deep forgings and castings.

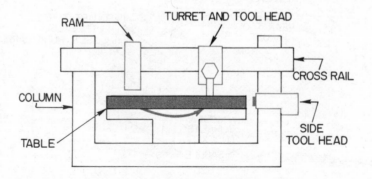

Fig. 5-138. Vertical boring and turning machines will machine work diameters up to 40 feet.

deep internal faces and recesses, deep internal tapers or threads, multiple bores and faces. Turret lathes will produce large volumes of work.

Vertical Turret Lathes. The **vertical turret lathe** has a large chuck mounted so that its face is parallel with the floor, thus the lathe spindle stands vertically, Fig. 5-137.

The tool is very rigid and will turn large castings and forgings that would be difficult to mount in a horizontal position in a lathe on centers, Fig. 5-138.

The size of the lathe is indicated by the size of the chuck, and it usually ranges from 30″ to 46″. The machine may have a number of toolheads attached to the cross-rail and the column. Frequently it has a main swiveling turret of six stations, a ram head to the left of the main head, and a side head which operates as a square or a hexagonal turret on a horizontal lathe.

The vertical turret lathe can turn inside diameters; turn outside diameters; face; and cut grooves, chamfer, and radii, tapers, and multiple diameters, Fig. 5-139. All of the vertical's work is done in the horizontal chuck.

Automatic Horizontal Turret Lathe (Chucker). The automatic chucker is a high production machine, Fig. 5-140. It is designed somewhat like the saddle turret lathe, but it is fully automatic for the size control of the parts and changing of the turrets and tools from position to position.

Fig. 5-139.   Machining a gear for the turning assembly for a bridge.

Fig. 5-140. An automatic chucker has great power and production. Speeds, feeds, indexing, and length of stroke are controlled by adjustable trips in the selector drum.

Fig. 5-141. Cross slide moving tool in for a heavy cut on an automatic chucker with a 15½" swing, 11" turning length, and a 6" cross slide working stroke.

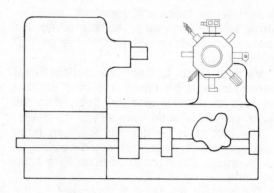

Fig. 5-143. Automatic screw machines are very fast and will produce volumes of small parts.

Fig. 5-142. A rigid machine capable of bar, shaft, and chucking work.

These machines have automatic controls which may be set to perform the spindle speed changes; reverse or stop; index the turrets; and start, stop or change the feed rates. If the machine is a ram-type bar machine, the work will automatically be fed into the machine, and the machine scarcely stops between operations and parts.

When the machine has the turret slanted vertically, the front and back cross slides may move simultaneously in front of the workpiece held in a chuck, Fig. 5-141. This lathe has great rigidity and can machine large diameter work easily. This machine will accept many shapes of castings, forgings, and preworked bar stock, Fig. 5-142.

Single-Spindle Automatic Screw Machines. The single-spindle automatic screw machine is a very high production tool, Fig. 5-143. It is a small cam-operated turret lathe which operates at very high speeds. Originally the machine manufactured small screws, bolts, and round parts such as pins, inserts, couplings, etc. The spindle had a collet capacity of 1/2" and a turning length of 1". Today the automatic screw machine is capable of producing larger products, including work up to 2" in diameter. Electronics has made recent application of screw machines for the manufacture of many microwave components which require great accuracy in manufacture and require many diameters and sealing surfaces.

The material supplied to the screw machine is usually bar stock of round, hexagonal, square, or tubular shape. Because these machines are completely automatic, it is possible for one operator to care for six of these machines running at maximum production. Very accurate and rapidly produced parts are available from the screw machine, Fig. 5-144 and 5-145.

Multispindle Automatic Screw Machine. The multispindle automatic screw machine may have up to six spindles which hold the workpiece in a collet or chuck, all rotating at machining speeds, Fig. 5-146. The stock is in a magazine at station one and is also loaded at station one. The spindle carrier with all the spindles rotates (index) one station. A roughing cut or drilling operation is performed in station two. When the cut is complete, the spindle carrier rotates all the spindles, and the part is moved to station three. Here the next operation is performed. When the machine is in operation, a part is completed and comes out of the machine each time the spindle carrier is indexed.

There are six parts in the machine in various machining stages and one part comes out per index. In reality, six parts are produced in the time it would take to pro-

Fig. 5-144. The screw machine will machine bar stock continuously with the addition of a bar feed magazine. An air-operated chuck may be added and square, hexagonal, and irregularly shaped castings and forgings up to 3″ in diameter may be made.

Fig. 5-145. An automatic screw machine machining a part from bar stock.

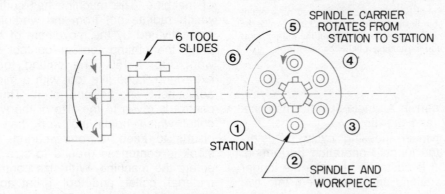

Fig. 5-146. The multispindle automatic screw machine has six parts being machined at the same time.

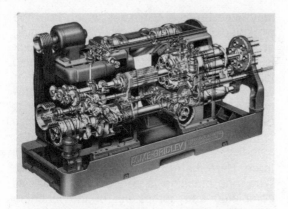

Fig. 5-147. A phantom view of a multispindle auto-matic screw machine. This type of machine has a voluminous production output.

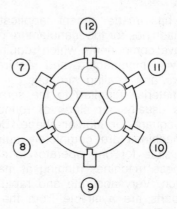

Fig. 5-149. Additional tool slides may be added to the outside of the automatic screw machine so that a total of 12 slides are available for tools.

Fig. 5-148. Small parts being manufactured on a multi-spindle automatic screw machine. Notice the coolant used for preserving the cutting tools and for gaining a better surface finish.

duce one part in a single-spindle machine, Fig. 5-147 and 5-148.

The individual indexing cycle is determined by the longest operation performed. Therefore, it is customary to take the longer cutting operations and break them into two stages. Such operations as drilling may be separated into two stations in order to reduce the cycle time. When tools are placed on the outside of the tool slide, the number of tool slides available may be increased to twelve, Fig. 5-149. Recently a multispindle numerically controlled chucking machine has been developed. The production of the chucker has been combined with the flexibility of numerical control. The tools can be programmed by tape to perform the many cutting operations and can use the same tools for another product by reprogramming the tool movement.

Swiss-Type Automatic Screw Machine. The Swiss-type screw machine is designed to produce small diameter parts. The various shafts in a wristwatch are made by this type of machine. The machine has cutting tools which radiate out from the workpiece and are activated by the movement of a cam to bring the cutting tool in contact with the work, Fig. 5-150. The cutting tool is an extra-long, lathe-like tool with a micrometer adjustment on the end of it. Thus, great control is given to the size of the work. The collet which holds the work has carbide inserts to keep it from wearing, and the stock is centerless ground to size before it enters the machine. With size control of the material, collet, and tool, great accuracies are possible.

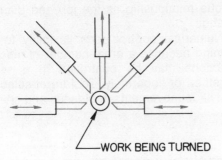

Fig. 5-150. Swiss-type automatic screw machine has a number of cutting tools that contact the work in a cam-controlled sequence. Fine watch parts are frequently made on this type machine.

Fig. 5-152. The technologies of control and machining come together in numerical control of machine tools.

Fig. 5-151. Stainless steel miniature precision components for use in aircraft, missiles, cameras, hydraulic controls, and watches are produced economically with the Swiss-type screw machine.

The cutting tools are activated by a series of cams which supply the movement through rocker arms to move the tool for the precise cut. Accuracies may be obtained on small diameter parts of 0.0005″ on a production run. These machines are particularly useful for making the small parts of clocks, electronic parts, calculating machines, meters, or any small instrument or machine part, Fig. 5-151.

Numerically Controlled Lathe. The numerically controlled lathe is the result of the merging of a number of technologies into another generation of machine tools. These tools are controlled by a 1″ punched tape that carries the information for controlling all the movements of the tool, Fig. 5-152.

In order to produce the rigidity and strength necessary for the numerically controlled lathe speeds and power, the bed of the lathes were moved to a slant position or to 90° to the horizontal position of the engine lathe, Fig. 5-153 and 5-154.

The lathes are designed with direct current spindle drives so that the spindle speeds can be changed while the cut is in progress. Constant cutting speed can be achieved as the work diameter changes. The greatest cutting efficiency is obtained by maintaining the machine's speed and torque relationship as the job and diameter change.

Numerical control adds flexibility to machining because a great number of machine functions may be controlled in a series, together, or separately. Tool turret selection, tool slide movements, tool feeds and speeds,

Fig. 5-153.   Heavy turned part is cut off with a parting tool on a cross slide. High-speed machining demands extreme power and machine rigidity.

Fig. 5-155.   With these 2 turrets, 12 tools are available for machining. Idle time is reduced, because both turrets can be programmed for rough and finish turning, facing, contouring, and threading.

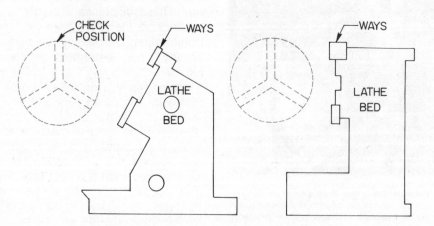

Fig. 5-154.   Lathe bed designs for heavy metal removal, easier part loading, and chip clearance with high accuracy.

and spindle speeds or work rotation are controlled by a tape, Fig. 5-155.

Numerically controlled lathes are turning centers, because they can machine such a broad range of work, Fig. 5-156. These tools have such a tremendous production output that frequently they have a chip removal and disposal system built into the machine, Fig. 5-157.

Electrical Chemical Separation

Electrical chemical separation occurs by energy release, rather than by the application of physical force of a cutting edge. This type of separation is performed by electrical discharge machining, electrochemical machining, and chemical machining.

Electrical Discharge Machining

In electrical discharge machining, a pulsating D-C electric spark jumps between the workpiece and the electrode. At that moment a very small particle of metal under the electrode is melted or vaporized and carried away, Fig. 5-158. The E.D.M. (electric discharge machining) machine is designed with a **servo-controlled** electrode feeding system. The electrode will advance into the workpiece and maintain the proper gap between the electrode tool and the disintegrating workpiece, keeping the electric spark jumping and removing metal.

The heat is controlled by submerging the entire cutting system in a dielectric fluid similar to kerosene. The solution removes the heat and flushes the eroded particles away.

The electrode tools are made from brass, mild steel, and carbon. They are usually made from a high quality carbon block, or graphite. These blocks may be machined easily with a profiling milling machine to produce an accurate carbon tool in a short time. Electrode tools may also be cut with tape or computer-controlled, numerically controlled machines in which no pattern or model is needed.

Electrical discharge machines have many applications in industry. One outstanding

Fig. 5-156. In addition to turning, these lathes will perform off-center drilling, contour milling, boring, tapping, and reaming while the work is held in the chuck.

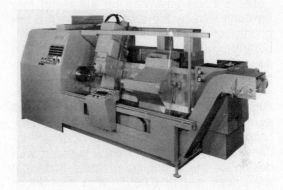

Fig. 5-157. In high-production machine tools, chip removal is a serious problem. Chip conveyors are built into the machine tool to dispose of chips.

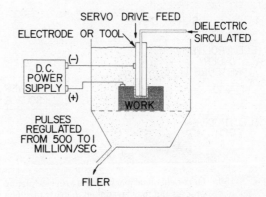

Fig. 5-158. Electrical discharge machining is performed by electrical sparks eroding away the metal under the tool.

use has been in the making of dies. Diesinking is a very expensive process when it is done with traditional machine tools. The electrical discharging machining has reduced this cost considerably.

E.D.M. is unique in that any material that is electrically conductive can be cut. Hardness has no relation to the cutting. Parts can be hardened before they are cut. Super alloys and cemented carbides which would be extremely difficult or impossible to cut by other means may be cut by E.D.M. Work cut by E.D.M. does not have stress or distortion put into the workpiece by cutting tool pressures. This is an important asset when thin or fragile sections are cut.

The surface finish left on the metal may be controlled by the amperage and the frequency of pulses used. Increasing the amperage produces a rougher surface, but this can be offset by a higher frequency to improve the surface.

This process is extremely useful for cutting odd-shaped holes and cavities, as well as narrow slots and small deep holes. The shapes will have a slight taper to them because of electrode tool wear. However, with design or the changing of tools, practical tolerances are $\pm 0.005''$, with tolerances as close as $\pm 0.0001''$ in special cases.

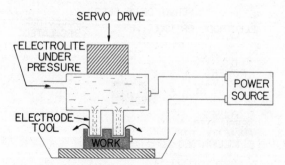

Fig. 5-159. Electrochemical machining is a reverse plating process and may be applied to many machine designs. The electrode tool does not contact the work. The electrolyte deplates the metal under the electrode tool.

Electrochemical Machining

Electrochemical machining makes use of a chemical process in dissolving metal from the workpiece. The process is the reverse of electroplating. The work is machined by being brought very close to the tool (cathode) and a high D.C. current is passed between the tool and the workpiece, Fig. 5-159. A salt solution (electrolyte) is pumped between the two parts at up to 350 psi at a temperature of 100° F. to 120° F. These machines apply up to 20,000 amperes to cause the metal on the workpiece to go into the solution.

The electrode tool does not come in contact with the workpiece, and therefore it keeps its accuracy. The tool may be made of copper, brass, stainless steel, graphite, or copper-tungsten. The accuracy and surface finish of the tool will be reproduced on the workpiece. The released metal particles are carried away by the solution. Round or square holes may be cut with sharp burr-free edges. Circular pieces may be turned, or complex contoured shapes formed.

The advantages of electrochemical machining (E.C.M.) are similar to E.D.M. There is no concentration of heat in the workpiece, so no stress or distortion is introduced into the work. The electrode tool does not touch the work and can be made of soft materials if convenient. Multiple shaping operations may be carried out at the same time thus machine operation time is saved. Surface finishes produced are excellent, and an accuracy of $+ 0.001''$ to $+ 0.002''$ may consistently be maintained. When tough materials are shaped, E.C.M. is capable of removing metal at a rate significantly faster than conventional machining.

Chemical Machining

Chemical machining is a process of removing metal by chemical action wherein the material is dissolved by an acid or base, as in etching. The process is used for an area, point, or line removal of metal. Its chief application is for the production of delicate thin parts with unlimited complexity

Fig. 5-160. Artwork that will be photo-reduced and printed on light sensitive metal sheets.

Fig. 5-161. The workpiece has been etched free from the blank. The part is covered with a photographic resist and is etched only on the line to produce the part.

of details and shapes. The process can be used on most metals.

Chemical Blanking. The copy of the part is prepared by artwork, Fig. 5-160. The copy will be carefully drawn and reduced by a photographic process to the correct size on a negative. The metal to be etched is thoroughly cleaned and coated with a photosensitive resist. The photographic negative image is contact-printed onto the photosensitive metal surface and the image is developed. The exposed and developed metal surface becomes an etchant-resist image and protects the metal while the other unexposed areas are cleaned with a solvent, leaving the exposed metal. An aligned negative can be printed on each side of the metal, thus reducing the part etching time by etching from both sides of the metal at once, Fig. 5-161 and 5-162.

Chemical Milling. Chemical milling is a process of milling usually applied to large

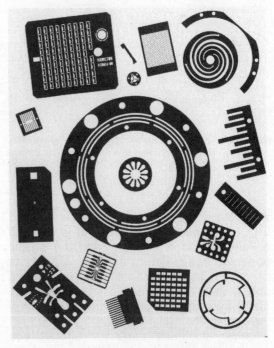

Fig. 5-162. Typical chemical blanked parts. Thin, detailed parts without stress or distortion are made by this process.

Fig. 5-163. Chemical milling of an aluminum part being lowered into a caustic tank. The parts not protected by a mask will be removed by the caustic, and the weight of the part will be reduced.

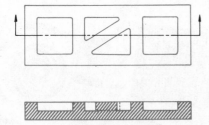

Fig. 5-164. Aircraft and spacecraft panels are lightened by the use of chemical milling in uncritical areas of parts.

workpieces, Fig. 5-163. The metal is sprayed with a thin plastic film. The film is cut with a knife around a template, and the plastic is peeled away, leaving the area which is to be etched bare. The bare metal is exposed and etched, while the plastic protects the metal from etchant on the masked parts of the workpiece.

Large aluminum parts for aircraft and spacecraft may be reduced in weight by removing uncritical areas, Fig. 5-164. Aluminum is etched with sodium hydroxide solu-

tions, while ferrous metals are etched with acid solutions.

Large areas may also be etchant-resisted with the use of a silk screening process. High volume part production can be done by this method, but it is less accurate than the photographic method of masking.

Advantages of Chemical Machining. There are four advantages of chemical machining:

1. A stress-free part is produced where a sheared or machined part would be heavily stressed.
2. Almost all metals may be shaped by this method, and it makes little difference if the metal is hard or soft.
3. Small and curved surfaces are easily shaped accurately.
4. The equipment required to do the work may be used for a wide range of work, thus it has advantages over other processes on short production runs.

The disadvantages of chemical milling is that precision work must not be more than 1/16″ in thickness. Also, any damage or surface imperfections tend to be enlarged in area as etching progresses.

Activities

Design and build a project which requires layout, hand tools, and small machine tools.

Related Occupations

These occupations are related to metal separation:
Tool-and-die maker
    Pressing die maker
    Die casting maker
    Drill jigs maker
    Special gages maker
Instrument maker
Setup technician
Machinist
Inspector

Machine tool operator
   Engine lathe operator
   Screw machine operator
   Milling machine operator
   Boring machine operator
   Threading machine operator
   Broaching machine operator
   Surface grinder operator
   Honing machine operator
   Chucking machine operator
   Shaper operator

   Planer operator
   Turret lathe operator
   Centerless grinder operator
   Multiautomatic turret lathe operator
   Profiling machine operator
   Keyseating machine operator
   Gear cutting and shaping machine
     operator
   Tapping and threading machine operator
   Contour saw operator
   Cold saw operator

Metal Separation by Shearing and Drilling

Chapter
6

Words You Should Know

Shearing — Separating metal across its width by pressure.

Cutting off — Separating metal along a line, not necessarily perpendicular to the edge.

Parting — Separating metal by removing metal from between two parts.

Blanking — Separating metal shapes from a sheet.

Punching or piercing — Small areas, usually in the shape of a hole, are separated from a sheet of metal.

Notching — Small areas are separated from a sheet of metal. The metal removed will be on the edge of the product to aid in bending and forming operations.

Slitting — Separating metal along its length.

Lancing — Separating metal partway across a sheet, from one edge or in the center of the sheet.

Trimming — Removing excess material to form correct shape and size of work.

Nibbling — A series of small holes or cuts connected together so that work may be shaped and parted.

Stylus — A pointed rod that fits into a series of holes and thus locates a position of a machine or tool.

Concentric position — A series of circles with a common center.

Drilling — Producing a hole with a tool with cutting edges on its end. Rotating the tool

causes the metal to be removed beneath the end of the tool.

Countersinking — Removing the metal from the top of a hole so that a conical shape is produced.

Counterboring — Removing the metal from the top of a hole so that a hole with a larger diameter is made with a flat shoulder at the bottom.

Boring — Enlarging an existing hole to make it round and to improve its accuracy.

Reaming — Enlarging an existing hole to make it straight and to improve its accuracy and surface finish.

Spot facing — Finishing the area around the top of a hole so that the flat surface will be at right angles to the center line of the hole.

Drill jig — A tool that accurately positions the work and precisely locates the hole by a bushing that guides the drill.

Metal Shearing

Shearing is the process of separating metals without loss of material by concentrating the stress in a metal between two sharp edges, Fig. 6-1. A narrow area of the metal is stressed past ultimate strength and the metal fractures, Figs. 6-2 and 6-3. If the pressure or stress can be controlled and the tool guided, metal separation may be either contoured or straight.

Shearing Operations of Sheet Metal

Sheet metal is separated by the following operations of shearing, Figs. 6-4 and 6-5:

1. Cutting off
2. Parting
3. Blanking
4. Punching (or piercing)
5. Notching
6. Slitting
7. Landing
8. Trimming
9. Nibbling.

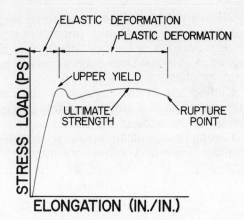

Fig. 6-2. Sheared metal is stressed beyond its ultimate strength to its rupture point. There the metal is parted without loss of material.

Fig. 6-1. Shearing is the concentration of stress between two sharp edges until the rupture point in the metal is reached and the parts are separated.

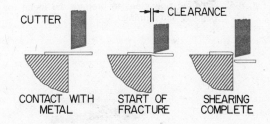

Fig. 6-3. The concentration of stress at the shear zone causes a narrow band of metal to rupture.

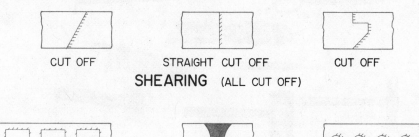

Fig. 6-4. Shearing operations used on sheet metal: cutting off, parting, blanking, and punching or piercing.

Shearing is generally considered a straight cut across the metal which brings the stock to semi-finished or finished length. It is designed to cut most stock such as round, bar, strip, or sheet.

Cutting off is the separation of metal along a single line, not necessarily perpendicular to the edge of the sheet.

Parting is a cutting operation in which material is removed between the metal parts, and radii or other features are formed.

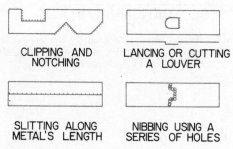

CLIPPING AND NOTCHING

LANCING OR CUTTING A LOUVER

SLITTING ALONG METAL'S LENGTH

NIBBING USING A SERIES OF HOLES

Fig. 6-5. Shearing operations to prepare the metal for bending and forming.

Blanking is the cutting process which makes metal shapes which are usually raw materials for other operations.

Punching (or piercing) is a similar operation to blanking, but the area is smaller and usually results in a hole in the workpiece.

Notching is the cutting operation which removes material from the metal, usually used to allow for folding or bending of the metal.

Slitting is the cutting of metal along its length. This operation may be applied to sheets or coils of metal.

Lancing is the cutting of metal partway across the sheet for a later operation. A louver is an example of lancing.

Trimming is the removal of excess metal after a forming operation to give the part the correct size.

Nibbling is produced by a series of small punched holes. It uses small, inexpensive dies and is very flexible. It may be used to follow almost any line on sheet metal.

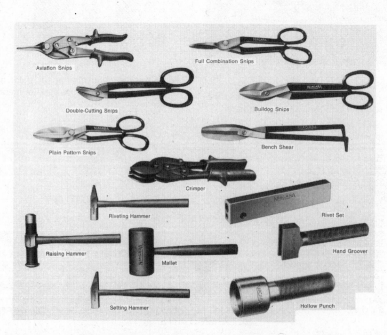

Fig. 6-6. Hand tools for sheet metal work.

Cutting Sheet Metal with Snips. Hand cutting of metal can be done by **tin snips** or **hand shears**. These tools cut straight lines or curves. One type of hand shears is similar to scissors. They are used to cut soft, thin metals. The other type of shears is more modern. They are called **aviation snips**. These snips contain a compound lever system, have short, serrated jaws with light handles and are designed to cut the harder and tougher metals such as stainless steels and duralumin.

The tin snips of the scissor pattern have been developed to perform many operations, and their shapes have been modified to do this cutting. The sizes vary, with cutting edges from 2″ to 4½″. They will accommodate different thicknesses of metal. The common types of tin snips of the scissor pattern are straight pattern, combination, bulldog, circular, slitting, hawkbill, double-cutting, compound lever, and bench shear, Fig. 6-6.

The most frequent use of hand shears is for clipping and notching and for straight cuts in sheet metal, Figs. 6-7 and 6-8.

Aviation snips cut most types of metals and cut thicknesses of metal up to 20 gage, Figs. 6-9 and 6-10. Aviation snips are made

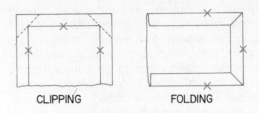

CLIPPING   FOLDING

Fig. 6-8. The clipping of metal corners is a simple and practical method of forming 45° miters for the reinforcement of edges of metal.

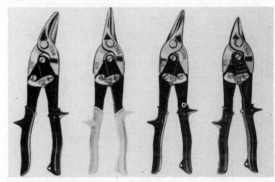

Fig. 6-9. Aviation snips: MIR cuts to the left, M3R straight-cuts, M2R cuts to the right, and M5R bulldog cuts heavier or tough metal.

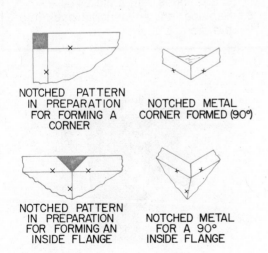

NOTCHED PATTERN IN PREPARATION FOR FORMING A CORNER

NOTCHED METAL CORNER FORMED (90°)

NOTCHED PATTERN IN PREPARATION FOR FORMING AN INSIDE FLANGE

NOTCHED METAL FOR A 90° INSIDE FLANGE

Fig. 6-7. Notching metal for a corner and a 45° inside flange.

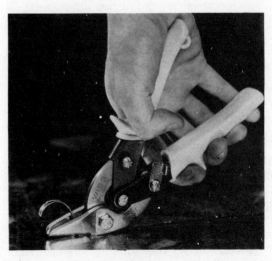

Fig. 6-10. Hand-cutting metal with compound action aviation pipe snips.

in these three styles: (1) cutting work curving to the right, (2) straight-cutting, and (3) cutting work curving to the left. These shears have a compound lever section which requires less hand effort to operate, so more energy is available for control while curves are cut. The snips are short, pocket sized, and thus very convenient for the mechanic.

Procedure for Using Hand Shears

1. Select hand shears that will produce the cut desired such as straight-cut, curving to the right, or curving to the left.
2. Gage the thickness of the material to find out if it is within the range of the shear.
3. Open the shears as far as possible and insert the metal, holding the cutting blade 90° to the sheet.
4. Close the shears slowly to follow the layout line, but stop cutting when the metal is within ¼″ to ½″ from the end of the blade. Reopen the shears and repeat the closing while carefully following the layout line. When the cut at the edge of the metal is being finished, the metal should be deep in the shears and the blades closed slowly. Cut metal should immediately be deburred with a file to prevent cuts on the hands.

Cutting with Squaring Shears. Squaring **shears** are frequently foot-operated machines designed to cut sheet metal across its full width and to square the edge with the guide, Figs. 6-11 and 6-12. The most frequent use of the squaring shear is for cutting large sheets into needed rectangular or square sizes. The squaring shears may be rigged with attachments which produce angular cuts.

Procedure for Operating the Squaring Shears

1. The bed and blade should be inspected so that no wire, tools, or rules are on the machine that could get under the blade and cause damage. The machine should be cleaned, oiled, and adjusted to be certain that it is serviceable.
2. The back gage is set to the dimension required for the length of the material and checked on both ends of the gage with a flexible rule.
3. The gage of the metal cut must be within the capacity of the machine, commonly 16 gage mild steel.
4. The metal is placed against the left-hand side gage and slipped under the hold-down until the metal strikes the back gage.
5. With fingers well back on the bed and the operator's foot well behind the treadle, press down on the foot treadle with the other foot.
6. When the metal has been cut free, release the foot treadle slowly back to the return position.
7. Remove the metal from the shelf behind the shears and deburr the sharp edges with a file.
8. Replace leftover material back into the storage area.

Fig. 6-11. Foot squaring shear is used for long, straight, and square cuts in sheet metal.

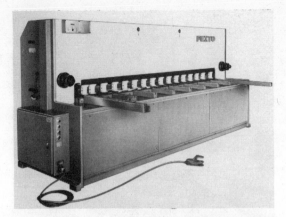

Fig. 6-12. A hydraulic under-driven shear with the cutting capacity of sheet metal 10 ft. by ¼" thick.

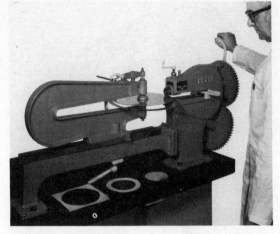

Fig. 6-14. Hand-operated ring and circle machine cutting a ring from a circle of sheet metal.

Fig. 6-13. The notcher will notch, clip, shear, and cope sheet metal into many shapes prior to bending the metal.

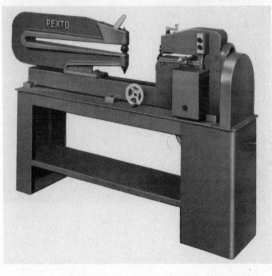

Fig. 6-15. A power-driven ring and circle shear with a capacity of 3½" to 44" circles up to 16 gage thick.

The squaring shears will rapidly cut sheet metal to the stretch-out size needed for sheet metal products.

Sheet Metal Notchers. The **tab notcher** is usually a hand-operated tool which greatly simplifies clipping and notching, Fig. 6-13. The notcher is a lightweight die that will cut a 6" x 6" corner notch in one stroke. It is also possible to take multiple strokes to produce irregular shapes. The tab notchers are frequently used for cutting the tabs and notches for box construction in sheet metal.

Ring and Circle Shears. The **ring and circle shears** can cut discs from sheet metal. By readjusting the radius of the machine, a second disc may be cut within the first disc, thus producing a ring, Figs. 6-14 and 6-15. These machines commonly cut circles from a few inches in diameter to 36" in diameter.

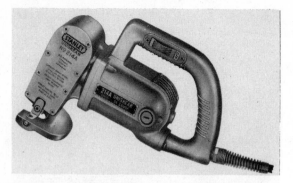

Fig. 6-16. Portable power shears will cut straight or irregular shapes in sheet metal up to 14 gage.

Fig. 6-18. A concave shear punch is capable of punching up to a 4"-diameter hole through ¼" thick material.

Fig. 6-17. A hydraulic-powered punch.

These machines have a guide that may be attached to the cutter head, and sheet metal slitting may be performed.

Portable Power Shears. Small motor-driven shears have a short blade which moves up and down very rapidly, cutting sheet metal up to 14 gage, Fig. 6-16. The unishear or portashear may be used in a shop or out in the field if 110 volt power is available. The shears will cut straight or will cut right or left curves because of their short blade and throatless construction.

Procedure for Operating Portable Power Shears

1. The work must be held flat on the edge of a support, table, or sawhorse.
2. The power cord must not be cut by the shears.
3. The cord should be kept out of water.
4. All parts of the operator's body and fingers must be kept away from the cutter.
5. Do not apply too much pressure. This will allow the cutting action to take place.
6. When cutting is completed, the tool must stop moving before it is set down or placed in its storage holder.

Punching Holes in Sheet Metal. Holes may be produced by many cutting operations. Their location and size control is very important in manufacturing. The precision of hole production ranges from the work of a cold chisel to that of a numerically controlled punch to a punch press, Fig. 6-17.

Examine a manufactured product and notice the number of holes that are in the work, whether they are for fasteners, ports,

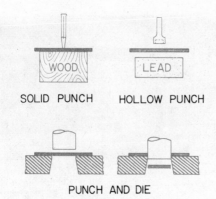

SOLID PUNCH   HOLLOW PUNCH

PUNCH AND DIE

Fig. 6-19. Rough openings in sheet metal may be produced by a solid punch, chisel, or a hollow
punch. A precision hole is produced by a punch
and die set.

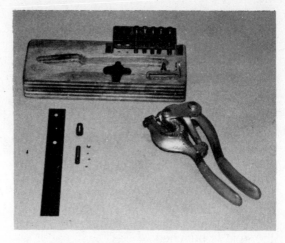

Fig. 6-20. A hand punch for work on sheet metal.

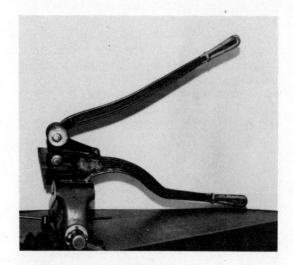

Fig. 6-21. Band iron punch for larger work.

or structural parts. The cutting of holes is a
major problem for the metalworking industry, so much so that a great number of
methods have been designed to produce
holes and openings, Fig. 6-18.

A hole may vary greatly in its precision by
the method of production. A cold chisel
striking sheet metal held on the end grain
of hardwood will produce a rough opening
in the workpiece. Much more precision in
the hole is made by using a punch and die
set to produce a hole in sheet metal, Fig.
6-19. This principle of punching may be
applied to many tools.

This punch makes use of a punch and
die set as well as a lever system to supply
the power for producing holes. The hand
punch is used to punch small holes in most
grades of commercial sheet metal. The hand
punch produces holes from 1/16″ to 9/32″.
The dies and punches increase in size by
1/64″, thus producing tight holes for rivets
and screw fasteners.

The hand punch is light and portable
making it very desirable for sheet metal
work, Fig. 6-20.

A larger punch along the same design
is the iron hand punch or band iron punch,
Fig. 6-21. This punch has a longer lever
system and thus more power. Heavier metal

may be punched with it. The punches range
in size from 3/32″ to ½″ with easy change-
able dies. Both of these types of hand
punches use a punch which has a projection
in the center. This projection may be carefully fitted into the center punch mark on the
metal layout, thus accurately locating the
hole to be punched.

The **hand-operated turret punch** is made
up of a series of single punch and die sets

and arranged in a circle so that one mechanism operates all punches when indexed for use, Fig. 6-22.

The punches have a projection so that the layout center punch mark may be placed on the punch center and punched through the metal. In the manual type machines, the thickness of metal punched must not exceed the strength of the punch. For 2″ diameter holes, 16 gage is the capacity of a 4-ton press. For a ¼″ diameter hole, 3/16″ mild steel is maximum capacity.

If the turret is indexed, the punches and dies must be checked to see if the correct die is matched with the correct punch before the hole is punched.

The design is quite similar to the hand-operated press, except that the two turrets (one for the punches and one for the die) are geared together and locked in exact alignment.

In front of the machine is a **cross slide,** which offers a convenient means of locating the sheet metal under the punch. The material on the cross slide follows the stylus when it is moved. When the stylus point enters a color-coded hole on the template,

Fig. 6-23. Workpieces may be located under the punch by the use of a template and stylus or located automatically.

Fig. 6-22. The hand-operated turret punch may be rotated so that the needed punch and die is aligned for punching the hole of the desired size.

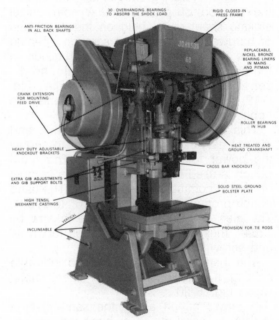

Fig. 6-24. Both high-volume work and accurate work are done in a punch press.

the position of metal has been reached. The punch is automatically tripped and the hole is punched, Fig. 6-23.

When a high volume of workpieces is produced, holes and other features are produced on a punch press, Fig. 6-24. The power is supplied by the press and the holes located and punched by the dies. A common type of punch press is the **C-type** frame which is inclinable. This inclinability makes it easier to remove the work.

The **die set** is the tool that is mounted in the press, and it does the actual punching, Fig. 6-25. Dies may cost from several hundred dollars for a simple die to tens of thousands of dollars for a complicated die, such as that used in automotive part production.

A die may be built with many punches and dies in the same die set so that all of the holes in a product can be made with one stroke of the press, Fig. 6-26.

The punch press may be power-fed. The raw stock feed is frequently automated, Fig. 6-27. In this case the punch press is strip

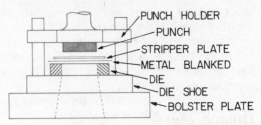

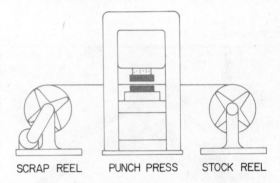

Fig. 6-25. A die set is made of many parts, but the punch and die do the shearing.

Fig. 6-27. Strip stock is frequently fed automatically into the machine for high-volume production.

SCRAP REEL   PUNCH PRESS   STOCK REEL

Fig. 6-26. A simple die set for making washers.

stock-fed off a reel and is automatically drawn through the machine.

The parts are ejected with a jet of air and fall to the back of a storage bin. The usable part is punched out of the strip. The remainder becomes scrap.

Drilling Metal

Metal can also be separated by drilling processes. As in other metal-working opera-

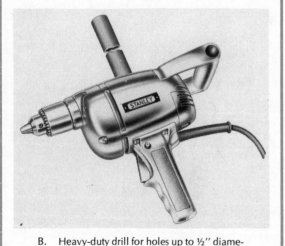

A. Light-duty, speed-controlled drill for holes up to ¼″ diameter in steel.

B. Heavy-duty drill for holes up to ½″ diameter in steel.

Fig. 6.28.   Different drills are used for making holes of different diameter.

tions, safety procedures should be learned before drilling is begun.

Portable Drilling

Light-duty drilling has been given flexibility with the development of a lightweight drill motor, Fig. 6-28A. This tool is extensively used in construction and manufacturing. However, it is most frequently used in repair work. The tool may be brought to the work and the hole drilled.

The electric portable drill is available in a series of sizes: ¼″, ⅜″, ½″, and larger, Fig. 6-28B. These drills allow holes to be drilled in almost any position or place, except in explosive gas areas. The size of the drill is indicated by the largest drill it will power through mild steel.

Safety Procedure for Using a Portable Drill

1. This is a hand-held tool and is subject to all types of working conditions. The drill should always be grounded. If any of the electrical insulation should fail, the circuit breaker or fuse will go out, preventing an electrical shock to the operator.
2. When preparing to drill the hole, a center punch mark is made so that the drill will not wander when the hole is being started.
3. Select the correct size drill, place it in the drill chuck, and tighten the chuck. Remove the chuck key from the chuck and replace it in its storage position.
4. Check to see that the electric drill is off, the button is released, and the switch is off. Plug the cord into the proper voltage plug.
5. Observe all safety rules. The electric cord should be kept dry and safe from falling objects which may cut it. Hands and all parts of the body should be kept away from the spindle and between the drill motor

and any structural member in case the drill should bind or lock.

6. Apply a steady pressure to the drill so that it will cut. The drill must cut or its cutting edge will be destroyed.

7. Release the pressure on the drill as the drill point breaks through the metal so that the drill will not be overfed and lock.

8. When the hole is drilled, shut off the power and back the drill out of the hole while it is turning slowly. When the drill has stopped, unplug the power and remove the drill. Return the drill and motor to their storage area.

Drill Press Processes

The drill is one of the earlier hand tools used, since people had to drill holes in objects such as shells, bone, stone, and metal. Today the most common method of producing holes is with the drill in a drill press. The work done by the drill press has been expanded to include related operations of countersinking, counterboring, boring, reaming, and spot facing. These operations enlarge, straighten, and bring the surface of holes to a finish and to the correct size.

In **drilling**, a drill is forced against the metal as it rotates. The cutting edges cut the metal beneath the drill and produce a hole, Fig. 6-29. A **countersink** hole has the top of the hole enlarged to a 60°, 82°, or 100° included angle so that a fastener will fit into the recess, Fig. 6-30. A countersunk hole allows screws and rivets to be flush with the surface.

A **counterbored hole** has the top of the hole enlarged with a flat shoulder in the bottom of the hole, Fig. 6-31. In a counterbored hole, the bolt heads are flush with the metal's surface.

A **bored hole** is a hole which is enlarged after drilling for the purpose of straightening, rounding, and improving the size accuracy of the hole, Fig. 6-32. **Reaming** is the op-

COUNTERSINKING

Fig. 6-30. Countersinking cuts a conical hole to fit bolts, rivets, and screws so that the top of the fastener will be even with the metal surface.

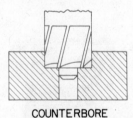

COUNTERBORE

Fig. 6-31. In counterboring, the large diameter hole is cut down into the metal and provides a flat surface for a bolt head to contact under the suface.

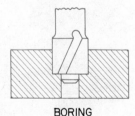

BORING

Fig. 6-32. Boring is performed after drilling to bring the hole to size, straighten the hole, and to produce a round hole.

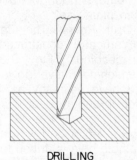

DRILLING

Fig. 6-29. Drilling is a practical method of removing metal with a cutting edge to make a hole.

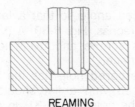

REAMING

Fig. 6-33. Reaming removes a small amount of metal from a drilled hole and brings the hole to size. It makes the hole round and straight, while improving the surface finish of the hole.

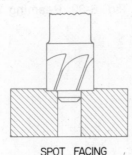

SPOT FACING

Fig. 6-34. Spot facing is done on castings or surfaces that are uneven. The spot-facing tool machines a surface that a nut or other fastener will sit on squarely.

Fig. 6-35. A bench drill press is designed for small, light work and will commonly drill holes in steel up to ½″ in diameter.

eration most frequently used after drilling. It produces a round, straight hole and enlarges the hole a small amount to produce a quality surface finish and an accurately sized hole, Fig. 6-33.

Frequently on castings the drilled hole will not be at a right angle of the metal's surface. **Spot facing** finishes a round, flat surface around the top of a drilled hole so that the head of a bolt or nut will tighten without being tilted, Fig. 6-34.

Drill Presses

Drilling machines vary widely as to their size classification and construction. Manufacturing plants use these classifications of drill presses: sensitive drill presses, upright drill presses, radial arm drill presses, gang drill presses, multispindle drill presses, deephole drilling machines, and automatic production drill presses.

Sensitive Drill Press. The sensitive drill press is a lightweight, high-speed drill press. It is designed for drills from 0.004″ to 5/16″. Frequently the drive for this type of press has continuous speed regulation for the drill speeds. The drill is counterweighted so that the cutting action of the drill may be felt by the operator.

A bench or floor drill press is the most common drill press. It powers drills up to ½″, Fig. 6-35. The bench press usually has four spindle speeds powered by a fractional horsepower motor and a belt drive. The machine is inexpensive and very adaptive to home, school, toolroom, or light manufacturing plant use.

Upright Drill Press. The upright drill presses are vertical in construction and are heavy-duty machines, Fig. 6-36. These production machines have power feed on the drill and have as many as 12 spindle speeds driven through a gear head. A common upright drill press is a 26″ press with a 5 horsepower motor and with a hydraulic feed. The spindle hole is drilled to receive a number 4 Morse taper. The upright drill press has a very rigid construction of cast iron with a boxlike column. This added rigidity makes it possible to drill, bore, ream, and tap holes. This tool does general drilling in a machine shop on short run production products.

Radial Drill Press. The radial drill is a large machine that is movable through a large arc. Large work may be mounted on or beside the press. The radial arm swings to a position over the work, and the holes are drilled. The radial drill is a precision tool designed with a heavy base and a massive column which supports a 6′, 8′, or 10′ arm and a drilling head, Fig. 6-37. With the workpiece squared and leveled, the drill may be moved around the whole workpiece and the holes drilled. The tool may continue to work by reaming, boring, spot facing, and tapping the holes that have been drilled. All feeds are by power, including the raising of the radial arm. This machine will drill 1½″ holes in steel and 2″ holes in cast iron with precision results. The universal radial drill enables the drill head to be swiveled. This gives the added advantage of drilling angular holes off of the vertical plane.

This tool is designed for accurate large, heavy work. It is useful for repetitive drilling of a series of different sized holes and operations on moderately short production lot sizes.

Gang Drill Press. The gang drill press performs a series of operations on a workpiece as it is moved down the drill press table from spindle to spindle, Fig. 6-38. Any

Fig. 6-37. A radial drill press is designed to drill work mounted on the table or larger work setting on the floor within the radius of the arm and drilling head.

Fig. 6-36. The upright drill press is a production machine with power feeds and a selection of spindle speeds.

Fig. 6-38. Gang drill presses are set up so that the work may be mounted in a drill jig and each operation performed in a **separate** drill press as the fixture is moved down the line.

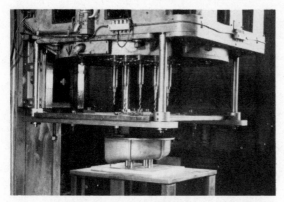

Fig. 6-39. A multispindle drill press will drill a large number of holes at the same time.

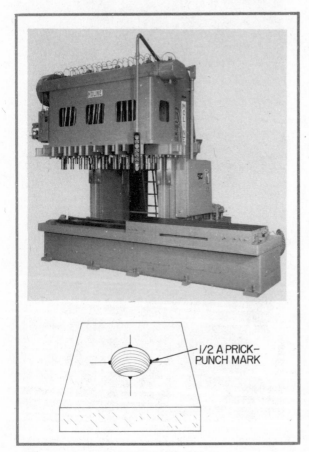

1/2 A PRICK-PUNCH MARK

Fig. 6-40. A multispindle, adjustable driller with 18″ x 72″ drilling area. These drills are used for large-volume production of workpieces.

number of drill press heads and spindles may be mounted on a single table, each with its own motor and speed and feed set. The workpiece may be mounted in a drill jig or fixture as required and taken through the sequence. Perhaps three different sizes of drills, counterbores, and tapping heads (with each tool in a separate drill press) may be used. The operator may take the work through all the operations without having to change tools or reset the machine.

Multispindle Drill Press. Multispindle drills have been developed so that several holes may be drilled simultaneously into the workpiece, Figs. 6-39 and 6-40. This type of drilling is for long-term mass production of parts for the mass market. The machines have power feeds and drill speeds related to the size of the drill and type of work. The drilling cycle involves rapid advance to the workpiece, proper feed during drilling, and rapid return to unloading position.

In most cases these machines drill accurately spaced holes without drill jigs because the press itself is designed to perform this function. The machines are specially designed for a single workpiece and are for large volume production. Over 100 holes may be drilled simultaneously.

Way Drilling. Way-type drilling machines are machines that drill multiple holes on different axes. These drills are used on production work such as engine block castings, transmission and differential cases, and like work. The drills come into the workpiece from different directions, or ways, Fig. 6-41.

This type of machine is built into a transfer machine and produces great volumes of parts necessary in today's automated production, Fig. 6-42.

Drill Press Operation

Drill press operation requires some knowledge of the cutting tool, the drill, before the press may be used efficiently, Fig. 6-43.

Drill Systems. Drills are sized by four systems: (1) by fractions, (2) by numbers, (3) by letters, and (4) by metric. The **fractional set** of drills is sized by a fraction of an inch—such as 1/16″, ⅛″, ⅜″.

These drills are frequently in sets of drill sizes from 1/16″ to ½″, increasing in sizes of 1/64″. The fractional drills go in size to 3¼″ in diameter (even up to 4″ on special order). These drills are the common drills which are used for the majority of the drilling jobs.

Another set of drills which are very commonly used for smaller holes are the **number drills**, or drills which are assigned a number. The number comes from the wire gage, and the drill size is equal to that particular wire diameter. These drills are used for the smaller sized holes such as those found in carburetor parts, etc. This set of sizes range from drill size 1 to 60 and

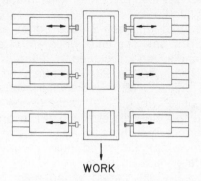

WORK

Fig. 6-42. Way machines may be combined into a transfer machine. The workpiece is moved under the cutting tools. The work is performed and then transferred to the next station.

Fig. 6-41. The way drill may have single or multiple tools or drills mounted on the head.

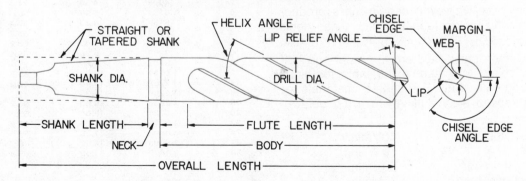

Fig. 6-43. The general-purpose drill is a very efficient tool and will commonly be supplied with a straight or tapered shank.

another set for very small holes range from size 61 to 80.

The third drill set is referred to as the **letter drill set**. The letter drills are sized by thousandths and assigned a letter from A to Z. The A drill is .2340″ in diameter, and the Z drill is .4130″ in diameter. The letter drills have two general uses: (1) to finish a hole to a specific size and (2) to provide the correct sized hole for a tap. The tap will require a specific sized hole in order to provide enough material around the tap to produce a 75% thread engagement of the two threads, thus the size of the hole is the controlling factor. The letter drills are sized so that there will be additional drill sizes between those in the fractional set of drills. When letter drills are used, they should follow a hole drilled by a fractional drill.

The **metric set** is the fourth set of drills and is sized directly in millimetres. The metric drills are frequently used in the manufacture of scientific instruments and the space industry, Fig. 6-44.

Drilling Speeds. Drills are made of various materials. Those made of carbon tool steel must be used with low cutting speeds. Others are made of high-speed steel, which doubles the cutting speed. They are more expensive than the carbon tool steel drills, but they are strong and will increase production because of their cutting speed. Cemented carbide-tipped drills are economical for high-speed production, but they require special care in use in order to prevent breakage. The expense of carbide drills is greater than that of the high-speed steel drills. However, certain mass production drilling operations make these tools economical, Figs. 6-45, 6-46, and 6-46A.

Carbon steel drills are turned at one half the speed indicated in the table for high-speed drills. Cutting speeds of materials for high-speed steel drilling are listed below:

Cast steel	40′/min.
Tool steel	50′/min.
Drop forgings	60′/min.
Cast iron (hard)	70′/min.
Malleable iron	90′/min.
Mild steel	100′/min.
Cast iron (annealed)	150′/min.
Brass and aluminum	300′/min.

Calculate the revolution per minute (RPM) of a drill by selecting the cutting speed from the table in Fig. 6-46 and multiplying it times four. Then divide the results by the diameter of the drill. The formula is given below.

$$RPM = \frac{C.S. \times 4}{D}$$

TAPPING DRILL DIAMETERS FOR METRIC SCREW THREADS

NOMINAL DIA. OF THREAD (IN MM)	DEPTH OF EXTERNAL THREAD (IN MM)	TAPPING DRILL DIA. (IN MM)	RADIAL ENGAGEMENT WITH EXT. THREAD %
1.6	.2147	1.30	70
1.8	.2147	1.50	70
2	.2454	1.65	71.3
2.2	.2760	1.80	72.5
2.5	.2760	2.10	72.5
3	.3067	2.60	65.2
3.5	.3681	3.00	68
4	.4294	3.40	70
4.5	.4601	3.90	65.3
5	.4908	4.30	71.3
6	.6134	5.20	65.4
7	.6134	6.20	65.4
8	.7668	7.00	65.2
9	.7668	8.00	65.2
10	.9202	8.80	65.1
11	.9202	9.80	65.1
12	1.0735	10.60	65.2
14	1.2269	12.40	65.2
16	1.2269	14.25	71
18	1.5336	16.00	65.2
20	1.5336	18.00	65.2
22	1.5336	20.00	65.2
24	1.8403	21.50	68
27	1.8403	24.50	68
30	2.1470	27.00	69.8

Fig. 6-44. Metric tap drill sizes.

Tap Drill Sizes

Sizes of Taps, Tap Drills[1], and Clearance Drills[2]

Size of Tap		Outside Diameter (Inches)	Root Diameter (Inches)	Size of Tap Drill (In Inches) For 75% Thread Depth			Clearance Drill (Inches)		Clearance (Inches)
UNC NC (USS)	UNF NF (SAE)			Number and Letter Drills	Fractional Drills	Decimal Equivalent	Size	Decimal Equivalent	
	#0-80	0.0600	0.0438	...	3/64	0.0469	#51	0.0670	0.0070
#1-64	...	0.0730	0.0527	53	...	0.0595	#47	0.0785	0.0055
	#1-72	0.0730	0.0550	53	...	0.0595	#47	0.0785	0.0055
#2-56	...	0.0860	0.0628	50	...	0.0700	#42	0.0935	0.0075
	#2-64	0.0860	0.0657	50	...	0.0700	#42	0.0935	0.0075
#3-48	...	0.0990	0.0719	47	...	0.0785	#36	0.1065	0.0075
	#3-56	0.0990	0.0758	45	...	0.0820	#36	0.1065	0.0075
#4-40	...	0.1120	0.0795	43	...	0.0890	#31	0.1200	0.0080
	#4-48	0.1120	0.0849	42	...	0.0935	#31	0.1200	0.0080
#5-40	...	0.1250	0.0925	38	...	0.1015	#29	0.1360	0.0110
	#5-44	0.1250	0.0955	37	...	0.1040	#29	0.1360	0.0110
#6-32	...	0.1380	0.0974	36	...	0.1065	#25	0.1495	0.0115
	#6-40	0.1380	0.1055	33	...	0.1130	#25	0.1495	0.0115
#8-32	...	0.1640	0.1234	29	...	0.1360	#16	0.1770	0.0130
	#8-36	0.1640	0.1279	29	...	0.1360	#16	0.1770	0.0130
#10-24	...	0.1900	0.1359	25	...	0.1495	13/64	0.2031	0.0131
	#10-32	0.1900	0.1494	21	...	0.1590	13/64	0.2031	0.0131
#12-24	...	0.2160	0.1619	16	...	0.1770	7/32	0.2187	0.0027
	#12-28	0.2160	0.1696	14	...	0.1820	7/32	0.2187	0.0027
1/4"-20		0.2500	0.1850	7	...	0.2010	17/64	0.2656	0.0156
	1/4"-28	0.2500	0.2036	3	...	0.2130	17/64	0.2656	0.0156
5/16"-18		0.3125	0.2403	F	...	0.2570	21/64	0.3281	0.0156
	5/16"-24	0.3125	0.2584	I	...	0.2720	21/64	0.3281	0.0156
3/8"-16	...	0.3750	0.2938	...	5/16	0.3125	25/64	0.3906	0.0156
	3/8"-24	0.3750	0.3209	Q	...	0.3320	25/64	0.3906	0.0156
7/16"-14	...	0.4375	0.3447	U	...	0.3680	29/64	0.4531	0.0156
	7/16"-20	0.4375	0.3725	...	25/64	0.3906	29/64	0.4531	0.0156
1/2"-13		0.5000	0.4001	...	27/64	0.4219	33/64	0.5156	0.0156
	1/2"-20	0.5000	0.4350	...	29/64	0.4531	33/64	0.5156	0.0156
9/16"-12	...	0.5625	0.4542	...	31/64	0.4844	37/64	0.5781	0.0156
	9/16"-18	0.5625	0.4903	...	33/64	0.5156	37/64	0.5781	0.0156
5/8"-11	...	0.6250	0.5069	...	17/32	0.5312	41/64	0.6406	0.0156
	5/8"-18	0.6250	0.5528	...	37/64	0.5781	41/64	0.6406	0.0156
3/4"-10	...	0.7500	0.6201	...	21/32	0.6562	49/64	0.7656	0.0156
	3/4"-16	0.7500	0.6688	...	11/16	0.6875	49/64	0.7656	0.0156
7/8"- 9		0.8750	0.7307	...	49/64	0.7656	57/64	0.8906	0.0156
	7/8"-14	0.8750	0.7822	...	13/16	0.8125	57/64	0.8906	0.0156
1"- 8	...	1.0000	0.8376	...	7/8	0.8750	1 1/64	1.0156	0.0156
	1"-14	1.0000	0.9072	...	15/16	0.9375	1 1/64	1.0156	0.0156

If you cannot get the size of tap drill given here, see Fig. 6-46 to find the size of drill nearest to it. Be sure to get a drill a little larger than the **root diameter** of the thread.

The drill that makes a hole so that a bolt or screw may pass through it is called a **clearance drill**. This drill makes a hole with a **clearance** for the **nominal diameter of thread**. The **clearance** equals the difference between the clearance drill size and the nominal diameter of the bolt or screw:

Clearance drill = Diameter of bolt or screw + Clearance

Example: The clearance for a 1/4" bolt or screw is 1/64". The size of the clearance drill should, therefore, be 1/4" + 1/64" or 1 7/64".

Fig. 6-45. A drill size is selected so that enough metal is left around the hole to cut the threads and produce a thread with a 75% area contact.

Drill Size	Decimal	Drill Size	Decimal	Drill Size	Decimal	Drill Size	Decimal
80	.0135	42	.0935	13/64	.2031	X	.3970
79	.0145	3/32	.0938	6	.2040	Y	.4040
1/64	.0156	41	.0960	5	.2055	13/32	.4062
78	.0160	40	.0980	4	.2090	Z	.4130
77	.0180	39	.0995	3	.2130	27/64	.4219
76	.0200	38	.1015	7/32	.2188	7/16	.4375
75	.0210	37	.1040	2	.2210	29/64	.4531
74	.0225	36	.1065	1	.2280	15/32	.4688
73	.0240	7/64	.1094	A	.2340	31/64	.4844
72	.0250	35	.1100	15/64	.2344	1/2	.5000
71	.0260	34	.1110	B	.2380	33/64	.5156
70	.0280	33	.1130	C	.2420	17/32	.5312
69	.0292	32	.1160	D	.2460	35/64	.5469
68	.0310	31	.1200	1/4	.2500	9/16	.5625
1/32	.0312	1/8	.1250	E	.2500	37/64	.5781
67	.0320	30	.1285	F	.2570	19/32	.5938
66	.0330	29	.1360	G	.2610	39/64	.6094
65	.0350	28	.1405	17/64	.2656	5/8	.6250
64	.0360	9/64	.1406	H	.2660	41/64	.6406
63	.0370	27	.1440	I	.2720	21/32	.6562
62	.0380	26	.1470	J	.2770	43/64	.6719
61	.0390	25	.1495	K	.2810	11/16	.6875
60	.0400	24	.1520	9/32	.2812	45/64	.7031
59	.0410	23	.1540	L	.2900	23/32	.7188
58	.0420	5/32	.1562	M	.2950	47/64	.7344
57	.0430	22	.1570	19/64	.2969	3/4	.7500
56	.0465	21	.1590	N	.3020	49/64	.7656
3/64	.0469	20	.1610	5/16	.3125	25/32	.7812
55	.0520	19	.1660	O	.3160	51/64	.7969
54	.0550	18	.1695	P	.3230	13/16	.8125
53	.0595	11/64	.1719	21/64	.3281	53/64	.8281
1/16	.0625	17	.1730	Q	.3320	27/32	.8438
52	.0635	16	.1770	R	.3390	55/64	.8594
51	.0670	15	.1800	11/32	.3438	7/8	.8750
50	.0700	14	.1820	S	.3480	57/64	.8906
49	.0730	13	.1850	T	.3580	29/32	.9062
48	.0760	3/16	.1875	23/64	.3594	59/64	.9219
5/64	.0781	12	.1890	U	.3680	15/16	.9375
47	.0785	11	.1910	3/8	.3750	61/64	.9531
46	.0810	10	.1935	V	.3770	31/32	.9688
45	.0820	9	.1960	W	.3860	63/64	.9844
44	.0860	8	.1990	25/64	.3906	1	1.0000
43	.0890	7	.0210				

Fig. 6-46. Decimal equivalents of common-sized drills.

Fig. 6-46A.
Drill Speed RPM Chart for High-Speed Drills

Drill Size	50 Ft./Min. Tool Steel	70 Ft./Min. Cast Iron	100 Ft./Min. Machine Steel	300 Ft./Min. Aluminum, Brass, Bronze
1/16"	3056	4278	6111	—
1/8"	1528	2139	3056	9170
1/4"	764	1070	1528	4585
3/8"	510	713	1019	3056
1/2"	382	535	764	2287
5/8"	306	428	611	1830
3/4"	255	357	509	1525
1"	191	267	382	1143

A 2″ diameter high-speed drill cutting cast iron would be found thus:

$$R.P.M. = \frac{70 \times 4}{2} = \frac{280}{2} = 140 \text{ R.P.M.}$$

Drills must be sharpened and repointed to operate efficiently. A drill point has a number of complicated angles and lengths which must be correct and equal on both sides of

Fig. 6-47. The drill is positioned on a grinding wheel, and cams control the clearances as the drill point is ground. Tip lengths and angles are equal within .0002″.

the drill if the metal is to be removed without wasting power and creating heat. Formerly drills were sharpened by hand on a pedestal grinder with the aid of a drill gage. This process required great hand skill. Now drill point grinders which produce precision points on drills have been developed. The motion of the drill point during grinding is controlled by cams; therefore, very accurate clearances and angles are produced, Fig. 6-47.

Drilling Procedure. The workpiece is laid out and prick-punched holes that are to be drilled are opened to receive the drill point with a center punch. The larger center punch mark makes it easier to start the drill in the correct position on the metal. A drill that is about three-fourths the size of the finished hole is selected and mounted in the chuck of the drill press. Be certain to remove the chuck key from the drill chuck when the drill has been tightened. Place the workpiece on the drill press table and directly under the drill. Allow the drill point to locate in the center punched hole. Clamp the work or vise holding the work to the drill press table so that it will not move during the drilling operation.

Select the proper drill speed from the chart and set the machine for that speed. Remove all tools and rags from the drilling area and start the drill. The drill is brought in contact with the work, and the point of the drill is fed into the work until a conical section is drilled. This conical section has the diameter of about one-fourth the diameter of the drill. The drill press is then stopped, Fig. 6-48.

LAYOUT CONICAL SECTION 1/4″ DIAMETER METAL CHISELED AWAY TO MOVE HOLE HOLE CONCENTRIC WITH LAYOUT

Fig. 6-48. A hole started off the layout may have the drill moved by chiseling away some of the metal in the direction the drill should move.

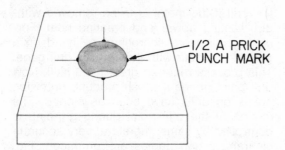

1/2 A PRICK PUNCH MARK

Fig. 6-49. The accuracy of drilling compared to the layout is checked by seeing if one-half of a prick-punched mark is left beside the hole. If present, the hole has been drilled according to the layout.

Fig. 6-51. In numerically controlled drilling, jigs are eliminated, and the complexity of fixtures is reduced.

Fig. 6-50. The drill jig provides a guide for the drill so that the drill is located in the same position as new workpieces are placed in the jig.

If the conical section is not concentric with the layout, the center punch is used to cut away some of the metal towards the direction that the drill point is needed to move. The work or vise is also moved slightly so that the drill will be moved into a concentric position.

The drill press is started, and the drill fed into the work until one-half or three-quarters of the drill diameter is drilled. Again, the hole is checked to see if the drill has moved over and the conical section is concentric with the layout. If the drill is now concentric, the hole may be drilled. If the drill needs further correction, the cutting with the punch and the movement of the vise are repeated.

When the hole has been accurately located and drilled, the drill press is stopped. The drill is removed and replaced by a finish drill of the size indicated on the blueprint. The drill press is started, and the hole is drilled to the finish size. If the hole has been drilled correctly, there should be one-half of a prick-punch mark left on all four sides of the hole. The four half prick-punch marks tell the inspector that the hole has been drilled according to the original layout, Fig. 6-49.

Fig. 6-52. A pressure vessel head for a nuclear reactor being checked after drilling, boring, and counterboring.

Drill Jigs and Fixtures. A **drill jig** is a tool that provides a guide for a drill. The jig holds the workpiece, locates the hole on the work, and guides the drill into the proper position. A jig is made up of a plate with precisely located holes fitted with hardened steel bushings to locate the drill. Frequently, drill jigs are built like boxes so that the workpiece may be located accurately, clamped under the bushed plate, and held precisely while the drilling takes place, Fig. 6-50.

Drill fixtures are clamping devices for locating and clamping the workpiece under a drill. A drill fixture does not have a drill guide.

Numerically Controlled Drilling. Numerically controlled drilling has changed the concept of drilling in that the location of the hole is not controlled by the operator or by a drill jig but rather by very accurate positioning of the work under the drill. A simple fixture mounts the work to the table, and the table is moved to a precise position by a coded tape or computer, Fig. 6-51. Consistently, the location of the table under the drill within a half-thousandth of an inch or less can be positioned under the cutting tool. This degree of accuracy can be attained on the entire working range of the table, Fig. 6-52.

With a machine available for production use that locates the holes this accurately, the old idea that accurately located holes cost more money is no longer true. Conventional tooling would require expensive jigs and presses to equal the tolerances that the numerically controlled tools can produce without a cost increase.

Activities

Build a project, using metal shearing and drilling.

Related Occupations

These occupations are related to shearing and drilling:

Sheet metal worker
Structural iron worker
Boilermaker
Ornamental iron worker
Machinist
Power shear operator
Blanking machine operator
Angle cutting machine operator
Bolt cutter operator
Material cutter
Parts trimmer
Squaring shear operator
Sensitive drill press operator
Upright drill press operator
Radial drill press operator
Gang drill press operator
Turret drill press operator
Multispindle head drill press operator
Way drilling machine operator
Tape controlled drilling machine operator
Gun drilling machine operator

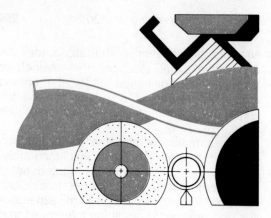

Practices of Metal Forming

Chapter 7

Words You Should Know

Press brake — A powerful machine that holds forming dies for bending and shaping metal sheets.

Air bending — An acute angle punch is closed partly into the die, and the bend is made. The deeper the punch goes into the die, the sharper the bend. Normally the die does not close.

Roll forming — A mill made up of a set of rolls. The metal strip passes through the mill with each set of rolls making part of the total bend.

Rubber forming — Rubber is restricted over a male die. The crushed rubber provides the energy to form the metal over the die.

Hydroforming — The use of oil pressures to apply the energy to form metal over a die.

Stretch forming — Metal is prestressed between two jaws, and the die is pushed up into the sheet to form the new shape.

Metal spinning — The metal is rotated, and pressure is applied to the spinning disc. The metal flows down to the surface of the chuck and takes on its shape.

Metal drawing — Metal is stretched between die surfaces and stretched or pulled through a die, as in wire drawing.

Coining — Squeezing metal in a closed die to straighten, shape, or finish the part.

Swaging — The reduction in size or shape of a metal part by pressure.

Intraforming — Forming a special shape on the inside of a part over a mandrel such as a spline in a gear hub.

Microsecond — A millionth part of a second.

Magnetic pulse — A strong attractive or repulsive force which occurs in a short burst.

Detonate — Explode with violence.

Shock wave — A progressive disturbance of high energy.

Extrusion — To push or thrust out through a die.

Compact — Pressed together until firmly united.

Sintering — The bringing together of metal by heating particles.

Forging — Heating, hammering, or pressing metal into a shape.

Drop forging — A heavy die half falls vertically upon the workpiece. The bottom half of the die is the anvil for shaping the metal.

Cold-Working Metal

The manufacturing processes that bend, squeeze, upset, and draw out metals are called **forming operations**. They compress or stretch metals into the desired shape. The work which is done with this type of work is commonly referred to as **cold-working** if the work is performed below the metal's recrystalization temperature, and **hot-work-**

ing if the work is performed above the metal's recrystallization temperature.

When metals are bent into a shape, both stretching and compression occur, Fig. 7-1. The metal above the center line of the material is usually stretching while the material below the center line is frequently upsetting or compressing. Theoretically there is little change at the center line.

Cold-Forming Sheet Metal

Angle Bending. A bar folder is a manufacturing tool used to form light-gaged sheet metal products. This machine is used to form edges, seams, and small structural parts, Figs. 7-2 and 7-3. The bar folder bends sheet metal to any angle from 0° to 180°, and to a depth of 1″ and a length of 36″. The bar folder is a hand-operated tool that forms the following edge finishes and locks, Fig. 7-4.

The **box and pan brake** is a variation of the cornice brake, which is designed to produce folds and bends on long pieces of sheet metal. The box and pan brake is designed to have adjustable and removable

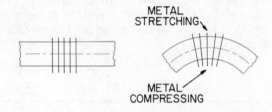

Fig. 7-1. When metal is formed, metal above the center line stretches, and metal below the center line is compressed.

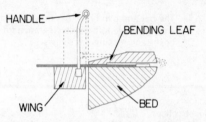

Fig. 7-3. The adjustable bar folder will make a sharp bend when the wing is level with the bed. A round or pocket fold is made when the wing is below and away from the bed.

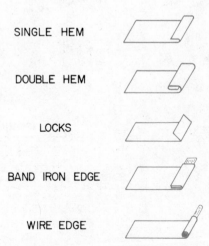

SINGLE HEM

DOUBLE HEM

LOCKS

BAND IRON EDGE

WIRE EDGE

Fig. 7-2. Angle bending may be done on a bar folder. Frequent bends are for making structural parts using single hems, double hems, or locks.

Fig. 7-4. The bar folder forms edges, seams, and small parts. Sharp and radius bends may be formed from 0° to 180°.

fingers of various sizes, Fig. 7-5. These fingers make the upper jaw adjustable so that clearance is provided and a box may be

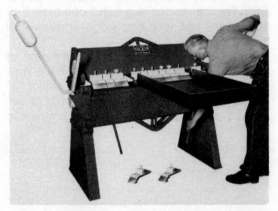

Fig. 7-5. The bar and pan brake has movable fingers that are adjustable to provide clearance for bending boxes and rectangular shapes.

formed. When the sides of a box have been bent up to 90°, the fingers will be adjusted to the length of the ends, and clearance will be allowed by removing the finger on either side of the end so that the metal will have clearance while bending (Fig. 7-6). Boxes will be made with odd dimensions such as ⅛″ or 3/16″ longer than an even inch dimension, because the fingers have widths of even inches, and the box must clear the ends of the fingers of the box for forming.

When metal 25 gage or less is bent, an allowance for the thickness of the metal is not required. There is enough stretch in the metal to do accurate work. Metals heavier than 26 gage, however, do require bending allowances equivalent to 3-5 thicknesses of the metal. For heavy metal a table of bending radii should be consulted for the correct allowance.

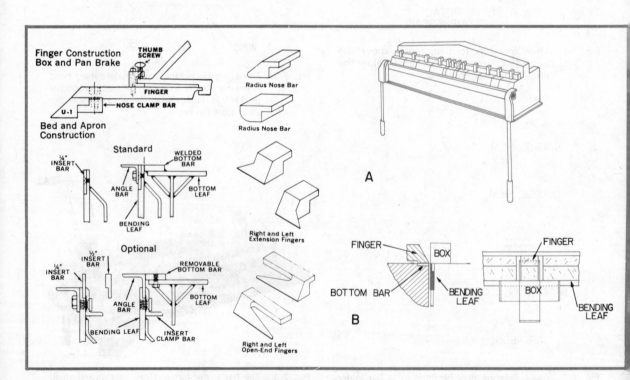

Fig. 7-6. The box and pan brake is constructed so that pre-bent shapes may clear the forming fingers. Sharp bends and radius bends are made with special attachments.

Bending radii should not be so short that cracks develop on the stretching side. The bending radius varies with the ductility of the material being bent. Materials that crack readily require larger bending radii for forming, Fig. 7-7. The additional material is added to the layout.

Metal seams may be made from the metal being formed. To allow for the seams, additional material is added to the pattern. The flat lock groove seam is often used to join sheet metal together, Fig. 7-8. This seam requires a dimension of 3 widths of the seam in additional metal added to the pattern. Different seams have different allowance requirements, Fig. 7-9.

The metal allowances needed may be found by determining the width of the seam and multiplying it by the number of thicknesses of metal used. Remember, one width of the seam is part of the product and is usually included in the size dimension. Accurate determination of the amount of material allowance can be made by taking two pieces of the same gage metal 1″ by 6″ and making a seam across the 1″ dimension, fastening the material end to end. By measuring the amount of metal and subtracting from 12″ (the original length of the two strips), the amount needed for the seam allowance is determined.

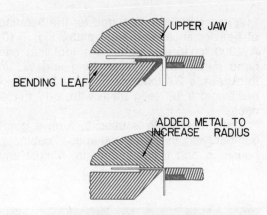

Fig. 7-7. Heavy metal or less ductile metal will require a larger bending radius and may be obtained by adding metal to the upper jaw to provide the radius.

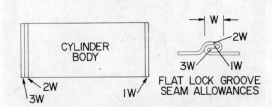

Fig. 7-8. The flat lock grooved seam requires three widths of the seam to make the metal lock. This is a strong and frequently used sheet metal seam.

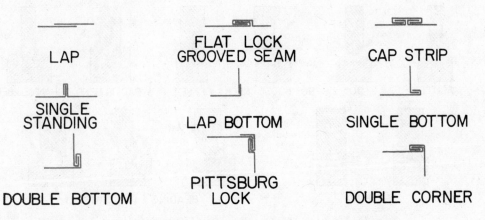

Fig. 7-9. Sheet metal seams used to permanently fasten metal parts by bending.

A **press brake** is required for the bending of heavier sheet and steel plate, Fig. 7-10. A 1000 ton press brake is a tool that can bend metal up to 30 feet long and ½″ in thickness, a 20′ long plate with a ¾″ thickness, and a 12′ long plate with a 1″ thickness.

The press brake is used to form a great number of structural shapes, cabinets, housings, and tubular sections. Almost any cross-sectional design may be bent if it is possible to get the metal into the press and out after the forming, Fig. 7-11.

The press brakes use a few standard dies which will produce a great number of bends and shapes. The common groups of dies

Fig. 7-11. A 200-ton press brake forming small ⅜″ thick metal parts.

Fig. 7-10. The press brake is a production tool for bending heavy, long steel plates and sheet. A 1000-ton brake will bend material up to 1″ thick.

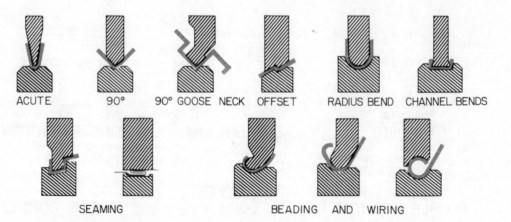

ACUTE   90°   90° GOOSE NECK   OFFSET   RADIUS BEND   CHANNEL BENDS

SEAMING   BEADING AND WIRING

Fig. 7-12. Press brakes with a few standard dies can produce many bends and shapes.

used on 14 gage metal or less are these: acute angle 90° bends, offset bends, radius bends, seaming, beading and wiring, and channel bends, Fig. 7-12.

In addition to performing bending, large press brakes perform a variety of notching, punching, blanking, straightening, trimming, and corrugating operations, Fig. 7-13.

The press brake is used to do air bending. **Air bending** is done by using a set of acute angle dies but not allowing the punch to close down tightly on the metal. Controlling the distance the punch moves into the die determines the amount of bend radius formed. An indefinite number of radii may be formed by air bending with a series of bends, Fig. 7-14. With the aid of the proper press brake tools, nearly all sheet and plate stock may be formed into round, curved, rectangular, straight, or tapered products.

Recently small sheet metal press brakes have eliminated the lower die from the forming operation by substituting a polyurethane pad. The urethane offers resistance to the metal as the punch is driven into the urethane, and the metal is formed, Fig. 7-15. Prefinished materials and materials which scratch easily can be formed without damage by this method. Many products are more economically produced if the finishing is done while the stock is in bulk form before it is shaped. Aluminum, stainless, copper-plated metal, and even painted steel can be thus formed without marks.

Roll Forming. Roll forming is based upon the concepts of air forming except that the support points are rolls whose distance is adjusted between their respective centers. The bending force of the metal is supplied by the relative position of the three rolls or the bending radii.

Light gage sheet metal may be hand roll-formed by using the same forming principles. Sheet metal is rolled, turned, burred,

Fig. 7-14. Air bending is done by controlling the distance the punch moves down. Moving the work under the punch makes it possible to form an indefinite number of radii.

Fig. 7-13. Tools may be added to the press brake so they may perform punching, notching, blanking, trimming, and straightening operations.

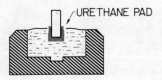

URETHANE PAD

Fig. 7-15. The urethane offers resistance to the metal and punch and forms the metal around the punch. A lower die is not needed for sheet metal.

Fig. 7-16. A hydraulic pinch-pyramid, plate-bending roll will form cylindrical, cone, and crimp-plate edges of cold steel plate which are 7″ thick by 12 feet wide.

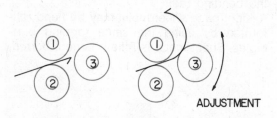

ADJUSTMENT

Fig. 7-17. Roll forming is done between three rolls that may be positioned to bend the metal into an arc or radius.

beaded, wired, and crimped by using these bending forces, Fig. 7-16.

The **slip roll-forming machine** is used on light gage sheet metal to bend the stock into cylindrical shapes. A lock usually is bent on each end of the sheet to be formed in a bar folder. Later it is closed into a seam, then the cylinder is formed. The two front rolls feed the metal into the third roll, bending the metal into a curve, Fig. 7-17. The curvature is controlled by the distance roll number 3 is from roll number 1. The roll numbers 2 and 3 are adjustable. Number 2 controls the space allowed for the thickness of the metal and the locks on the ends of the sheet. The bending takes place by pressure on all three rolls, but the adjustment of roll 3 determines the radius that the metal will be bent.

Cylinders from 2″ in diameter to a number of feet may be formed in this manner, Fig. 7-18. Sheet metal cylinders use the locks on the ends of sheet to make a flat lock groove seam down the length of the cylinder.

Turning, wiring, beading, crimping, and burring are performed on **hand rotary ma-**

Fig. 7-18. Sheet metal cylindrical and conical shapes may be rolled to shape and slipped out the end opening.

Fig. 7-19. Combination rotary machines will turn, wire, bead, crimp, and burr. The machine illustrates a crimping operation.

chines, Fig. 7-19. These operations may be done on separate machines, or they may be performed on a combination rotary machine. The combination machine has sets of rolls for each operation, and the various rolls may be selected and mounted on the machine to perform the above roll-forming processes.

The metal is placed between the rolls of the **turning machine** and against the gage. The top roll is lowered until a crease is made in the metal, and the rolls are turned until the crease runs completely around the product. The machine is turned and the turning roll is lowered after each revolution. As the pocket is being turned, the metal is slowly raised until the metal is nearly parallel with the upper roll. **Turning** is the operation which produces a pocket around the top of a funnel for a wired edge or other round sheet metal products which require a wire edge after the parts have been formed, Fig. 7-20. The wiring machine is another hand rotary machine for the closing of a pocket around a wire. The upper roll of the wiring machine closes the pocket around the wire, Fig. 7-21.

The **beading machine** is used to reinforce the sheet metal. The bead changes the direction of the metal and thereby increases the material's rigidity or stiffness.

There are three common beads for sheet metal: (1) the ogee bead, (2) the single bead, and (3) the triple bead, Fig. 7-22. These beads may be obtained in different sizes; the width of the single bead is the bead's size. The metal is placed against the gage and the top roll is gradually closed as the metal is rotated through the rolls. The bead is gradually rolled into the metal.

The **crimper** has the function of reducing the diameter of the end of a sheet metal pipe so that it will slide into a pipe of the same diameter. Crimpers are frequently a

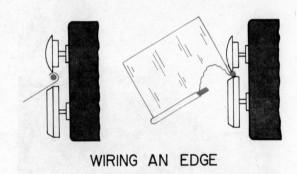

WIRING AN EDGE

Fig. 7-21. The wiring machine rolls the metals tightly around the wire.

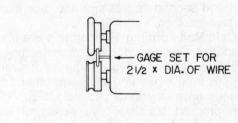

GAGE SET FOR 2 1/2 × DIA. OF WIRE

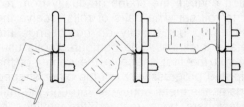

Fig. 7-20. The turning machine will form a pocket for a wire that is frequently installed to strengthen the edge of a product.

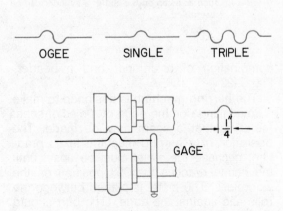

OGEE   SINGLE   TRIPLE

$\frac{1''}{4}$

GAGE

Fig. 7-22. The rolls are changed on the machine to produce the bead desired. Beading reinforces metal by changing the planes of the metal's surface.

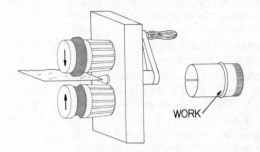

Fig. 7-23. Combination crimping and beading rolls perform two forming operations in one rotary machine. The crimper gathers the metal at the end of a pipe and makes its diameter smaller.

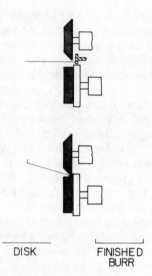

Fig. 7-25. Burring requires hand skill, experience, and proper placement of the fingers.

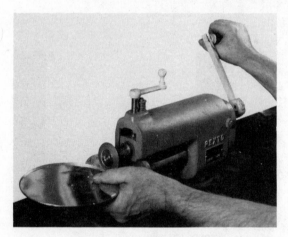

Fig. 7-24. Burring turns up a flange on metal so that a seam such as a cap on the end of a cylinder can be assembled.

combination of a crimper and a beader, Fig. 7-23.

The **burring machine** is designed to make a small flange on the edge of a disc of sheet metal so that a seam may be made. The operation of a burring machine takes practice, because the metal must be upset until the flange reaches the 90° position on the workpiece. The metal is placed between the rolls and against the gage. (The burr should be less than 1/8″ on small diameter discs). The first and second fingers of the left hand rest behind the lower roll and under the

metal. The thumb is on top of the metal, holding the metal between it and the first finger and offering resistance while holding the metal against the gage, Fig. 7-24. The crank screw lowers the upper roll until an equidistant line is scored around the disc. (This scoring is very important and should not have any crossover lines in its pattern.) The metal is rotated, and pressure from the first and second fingers spin the disc into a vertical position, Fig. 7-25.

Cold Roll Forming. Roll forming is a mass production process for the manufacture of standard stock such as garage door rails, architectural forms and trim, and other continuous cross-sectional products, Fig. 7-26. The process consists of a series of rolls which change the shape gradually from roll set to roll set. The pairs of rolls receive the metal strip but usually do not change the metal's thickness. The contoured rolls usually roll the first shape in the center of the strip and progressively roll the other shaped parts as the metal continues through the mill.

Designers have realized the savings of weight and increased strength provided by using alloy steels and aluminum rolls formed from strip metal.

Fig. 7-26. This rolling mill forms sheet metal into many continuous shapes for construction and manufacturing.

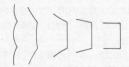

Fig. 7-27. Roll-formed shapes are continuously rolled to produce light, strong supports, and strip products.

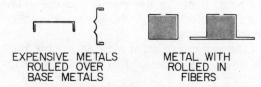

EXPENSIVE METALS ROLLED OVER BASE METALS

METAL WITH ROLLED IN FIBERS

Fig. 7-28. Expensive metals such as brass and stainless steels are rolled around base metals. Fillers such as fiber, felt, rubber, and fiberglass can be rolled into a product.

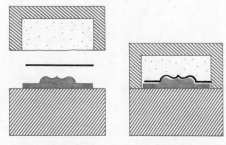

Fig. 7-29. Compressed rubber forms the metal part over the die. Aluminum sheet metal aircraft parts are frequently formed by this process.

The roll-forming process has been controlled to the point that the metal supports for acoustical ceilings have been roll formed with the finish coat of paint on the metal before it was rolled to shape. These machines produce at a rate of from 50 to 200 feet per minute. The small, light mills produce at the higher speed range. Roll forming also rolls fillers into a structural member, Fig. 7-27. Fillers such as rubber, fiber, felt, and fiberglass may be included in the roll-formed product.

Metal shields of expensive metals may also be rolled around a base metal during the forming to produce a heavy, gaged, inexpensive finished form, Fig. 7-28. The article may have a brass or stainless finish which is 0.004″ to 0.006″ thick.

Cold-Formed Metal Shaped by Stretching

Flexible Die Forming. These forming machines utilize rubber, oil, or water as the energy transmission source to form the metal.

Rubber that is put under pressure in a confined area will flow and fill up any existing cavities or voids. Because rubber of the correct hardness is flexible, it will respond much like an entrapped liquid. This characteristic is used in the forming of sheet metal products.

The male shape of the product form is made of aluminum, plastic, steel, kirksite, masonite, or almost any material that will not break down under the forming pressure. Small aircraft parts are frequently made by this process, Fig. 7-29. Such metals as aluminum, magnesium, and light stainless steels are formed by the compressed rubber.

To increase the life of the rubber and the die, sharp corners are removed from the die and a lubricant such as powered talc or graphite is used to help the rubber flow over the metal.

Products such as license plates, furniture, instrument cases, and aircraft structural parts are applicable to this process because of the limited production of parts. The rubber will form about 20,000 parts before it must be replaced.

Bulging is the reverse process of rubber forming, because the rubber is placed inside the product and the pressure applied from the press.

The workpiece must start as a tube, cup, or shell. It is expanded by pressurized rubber, liquid, or elastomers. Bulging will increase the diameter of the part up to 30% in one operation. Bulging requires frequent annealings of the metal in order to form the metal to larger diameters.

Products which are shaped by bulging are automobile hubcaps, various pot shapes, electrical fixtures, and ornamental fixtures, Fig. 7-30.

Hydroforming. Hydroforming is done by the use of hydraulic pressures applied to a flexible diaphragm that in turns forms the metal around the die. There are three pressures that are regulated so that the drawing process can be very carefully controlled:

1. Pressure 1 closes the system and produces a pressure on the blank holder.
2. Pressure 2 causes the diaphragm to wrap the metal around the punch, Fig. 7-31.

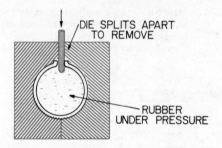

Fig. 7-30. Bulging expands the metal from the inside against a die. The pressure is supplied by rubber, oil, or water.

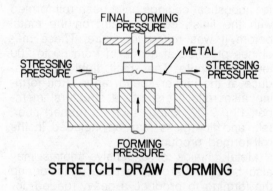

STRETCH-DRAW FORMING

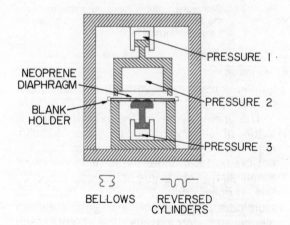

Fig. 7-31. Oil pressures and a diaphragm form difficult parts with complicated shapes. Circular bellows and reverse bent cylinders are formed by this process.

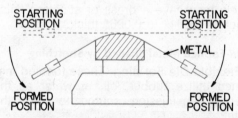

STRETCH-WRAP FORMING

Fig. 7-32. Compound and tapering curves are formed by stretch forming. Curved sections of airliners' cabins, cowling, and wing sections are formed by this process.

3. Pressure 3 causes the punch or die to push into the metal.

Very difficult parts such as circular bellows and reverse bent cylinders may be formed, because this forming process gives the operator so much control.

Stretch Forming. Stretch forming is used when a limited number of very carefully contoured sheets of metal are needed, such as special compound carved pieces of airplane skin, truck, and trailer bodies, Fig. 7-32.

The metal is held between two jaws and is stressed or stretched as a hydraulic cylinder lifts the die into the metal and forms the material. Because the whole thickness of the metal is under tension when the metal is shaped, there is no springback and little thinning of the metal.

The stretch process can be used on metals difficult to form, because the work is prestressed by a pulling force until the tension reached is nearly the yield point. While the work is being stressed, it is formed by dies that contact the metal and form the metal with the additional energy necessary to plastically deform the work.

The forming takes place over the entire surface of the workpiece, and the result of this is fewer wrinkles and buckles. The stretch forming conditions and increases the tensile strength of the metal by about 10%. Because so much surface is used in the forming process, springback is greatly reduced and can be removed by slightly overforming the part.

An important part of the stretch-forming process is the inexpensive materials that may be used for the dies: wood, zinc alloy, cast plastic, kirksite, masonite, cast iron, or low-carbon steel. These materials can easily be shaped and formed locally for the production run. To prevent damage to the metal during forming, the die may be polished or covered with thin rubber sheets to avoid scratches and the need for lubrication.

In **stretch-wrap forming** the metal is drawn around the forming die. Stretch form-

ing is most frequently used for short runs of large, difficult - to - form, double - curvature parts of aluminum, titanium, and stainless steel sheets.

Metal Spinning. In metal spinning, a metal disc is clamped against a chuck, and both the disc and chuck are revolved on the spindle of a lathe. A tool is pressed against the revolving disc and the disc is forced to take the shape of the chuck.

Manual spinning is limited to about 1000 workpieces, because the chuck wears out. Doing more parts than this requires other processes such as mechanical spinning, drawing, or stamping.

Spinning operations are used on odd shaped cylindrical parts such as bowls, pans, food machinery, small tanks, antenna reflectors, floats, dust covers, trays, and missile parts. Aluminum is the most frequently spun. Copper, steel, stainless steel, and brasses are also frequently spun.

Spinning has great flexibility, because the forms or chucks may be made with few tools and shaped to the desired shape or changed to a new shape with wood-turning tools or metal-turning tools. Chucks are made of various woods and metals. Maple, cherry or woods which are hard and possess a close grain are best short-run chucks. These chucks may be changed so that the correct dimensions on the work may be produced after making a few test pieces.

Metal spinning chucks are also made of cast iron, brass, and aluminum. Cast iron chucks will produce a large number of workpieces if given the proper care and use.

Metal spinning is a metal flowing process. The metal will have the same thickness after it is spun as it had before if spinning is done correctly. A skilled metal spinner can actually thicken the metal in certain areas above the basic metal thickness when required.

The metal is spun between the follow block, which is mounted on a spinner's center, and the chuck. The metal rotates toward the operator, and the spinning tool contacts the metal in the third quarter of

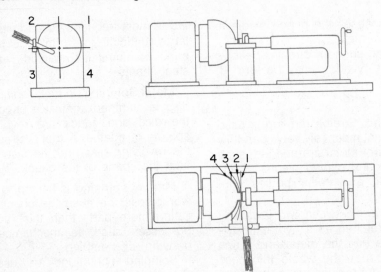

Fig. 7-33.   The spinning tool should contact the metal in the third quadrant. The metal is caused to flow under the tool.

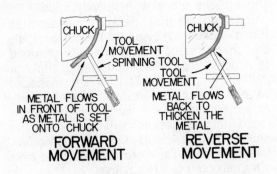

FORWARD
MOVEMENT

REVERSE
MOVEMENT

Fig. 7-34.   On the forward stroke of the spinning tool the metal flows ahead of the tool. On the reverse stroke the metal is set down on the chuck and smoothed out by thickening the metal in its new shape.

the disc, Fig. 7-33. The metal is caused to flow both by the tool's upward pressure upon the chuck and again as the tool is moved back down. This keeps the metal from thinning out. When the metal has received the contour of the chuck, the edge may be trimmed to remove any irregularities, burnished, or rolled over into a bead and polished, Fig. 7-34.

Complicated spinning of several contours will require a series of chucks, as well as annealing of the metal. A break-down chuck, a rough contour, and a finished segment may be required of an article which contains reverse curves in its surface.

Power Spinning. Metal up to 12 feet in diameter and up to ¼″ thick may be spun manually in soft, ductile metals. Mild steel may be spun in thickness of 3/16″ at the same diameter. Work of a greater thickness and diameter requires special power-driven equipment, Fig. 7-35.

Flow turning, shear forming, or power spinning uses the concepts of spinning, but enough power is applied so that forward extrusion of the metal takes place. The process begins with a thicker blank than used for manual spinning and the metal is thinned out during the roll-forming process.

A large 62″ hemisphere may be shear formed. On a large break-down chuck, the 1″ thick aluminum sheet is preformed or shaped by shear forming. The 1″ aluminum is placed on the mandrel and formed by hydraulic-powered wheels. The second wheel follows but has a primary function of keeping the metal in contact with the chuck. The action of the roll causes the metal to flow ahead, producing a longer and thinner hemisphere, Figs. 7-36 and 7-37.

Fig. 7-36. Shear forming on one wheel causes the metal to flow ahead of the roll, while the other is supporting the work in the mandrel.

Fig. 7-35. For heavy work, power spinning is done by power-driven equipment.

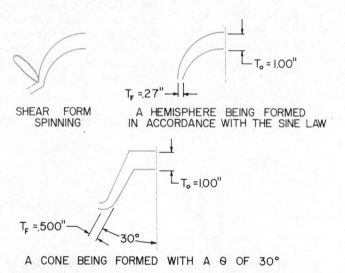

$T_F = .27"$

$T_o = 1.00"$

SHEAR FORM SPINNING

A HEMISPHERE BEING FORMED IN ACCORDANCE WITH THE SINE LAW

$T_o = 1.00"$

$T_F = .500"$

30°

A CONE BEING FORMED WITH A Θ OF 30°

Fig. 7-37. This type of spinning follows the sine law ($T_F = T_o \sin \Theta$). T_F (the final thickness of the metal) equals T_o (the original thickness of the metal).

Products such as rocket fuel tanks, missile cones, thin-walled seamless tubing, television cones, and light fixtures are made by shear forming.

Frequently the power-spinning operation starts with a pre-shaped workpiece produced

Fig. 7-38.   Deep-drawn parts frequently start from a disc of metal and are stretched during the drawing process into a new shape.

in a deep-drawn cup or forging and finished by power spinning. The preforming saves considerable manufacturing time, thus a reduction in cost.

Metals which may be power spun are aluminum alloys, copper alloys, mild steel, stainless steel alloys, magnesium, and titanium. Other metals may be power spun but with less success.

Rigid Die Forming. When drawing processes are used in manufacturing, a punch is used to contact the flat metal and force it to flow between the punch and a die. The punch and die cause the metal to flow into a shape of a cylinder, cup, or box, depending on the shape of the die.

Deep drawing is a stretching process, Figs. 7-38 and 7-39. The blank is placed in the draw-press die. The blank holder places pressure on the blank to hold it tight enough so that it will slide into the die while keeping the metal flat and free from wrinkles.

The pressure of the blank holder during the drawing of cylindrical shells will vary with the thickness of the blanks. The thinner the blank, the greater the holding pressure on the metal from the blank holder. The pressure on the blank holder will vary from very little to one-third the drawing pressure for a 0.030″ blank. Thicker blanks may require less blank holder force; the thicker metals tend to wrinkle less than thin metals.

Deep drawing produces straight walled cups (shells) from flat materials, Fig. 7-40. The shells become the raw material for

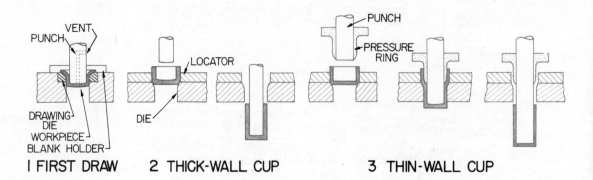

I FIRST DRAW   2 THICK-WALL CUP   3 THIN-WALL CUP

Fig. 7-39.   In deep drawing the cup may be polygonal, stepped-down, hexagonal, square, or circular in shape.

drawing processes. Such products as metal tumblers; cooking utensils; tubular products; automobile bodies; and structural shapes for stoves, sinks and refrigerators are made from deep-drawn shells, Fig. 7-41.

As metal is drawn into the dies, the metal is stretched and work hardened, Fig. 7-42. For deep drawing and angular shapes, the workpiece may be partly drawn, annealed, and redrawn to the final shape.

Deep drawing may be followed by another process — **ironing**. Different thicknesses of metal on the workpiece are eliminated when the material has gone through the ironing process. Ironing provides precision control for wall thickness, Fig. 7-43. A shell is pushed through a die, which sizes and thins the walls. The reduction in wall thickness is usually less than 50% of the original thickness. A heavy top section may be left on the shell so that machining operations may be carried out on the final product.

The manufacture of products which require an accurate size and a good surface finish may be done by drawing the metal

Fig. 7-40. This electrical circuit breaker cover is flanged, dished, pierced, and embossed in one press operation. The cover is made of ¼″ stainless steel.

Fig. 7-42. A rigid die draw-forming an industrial lift truck cover of steel ⅛″ thick.

Fig. 7-41. Draw presses produce industrial parts and heavy parts for passenger cars, trucks, and farm machinery.

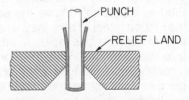

IRONING OR THINNING

Fig. 7-43. Ironing or thinning is a means of precision control over wall thickness.

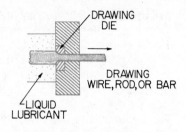

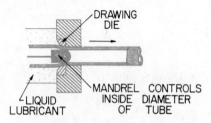

Fig. 7-44.   Steel wire, rod bar, and tubing are cold-drawn to size through dies to produce a product of high quality.

through a die cold, Fig. 7-44. This process is called **cold drawing**. Bars, tubes, wire, and round, square, rectangular, hexagonal, and octagonal stock may be produced by drawing the material through the proper die.

The drawing operation improves the physical characteristics of the metal, orienting the grain and work hardening the material, thus increasing its tensile strength.

Steels are frequently drawn by swaging down the end of the rod or tube until the end passes through the die and is locked to a gripper, which pulls the material through the die. Its cross-sectional area is reduced by 40% or less.

To prepare them for drawing, steels are cleaned with acid to remove the scale from the hot-rolled steel. Then they are washed in lime to remove this acid. A lubricant of soap unites with the lime and provides a lime coating that reduces wear on the die.

The dies that shape the material as it is drawn through must be very strong, hard, and resistant to abrasion. The materials used for these dies are hardened alloy tool steel, tungsten carbide, and diamond.

The power for drawing the material through dies is supplied by a chain, which pulls the stock from the die onto a draw bench. The bar, rod, or tubing is held by a gripper. This gripper pulls the stock down the length of the draw bench, where it is detached. The gripper is moved and attached to a new stock coming through the die. Large amounts of cold-drawn steel are finished by this process. In the drawing process, the dimension of the steel is reduced a few fractions of an inch.

Countless miles of wire are drawn through dies that produce the various sizes and characteristics of the wire. Wire is applied as stock for many manufactured products — from springs, cables, and clips to electrical wire. Each wire will have its own characteristics of hardness and tensile strength. Hardness and tensile strength are related to the composition of the material and the work hardening obtained during drawing. Steel wire may run from .3% carbon for soft, strong wire for general use to very strong, hard wire of up to 1.2% carbon for springs and music wire used in pianos and string instruments.

Cold-Forming Metal by Compression

Squeezing Processes. Squeezing processes that form and mechanically condition metal are cold-rolling, sizing, swaging, cold-forging, coining, stamping, impact extruding, and hobbing, Fig. 7-45. Some of the processes also finish the product in the same forming conditioning operation.

A closed die is used in **coining**. Metal with a larger volume than the die capacity is enclosed and restrained in all directions. When high pressure is applied to the die, the metal flows to the configuration of the die surfaces. Soft metals such as copper, silver, and gold are usually coined. However, other metals such as steel and stainless steel may be struck if they are annealed. Annealed forging and powdered metallurgical products may be coined to finish the size within tolerance and to improve the physical properties by work hardening. Metal coins, silver forks and

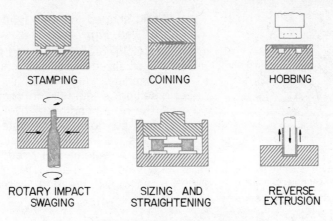

STAMPING   COINING   HOBBING

ROTARY IMPACT SWAGING   SIZING AND STRAIGHTENING   REVERSE EXTRUSION

Fig. 7-45.   Squeezing processes.

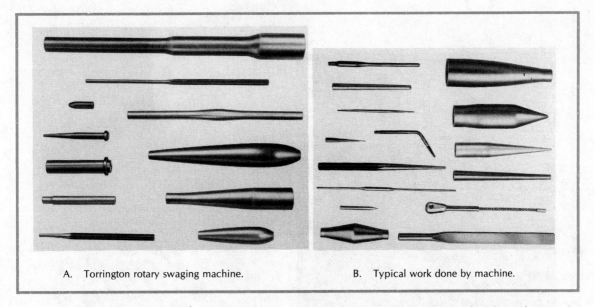

A.   Torrington rotary swaging machine.   B.   Typical work done by machine.

Fig. 7-46.   Rotary swaging is a hammering action that reduces the diameter and increases the length of the work.

spoons, and metal buttons are coined products.

Swaging squeezes the piece in a crushing operation. The diameter or size of the piece is slightly reduced. Consequently, either the piece is elongated, or the walls of the material become thicker. Swaging is used to straighten the work or to size it to a dimension, Fig. 7-46. It often is used as a finishing process, because it improves the appearance of the part.

Rotary swaging is a manufacturing process. The swaging is done in a rotary machine. Long tapers on tubes and such work as round metal furniture legs, mechanical pencil points, speedometers, and sewing machine parts are produced. The ends of bars, rods, or tubes are reduced

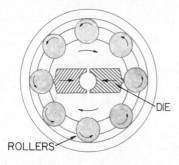

Fig. 7-47. Rotary swaging is done by the dies revolving. The rollers cause the dies to move in each time a roller is passed.

and shaped, and the metal condition is improved, Fig. 7-47.

Cold-forging is carried out by high-speed specialized machines called cold-heading machines, Fig. 7-48. It usually is an up-setting, swaging, and conditioning process. The cold-heading of rivets, bolts, and nails can be done at the rate of 400 or more parts per minute. Raw stock in the form of rods is automatically fed into the machine, cut, headed in a die, removed, and (in some cases) roll-threaded all in this one machine.

A. Flat dies for thread rolling.

B. Round dies for thread rolling.

C. Group of typical threaded parts.

Fig. 7-48. Cold-headed parts are produced automatically by feeding stock into the machine, cutting and heading the part, and frequently by threading.

Many screw fasteners and ball bearings are made in this type of machine which forms and mechanically conditions the metal at the same time.

Intraforming is the squeezing of the metal with a forming mandrel inside the workpiece, Fig. 7-49. The work is crushed around the mandrel. Uniformly spaced ridges (splines) may be formed on the inside of a tube, or internal gears may be shaped by intraforming.

Magnetic pulse forming conditions metal, joins assemblies, and forms metals in microseconds. The material being formed is a conductor of electricity and becomes a single turn of a short-circuited coil. The energy is supplied by a large charged capacitor that is discharged into a shaped coil around, over, or within the workpiece, Fig. 7-50. On the other side of the workpiece is the shape-forming die or component to be swaged. The discharge through the coil creates a strong magnetic field that induces a heavy current into the workpiece conductor. The induced current in the workpiece interacts with the field of the coil, producing a force which slams the workpiece away from the coil and thus onto the forming surface.

The magnetic pulse conditions the metal at the same time the parts are swaged to-

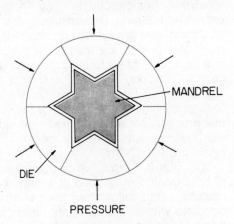

Fig. 7-49. Intraforming work is swaged onto a mandrel. The mandrel may be many shapes, but it must be able to slide out of the work.

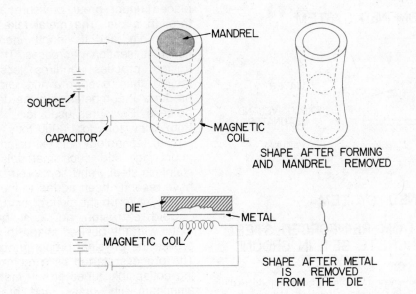

Fig. 7-50. The capacitor bank is discharged into the magnetic coil. It induces a magnetic field that forces the metal into the die cavity or mandrel.

gether into a shape. Tubing may be joined by swaging the ends together and sealed by placing a neoprene O-ring on the tube before swaging. Cables may be crimped tightly into a terminal or connector. Many small components such as electric motors, loudspeakers, or transformers may be joined by magnetic pulse forming. The process can be used to seal containers of radioactive or sterilized materials to prevent contamination.

Explosive forming is a process in which a large amount of energy is released over the workpiece to be formed. A shock wave then forces the metal into a die cavity or other shape. This is done by using high explosives such as T.N.T., C-4, dynamite, or primacord that detonates and produces a

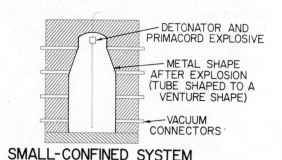

SMALL-CONFINED SYSTEM

- DETONATOR AND PRIMACORD EXPLOSIVE
- METAL SHAPE AFTER EXPLOSION (TUBE SHAPED TO A VENTURE SHAPE)
- VACUUM CONNECTORS

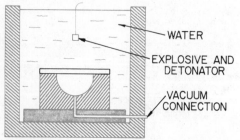

UNCONFINED SYSTEM

- WATER
- EXPLOSIVE AND DETONATOR
- VACUUM CONNECTION

LARGE TANK REINFORCED STEEL OR CONCRETE SET IN GROUND

Fig. 7-51.   High-energy rate forming quickly produces difficult forms, because forming velocities above 100 ft./sec. make many metal slip planes available to move.

shock wave. The shock wave may be transmitted to the workpiece through air, water, or (in the case of heated forming), molten aluminum. Solid materials such as rubber sheets, plastic, and other metals are used to transmit the shock to the metal for special treatments.

When the explosion occurs, the metals receive a tremendous forming force for a very short time. Because of this overwhelming energy which is applied, many more slip planes are moved than would be in a slow traditional forming process. Metals such as titanium, tungsten, and large spherical sections of aluminum which are difficult to form by other methods can be shaped by this process. Most unconfined, formed parts will be large and will be portions of a sphere in shape, Fig. 7-51.

The dies for explosive forming are made of heat-treated alloy steel, low-carbon steel, castable zinc alloy, reinforced concrete, concrete with an epoxy facing, and (for one part,) plaster dies. The material selected for the die is dependent on the size, accuracy, number of parts to be made, and size of the charge which will be used.

In the **extrusion process**, the metal is placed under great pressure and squirted through a die. The metal takes the shape of the die, and its length increases while its cross section decreases. The metal extrusion process requires less expensive tooling than deep drawing, and on many products it competes with the deep-drawing process. Pressures used for extrusion vary from a few tons to 14,000 tons.

The technology of extrusion has been improving, and such materials as steels, stainless steels, and some refractory metals have recently been added to the soft metals which make up the bulk of extrusions.

Cold extrusion not only provides the metal with the desired shape, but it also improves the physical properties of the metal. The process started as a method for forming collapsible tubes of soft metals such as aluminum, tin, brass, and bronze. These tubes are used for such items as toothpaste tubes. In the manufacture of these items a

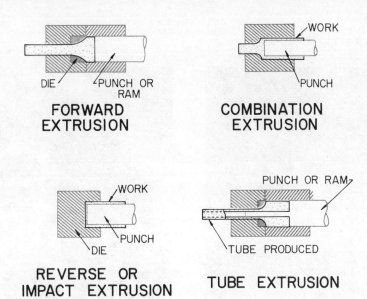

FORWARD EXTRUSION

DIE — PUNCH OR RAM

COMBINATION EXTRUSION

WORK — PUNCH

REVERSE OR IMPACT EXTRUSION

WORK — PUNCH — DIE

TUBE EXTRUSION

PUNCH OR RAM — TUBE PRODUCED

Fig. 7-52.   All forms of extrusion force metal through a die to produce a new shape. Forward, reverse, combination, and tube extrusion are the principal processes.

metal slug is dropped into the die, and the punch is moved forward with great impact. The punch is moved into the die, and the metal is pushed through the orifice.

The common mill products of tubes and rods can be made by this process, and complex shapes can also be extruded. Structural and architectural shapes, molding, ornamental trim, flashlight cases, cartridge cases, radio condenser cans, automotive aluminum pistons, rocket motors, shock absorber cylinders, and fire extinguisher cases can be extruded.

There are four primary methods of cold extrusion forming: (1) forward extrusion, (2) backward extrusion, (3) combination extrusion, and (4) tube extrusion, Fig. 7-52.

In the **forward extrusion,** the metal flows in the same direction as the punch. The metal receives its shape from the die and is considerably lengthened.

The **backward extrusion** is made by placing the slug in the die and closing the die. The metal is extruded back over the punch. The parts that are formed by this method are tubular, such as cans, tubes, and shells with forged bottoms.

In **combination extrusion**, these parts flow two directions at the same time. Part

of the slug is extruding forward through the die while another part of the slug is extruding backward up the punch side.

Tube and architectural extrusion is an adaption of forward extrusion, Fig. 7-53. It

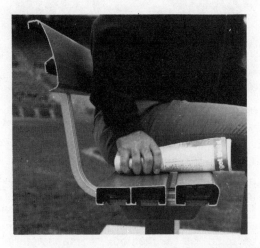

Fig. 7-53.   Extrusions may be any shape, just so the shape is continuous. Architectural shapes, rails, and trim of complicated design are frequently extruded.

281

Fig. 7-54.   Alloy steel powders with an atomizing process disperse additive metals uniformly within the power granules.

Fig. 7-56.   Powdered metal technology has made the combining of different materials and densities to provide the exact physical characteristics demanded of the part.

Fig. 7-55.   Metallic ores are processed to supply the different metals used in powdered metal parts. These may vary from copper to stainless steels.

uses a mandrel to produce a hollow extrusion.

Powdered Metal (Powder Metallurgy). Powder metallurgy is dependent upon compression as a method of manufacturing metal parts from powders, Fig. 7-54. These metal powders are frequently blended with a metallic stearate, which acts as a die lubricant, Fig. 7-55. The powder is compacted by punches pressing from the top and from the bottom to form a compacted briquetted part. During pressing the powder will receive pressures from 5-50 tons per square inch. A "green" product results that can be handled and moved to a furnace for sintering at a temperature slightly below the lowest melting point of the powders. After cooling, the parts are sized or coined. The sizing process merely burnishes the surface as it passes through the die.

The physical properties of a wide range of materials can be combined because of the unique technology of powder metallurgy. Metals may be combined with ceramic materials of alumina or silica base or with organic polymers (plastics) and with metal oxides and carbides.

This combining of materials offers control over density, hardness, strength, porosity, and toughness. Parts may be designed to wear evenly, resist heat, or conduct electricity, Fig. 7-56. Industrial items such as electrical circuit breaker points are required to resist damage when an electrical spark is drawn between them. With powdered metallurgy, the points can be made with a combination of tungsten powder and silver powder on metals whose melting points are thousands of degrees apart. The silver conducts the electrical current, and the tungsten resists the electrical spark erosion.

Metals commonly used for powdered metals are steels, copper, copper-based alloys, stainless steels, aluminum, molybdenum, tin, nickel, tantalium, chromium, and tungsten. Powdered metals, metal oxides, metal carbides, and other materials are mixed to manufacture products with unique properties. Powdered metals are obtained from metallic ores by chemical reaction, atomization of liquids, electrolytic deposition, and mechanical milling.

Copper powders are frequently chemically displaced from a copper solution by precipitation. Metals may also be liquefied and passed through small holes, broken up, and solidified quickly by dropping the metal particles in water (atomization). High-purity powders of copper, beryllium, and iron may be made by electrolysis. The mass is spongy and must later be pulverized. Metals may also be reduced to powdered flakes by grinding them in a ball mill. This process is called **mechanical milling**.

When metal powders are compacted, the pressing operation crushes the powders together. This action makes particles rub each other. The oxide film on the metal is scratched away, and a cold weld is produced at the particle contact points. The slip or abrasion of the particles instead of the pressure produces the weld points. For this reason, punches enter the dies from the top and from the bottom so that the pressures will cause the slippage of particles throughout the piece.

To obtain high densities, presses and dies must be designed and built so they are very strong. The size of work is limited by the

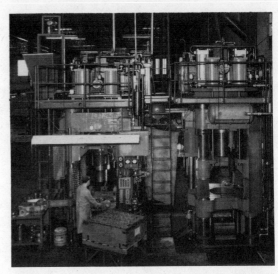

Fig. 7-57. The operator is changing the dies in a 1000-ton powder metallurgy press. The press on the right is a 1500-ton press with the tooling removed.

die's strength and the press's capacity, Fig. 7-57. At the present time parts as large as six inches in diameter can be made.

Sintering or heating the compacted part allows molecules in the cold-welded areas to return to a normal condition of the original solid metal. Considerable diffusion takes place, the weld area is increased, and many of the pores between powder particles close.

The parts are heated during sintering to a high temperature, usually about two-thirds of the lowest powdered metal's melting point.

Sizing, coining, and **oiling** are finishing processes of powder metallurgy. Sizing and coining finish the parts to precise size and straightness. Coining also adds density to the parts and improves the physical characteristics.

Oils are infiltrated into the coined parts for oil-less bearing material, Fig. 7-58. In special cases metals such as copper and silver are infiltrated through a product with a high melting point. This is done by heating the metal in an oxygen-free furnace. The melted metals fill the pores of the powdered metal parts.

Fig. 7-58.   Sintered parts are frequently sized or coined to provide accurate dimensions and density.

Fig. 7-59.   Large forged truck crankshaft ready for machining.

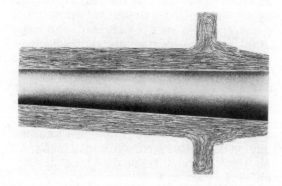

Fig. 7-60.   The grain flow of a forged part is called fibering.

Hot-Working Metal by Compression

Forging is a very old process and is successful because of the strength and toughness that forging gives to the part. Forgings are used where highly stressed parts — such as wheel spindles, connecting rods, axles, and front-end steering parts — are required. Forging produces the lightest and strongest parts because of soundness and the orientation of the grains within the metal, Fig. 7-59.

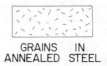

GRAINS IN ANNEALED STEEL

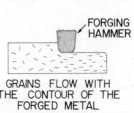

FORGING HAMMER

GRAINS FLOW WITH THE CONTOUR OF THE FORGED METAL

Fig. 7-61.   The forged part is shaped so that the grain flow is with the contour of the part.

Squeezing Processes

When a part is forged, the grains of the metal turn and orient themselves perpendicular to the force applied. They then grow longer. The grain flow of a forged part is called **fibering**. With this flow and orientation of the grain, great strength of the part is achieved, Fig. 7-60.

The shaping of the part during forging allows the operator to thin some sections and to thicken others. The operator can take advantage of the grain fibering to produce a part with high strength-to-weight ratio characteristics, Fig. 7-61.

Smith forging. This is the oldest process of working iron and steel into the shapes required for products which were used in a growing country. Wagon parts, bolts, window guards, hinges, fences, and iron and steel products were made by this process. The work of the blacksmith was important to the early settlers, because the blacksmith's work kept tools and equipment in working order. This was important, since they were many miles from the manufacturing plants.

In smith forging, a few basic tools are used to shape the metal. A **forge** heats the metal to a bright red heat, and an **anvil** provides a base upon which to rest the hot metal when it is struck. A series of **hammers** of various sizes are used to strike the metal into shape. **Tongs** hold the hot metal, and **hot and cold cutters** separate the metal. Other tools are needed for special operations, Fig. 7-62.

The operations of drawing metal out, upsetting metal, swaging metal, fullering metal, and bending metal are required to complete most products.

The drawing out of metal — making it longer and thinner — is done by placing the hot metal over the round edge of the anvil and striking the metal with a heavy hammer. The metal is squeezed between the anvil's edge and the hammer face. This squeezing process causes the metal to grow longer and thinner. The metal then is turned 90° and forged down on the anvil face until the original or desired width is produced. The operation is repeated with the other side of the metal on the round edge of the anvil. This causes the metal to flow out. A point may be drawn out by forging all sides of the stock until the point is drawn out over the round edge of the anvil, and is finished on the face of the anvil, Fig. 7-63.

Metal may be rapidly drawn out by using a top and bottom fuller and a sledge hammer. The metal is drawn out in the direction perpendicular to the fuller, Fig. 7-64A.

The top and bottom fullers squeeze the metal between them, making the metal thinner and at the same time drawing it out.

A **flatter** is a finishing tool that is used to smooth out rough surfaces such as those created by the fullers, Fig. 7-64B. The flatter is worked on cooler metal and is struck with a sledge hammer to remove rough surfaces.

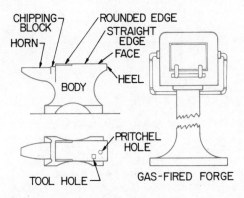

Fig. 7-62.   The anvil, forge, hammers, and tongs are basic tools needed for hand forging.

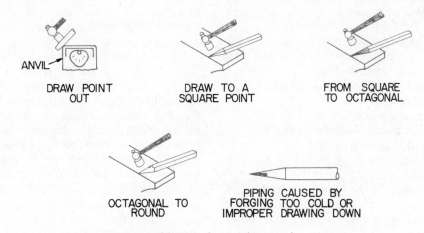

Fig. 7-63.   Hand forging: drawing down steel to a point.

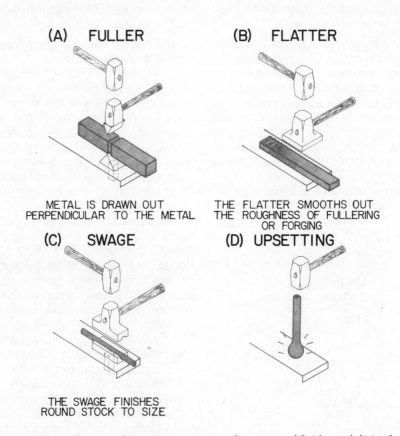

(A) FULLER

METAL IS DRAWN OUT
PERPENDICULAR TO THE METAL

(B) FLATTER

THE FLATTER SMOOTHS OUT
THE ROUGHNESS OF FULLERING
OR FORGING

(C) SWAGE

THE SWAGE FINISHES
ROUND STOCK TO SIZE

(D) UPSETTING

Fig. 7-64. Fullering, flatting, and swaging are processes that move and finish metal during forging.

The **swage** is used to finish round pieces as the flatter does, except the swage squeezes the metal into its finished shape and size, Fig. 7-64C.

The round stock is forged by hand on the anvil face with a hammer. The swage that produces the correct finish diameter is selected and struck to smooth and round out the bolt or rod.

Upsetting processes force the metal to thicken in sections. The thickness can be evened out by hammering the metal.

Upsetting may be done in the center of a bar by heating the center and by dipping the two ends in water, Fig. 7-64D. When the end of the steel is struck with a sledge hammer, the metal will be shortened, and the center will enlarge.

In smith forging, the bending of the metal must be done while the metal is hot. Thick materials may be bent into short radii when the metal is hot to produce bends, rings, eyes, and twists, Fig. 7-65.

Large bends are frequently done over the face of the anvil by delivering a lever blow to the metal. Bends with less arc may be formed over the horn of the anvil to form the first bend. In any case, the hammer should hit the metal and not the anvil face or horn.

Still smaller bends may be started over the horn and finished on the anvil face. An eye is finished by rounding up the circle by placing the eye over a tool in the pritchel hole and rounding up the eye.

A zero-radius bend may be made by first upsetting the area of the bend, shaping the

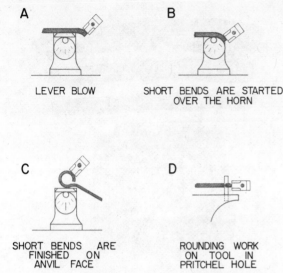

LEVER BLOW   SHORT BENDS ARE STARTED
OVER THE HORN

SHORT BENDS ARE
FINISHED ON
ANVIL FACE   ROUNDING WORK
ON TOOL IN
PRITCHEL HOLE

Fig. 7-65. Thick metal is bent into short radii by the anvil and simple tools. Bends, rings, eyes, and twists are produced.

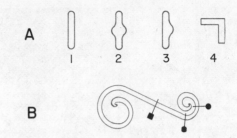

Fig. 7-66. Forging to produce a bend.
A. Metal must be gathered by upsetting so that enough metal will be available for the outside radius of an inside zero-radius bend.
B. Forging with a changing cross section produces quality art forging.

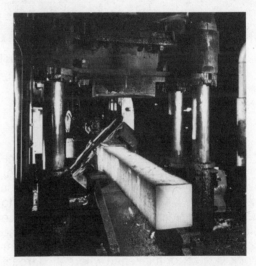

Fig. 7-67. A skilled operator and simple tooling are required for the operation of an open-die forge.

upset material, pushing it to the outside of the bend, and finally making the bend, Fig. 7-66. This method of making a bend on the anvil will allow the metal to be the same thickness throughout the bend area.

Fancy hand forging may be done by having the work change its cross-sectional area as it goes through the form of the part. The forging is done on the anvil face while the scroll is formed on a scroll former.

Open-Die Forging is an outgrowth of smith forging. The forge finishes the metal, and the work is done between two flat dies or V-shaped, half-round, or half-oval dies. Small open-die forgings may be produced in either hammer or presses. A steam, hydraulic, or air-driven hammer substitutes for the power, hammer, and anvil of the blacksmith, Fig. 7-67. This type of work requires a skilled operator. This operator must decide how much reduction in the part must be made with each blow, where the blow will be struck, and when and where the part needs heating. Because of these individual decisions, the production rate is generally

low. The advantage of this forging method is that simple tooling is required, and a great flexibility of forged parts and size ranges may be obtained.

Closed-Die Forging. This method is also called drop forging. It is done in a press which has the top half of the forging die on the upper or drop hammer and the lower half of the die mounted to the anvil, Fig.

Fig. 7-68.   A 2000-pound board hammer drop-forges small steel parts.

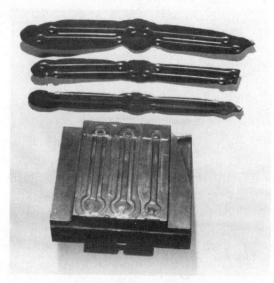

Fig. 7-69. The work is progressively moved to these various forging stations in the die: (1) edging, (2) drawing out, (3) blocking, (4) finished shape, and (5) trimmer to complete the connecting rod.

7-68. The hammer rises to the top of the press travel and falls, striking the work between the dies at the bottom of the press travel. A number of stages of the shape to be formed are cut into the surface of the dies. The white hot metal is placed in one position of the die and struck. The metal bar stock is upset or thickened where the largest amount of metal will be needed. This is the "edger" which forms the stock to fill the die. The work is moved to the next station. There the work is drawn out in the die cavity, producing a narrow section in the part. The third stage blocks out the shape of the article in a rough form. In the fourth position the part is struck and a shape is produced that is near the configuration of the finished workpiece. A "blocker" forms a rough contour of the work. A fifth station brings the part to finished size and contour. The extra metal showing around the work is called **flash**. Later it is trimmed off. As the work is

formed, it is progressively moved to the various stations, Fig. 7-69. In the forging operation, more than one blow is struck while the work is in a station.

An advantage of drop forging is the speed of the forging. Rapid blows complete the part with only one heating of the metal. Numerous shapes can be formed.

Drop forgings have advantages over open die forgings, because less skill is required of the operator. The shaping is done by the die, so a higher production rate is possible with a drop forge, because less manipulation of the workpiece is required. The drop forge delivers a consistent piece of work to much closer tolerances because of the die, Fig. 7-70.

Closed die forgings can produce a considerable range of parts. The parts may range from less than an ounce to several hundred pounds.

Press Forging. This method is similar to drop forging. However, the forging is pro-

Fig. 7-70. Automatic hot-forging of a 2½ pound wheel hub produced from 1¾" bar stock. The forge automatically transfers the parts to the various dies.

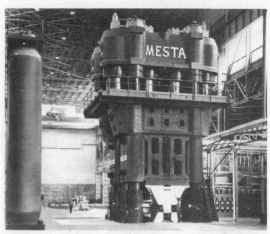

Fig. 7-71. This 50,000-ton hydraulic closed-die forging press was designed and built by Mesta Machine Company for forging aluminum and magnesium alloy aircraft parts.

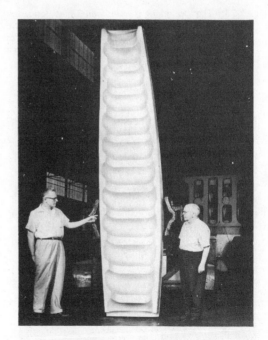

Fig. 7-72. Closed-die forgings are made of high-strength aluminum and titanium shaped parts. This aluminum forging weighs 3,000 pounds, and it is 156" long, 36" wide, and 12" thick.

duced by squeezing rather than by hammering. The pressure exerted on the workpieces may be as much as 50,000 tons, Figs. 7-71 and 7-72. The presses make a few strokes down on the workpiece and force the metal into the die. These large presses forge aluminum and titanium alloys for transportation and other industries in one or a few strokes of the press, Fig. 7-73.

Press forgings are equal to drop forgings in metallurgical quality. The slow squeezing

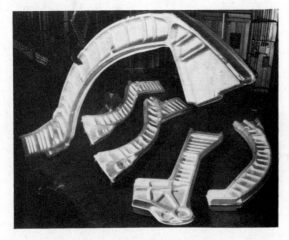

Fig. 7-73. The top forging is more than 23 feet long, and weighs 27,000 pounds. This closed-die aluminum forging supports the forward unit of the Air Force C-5A transport. It was forged on a 50,000 ton press.

of the metal under these great pressures causes the metal to flow and produce the fibering which makes forging a superior production method. Because few strokes are needed to produce the parts, they can be made quickly and there is less wear on the dies. As a result, they last longer. Press forgings may be large, but they are comparatively simple, Fig. 7-74. Parts which have irregular or complicated surfaces and shapes are done better and cheaper by the drop forge method.

Forging presses are used to finish the work done by other presses. Drop-forged work may be passed on to the forge press for finish sizing or coining. The coining does not move a large amount of metal, but it shapes the part in a closed die exactly the way the edge and face are produced upon a quarter.

Draftless Forging. No draft is the slope on the sides and ends of a die cavity which allows the part to be released from the die when it has been forged. The draft in press forging is normally 1° to 5°, but in no-draft forging, it is reduced to nearly zero, Fig. 7-75.

This precision forging drastically reduces the amount of metal in the forging.

Fig. 7-74. A large steel disc-type forging coming out of a heat furnace.

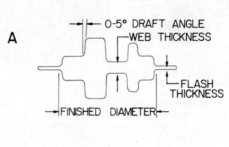

Fig. 7-75. No-draft precision forgings eliminate weight without the loss of strength in aerospace industries.

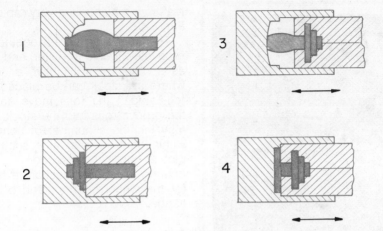

Fig. 7-76. The upsetting of bar stock into a forging for an automotive transmission cluster gear.

Consequently, the weight is reduced. The web and wall thicknesses are thinned. No-draft forgings are used on aluminum and titanium aircraft parts in which all excess weight must be eliminated.

Upset Forging. Upset forging machines are sized by the largest bar that they are capable of working, Fig. 7-76. The largest commercial machine will forge bars up to 9″ in diameter. Parts weighing from a few ounces to 500 pounds can be made by these machines. The upsetter forges products made of round, square, hexagonal, or tubular stock. Hollow or tubular stock is used in the manufacture of such products as axle housings and like products, Fig. 7-77.

This forging process is very fast. It will produce enlarging diameters up to three times the original bar. It produces a strong uniform forged structure throughout the work, thus making it very strong and durable.

Roll Forging. Roll forging is designed to roll the product between rollers with the contour of the part cut into the surface of the roll so that the shape of the metal is formed when the metal is forged. The surface of the roll die has a series of grooves on its surface. Each groove produces a reduction in the size of the metal and shapes

Fig. 7-77. In upset forging the material is gathered so the gears, flanges, or other shapes may be made on the end of the stock.

it until the last groove of several gives it its final shape and surface configurations.

Forge-rolling machines produce tapers, channels, bosses, and shaped parts in the rolling. This process is frequently used to produce the rough stock for drop forging, upset forging, and closed die forging.

Such products as levers, shafts, cam shafts, axles, crowbars, knife blades, and

A.   Roll-forging machine forming an axle shaft.

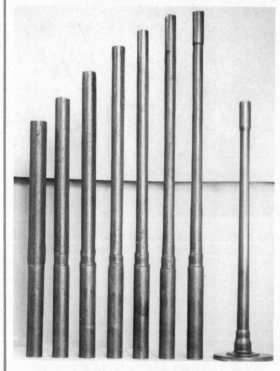

B.   Typical parts formed by roll forging.

Fig. 7-78.   Roll forging is done with a pair of rolls with the shape to be forged machined into their surface.

leaf spring blanks are very rapidly produced, Fig. 7-78.

The rolls are geared together so that the grooves will hold, register, and roll the metal as a pair. The rolls have an open section where the stock can be placed between the rolls. When the rolls move, the contour of the groove rolls into the stock. The grooves may include a number of contour changes within them. A rear axle shaft of an automobile has a thick section for a spline and bearing, a tapered section, a heavy section for the wheel bearing, and a section for a flange.

Activities

1. Design a forging project and heat treat as necessary.
2. Design a sheet metal project and construct it.

Related Occupations

These occupations are related to metal forming:
Sheet metal worker
Forge operator
Press operator
Blacksmith
Boilermaker
Layout technician
Brake operator
Twisting machine operator
Four-side bending machine operator
Rolling machine operator
Iron worker machine operator
Power brake operator
Power hammer operator
Power shear operator
Profile cutting machine operator
Tube bender
Riveter
Welder
Painter
Plater
Heat treater
Spot welder operator
Spinner
Four-column hydraulic press operator

Hammersmith
Hammer operator
Cold header operator
Heater
Inspector

Trimmer
Grinder
Shot blaster
Pickler

Forming Metal by Casting

Chapter
8

Words You Should Know

Solidification range — The temperature range in which liquid metal changes to a solid.

Nucleus — The forming of a fundamental particle within the liquid metal.

Dendritic — A branching figure in a tree-like form.

Eutectic composition — A composition of metal that has the lowest melting point of any alloy or metals in the alloy.

Pattern — A form designed to produce a cavity equal to the desired cast shape.

Cope — The top part of a foundry flask.

Drage — The bottom part of a foundry flask.

Dielectric heating — A nonconducting, high-frequency electrical heating process.

Machine molding — The mechanization of heavy work by the use of air and hydraulic devices.

Sand slinger — A movable machine that accelerates the speed of sand by an impellor. The sand is thrown into the flask with such velocity that the sand is compacted on contact.

Impellor — A rotor that transmits motion to a surrounding media. An example is driving water with a pump impellor.

Feldspar — A mineral occuring in pegmatite dikes and granite rocks. It weathers into fire clay.

Permeability — The ability to pass through a substance or mass.

Permanent molding — A mold that is used repeatedly and is made of cast iron.

Centrifugal casting — Rotating the mold so that the forces move the metal away from the central axis or point of rotation to produce a dense casting.

Die casting — Molten metal is forced into a die mold by a plunger. The metal is under great pressure as it solidifies.

Shell molding — A thin shell made of sand and resin binders 1/8″ to 3/8″ thick. The mold may be backed by other materials and poured full of metal to produce a casting.

Precision casting — Investment or lost wax casting is done by making a wax pattern and imbedding it in plaster of paris refractory mix. The wax is later burned out, and metal is poured into the cavity.

Theory of Casting

Metal may be shaped and given form by heating it until it becomes a liquid and then pouring it into a mold. The mold provides the shape of the product and holds the metal in this shape until it has solidified, Fig. 8-1.

The melting of metals is accomplished by applying the energy of heat to the solid metal. Metal is made up of atoms and molecules that are in motion even when the metal is in a solid state. The atoms are confined to a definite space within the metal's space lattice system. When heat is added to the metal, as it is in a melting furnace, the molecules absorb the heat. This heat does not increase the vibration

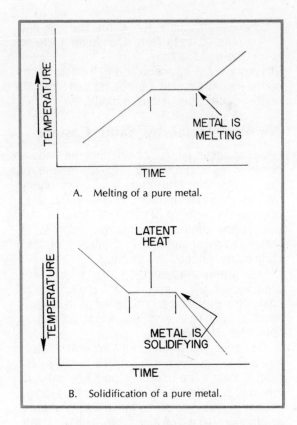

A. Melting of a pure metal.

B. Solidification of a pure metal.

Fig. 8-1. Metal can be given form by being heated and then allowed to solidify in a specific shape.

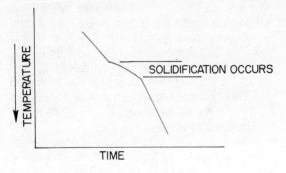

Fig. 8-2. Solidification of an alloy.

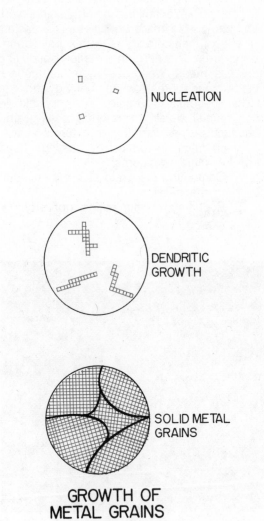

GROWTH OF
METAL GRAINS

Fig. 8-3. Growth of metal grains from nucleation (seeds), dentritic growth, to grains of solid metal.

of the atoms. However, the heat removes the bonds which hold the molecules together and when the bonds are broken, the metal melts. After melting is completed, the vibrations speed up in the molecules as the temperature continues to rise.

Melting and Freezing of a Pure Metal

Metals which are made up of two or more metals are called **alloys.** Alloys do not have a freezing or melting temperature. Instead, they have a **solidification range** and a curve that will be related to the percentage of the various metals in the melt. Crystal growth in the liquid metal is dependent upon the time involved and the temperature of the metal, Fig. 8-2. While the liquid is cooling, nucleus or seed particles are formed. The nuclear particles are distributed within the liquid in a random position in the total mass. Solidifying material forms in different direction on each crystal axis into a fernlike or a **dendritic** structure. As solidification occurs, the solid grains of metal are removed from the liquid melt, thus the chemical composition of the remaining liquid metal is changed, Fig. 8-3.

The remaining liquid metal cools further and eventually solidifies. When the alloy reaches its lowest liquid temperature, the eutectic composition (composition with the lowest possible melting point) is achieved, and the whole mass solidifies. The easiest or lowest melting alloy is in the grain boundary area or where the earlier solidifying crystals meet. When the metal is melted, these bonds break when melting and the metal becomes a liquid. The crystals in commercial metals are referred to as **grains** because of their variation in external shape.

When metal is formed by casting, these are the major concerns:

1. Controlling the **grain size** of the metal in the casting.
2. Reducing or eliminating **porosity.**

Fig. 8-4.   Pouring the right side of a hoist drum in a large green-sand mold.

3. Compensating for shrinkage as the metal passes from the liquid state to the solid state.

The primary purpose of heat treating is to control the grain size, hardness, and strength of the metal after it has solidified.

Forming Metal by Sand Casting

Sand casting is a convenient and inexpensive method of changing the shape of metals. Metals that can be commercially melted may be poured into a mold and allowed to solidify. They will assume the new shape of the mold. Large volumes of metal used for structural members of machines are shaped by casting.

The green-sand method of casting produces over 90% of the tonnage of metal cast. The processes connected with green-sand casting are very important to manufacturing, Fig. 8-4.

The cost of raw materials for sand casting is low; however, it is related to the metal that is cast. The factor which is costly is the direct labor cost. Much of the sand-casting process requires skilled labor. Large production runs reduce the skilled labor costs by automating large foundries, such as the automobile engine foundries. Scrap loss in a foundry is low, because the metal in damaged parts, risers, gates, and the like may be recycled by the remelting process.

Cast products have some beneficial physical characteristics that other types of formed products do not contain. Cast products have the property of "no laminated" or "segregated structure" such as in forged products, Fig. 8-5. Certain products such as engine cylinders, piston rings, and gears best utilize the random metallic structure. The metal has the same strength in all directions, and has even wearing characteristics.

Most cast materials have the ability to flow. **Fluidity** is the ability of the materials to form the shape of the mold before the metal solidifies. Fluidity is important for the casting of thin sections.

Cast iron is unique, because it dampens vibration which may be introduced into a

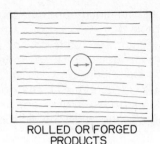

ROLLED OR FORGED
PRODUCTS

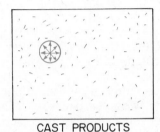

CAST PRODUCTS

Fig. 8-5. Gray iron castings have a random metallic structure, thus they will have the same strength and wearing characteristics in all directions when under load.

TIME

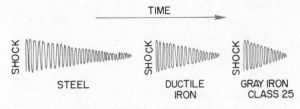

SHOCK

STEEL

SHOCK

DUCTILE
IRON

SHOCK

GRAY IRON
CLASS 25

Fig. 8-6. Vibration dampening characteristics of metals. Gray cast iron dampens shock quickly.

Fig. 8-7. Cast iron castings are rigid and stable. They machine easily and provide an excellent surface finish.

structural member. This dampening (vibration effect) characteristic makes cast iron desirable for machine bases, engine blocks, and frames, Fig. 8-6. Also, cast iron is rigid and thus useful as the structural material for heavy equipment and machine parts which require stable alignment. Cast iron can be machined to produce a smooth, fine surface finish without additional finishing processes such as grinding or honing, Fig. 8-7.

Foundry Patterns

It is necessary to produce a **pattern** before the shape of the casting can be molded sand. The pattern is surrounded with compressed molding sand to form the shape. Then the pattern is carefully withdrawn from the sand to produce an accurate cavity for the casting. At a later time, molten metal is poured into the cavity and allowed to cool. The shape of the metal will be the same shape as that of the pattern, Fig. 8-8. How-

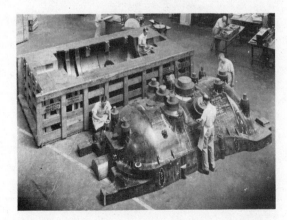

Fig. 8-8. When the pattern is removed from the sand, it leaves an accurate cavity into which the molten metal is poured. The core box for the pattern is being constructed behind the pattern.

ever, as the metal cools, it shrinks (contracts). The finished casting will be slightly smaller than the pattern.

Patterns are a very important part of the casting industry, because they produce the many intricate shapes for industrial component parts. The patternmaker is a very skilled craftsman. Patterns are made of three materials: wood, metal, and plastic.

Wood Patterns. The easiest material to work with in patternmaking is wood. A pattern may be made of Honduras Mahogany, Eastern White Pine, Western Sugar Pine, or Maple, Fig. 8-9. The wood must be clear, closed grain, and hard. Patterns are frequently made of mahogany because of its hardness and its workability. Maple is an excellent pattern material for carvings which have intricate detailed designs. Frequently a wood pattern will be converted into a metal pattern to reduce pattern wear. The master pattern is usually made of wood.

Metal Patterns. Patterns made of metal are durable and will resist shrinkage or swelling caused by moist sand. Metal patterns produce many castings before they must be replaced, Fig. 8-10. Navy bronze metal patterns are the best. Cast iron is good, but aluminum and magnesium are less durable.

Permanent metal patterns are expensive to construct because the master pattern must first be made of wood or machined from a solid piece of metal. The wood master pattern is then cast into a permanent metal. The roughcasting must be finished by removing the gates and flash. Then it is smoothed and polished. Any pits or damage to the rough pattern casting must be filled, repaired, and machined. These extra processes result in a superior permanent pattern, but the cost also has been increased when compared to a wooden production pattern.

Plastic Patterns. Plastic patterns have come into wide use because of their excellent dimensional stability in moist conditions. They may be made by craftsmen with less skill than those who make other patterns.

Patterns made of plastic are of three general classes: (1) solid cast pattern, (2) cast with a core, and (3) wet layup. Solid plastic patterns of epoxy resins with aluminum or mineral fillers may be cast with a cross section as small as 3/8" thick if the pattern is uniform in thickness. Small parts

Fig. 8-9. Wood patterns are frequently made of mahogany or clear pine.

Fig. 8-10. Metal patterns will better withstand the wear and rough handling given the pattern during molding in the foundry.

of the total area of a plastic pattern may have thickness up to 3/4" in thickness. Plastic patterns with a variety of thicknesses must be cured evenly and carefully to prevent distortion of the shape.

Plastic patterns are strong and flexible under pressure. This is an advantage since brittle patterns break under stress.

Pattern Allowances. Patterns require a number of allowances: draft, shrinkage, distortion, and machining. It is necessary to be able to remove a pattern from the sand mold. The vertical surfaces of a pattern are tapered on the pattern to allow withdrawal from the sand without damaging the molded walls of the cavity. This taper is called **draft.** A pattern may have 1/8" to 1/4" per feet of draft so that the pattern can be withdrawn from the sand, Fig. 8-11.

The draft must be added to the basic dimension of the part. It must be consistent from the side of the pattern so that all surfaces will release from the sand at the same time when the pattern is removed from the cavity. If the pattern has a large hole in the pattern, the hole must have draft also.

When metals change from a liquid to a solid state, there is a reduction in size. Different metals shrink different amounts. For example, cast iron shrinks 1/8" per foot. Every foot of cast iron castings dimensions will be less by 1/8". To allow for this metal shrinkage, the patternmaker must enlarge the original pattern dimensions of the drawing. The patternmaker uses a shrink rule to enlarge the pattern accurately. The shrink rule has additional length added throughout its length. A 1/8" shrink rule for cast iron patterns would measure 12⅛" to provide the final cast size of 12". It looks exactly like a 12" rule, except each division is slightly larger so that the dimensions total 12⅛". If the steel rule has "Shrink 1/8" to ft.," this rule should be used only for pattern work, not for general metalworking layout work, Fig. 8-12.

Each pattern will be built with a shrink rule to compensate for the average shrinkage of the metal. For example, the following is the average shrinkage of casting of these metals:

1. Cast iron 1/8"/ft.
2. Malleable iron 1/8"/ft.
3. Steel 1/4"/ft.
4. Brass 3/16"/ft.
5. Copper 3/16"/ft.
6. Aluminum 5/32"/ft.
7. Magnesium 5/32"/ft.
8. Lead 5/16"/ft.
9. Zinc 5/16"/ft.
10. Tin alloys 1/12"/ft.

When a wood pattern is to be used to produce a metal pattern, a double shrinkage must be allowed. For instance, if a wooden pattern is used to make a cast iron pattern which in turn will form the mold for a brass castings, the two shrinkages must be added. The first shrinkage is 1/8"/ft. for the cast iron, and the second shrinkage is 3/16"/ft. for the brass. The patternmaker would select

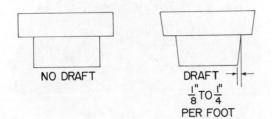

Fig. 8-11.   Draft will aid in the drawing of the pattern from the sand.

Fig. 8-12.   Shrink rules indicate how much shrinkage the pattern will undergo.

a 5/16″ shrink rule to build the wooden pattern.

The patternmaker will carefully examine the drawings for the draftsman's finish marks. These marks indicate that the metal will be cut by a machining operation to the finish dimension of the part.

Castings which have machined areas or bosses to be machined will have an additional allowance of metal provided by adding 1/8″ to the height or thickness of the pattern. This supplies the metal for machining. If the casting has a tendency to warp so the whole surface does not "clean up" or is not machinable over the entire area, additional allowance will be made. When required, the patternmaker will add metal cylinders or pads to the casting so the part may be held in a chuck or mounted in a machine.

Metal of irregular shapes may shrink at different amounts when cooling, thus causing distortion in the workpiece. The pattern is enlarged in thick parts of the casting (off square) so that as the heavy intersection area is cooled to room temperature, the legs will straighten out to make a square workpiece, Fig. 8-13. Allowing for distortion requires the technical experience and knowledge of a patternmaker for accurate construction of patterns for complicated castings.

Types of Patterns. There are all kinds of patterns used for castings. They include solid, split, match plate, cope and drag, and special patterns.

The **solid** pattern is a simple pattern — a slightly larger duplicate of the final part — with the shrinkage allowances added to its dimensions, Fig. 8-14. The solid pattern does not have any gates, sprues, or risers attached to it, Fig. 8-15. Thus, it is simple and inexpensive to construct. The molding cost will increase because the mold, the gates, runners, sprues, and risers are formed by handwork by the molder so that the metal may reach the mold cavity. This type of pattern is used to form prototype parts, and it is a limited production technique. It may be used in school industrial courses or job shops (special production shops that serve a larger production plant).

The **split pattern** separates into two sections so that half of the pattern is molded in the **cope** and the other half the **drag** of

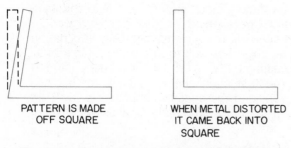

PATTERN IS MADE OFF SQUARE

WHEN METAL DISTORTED IT CAME BACK INTO SQUARE

Fig. 8-13. The pattern is made so that it is off square. When the metal shrinks and distorts, it moves back into the correct position.

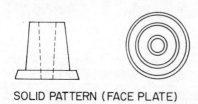

SOLID PATTERN (FACE PLATE)

Fig. 8-14. Solid, loose, or single patterns are slightly larger duplicates of the final part to allow for shrinkage of the metal.

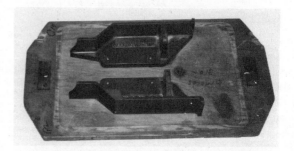

Fig. 8-15. A solid pattern mounted on a molding board.

the flask. The parting line of the pattern and the flask will usually be on the center line of the pattern.

Complicated patterns may be formed with a split pattern because the draft will slope away from the parting line of the pattern, Fig. 8-16. The flat parting line surface may be used by placing one-half of the split pattern on the mold board to establish the parting line of the flask and pattern in the same plane. Using this method, detailed configurations of shapes may be on both sides of the pattern.

The split pattern has the same molding limitations as the solid pattern in that the molder must cut the gates, runners, sprues, and risers into the molding sand by hand, so molding costs will increase.

Match-plate patterns are used for making large quantities of small castings, Fig. 8-17. Production is increased by mounting one-half of the pattern on one side of a board or plate and the other half of the pattern precisely on the opposite side of the plate. The surfaces of the plate become the parting line of the pattern, and more importantly, the parting line of the flask, Fig. 8-18.

The match plate also has the gates, runners, sprues, and risers all mounted on the plate so the molder merely has to ram, jolt, squeeze, or sling the mold with sand; separate the flask; and draw the mold plate,

Fig. 8-16. Split pattern constructed of wood. The parting surface runs through the center of the pattern.

Fig. 8-17. The aluminum match plate has one-half of the pattern on one side of the plate and the other half of the pattern on the bottom side. The match has the runners and gates also mounted on the plate.

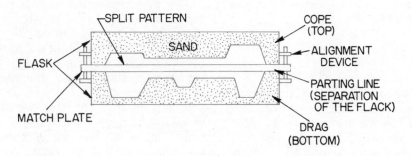

Fig. 8-18. When the match plate is rammed up in a foundry flask, the plate is mounted at the parting line of the flask.

Fig. 8-19.   A match board illustrating the pattern, runners, sprue, and riser shown on an end view of a match board.

Fig. 8-20.   Cope and drag patterns.

Fig. 8-22.   Patterns with irregular parting surfaces are easily molded by using a follow board. The board provides a simpler parting surface.

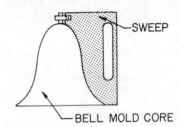

Fig. 8-21.   A simple circular shaped sand mold may be made by rotating a board with the desired cross section in the molding sand.

Fig. 8-19. Molding efficiency is greatly increased by the use of patterns mounted on match plates. The match plates cost more when they are constructed, but the ease and efficiency in manufacturing castings quickly offsets the cost of the patterns.

Foundries that produce several large castings may mount the **cope** pattern on one plate and the **drag** pattern on another plate, Fig. 8-20. The molding process may be speeded up by having one person mold the cope pattern on one molding machine while a second molder and machine molds the drag half of the pattern. Later the cope and drag molds are put together for pouring. By specializing the work, greater productivity is achieved in molding.

The ingenuity of foundrymen and pattern-makers has produced many **special devices and patterns.** Some of the common devices used for molding are the sweep, follow boards, molders parting, and Styrofoam patterns.

The **sweep** is used to make concentric patterns such as bells, fly wheels, and pulleys. The pattern is cut to the half cross section and pivoted about a central point to generate a circular pattern, Fig. 8-21. The configuration is cut into the foundry sand by the sweep board.

Follow boards are used with loose patterns which have an irregular parting surface, Fig. 8-22. The follow board cradles the pattern so that the follow board establishes a new parting line. The drag is

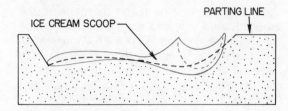

Fig. 8-23. The molder may remove the sand from the part until a natural parting line is reached.

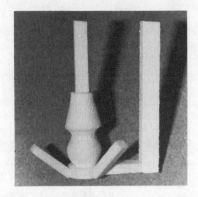

Fig. 8-24. The Styrofoam vaporizes as the molten metal comes in contact with the plastic.

rammed around the pattern while the pattern is held level by the follow board. The drag is turned over and the follow board is removed, leaving an irregular parting line that has part of the cope sand occupying the area of the follow board.

Molder's parting is sometimes used when a solid pattern or a broken part is used as a pattern, Fig. 8-23. The part is rammed in the drag without concern for draft in the drag section below the parting line. The drag is turned over, and all sand is cut away to the parting line with a trowel and packed tightly. The parting compound is added to the surface. The cope is placed on top of the drag and rammed, completing the parting line.

Styrofoam patterns or disposable pattern molds are made from expanded polystyrene beads pressed into thick sheets or blocks. A pattern may be sculptured out or glued together, since this material is very quickly shaped with hot tools, hot wires, or woodworking tools. The beads may also be molded into a pattern form for ramming.

The advantage of this method is that no parting line is needed, since the pattern vaporizes in the mold. Also, undercuts, draft, and loose pieces are no problem and are no longer needed. The Styrofoam patterns are given a wash coat of zirconite sands to reduce metal penetration and are rammed up with the regular foundry sand.

The hot liquid metal is poured into the Styrofoam mold. The metal heat causes the polystyrene to vaporize and generate a gas, Fig. 8-24. While the metal fills the mold the

Fig. 8-25. The core is made of sand and a binder that produces the hole in the center of the casting. These are manifold cores.

gas holds pressure on the sand and does not allow the sand to collapse. The gas is finally vented out through a riser or through the sand.

Styrofoam casting makes both pattern-making and molding a very simple operation, but a new pattern is required for each pouring of the metal. This property makes Styrofoam an attractive method for artistic castings and large sized limited production parts.

Coremaking

A core provides a shaped sand filler which is placed into the mold. The liquid metal flows around the core to form a hole in the roughcasting, Fig. 8-25. The core should lose its strength after the metal has solidified so that the core sand will crumble and shake out easily. This leaves a clean hole in the casting.

Cores may be any shape and may be enlarged by cementing several pieces to-gether. They may be made by ramming the special sand mix into a core box the shape of the core desired, Fig. 8-26. Core sands are commonly bound together by cereal and linseed oil, natural resins, synthetic resins, or plastics. The sand used is a washed silica which is mixed with the binders.

A large foundry will utilize a core-blowing machine, Fig. 8-27. This machine uses pressure of about 100 pounds per square inch to fill a core with sand in a few seconds. The cores are removed from the core boxes and placed on a conveyor belt that passes the cores through a dielectric (radar-heating) oven which bakes them hard in a few minutes, Fig. 8-28.

Types of Cores. Cores are defined by the material they are made from and the way they are used in the mold. The **green-sand core** is made in the mold by ramming foundry sand into large holes in the pattern. When the pattern is withdrawn from the sand, a core of foundry sand is left in place, Fig. 8-29.

Fig. 8-26. A large core is being lowered into the mold for a machine tool.

Fig. 8-27. Small core-blowing machine.

Fig. 8-28. A large dual-station, hot-box core machine.

Dry sand cores are made in a core box or are extruded in common diameters. These cores are made of sand and binders. The carbon dioxide cores combine sodium silicate with the sand. After the core is shaped, it is exposed to carbon dioxide gas. The gas reacts to form an inorganic silicate bond holding the sand and core in a shape that can be handled and later set in a mold, Fig. 8-30.

Other types of cores are made with the use of phenolic resins mixed with the core sand and bound together by heating to about 450° F. to produce a shell core. A cold box-type of core is made by mixing phenolic resin with sand and exposing it to an active organic gas such as carbon dioxide to harden the resin and thus the core.

The principle types of dry sand core supporting are parting line, balanced, vertical, drop, and hanging cores, Fig. 8-31.

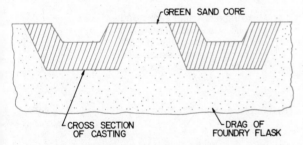

GREEN SAND CORE

CROSS SECTION OF CASTING

DRAG OF FOUNDRY FLASK

Fig. 8-29.   A green-sand core is made of the foundry sand and rammed as part of the pattern.

Fig. 8-30.   A dry-sand core made by the carbon dioxide processes for an automotive differential case.

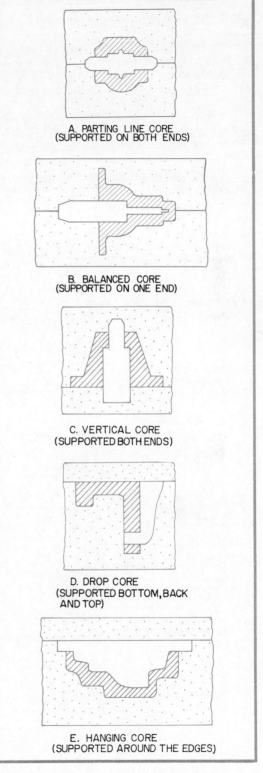

A. PARTING LINE CORE
(SUPPORTED ON BOTH ENDS)

B. BALANCED CORE
(SUPPORTED ON ONE END)

C. VERTICAL CORE
(SUPPORTED BOTH ENDS)

D. DROP CORE
(SUPPORTED BOTTOM, BACK AND TOP)

E. HANGING CORE
(SUPPORTED AROUND THE EDGES)

Fig. 8-31.   Dry-sand core types and how they are supported.

1. Place drag pins down on a patternboard.

2. Add a pattern, flat side down.

3. Sprinkle pattern surface with parting material.

4. Use riddle to sift or riddle sand to one inch depth over pattern. Tuck with fingers.

5. Fill drag with sand, using shovel.

6. Ram or pack down sand with bench rammer. Use peen end first.

7. Strike off excess sand with striking bar.

8. Vent the drag sand to within 1/4" of pattern.

9. Bed slightly with loose sand and place molding board on top.

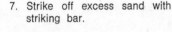

10. Turn drag over and remove patternboard, exposing pattern.

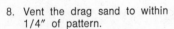

11. Sprinkle entire surface with parting.

12. Slide the cope in place. Now repeat Steps 4 through 8.

13. Use sprue cutters to cut sprue (metal entrance) and riser. This combats shrinkage.

14. Lift cope from drag; form pouring basin with slicks and trowels.

15. Use gate cutter to cut passage from sprue and riser to the pattern.

16. Loosen and draw (lift out) the pattern with pin or screw.

17. Make repairs using slicks and trowels. Use wet brush on delicate edges, dry brush for removing excess sand, and bellows to remove loose sand.

18. Replace cope on drag.

19. Heat metal in crucible furnace, then test temperature with pyrometer. Aluminum pouring temperature is 1300° F. Brass pouring temperature is 1900° F.

20. Remove crucible, flux, and skim refuge. Pour, using crucible tools.

Fig. 8-32. Procedures for bench molding using hand processes.

Sand Mold Characteristics

A sand-casting mold will have these characteristics:

1. The sand and binder must result in a strong mold, a mold that will hold the weight of the metal in the liquid state without washing, cracking, or collapsing.
2. Sand molds will result in surfaces and shapes that will not be eroded or wash away by the rapid and heavy flowing metal as it fills the mold.
3. The mold will give off very small amounts of air and gas when the hot metal comes in contact with the mold, thus gas inclusions and porosity are reduced.
4. A properly designed and constructed mold will have permeability — the ability to pass gases through the mold to a vent, thus improving the work.
5. High temperatures are a problem for sand molds. The sand and mold parts must be utilized to withstand high temperatures and allow for easy cleaning of the casting upon cooling.
6. In addition to molds, cored parts must be strong enough to withstand the forces developed within the mold. However, when the metal has solidified, the core material must collapse and shake out easily and clean when the metal has cooled and contracted.

Green-Sand Molds

Green-sand molding processes are classified by the method by which the mold is made. There are five of these processes: 1. bench molding, 2. floor molding, 3. pit molding, 4. machine molding, and 5. sand slingers.

Bench Molding. Bench molding is performed by a molder who uses hand tools. The work is performed at a bench which is designed for the height and convenience of the molder. Much small work and handcrafted work for special prototype development of job shops and schools utilize the flexibility and freedom development that the

Fig. 8-33. Bench molding of a medium-sized part.

Fig. 8-34. Floor molding is used where large, heavy castings are produced.

craftsman can produce with the skilled use of the hand tools, Figs. 8-32 and 8-33.

Floor Molding. Floor molding is used where larger castings are produced, Fig. 8-34. The amount of sand in a flask may rapidly become too heavy for a worker to handle or safely turn over with tools so that the foundry floor becomes the place to work instead of the bench.

Floor molding will be done on large castings weighing above one ton and will be cast on the floor of the foundry set aside for large castings. Mold construction for large castings require extensive engineering control to assure correct setting of cores, gating, pouring, and cleaning of the castings. Vast amounts of labor, time, and materials go into the makings of a large casting which can only be poured once. Either the cast is successful, or it will be scrap.

Pit Molding. Pit molding is used for very large castings. This process is used to produce castings as large as 100 tons when finished. The weight and hydraulic pressures are so great that the earth itself is used as part of the flask.

Large holes are dug into the foundry floor, and the sides of this large box-like hole are lined with brick or concrete walls, Fig. 8-35. The bottom of the pit is covered with a thick layer of cinders with vent pipes rising from the cinder area to the surface outside of the mold. The pattern is loaded into the pit and the molding sand is packed under and around the pattern by the molder. The molder frequently uses a ladder to climb down into the mold to do finishing work.

The molder may do finishing work in the mold by placing cores, building up parts of the mold with bricks, shovelfuls of sand, sweeps, and trowels to supply the finish to the inside of the mold. The floor mold must be strong and able to resist the pressures developed by the hot gases and 100 tons of liquid metal. This large amount of liquid is much safer to cast with the metal at the lowest point in the foundry in case the mold should leak or fail.

Machine Molding. Machine molding makes possible the mechanization of much of the heavy labor necessary for making molds. The molder's production may be increased by applying a machine that will ram the sand evenly around the pattern, turn the flask over, squeeze the sand tight on the other side of the flask, lift the cope of the

Fig. 8-35. Pit molds are for very large castings. The metal forces are so large that the earth is used as a flask.

Fig. 8-36. A molding machine performs match-plate molding operations.

flask, and vibrate and remove the pattern, Fig. 8-36. With the application of machinery to the repetitive and heavy work in molding, greater production with less human energy may be achieved.

Molding machines are designed from three molding principles. The first design is the **jolt machine,** Fig. 8-37. The mold is produced by placing the pattern in the bottom of the flask and filling the flask with sand. The flask, sand, and pattern are raised a short distance and dropped. When the machine strikes bottom, the sand is jolted down upon the pattern, thus packing or ramming the sand around the pattern, Fig. 8-38. The number of strokes determines how dense the sand is compacted around the pattern. At the same time it produces a uniform ramming throughout the flask.

The **squeeze machine** makes use of pressure on the sand to compact the sand around the pattern. The pattern is placed on the mold board and the flask around the pattern. The flask is filled with sand, and the squeeze head is brought into place over the flask. Pressure is applied to the sand. The squeezing packs the sand around the pattern to provide the correct mold density. The pattern is drawn, and the mold is closed for pouring. The density of the sand varies with the distance or the thickness of the sand above the pattern. For this reason, the squeeze machine is used to compress only a few inches of sand. The foundry industry has learned that these two operations may be combined into one efficient machine. The two combined become the jolt rollover squeeze-pattern draw machine. This type of machine is frequently used for high pro-

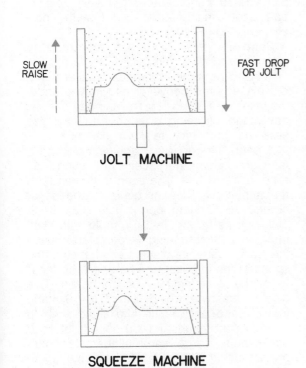

Fig. 8-37. The drag of the flask is rammed by jolting or raising and dropping to pack the sand. The flask is rolled over and the cope rammed by squeezing the sand.

Fig. 8-38. Jolt rollover squeeze machine with match-plate between cope and drag. Note pneumatic vibrator mounted on match-plate to aid in drawing the patterns.

Fig. 8-39.   Automatic molding units using hydraulic and pneumatic high-pressure molding.

Fig. 8-40.   Automatic ramming of flasks with high pressure on high-production runs.

Fig. 8-41.   Sand slinger ramming a large flask.

duction specialized molding, Fig. 8-39. A machine to handle the sized flask for the part under manufacture can thus be purchased, Fig. 8-40.

Sand Slingers. Sand slingers are used when the molding flasks are too large and difficult to be moved or when smaller flasks are placed on a conveyor and moved under the slinger, Fig. 8-41. The slinger has an advantage of being able to ram any size flask that the machine's arm can reach.

A **sand slinger** is a machine which contains an impellor that receives the sand from a hopper and belt. The sand is thrown from the impellor blades or cups at speeds of around 10,000 feet per minute, Fig. 8-42. The sand is thrown into the mold with such force that the ramming is completed when the sand lands in place. By changing the speed of the impellor, the hardness or density of the ramming may be controlled. This type of machine will deliver 7 to 10 cubic feet of sand per minute into a mold which is equivalent to about 1000 pounds of sand per minute. Large sand slingers for high production are capable of delivering up to 4000 pounds of sand per minute. This is considerably more than workers could shovel and ram by hand in half of a working day. When starting to fill a flask, the velocity of

Fig. 8-42. Impellor of a sand slinger. The impellor head cap has been removed for the photograph.

the sand is so great that the sand slinger operator will frequently bounce the sand off the edge of the flask to reduce the wear on the pattern. Sand slingers are very efficient molding tools that lend themselves well to numerical control and automation, Fig. 8-43.

Composition of Molding Sands

Molding sands are the materials which provide the mold shape into which the liquid metal is poured. Therefore, it is important that the raw material be controlled to produce the best possible mold. The raw materials are analyzed and compounded to provide the sand for the type of mold the foundry castings require.

Sand molds are classified as green-sand molds, skin-dried molds, dry sand molds, and other sand molds. Green-sand molds are the most widely used; therefore, that method will be discussed here.

Silica Sand. Silica sand is a natural sand which occurs in many parts of the world. In

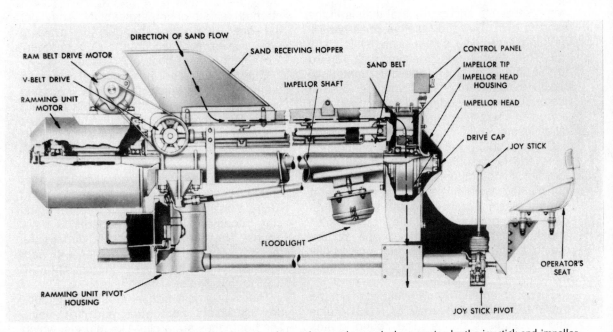

Fig. 8-43. On a large sand slinger an operator rides the machine and controls the ramming by the joy stick and impellor speed.

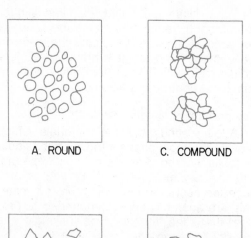

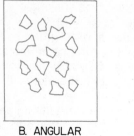

Fig. 8-44. Grain shapes of molding sands.

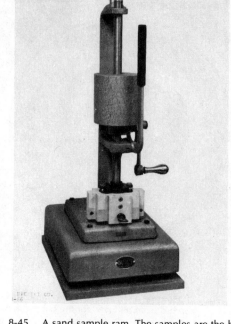

Fig. 8-45. A sand sample ram. The samples are the basis for foundry sand testing and core sand testing.

some sand pits enough natural clay is in the sand to bind the sand together for a natural molding sand. The sands used by industry must be reliable and constant with regard to the amount of binder and other materials. Most foundries test sand for the correct material content. Additional ingredients are added to bring the sand up to proper specifications.

Silica sand is the major part of molding sand and is important because it will resist the high temperatures of molten metal without melting itself. Silica sand is available in a range of sand grain sizes and shapes, Fig. 8-44. The grain shapes are round, subangular, and angular. The shape of the grains affect the sand's interlocking ability and thus the sand's strength when it is rammed around the pattern.

Sharp or angular sand grains cannot be packed as closely together as round grains and subangular grains. The round grains and subangular grains produce higher sand

strength and higher permeability. The finer the grain size, the stronger the molding sand. However, when grain size is reduced, the ability to vent gases is reduced.

Clay. Silica sand by itself will not hold together, so the silica sand grain is coated with clay. The clay is the binder in molding sand. The amount of clay in the mixture will vary from 2% to 15%. Molding sands will frequently be found with 4% or 5% clay in the mixture.

Clay is mulled or mixed with the sand until clay covers all surfaces of the sand grain. Water is added to the mix, and the clay becomes plastic and adheres to other grains. The amount of water added to the mix to hold the grains together will vary with the type of clay used. It is usually from 3% to 5% water.

The clays most frequently used for foundry sands are **bentonites** and **fire clays**. The bentonites are a clay-like material which was formed by the alternation of volcanic ash. Fire clays or kaolinite has formed by the weathering of feldspars (minerals occurring in granite rocks) and other aluminum silicates from granite rocks.

Sand Control

Sand control provides the greatest opportunity to bring about the control of the product quality being cast. A sand sample is weighed and rammed to be used as major sand tests, Fig. 8-45. The factors which must be controlled by the raw material selection and processing are these: (1) permeability, (2) compression strength, (3) clay content, (4) moisture content, (5) grain fineness, and (6) hardness of the mold.

Permeability. The test for a sand's permeability is the measurement of a standard pressure of air that is forced through the sand specimen, Fig. 8-46. A standard pressure of 10 centimetres of water (air pressure) is maintained above the specimen. The time necessary for 2000 cubic centimetres of air to pass through the sand will yield a permeability number or rating. This is a rating of the sand's ability to vent foundry casting gases through the sand and away from the molten metal.

Compression Strength. Compression strength is measured by the use of a universal sand-strength machine that gradually applies a load to the sand specimen, Fig. 8-47. When the sand sample fails, the pointer will indicate the compression strength of the sand in pounds per square inch, the strength of the foundry sand sample. This sand strength indicates how well the molding sand will hold the liquid metal.

Clay Content. Clay content may be determined by taking a dry-weighted sample of foundry sand and adding it to a blender-type agitator full of slightly alkaline water. The clay is washed free of the sand by agitating the sand in the water, allowing the large grains of sand to settle to the bottom while the mixture stands for five minutes. The clay in suspension in the water is siphoned off the top of the sand and the water discarded, Fig. 8-48. This process

Fig. 8-46.   A permeability test gives a sand sample a rating as to its ability to vent foundry gases through the sand.

Fig. 8-47.   Universal sand strength machine will test the compression strength of sand samples, as well as the tensile strength of core samples.

Fig. 8-48.   Clay content is measured by mixing the sand with water. After the mixture stands, the clay is poured off with the water.

may be repeated up to 15 times, until the water is clear after the five-minute settling period. The molding sand sample is dried and reweighed. The loss of weight of the sample — the difference in weight between before and after washing — represents the weight of the clay in the foundry sand and is expressed as the percentage of clay in the sample.

Moisture Content. Moisture content of the foundry sand is obtained by weighing a sample of sand, drying the sand sample and reweighing the sample so that there is a moist weight and a dry weight.

$$\frac{100 \text{ grams of moist sand} - 95 \text{ grams weight of dry sand}}{\text{weight moist sand sample}}$$

$$\frac{100\text{-}95 \text{ grams}}{100} = \frac{5}{100} = 5\% \text{ of moisture}$$

This test indicates the moisture in the foundry sand sample and is an important test to the quality control of castings, Fig. 8-49.

Grain Fineness. The grain fineness test is useful in analyzing the percentage of the various size of sand grains in the foundry sand sample. The washed and dried sand sample is weighed and placed in the coarse-sized sieve in a stack of eleven sieves with a pan. The stack of sieves is placed in a motor-driven shaker and shaken for 15 minutes, Fig. 8-50. The various sized grains

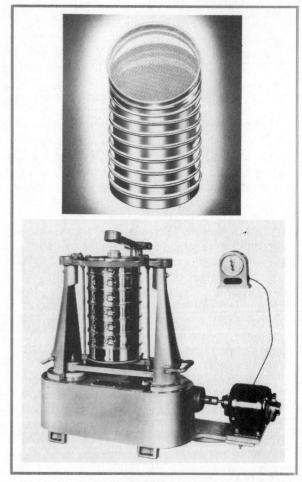

Fig. 8-50. A stack of sieves separates the various sizes of sand in the sample. The sand from the various sieves is weighed and the percentage calculated.

Fig. 8-49. Moisture teller dries out the foundry sand so that the amount of water in the sample may be calculated.

pass through the sieve openings corresponding to their size and lie upon the next smaller size screen.

The percentage of sand grain size retained on each sieve size may be calculated and yield the distribution or percentage of grain size in the total sample.

Mold Hardness. Mold hardness is another test. It determines how hard the sand has been rammed in the mold, Fig. 8-51. The hardness is related to the strength of the mold, but too hard a mold reduces the mold's permeability and causes gas inclusions and other problems in the castings.

A spring-loaded ball section is pressed into the sand, and the amount of penetration the ball makes into the sand is calibrated. It is possible to check very rapidly and easily the molds for ramming hardness and for uniformity of ramming in the various parts of the mold. The mold hardness tester helps to provide quality control during the foundry processes.

Forming Metal by Permanent Mold Casting

Permanent molding is the process of casting by using metal molds. The mold is filled by the liquid pressure or hydrostatic head of the mold. The molds are made of dense cast iron. These molds have all sprues, gates, runners, and risers machined into the molds.

The molds may be made of two or more parts and are hinged, clamped or toggled so they may be opened rapidly to remove the new parts. Cores are made of steel and are operated with cams or levers. The cores are withdrawn from the metal before the metal is completely solidified, reducing the possibility of the metal cracking from chilling and shrink cracking.

The typical metals which are cast in permanent mold products are lead, zinc, aluminum alloys, magnesium alloys, brasses, bronzes, and cast iron.

Products which are made by permanent molding include brass plumbing fittings, automotive pistons, aluminum typewriter parts, gear blanks, refrigerator cylinder blocks and heads, hydraulic cylinders, and water pumps.

Permanent molding processes become competitive in cost with sand casting if more than 200 simple castings are required. The average life of a die is near 100,000 parts per cavity. With a multiple cavity die a great number of parts may be produced, Fig. 8-52.

Fig. 8-51. Mold hardness tester indicates how hard the sand has been rammed into the mold.

Fig. 8-52. Permanent molding machine used for making gray iron castings.

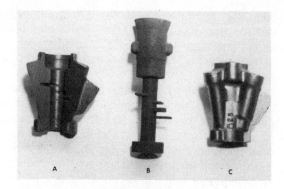

Fig. 8-53.   Permanent mold gray iron casting with a sand
core with intricate openings.
A.   Sectioned part
B.   Core
C.   Part

Fig. 8-54.   Pouring centrifugally cast pipe. Note the pipe
being rotated and the pipe to the left cooling.

Parts weighing from a few ounces to 100 pounds can be molded. Special castings using sand cores have been made as large as 500 lbs. Because of the additional heat necessary to melt gray iron, gray iron parts that weigh from about 8 ounces to 15 pounds are made.

The advantage of permanent mold castings is that the metal mold produces a fine grain casting with excellent surface finish. These castings also have uniformity and soundness of the material. Because of the fine grain, the castings may be up to 20% stronger than sand castings of the same material, Fig. 8-53.

Forming Metal by Centrifugal Casting

Centrifugal casting is done by adding molten metal to a mold that is spinning around on its axis. The centrifugal force adds pressure to the liquid, forcing dirt, sand, clay, and gases to the center surface, thus producing a dense high-quality casting.

Centrifugal castings are made in a sand, metal, or graphite mold that is rotated in a horizontal or vertical position. The metal is poured into the mold and revolved until the metal has solidified.

The type of work produced by centrifugal casting involves large, hollow, cylindrical forms, Fig. 8-54. The outside surface of the casting may be modified by flanges, bands,

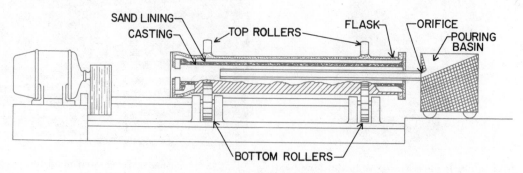

Fig. 8-55.   Centrifugal casting provides dense castings for large cylindrical shapes.

flutes, or projections. However, the total shape must be symmetrical around its central axis, otherwise the casting machine would be out of balance. Typical products are pipe, tubes, cylinder sleeves, liners, piston ring stock, bearings, rolls, pump parts, gear blanks, and oil well equipment.

The advantages of centrifugal castings are that they produce good dimensional accuracy; they have dense, clean metal; they have a rapid rate of production; and they are capable of being produced in large cylindrical shapes, Fig. 8-55. When the castings have large bores, this method of casting will produce superior castings that would be difficult to make by other casting methods.

Forming Metal by Die Casting

Die casting is the filling of a steel die cavity with liquid metal under great pressure. Because of the high pressures, the metal freezes, and very accurate parts are made with great detail reproduced on their surfaces.

The dies are made in two sections that may be locked tightly together by toggle action, by hydraulics, or by a combination of air and toggle-locking actions. The liquid metals are forced into the mold with pressures varying from 1500 to 30,000 psi. Experimental machines have increased the pressure to 50,000 pounds.

Metals with the lower melting points are used for die casting in order to obtain a reasonable life from the die cavity. High temperature melting alloys will erode and crack the special steel dies, thus rapidly destroying the expensive dies. The most common die casting alloys are zinc, aluminum, magnesium, and copper. These metals form well and will allow for a reasonable die wear, Fig. 8-56. The life of the mold will depend on the metal cast.

Zinc alloys which are cast from 750° to 800° F. will produce several million castings before the die will need to be replaced. If the metal used is a copper alloy, 10,000 castings may be produced before die repair

Fig. 8-56. Zinc die castings are used for enclosers and parts of all types. Electrical and blower cases are typical part-housing uses.

or replacement is required. The other metals will fall between these die production rates, depending on the die material used and the design of the die.

Zinc alloys are predominantly used for die casting because they have excellent dimensional accuracy. Tolerance of ±0.002″ can be achieved on articles up to 1″ thick. An additional 0.001″ per inch may be expected on articles over 1″ thick. Castings with long distances between holes or aligning surfaces which are considered uncritical areas will commonly have tolerances of ±0.010″/in.

Die casting requires high-grade zinc alloys, because impurities in the alloy such as tin, lead, and cadmium cause serious aging defects in the castings. Zinc alloys of high purity (99.99% zinc) are used to avoid intergranular corrosion. Aluminum (up to 4%) and copper (from 0.10% to 1.3%) may be added as alloys to improve the workability during casting and to improve the tensile strength and hardness of the cast part.

Hot Chamber Machine

Die casting machines are of two types, the hot chamber and the cold chamber. In the hot chamber machine the plunger system is in the molten metal and pushes the

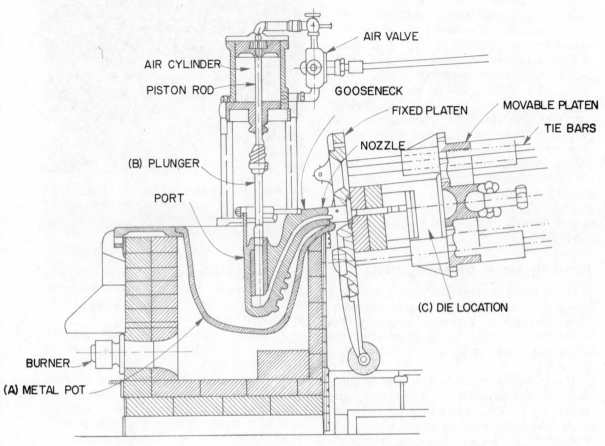

AIR VALVE

AIR CYLINDER

PISTON ROD

GOOSENECK

FIXED PLATEN

MOVABLE PLATEN

TIE BARS

(B) PLUNGER

NOZZLE

PORT

(C) DIE LOCATION

BURNER

(A) METAL POT

Fig. 8-57.   The hot chamber die casting machine for use with zinc, lead, or tin.

metal up through a gooseneck to inject the metal into the die. This design is fast and can be automated easily, Fig. 8-57. The disadvantage is that this machine will not operate at pressures over 6000 psi. However, for zinc alloys this is very serviceable. The die casting of aluminum with this type of machine is less successful, because the aluminum is contaminated by the iron in the long gooseneck, and the quality of the metal injected into the mold is changed.

Cold Chamber Machines

The cold chamber machine is designed so that the molten metal is ladled directly into the plunger ahead of the die, and the metal is forced into the die, Fig. 8-58. The advantage of this machine design is that the metal is not in contact with the steel plunger more than a few seconds and very high injection pressures are available from die or hydraulic plunger drives.

The products made by die castings are found in nearly all manufactured products of mass consumer use. T.V. picture tube frames, typewriter parts, sewing machine castings, carburetors, fuel pumps, front grill and trim of cars, camera parts, and many small parts for industry are common applications, Fig. 8-59.

Fig. 8-58. The molten metal is ladled directly ahead of the plunger and is quickly forced into the die.

Fig. 8-59. Workers removing an automobile grill from a die casting machine.

The advantage of manufacturing by die casting is that extremely smooth surfaces with excellent dimensional accuracy are possible at mass production rates.

Forming Metal by Plaster Mold Castings

Plaster mold-making for metal casting is a manufacturing process in which non-ferrous metals are cast into a plaster of paris mold, Fig. 8-60 (page 320). The mold is made of plaster of paris, fibrous talc, silica sand, and other chemicals to reduce expansion of the plaster of paris.

The plaster of paris mixture is added to water, mixed to a cream-like slurry and poured over a pattern made of polished brass, aluminum bronzes, or magnesium alloys. The patterns are coated with a thin coating of soap, wax and polish or light oil to aid in separating the pattern from the plaster of paris. The cope and drag are poured with the pattern, gates, and risers in the central parting line. The pattern is removed from the plaster of paris when the plaster is semi-set and contains some flexibility in the mold. This produces a plaster of paris flask with the mold cavity in the center.

The mold is carefully dried out to remove free moisture that would cause steam when the metal is poured into the mold.

A.   Preparing the pattern surface.

B.   Slurry being poured over a pattern.

C.   The poured pattern undergoes curing for surface hardness, etc.

D.   The cope half (top) is placed over the drag ceramic pattern half.

E.   The casting is ready for shipment to the customer.

Fig. 8-60.   Plaster stays fluid a sufficient length of time so that detailed and thin section castings are possible.

One of the chief advantages of plaster molding is that the plaster has a low conductivity of heat. Therefore, the metal stays fluid in the mold for a longer time. Very thin sections of castings can be produced because of this extra fluid state. Parts with webs or vanes as thin as 0.040″ may be cast.

The plaster casting will usually not require machining when it is completed. Its surface finish will have a fine finish equal to the pattern's finish. Accuracy in the casting will vary from 0.005″ to 0.010″ from the dimensions of the pattern when carefully cast. The manufacturing process of plaster casting will make articles of aluminum and its alloys; beryllium; copper; and brasses and bronzes and their alloys.

Pump propellors, vanes, diffusers and fans, gears, levers, handles, tire molds, manifold tube connectors, and other small, intricate parts are examples of parts manufactured by the plaster mold process.

The products of plaster casting are done in job lot numbers and run in weight from a few ounces to around 20 lbs. Larger castings have been made on special occasions, but this is not a characteristic of plaster casting.

Shell Molding

Shell molding is a manufacturing process used for casting increasing amounts of metal products. In this process a mold is made of a thin shell of sand and resin binders, which are baked to set the resin. The two half-mold shells are backed up with foundry sand or shot to support the sand shells, and the molten metal is poured directly into the cavity of the shells, Figs. 8-61 and 8-62.

The shells are made by heating the pattern to 450° F. and dumping the mixture of silica sand, urea, or phenol formaldehyde resin into a hot pattern. When the heat contacts the sand resin mixture, a shell of sand and resin is bonded together, forming a 1/8″ to 3/8″ thick shell. After a 20- to 25-second heating time, the pattern is turned over and the unbonded sand and resin fall

Fig. 8-61. Two shells are clamped together to make a cavity for pouring the metal. These shells are for making stellite welding rod.

Fig. 8-62. Stacks of shell molds being poured with a special cobalt-base alloy.

back into the dump box and may be used for the next shell.

The soft shell bakes from 30 to 60 seconds, depending on the shell thickness, with the pattern at about 450° F. The cured shell is carefully removed from the pattern with the aid of ejection pins. After cooling, two of these shells will be clamped, taped, or fastened with adhesives to assemble the mold for pouring.

Shell molding is gaining wide application as a manufacturing process, because it produces a casting equal in quality to the pattern. It is possible to hold dimensional tolerances of 0.002″ to 0.005″/inch of metal, Fig. 8-63. Smooth surfaces with detail and ease in cleaning the castings after pouring are characteristic of this process. Little skill is needed to produce these accurate molds, thus labor costs are reduced. The molds are light in weight and easily stored, Fig. 8-64.

Because of the thin shell next to the metal, gases may pass through the shell easily, reducing the number of castings which are

discarded because of blowholes, gas pockets, or porosity in the metal. The shell molding process is simple enough that the shells may be mass-produced by automatic machines in a production line, Fig. 8-65.

Shell molding does have disadvantages. A major one is that expensive metal patterns must be used. The use of resin is an added cost as is the heating equipment and the specialized dumping or shell blowing machines.

Because of the accuracy and light weight of the shell, the process is applied to the making of cores as well as to molds, Fig. 8-66. Detailed cores such as those required for automotive engine blocks and heads and other difficult coring problems are made by using the principles of shell molding, Fig. 8-67.

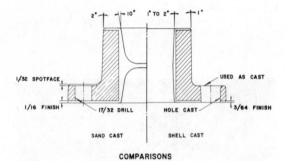

Fig. 8-63. With shell molding the finished shape and size of part may be reached because of the strength and finish of the shell. The shapes and metal needed in the two processes are compared.

Forming Metal by Carbon Dioxide Molds and Cores

Another method of producing mold cavities and cores for molds was developed which requires no water for the mold. Dry, clean silica sands are mixed with liquid

Fig. 8-64. The thin shell makes a considerable weight difference in the two casting processes. Less skill and labor are required in the shell-molding process.

Fig. 8-65. Two four-station shell-molding machines. Each can produce 180 shell molds per hour.

sodium silicate (water glass) and are packed into a mold around the pattern or in the mold box. A cover is placed over the sand in the core box or mold, and carbon dioxide gas is shot into the sand for about 15 to 30 seconds, Fig. 8-68. The carbon dioxide chemically reacts with the sodium silicate and converts it to sodium carbonate, which hardens and cements the sand together into a rigid mold ready to receive molten metal.

This process is widely used to make cores to go into many different types of molds and may be used with all types of metals.

Fig. 8-66. Hollow shell cormatic that produces accurate, lightweight shell cores.

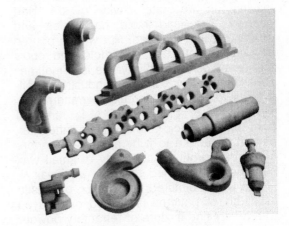

Fig. 8-67. A typical selection of hot-box cores.

Fig. 8-68. Carbon dioxide core and carbon dioxide mold for an experimental holding vessel.

Precision or Investment Casting

Investment casting of lost wax casting was applied by the early Chinese and Roman artists and craftsmen. The processes starts by either sculpturing the desired part out of wax directly or by making the exact part of brass or steel. If the part is made of steel, a split mold of lead alloy is cast around the steel part, making a mold. The steel part is removed from the lead mold, leaving an accurate cavity of the part needed. The lead mold has gates and sprues cut into the lead. Hot wax is poured or pumped into the lead mold and cools into a wax pattern. Polystyrene is sometimes substituted for wax in some industrial applications to make the precision patterns.

The wax or plastic pattern is removed from the lead mold and will have other wax patterns attached to it by means of runners and sprues until an assembly of small wax patterns are fastened together with wax.

The wax-sculptured piece, wax, or plastic assembly of patterns is dipped or sprayed with a very fine mixture of refractory powder mixed with water or alcohol, which becomes the metal contact surface of the mold, Fig. 8-69. The assembly cluster may be dipped several times, each time in a coarser refractory slurry mixture.

The assemblies are placed in a flask, and plaster is poured around the clusters until the flask is full. The plaster flask is dried and then placed upside down in a furnace. The wax runs out of the mold and all moisture and wax are burned out by the time the mold reaches 1000° F. The hot metal is poured into the mold cavity. Pressure is sometimes supplied to the metal in the mold by applying air pressure, spinning the mold to produce centrifugal force, or by placing the mold on a vacuum plate so that the atmospheric air pressure will push down on the metal. When it is cool, the plaster is broken, the parts are cleaned and the runners, gates, and sprues are removed.

Investment will produce our most accurate, intricate, and precise castings, Fig. 8-70. Investment casting produces very smooth castings that may have undercuts and no parting lines. An advantage of this process of manufacturing is that unmachineable metals may be cast to finish size and formed by this method.

Investment casting is a job-sized production and does not lend itself to mass manu-

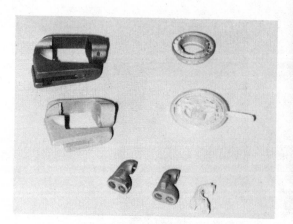

Fig. 8-69.   Investment castings and the investments. The patterns on the left are wax and the patterns on the right are injection-molded plastic.

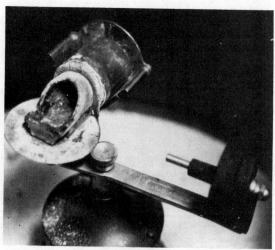

Fig. 8-70.   Centrifugal casting machine used with the investment casting of jewelry and small precision parts.

facture. A large amount of hand labor is required. The size of the casting that may be produced by investment casting is limited.

Jewelry, intricate machine parts, small special shaped gears and precision castings are examples of the type of work produced by this method.

Activity

1. Make a simple pattern for a name plate. Make the baseboard of clear soft pine or masonite. Cut letters out of balsa wood with an Xacto® knife and cement them to the baseboard. Sand a small amount of draft on all edges and around corners. Put two coats of shellac on the pattern. Allow shellac to soak well into the balsa wood and dry hard. Ram pattern in sand and pour with aluminum or brass.
2. Cast a blank plaque for use in the numerical control activity.

Related Occupations

These occupations are related to casting:
Patternmaker
Molder
Foundry worker
Molder's helper
Sand mixer
Machine molder
Shell mold machine operator
Coremaker
Core oven tender
Core assembler
Coresetter
Melter
Pourer
Shake-out worker
Shot blaster
Tumbler operator
Sand blaster
Chipper
Grinder
Heat treater
Casting inspector
Maintenance worker
Engineer and metallurgist
Quality control technician

Combining and
Conditioning Processes

Section
3

Bonding and joining are processes of combining materials. Cohesive bonding is bonding by heat and is performed by a number of welding processes: shielded arc welding, gas metal arc welding, gas tungsten arc welding, submerged arc welding, plasma arc welding, electron beam welding, and laser welding.

Bonding with heat and pressure are processes classified as electric resistance welding. These welding processes include spot welding, roll spot welding, seam welding, projection welding, flash butt welding, and upset welding.

Bonding with chemical heat is done by gas welding. Acetylene and oxygen gases are burned to produce the temperature required for fusion of the metals. The process is done manually, so the operator must have considerable skill.

The bonds of metal can be separated with a flame or heat, frequently an oxyacetylene flame. Flame cutting may be performed by hand or by machine. Machine cutting employs a number of methods of control such as mechanical control with cams and templates, photoelectric control with the cell following a line or pattern, and numerical control, which is activated by a programmed tape. Because of its great heat, plasma cutting can be used for separating metals, stainless steels, refractory materials, and carbon steels. It actually cuts by melting the metal at speeds up to three times as fast as conventional cutting processes.

Adhesive bonding joins metal with intermediate alloys or with polymer adhesives. Intermediate metal bonding is accomplished by bronze welding, brazing, or hard soldering and soft soldering. Polymer adhesive bonding of metal is done by thermoplastic, thermosetting, or elastomeric adhesives. Joint design is of prime importance in adhesive bonding.

Conditioning metal improves the metal characteristics by changing the structure of the metal. Methods of changing the properties of the metal are thermal or mechanical. Ferrous alloys are conditioned primarily by allotropic changes in the metal, while nonferrous alloys are conditioned primarily by precipitation or cold working. Some metal will utilize both thermal and mechanical conditioning processes.

Bonding-Processes of Joining

Chapter 9

Words You Should Know

Cohesion bonding — Welding in which the molten metal of two pieces of metal intermingles and solidifies as one metal.

Adhesion bonding — Metals are fastened by additional material which diffuses into the base metal. The metals are not melted.

Capillary action — The movement of liquids by the attraction or repulsion between the liquids or solids in a confined space.

Diffusion — The intermingling of molecules and penetrating of a surface.

Polarity — The direction of current flow in a welding circuit.

Penetration — The depth of fusion during welding.

Root pass — A weld in the bottom of a welded joint.

Distortion — Deforming of metal because of unequal heat in the various parts of the metal.

Dissociates — Chemically comes apart into simpler compounds.

High-current density — A high ampere flow in a small, compact area.

Constant voltage power supply — A welding power source that maintains a constant potential or voltage when welding or amperage is changed.

Wire feeder — A motor-driven set of rolls that push the wire into the weld area.

Metal transfer — The transporting of metal from a solid wire through an electric arc into the molten pool.

Ionization potential — The electrical energy needed to transform a gas atom into an ion of the gas.

Thermal conductivity — Low absorption of heat by the gas from the arc. Low thermal conductivity allows more heat to reach the metal, thus greater penetration is also achieved.

Inert gas — A gas that does not react chemically with surrounding metals and materials.

Constant current power supply — A welding power source that maintains a constant amperage as welding takes place.

Circumferential (girth seamer) — A motorized carriage that carries an automatic welding unit around the circumference or the out-seam of a cylinder.

Longitudinal (seamer) — A movable unit that will carry the welding unit down long, straight butt joints.

Pedestal boom — A vertical post and cross member that may be moved to various positions to hold an automatic welding torch head.

Manipulators — A large steel table that may be moved to any angle and also may be rotated under power. Positioners sometimes are very large.

Submerged arc welding — Heavy electrical arc welding under a mound of granular flux.

Electroslag welding — A vertical weld of heavy sections with electrode wires melting in the weld area under a heavy molten slag.

Plasma welding — Very high-temperature welding process that uses an electric arc to energize the gas in a confined chamber. This produces a plasma jet which is applied to melt metals.

Keyhole welding — The melting of a hole completely through the metal and the liquid metal rejoining behind the heat source.

Hard surfacing — Melting of very hard metal powders and welding them in the surface of the base metal.

Electron beam welding — A beam of electrons from an electron gun provides the heat necessary to melt a narrow path through the metal.

Kinetic energy — The energy of motion. Electrons traveling at high speeds strike the weld metal and slow down or stop. In doing so, the energy is transformed to heat.

Laser welding — A burst of light which is concentrated in a small area. It is so intense that it will melt thin metals.

Theory of Bonding

Bonding is divided into two large categories: cohesion bonding and adhesion bonding. Cohesion bonding is bonding by fusion welding, Fig. 9-1. Adhesion bonding is bonding by diffusion of solder alloys between metals.

Cohesion Bonding

In **cohesion bonding,** the edges of two pieces of metal are heated until the metal becomes fluid, Fig. 9-2. The liquid metals are caused to intermix, and upon cooling, a weld is made. This is a fusion weld, because the two metals (as liquids) have interchanged and the metals have been joined in the liquid condition. The liquid metal is now one metal and all its molecules are held together by the cohesion of the metal's atoms. The metal cools into one solid piece, and the bonding or welding has taken place.

Adhesion Bonding

Adhesion bonding is the fastening of two pieces of metals by diffusion, Fig. 9-3. The

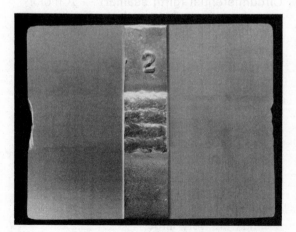

Fig. 9-1. Cohesive bonding: fusion welding of a cast iron block. The arc weld was done with a rod containing nickel. Notice the metal match where the part is machined.

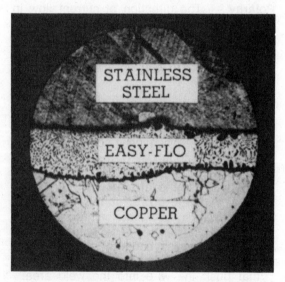

Fig. 9-3. Adhesion bonding is the fastening of two pieces of metal by diffusing a melted alloy between pieces that are not melted. The bonding alloy acts much as a solvent when melted.

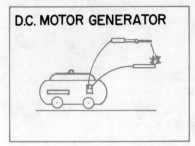

D.C. MOTOR GENERATOR

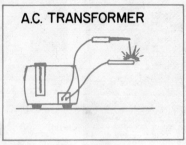

A.C. TRANSFORMER

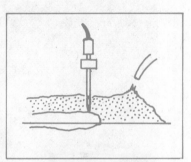

TUNGSTEN ELECTRODE
ELEC- HOLDER
TRODE
ELEC-
TRICAL
CONDUC-
TOR
WELDING
MACHINE
WORK- INSULA-
PIECE TING SHEATH

SHIELDED METAL ARC WELDING-MANUAL

INERT GAS-TUNGSTEN ARC WELDING-MANUAL

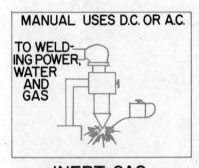

MANUAL USES D.C. OR A.C.

TO WELD-
ING POWER,
WATER
AND
GAS

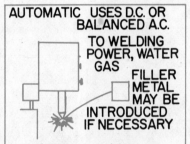

AUTOMATIC USES D.C. OR
BALANCED A.C.
TO WELDING
POWER, WATER
GAS
FILLER
METAL
MAY BE
INTRODUCED
IF NECESSARY

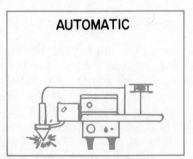

INERT GAS-TUNGSTEN ARC WELDING-MECHANIZED

INERT GAS-TUNGSTEN ARC WELDING-AUTOMATIC

SUBMERGED ARC WELDING

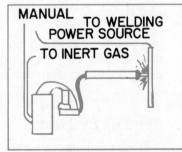

MANUAL TO WELDING
POWER SOURCE
TO INERT GAS

AUTOMATIC

INERT GAS-METAL ARC WELDING WITH CONSUMABLE ELECTRODES

Fig. 9-2. Arc welding processes and equipment.

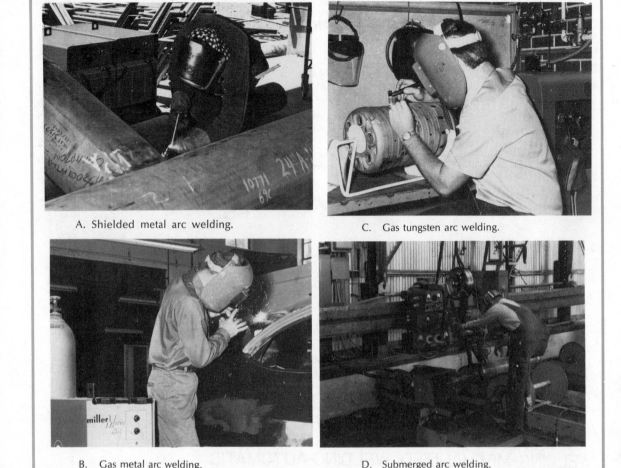

A. Shielded metal arc welding.

C. Gas tungsten arc welding.

B.   Gas metal arc welding.

D. Submerged arc welding.

Fig. 9-4. Bonding by heat

metals being joined are not melted, but they are heated and an additional material is added. The materials added are usually alloys which act much as solvents do. The metals are carefully cleaned, fluxed, and heated. A brazing or soldering alloy may be added before or during the heating. When the temperature is reached where the alloy will flow, it will proceed by capillary action through a correctly fitted joint. The alloy will act similarly to a solvent and allow some of the low temperature alloy to diffuse into the metal that it is contacting. The alloy diffuses into both pieces of the metal, thus adhering or holding the surfaces together.

Bonding by Heat

Bonding metal by heat is performed by a number of welding processes, Fig. 9-4. The common types are these: 1. Manual arc welding with consumable electrodes (shielded metal arc welding). 2. Gas metal arc welding or metal inert gas welding (M.I.G.). A continuous bare electrode wire is mechanically fed into the welding area and shielded from the atmosphere by carbon dioxide, argon, helium, or mixes of these and other gases. 3. Gas tungsten arc welding or tungsten inert gas welding (T.I.G.). A nonconsumable tungsten forms the electrode and establishes the arc for the welding temperature. A separate filler metal is added while being protected by a helium or argon shielding gas. 4. Submerged arc welding. The welding wire is automatically fed into the weld to produce the arc and supply the filler material. The whole weld area is covered with a granulated (powdered) flux similar to sand. The arc burns under the flux, thus the name "submerged arc." 5. Electroslag welding. Vertical welds are performed with the use of a heavy flux 1″ to 1½″ floating on top of the rising weld. Starting at the bottom of the weld, electrode wire or wires are added to the molten metal. These melt and fill the space between the two surfaces being joined. The ends of the weld are held in position by large water-cooled copper sliding molds. 6. Electrogas welding. This is a modification of electroslag welding except that the heavy slag is replaced by a shielding gas of usually 80% argon, 20% carbon dioxide. The heavy welds have an automatic wire feed for the filler and an automatic oscillation of the wire.

Arc Welding Safety

Arc welding safety is primarily concerned with preventing burns, electrical shock, and radiation. To prevent injury, the welder must wear and use protective equipment, Fig. 9-5. The arc welding processes generate large amounts of heat, ultraviolet rays, infrared rays, and flying sparks.

Welder's Equipment. Welding safety starts with the correct electrical installation of the welding power source and its equipment. The machine must be installed by a qualified electrical contractor so that proper equipment grounding is achieved. The

welder is grounded to protect the operator from a short developing within the machine because of any breakdown of insulation within the machine. When the ground is properly installed, a short in the welder will cause the circuit breaker to trip. The operator would be protected.

Welding cables should be periodically visually inspected for wear or cuts. Cables are checked after they have been in use by turning off the power source and running the cable through the hands to check for hot spots. A **hot spot** may indicate that the wires within the cable are breaking, due to the constant flexing during the welding process. If a hot spot is found, the spot is marked with chalk and taken to a technician to be cut and repaired. These cable breaks occur most frequently adjacent to the electrode holder.

Cable lugs and nuts should be periodically tightened when the machine power is off. The vibration of some welding machines cause cable fasteners to work loose with use.

Welder's Tools. These tools should be available to the welder:

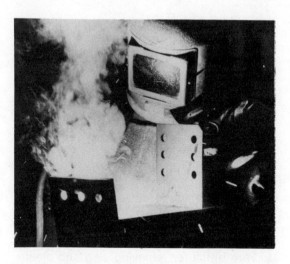

Fig. 9-5. Proper equipment protects the welder from the hazards of radiation, heat, and molten metal spatter.

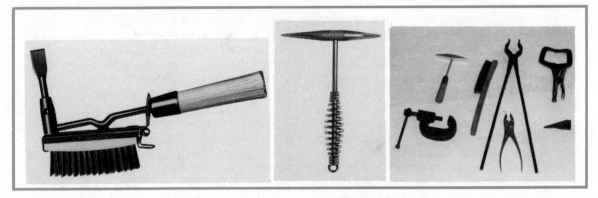

Fig. 9-6. Welder's tools for cleaning and locating work.

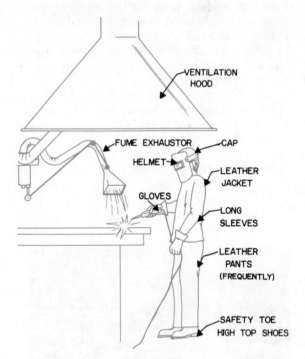

Fig. 9-7. Special welding safety equipment and clothing are needed for arc welding.

1. Pliers or tongs to move hot metal during the welding process.
2. Chipping hammer for removing slag and for aligning work.
3. Wire brush for removing slag and cleaning work.
4. Clamps and wedges for holding and positioning work.

5. Framing square and tape measure for squaring and locating work.

Figure 9-6 shows examples of these welding tools.

Protecting the Welder. Operators must have all skin covered so that their bodies will be protected from radiation and heat. Frequently, burns may occur at the V-neck of a low or unbuttoned shirt or on the forearm between the glove and sleeve of the shirt. During welding, sparks as well as globules of molten metal will scatter. A leather jacket, pants, and gloves are worn to offer protection from this hazard, Fig. 9-7.

The face and eyes are protected from the ultraviolet flash by a helmet with colored lens. At no time should an arc be looked at with bare eyes. A painful flash may damage the eyes, Fig. 9-8.

Another potential hazard to the welder is the fumes from welding fluxes and metal oxides. Welding must be done in areas which have proper ventilation and equipment to pick up and remove all fumes and gases from the welding area. These metal oxides are especially harmful to health: lead oxide from paint and leaded steels, zinc from galvanized coatings, and cadmium from plated steel and silver solders.

Welding is not dangerous when proper safety equipment and rules are followed.

Shielded Metal Arc Welding

Shielded metal arc welding is better known as manual arc welding, Fig. 9-9. It

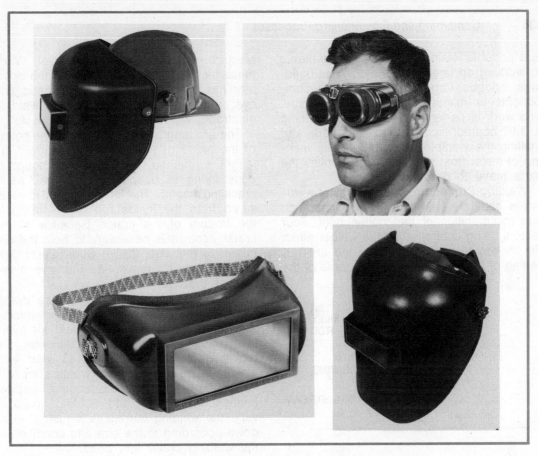

Fig. 9-8. Welder's protective equipment.

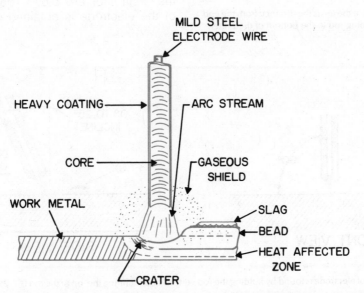

Fig. 9-9. The arc stream generates the heat and melts down into the base metal, making all sides of the joint liquid metal. The metals flow together, making a bond when the area cools.

is a portable joining process very adaptable to working on large projects such as buildings, bridges, large tanks, and other components or manufactured parts on a site. The work is performed by a welder striking a 14″ coated electrode to the work and melting the workpieces together. The melting of metal from the electrode supplies the extra metal to produce the bead.

Welding Procedure. The process of welding is dependent primarily on the condition and manipulation of the molten pool. Four factors being controlled by the welder affect the molten pool: (1) length of the arc, (2) setting of the power source, (3) angle of the

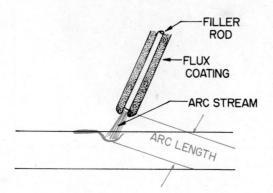

Fig. 9-10.   The length of the arc is the distance from the end of the welding rod to the bottom of the crater.

electrode, and (4) speed of travel. When these factors are coordinated and the operator has the necessary physical hand and eye coordination to manipulate the welding electrode, welding should be successful.

The **arc length** is determined by the sound of the arc, Fig. 9-10. When the correct arc length is reached, the arc has the sound of eggs frying in deep fat. There will be a crackling sound. The length of arc is measured from the inside rod or electrode to the bottom of the crater. Because of the crater depth it is necessary to hold the rod nearer to the work than it would seem. The arc length will approximate the wire diameter of the rod.

The setting of the power source determines the available amperage for producing the heat. One amperage is used for every thousandth of an inch of rod diameter. For example, a 1/8″ E6013 would start with 125 amperes, a 5/32″ rod would start with 156 amperes. Because the length of cables, connections, power, and welding positions vary, the welding current is adjusted up or down according to the size and condition of the welding crater and bead.

The angle of the electrodes is a factor which aids in the formation of the bead and the control of arc blow. The correct angle of the electrode is obtained when it is held

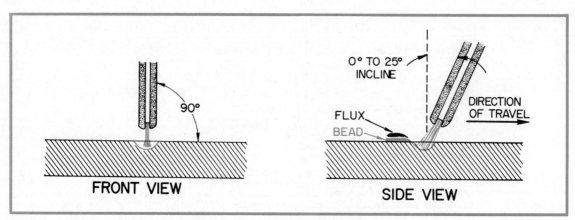

Fig. 9-11.   The angle of electrode is found by holding the rod vertical and inclining the top of the rod 0° – 25° in the direction of travel.

vertical to the plate lying on the welding bench and inclined toward the direction of travel from 0° to 25°, whatever is comfortable for the operator and makes it possible to see into the molten pool or crater, Fig. 9-11.

The correct speed of travel will produce a bead that is about one-half the rod diameter high and one and one-half times the rod diameter in width, Fig. 9-12.

Types of Joints. A number of different types of joints can be welded, Figs. 9-13

OVER-WELDED UNDER-WELDED CORRECT BEAD

Fig. 9-12.   The speed that the electrode is moved controls the size and shape of the weld bead.

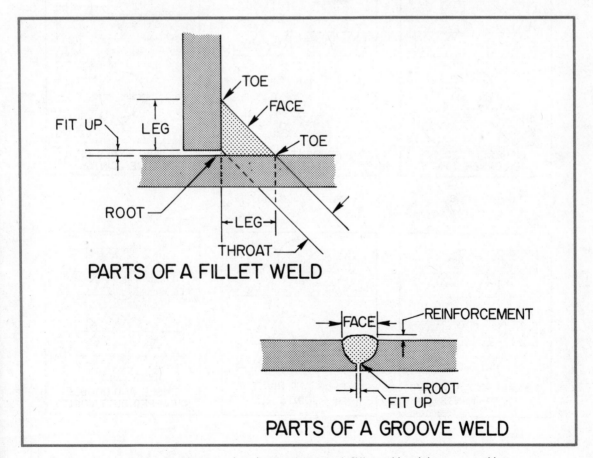

TOE
FACE
TOE
FIT UP
LEG
ROOT
LEG
THROAT
PARTS OF A FILLET WELD

FACE
REINFORCEMENT
ROOT
FIT UP
PARTS OF A GROOVE WELD

Fig. 9-13.   Weld joints are based on two designs — the fillet weld and the groove weld.

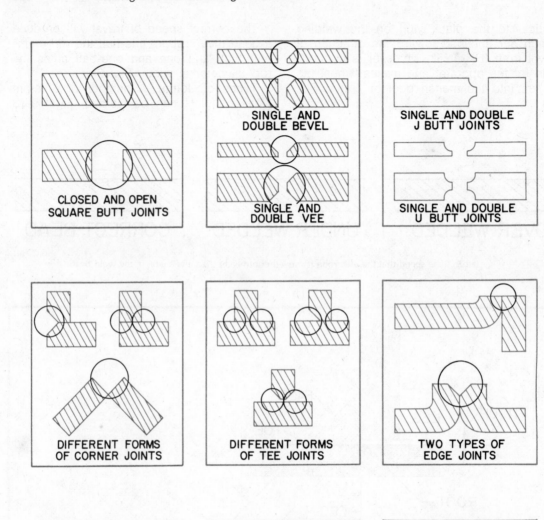

CLOSED AND OPEN
SQUARE BUTT JOINTS

SINGLE AND
DOUBLE BEVEL

SINGLE AND
DOUBLE VEE

SINGLE AND DOUBLE
J BUTT JOINTS

SINGLE AND DOUBLE
U BUTT JOINTS

DIFFERENT FORMS
OF CORNER JOINTS

DIFFERENT FORMS
OF TEE JOINTS

TWO TYPES OF
EDGE JOINTS

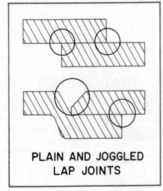

PLAIN AND JOGGLED
LAP JOINTS

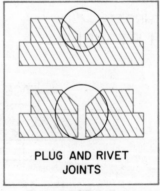

PLUG AND RIVET
JOINTS

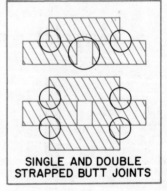

SINGLE AND DOUBLE
STRAPPED BUTT JOINTS

Fig. 9-14.   Types of weld joints made by fillet or groove construction.

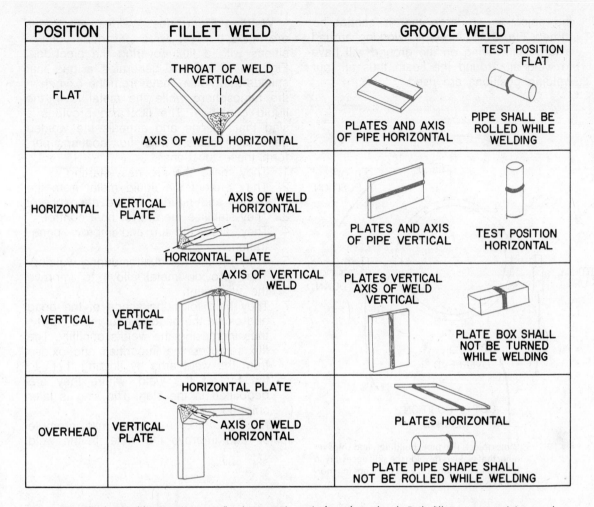

POSITION	FILLET WELD		GROOVE WELD	
FLAT	THROAT OF WELD VERTICAL / AXIS OF WELD HORIZONTAL		PLATES AND AXIS OF PIPE HORIZONTAL	TEST POSITION FLAT / PIPE SHALL BE ROLLED WHILE WELDING
HORIZONTAL	VERTICAL PLATE / AXIS OF WELD HORIZONTAL / HORIZONTAL PLATE		PLATES AND AXIS OF PIPE VERTICAL	TEST POSITION HORIZONTAL
VERTICAL	VERTICAL PLATE / AXIS OF VERTICAL WELD		PLATES VERTICAL AXIS OF WELD VERTICAL	PLATE BOX SHALL NOT BE TURNED WHILE WELDING
OVERHEAD	VERTICAL PLATE / HORIZONTAL PLATE / AXIS OF WELD HORIZONTAL		PLATES HORIZONTAL / PLATE PIPE SHAPE SHALL NOT BE ROLLED WHILE WELDING	

Fig. 9-15.   The four welding positions are flat, horizontal, vertical, and overhead. Both fillet or groove joints can be welded in all positions.

and 9-14. All joints are built from two basic types: **fillet** or **groove.** A fillet weld is made by welding the intersection of two surfaces at 90° or other angles to each other. A groove weld is made by welding two edges or surfaces that are placed beside each other. The groove weld may be made by spacing the metal, cutting a bevel on one or both plates or by cutting a gouge on one or both plates prior to welding.

Welding Positions. There are four common positions of welding, Fig. 9-15. In the flat position, the welded seam is parallel with the earth's surface, and the work is flat. In the **horizontal** position, the welded seam is also parallel with the earth's surface, but it is on a vertical plane. In the **vertical** position, the welded seam is perpendicular to the earth's surface. In the **overhead** position, the welding is parallel to the earth's surface, but it is over the operator's head. An operator who can work in all four positions is important in manufacturing and construction welding, because in many cases the object is so large that it cannot be turned, Fig. 9-16. A welder working around a large pipe lying on the ground will have to weld all around the seam thus all four welding positions are used.

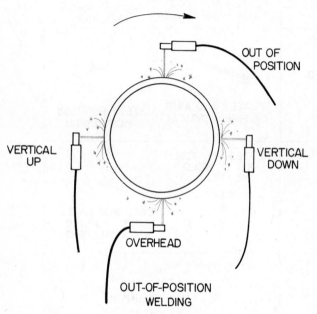

Fig. 9-16. Large objects like pipe, buildings, and ships require that welders develop the skill necessary to be able to weld objects which are not in a normal position.

Welding Electrodes. Welding electrodes are made of steel wires of various compositions with a flux covering the electrode, Fig. 9-17. The flux generates a gas that shields the molten metal from the oxygen in the atmosphere while the metal is in the liquid condition. The flux also provides a slag that shapes and anneals the welded bead as it cools. These flux coatings perform these functions.

1. They make the arc easy-starting.
2. They protect the liquid metal from the oxides and nitrides of the atmosphere.
3. They stabilize the arc for heat control.
4. They direct the arc and improve penetration.
5. They reduce the spatter during welding.
6. They provide metal alloys to improve the weldment.
7. They help form the shape of the bead and cause the bead to cool more slowly, thus increasing the weld's ductility. The flux removes the impurities and oxides from the weld area by floating them to the top of the weld, where they are deposited in the slag. The slag is later chipped off.

The flux adds materials that promote fusion and improve the metal in the weld

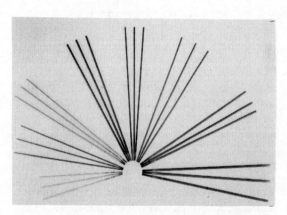

Fig. 9-17. The steel wire in the rod provides the filler metal, while the flux provides control and improvement of the weldment's chemistry.

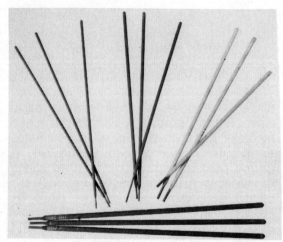

Fig. 9-18. The addition of iron powder to the flux improves the rods' penetration and metal deposition rate.

area. The flux coatings are made of inorganic materials such as feldspars, titanium dioxide, magnesium and calcium carbonate, asbestos, and aluminas. The organic fluxes produce a neutral or reducing gas shield. The gas shield is made of cellulose, carbon, hydrates, wood flour, cotton, starch, and paper. Metal powders and salts are added to produce alloy metal for the weld area.

Welding rods with heavy coatings of iron powder supply an additional amount of filler metal to the weld. The high temperature converts the powdered iron to steel in the weld and produces a deep penetrating welding action with a high deposition rate of weld metal, Fig. 9-18. Because the thick coatings produce a deep shield on the elec-

trode, it is possible to use a dragging electrode technique in the flat position. Higher amperage and an even arc length produce a sound, high quality weld.

Many electrodes are available to the welder. The correct electrode must be selected to meet the specifications of the welding job. The American Welding Society formulated a numbering system that identifies and classifies various welding electrodes. This classification system has been adopted by the American Society for Testing Materials.

A classification system using four numbers is used to identify the large number of welding electrodes used in manufacturing, Fig. 9-19. The shielded metal arc electrodes

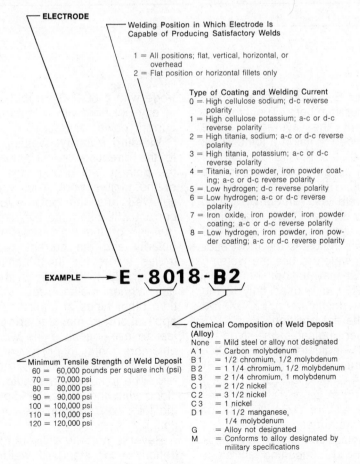

ELECTRODE

Welding Position in Which Electrode Is
Capable of Producing Satisfactory Welds

1 = All positions; flat, vertical, horizontal, or overhead
2 = Flat position or horizontal fillets only

Type of Coating and Welding Current
0 = High cellulose sodium; d-c reverse polarity
1 = High cellulose potassium; a-c or d-c reverse polarity
2 = High titania, sodium; a-c or d-c reverse polarity
3 = High titania, potassium; a-c or d-c reverse polarity
4 = Titania, iron powder, iron powder coating; a-c or d-c reverse polarity
5 = Low hydrogen; d-c reverse polarity
6 = Low hydrogen; a-c or d-c reverse polarity
7 = Iron oxide, iron powder, iron powder coating; a-c or d-c reverse polarity
8 = Low hydrogen, iron powder, iron powder coating; a-c or d-c reverse polarity

EXAMPLE → **E - 8 0 1 8 - B 2**

Chemical Composition of Weld Deposit (Alloy)
None = Mild steel or alloy not designated
A 1 = Carbon molybdenum
B 1 = 1/2 chromium, 1/2 molybdenum
B 2 = 1 1/4 chromium, 1/2 molybdenum
B 3 = 2 1/4 chromium, 1 molybdenum
C 1 = 2 1/2 nickel
C 2 = 3 1/2 nickel
C 3 = 1 nickel
D 1 = 1 1/2 manganese, 1/4 molybdenum
G = Alloy not designated
M = Conforms to alloy designated by military specifications

Minimum Tensile Strength of Weld Deposit
60 = 60,000 pounds per square inch (psi)
70 = 70,000 psi
80 = 80,000 psi
90 = 90,000 psi
100 = 100,000 psi
110 = 110,000 psi
120 = 120,000 psi

Fig. 9-19. Standard American Welding Society code for welding electrode designation.

FIG. 9-20.
Usability Characteristics of Mild Steel Electrodes

	Type of Coating	Position of Welding	Type of Current* Used	Pene-tration	Rate of Deposition	Appearance of Bead	Spatter	Slag Removal	Minimum Tensile Strength	Yield Point	Minimum Elongation in 2 in.
E6010	High Cellulose Sodium	All Positions	DC, Reverse	Deep	Average Rate	Rippled and Flat	Moderate	Moderately Easy	62,000 psi	50,000 psi	22%
E6011	High Cellulose Potassium	All Positions	AC DC, Reverse	Deep	Average Rate	Rippled and Flat	Moderate	Moderately Easy	62,000 psi	50,000 psi	22%
E6012	High Titania Sodium	All Positions	DC, Straight AC	Medium	Good Rate	Smooth and Convex	Slight	Moderately Easy	67,000 psi	55,000 psi	17%
E6013	High Titania Potassium	All Positions	AC DC, Straight	Mild	Good Rate	Smooth and Flat to Convex	Slight	Easy	67,000 psi	55,000 psi	17%
E7014	Iron Powder Titania	All Positions	AC DC, Straight	Mild	High Rate	Smooth and Flat to Convex	Very Slight	Very Easy	72,000 psi	60,000 psi	17%
E7016	Low Hydrogen Potassium	All Positions	AC DC, Reverse	Mild to Medium	Good Rate	Smooth and Flat to Convex	Slight	Moderately Easy	72,000 psi	60,000 psi	22%
E6020	High Iron Oxide	Flat Hor. Fillets	Flat—DC, AC Hor. Fillets— DC Str., AC	Deep	High Rate	Smooth and Flat to Concave	Slight	Easy	62,000 psi	55,000 psi	25%
E7024	Iron Powder Titania	Flat Hor. Fillets	DC AC	Deep	Very High Rate	Smooth and Flat to Convex	Very Slight	Very Easy	72,000 psi	60,000 psi	17%
E6027	Iron Powder Iron Oxide	Flat Hor. Fillets	Flat—DC, AC Hor. Fillets— DC Str., AC	Deep	Very High Rate	Flat to Concave	Slight	Very Easy	62,000 psi	50,000 psi	25%
E7018	Iron Powder Low Hydrogen	All Positions	AC DC, Reverse	Mild	High Rate	Smooth and Slightly Convex	Very Slight	Very Easy	67,000 psi	55,000 psi	22%
E7028	Iron Powder Low Hydrogen	Flat Hor. Fillets	AC DC, Reverse	Mild	Very High Rate	Smooth and Slightly Convex	Very Slight	Very Easy	67,000 psi	55,000 psi	22%

*DC Reverse means DC, reverse polarity (electrode positive).
DC Straight means DC, straight polarity (electrode-negative).

start with the letter E (for electrode), which is followed by four numbers. The first two numbers represent the average tensile strength of the welded metal in a joint. Thus, E60XX indicates that a tensile strength of 60,000 lbs. would be developed per square inch of weld metal. The 60 is multiplied by 1000 to yield a weld test sample of 60,000 psi. A welding rod with an E80XX classification would yield a test sample of 80,000 psi.

The third number indicates the most successful position in which to use the welding rod. An EXX1X rod may be applied in any position. An EXX2X rod is used only in the horizontal and flat position. An EXX3X rod may be used in the flat position.

The last number indicates the type of power supply: (1) indicates A.C. or D.C. (+) reverse polarity; (2) A.C. or D.C.; (3) A.C. or D.C.; (4) A.C. or D.C.; (5) D.C. (+) reverse polarity; (6) A.C. or D.C. (+) reverse polarity; (7) A.C. or D.C.; (8) A.C. or D.C. (+) reverse polarity.

The complete classification of an electrode would be E6013, meaning an electrode with a 60,000 psi that may be welded in any position with an alternating or direct current power source.

Welding Polarity. Welding polarity refers to the direction of current flow. Changing the polarity will change the characteristics of the molten pool. The type of electrode selected and the polarity of the welding current will determine the penetration depth (fusion) of the weld area, Fig. 9-21. The polarity and the current setting are controlled by setting the machine.

In shield arc welding, **straight polarity** has higher melting and deposition rates than other types of current. Thus, in certain applications in manufacturing it is economical to use, because the welder is able to put more metal in place per hour. Straight polarity delivers a shallow penetration pattern. The joint design should be considered along with the choice of polarity, because the joint can compensate for less penetration.

Reverse polarity delivers maximum penetration under standard welding conditions

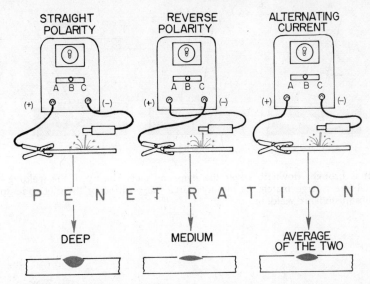

Fig. 9-21. The welding polarity is one factor which controls the penetration and characteristics of the metal deposit.

and gives an advantage for root passes in groove welding or where fit-up of the parts demands high heat for fusion. Reverse polarity is frequently used for out-of-position welding (horizontal, vertical, or overhead welding).

In **alternating current** the polarity reverses. The 60-cycle current used in American industry changes its polarity 120 times a second, producing the effect of both a straight and reverse polarity arc. Alternating current is excellent for welding thick sections with large diameter electrodes. Small alternating welders for repair and farm use have become popular because of their low initial cost and the availability of excellent small diameter electrodes.

Starting to Weld. Examine the thickness of the metal to be welded, the position of the weld, and the material to be welded. Study the drawings to determine the type of joint requirements. Will it be a fillet or groove weld? Is the weld bead to be convex, flat, or concave? What size is required? Select the electrode that best meets or exceeds the specifications.

The size of the electrode and the position of the weld determine the setting of the welding machine. The setting of the machine in most welding machines is controlled by a dial or by a dial and a switch.

Before any machine is used, the proper safety equipment must be put on. Then the electrode is placed in the electrode holder, the welder is started, the helmet turned down, and the electrode scratched against the weld area, Fig. 9-22.

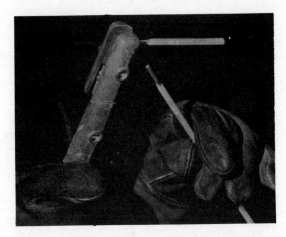

Fig. 9-22. Position of the electrode in the welding electrode holder. The bare end of the electrode is positioned flush with the metal parts of the holder.

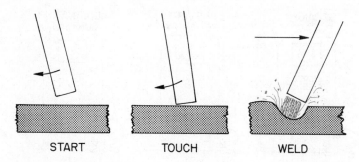

START TOUCH WELD

Fig. 9-23. The helmet is brought down to cover the eyes and face. The tip of the welding rod is scratched on the metal like a large match. As the arc starts, the position of the electrode and the length of the arc are adjusted for welding.

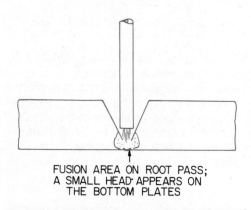

FUSION AREA ON ROOT PASS;
A SMALL HEAD APPEARS ON
THE BOTTOM PLATES

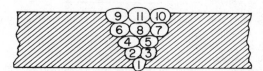

Fig. 9-25. Multipass welds will be needed on the welding of thick joints. The flux is chipped and the weld is carefully cleaned between each welding pass.

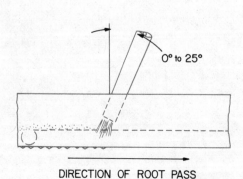

0° to 25°

DIRECTION OF ROOT PASS

Fig. 9-24. The arc is struck on the bottom of the double Vee opening. The root pass fuses the general area and produces a small bead on the bottom of the plates.

Fig. 9-26. A fabricated crosshead for a large press being welded together.

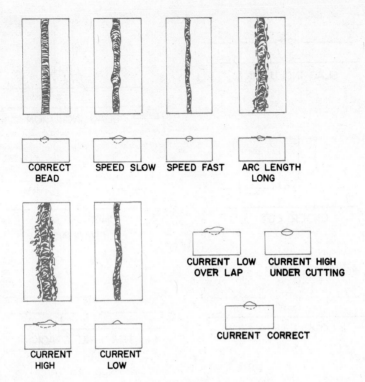

Fig. 9-27. Examine the weld and compare it to the most representative illustration. This indicates what is needed to improve the weld.

The arc is struck, Figs. 9-23 and 9-24. The distance between the bottom of the crater and the end of the electrode wire is adjusted by moving the electrode until a "deep fat frying" sound is heard. This sound indicates the correct arc length.

Metals which are thick may require a **multipass weld** and thus are joined with a series of welds, Fig. 9-25. Each pass must be cleaned by being chipped and wire brushed before the next weld pass is started. Beads may also be completed by weaving the electrode in a side-to-side motion during welding.

In order to get maximum penetration with a high deposition rate, two different electrodes may be used on heavy welds, Fig. 9-26. The root pass may be made with an E7016 electrode, and the remainder of the passes with an E7024. A high deposition iron powder electrode with smooth bead contour may be used to complete the weld.

After the weld is made, the operator checks to see whether it was made correctly. Figures 9-27 and 9-28 (page 346) show methods for checking the accuracy of a weld.

Preventing Welding Distortion. Distortion is caused by unequal heat in the work, which in turn produces stress in the weld area. Stress in the work causes the work to warp or distort. Overwelded work should also be avoided. More tensile strength than necessary is created by overwelding, and with it is the serious problem of heat distortion. This can be controlled by using a single pass or a stringer bead, by sequence welding, back stepping, pre-positioning, clamping, or pre-heating, Fig. 9-29 (p. 347).

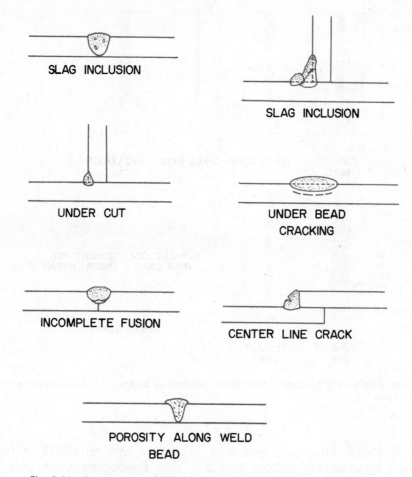

SLAG INCLUSION

SLAG INCLUSION

UNDER CUT

UNDER BEAD
CRACKING

INCOMPLETE FUSION

CENTER LINE CRACK

POROSITY ALONG WELD
BEAD

Fig. 9-28. Inspect, saw in half, or break a weld and examine it for defects.

Gas Metal Arc Welding

Gas metal arc welding (G.M.A.) is performed in a controlled atmosphere around a bare wire electrode which melts in the arc and provides filler metal to the weld joint, Fig. 9-30. The process in the past has been called M.I.G., or metal inert gas welding. When carbon dioxide gas was applied to this process for the inexpensive welding of mild steel, the shielding gas was no longer inert. The chemical activity of the oxygen in the weld zone improved the cleaning and wetting out of the base metal. The change to chemical activity made the name change necessary, since the gas was no longer inert.

Advantages of Gas Metal Arc Welding. One advantage of gas metal arc welding over the shielded metal arc is the great speed of welding which results from the continuous wire feed of electrodes, Fig. 9-31. With a gas shield, there is no slag to be chipped off to complete the weld in order to make a second pass.

The wire electrode diameter is a small area, thus a high current density is created. The wire therefore has a higher burn-off rate, and high metal deposition rate results.

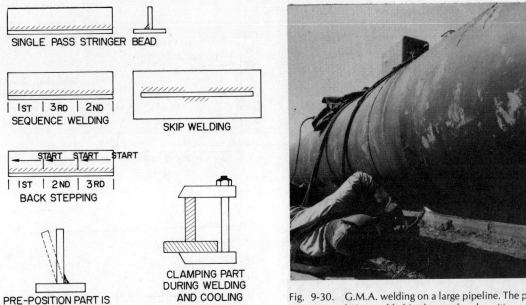

SINGLE PASS STRINGER BEAD

| 1ST | 3RD | 2ND |
SEQUENCE WELDING

SKIP WELDING

| 1ST | 2ND | 3RD |
BACK STEPPING

PRE-POSITION PART IS
STRAIGHT WHEN COOL

CLAMPING PART
DURING WELDING
AND COOLING

Fig. 9-29. Welding distortion is controlled by keeping the heat concentration low, by balancing the heat in the workpiece, by
mechanically overpowering the stress
forces with clamps and fixtures, or by
post-heating or shot peening.

Fig. 9-30. G.M.A. welding on a large pipeline. The pipe is
being welded in the overhead position.

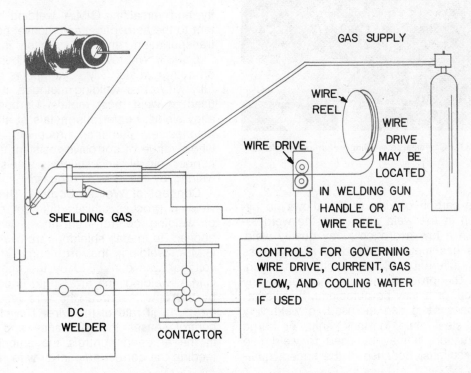

GAS SUPPLY

WIRE
REEL

WIRE DRIVE

WIRE
DRIVE
MAY BE
LOCATED
IN WELDING GUN
HANDLE OR AT
WIRE REEL

SHEILDING GAS

CONTROLS FOR GOVERNING
WIRE DRIVE, CURRENT, GAS
FLOW, AND COOLING WATER
IF USED

DC
WELDER

CONTACTOR

Fig. 9-31. Equipment necessary for making a gas metal arc weld.

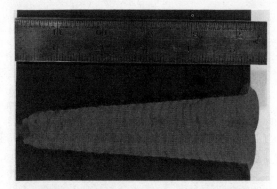

Fig. 9-32.   Test weld specimen 5¼″ thick welded with the gas metal arc process.

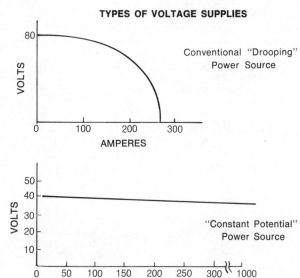

TYPES OF VOLTAGE SUPPLIES

Conventional "Drooping" Power Source

"Constant Potential" Power Source

Fig. 9-34.   A conventional or "drooping" power source drops in voltage output as the amperage increases.

Fig. 9-33.   Gas metal arc welding of aluminum yacht bulkheads.

The carbon dioxide shield allows no hydrogen in the weld area, and the weld results in a low hydrogen X-ray quality weld.

The gas metal arc welding process has many different welding applications, Fig. 9-32. The process may be manual in any position, or it may be automatic. It is versatile, because it can be used to weld very heavy steel plate for tank, ship, or barge construction. It may be used to weld thin steel sections. Because of the speed, quality, and versatility, G.M.A. welding is important to the aerospace, automotive, container, transportation, and construction industries, Fig. 9-33. Wherever welding is being done on a large scale, G.M.A. competes successfully with other welding methods. It may be used to weld most metals. Carbon steels; alloy steels; nonferrous metals of aluminum, magnesium, titanium, zirconium; and the whole range of corrosive-resistant metals of copper, nickel, and stainless steels can be welded by the G.M.A. process.

Concept of Welder Control. The G.M.A. welding process utilizes different concepts of welding control from those which are applied to metal shielded arc welding. In G.M.A. welding, the arc length (thus the voltage) was controlled by the operator. In G.M.A. welding, the arc length is controlled by the power source, the wire feeder, and the burn-off rate of the wire. The electronic control senses the resistance and voltage in the arc and controls the amperage by feeding the correct amount of wire. To make

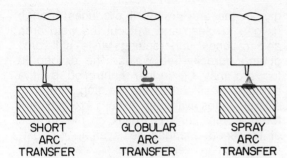

SHORT ARC TRANSFER   GLOBULAR ARC TRANSFER   SPRAY ARC TRANSFER

THREE TYPES OF METAL TRANSFER

Fig. 9-35.   Gas metal arc transfer changes its characteristics at different energy levels.

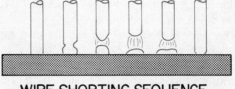

WIRE SHORTING SEQUENCE

Fig. 9-36.   Short-circuit transferring of metal takes place when the welding wire electrically shorts, pinches off, arcs to fuse metal to the work, and reshorts.

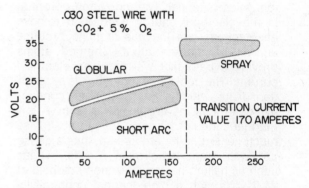

METAL TRANSFER ENERGY RANGES

Fig. 9-37.   The voltage and ampere ranges where the various metal transfer methods occur.

use of these electronic controls, it is necessary to develop a constant potential or constant voltage power supply, Fig. 9-34.

The constant potential power source has a flat volt-ampere characteristic which permits a wider range of arc self-correction by the electronic control. The welding arc amperage is controlled by the wire feed speed. Because of the constant voltage power source, the operator may vary the torch nozzle distance from the work within limits and not send the amperage or wire feeder out of control.

G.M.A. Arc Metal Transfer. There are three methods of metal transfer in G.M.A. welding, Fig. 9-35. They are primarily related to the energy supplied.

The **short circuiting method** uses a small diameter wire (.020″ to .045″) that comes in contact with the weld metal and electrically shorts, Fig. 9-36. When the short occurs, the electromagnetic pinch force squeezes the metal from the end of the wire, and the molten metal is left on the work as the short is opened. The arc restarts and generates heat in the metal and wire. The short arc method of transfer is valuable for welding in all positions. It gives moderate penetration.

Globular transfer occurs when the current density is low but above the short arc transfer level. The metal on the end wire melts and forms a liquid sphere which erratically transfers across the arc stream to the metal. The droplets crossing the arc stream are few, and they are affected by gravity. Thus globular transfer results in poor arc stability, excessive spatter, and shallow penetration.

When the transition current is reached (155 to 265 amperes for a 1/16″ steel wire with argon shielding), the number of drops crossing the arc stream increases dramatically from about 15 drops per second to about 240 drops per second. The transfer has then moved from a globular transfer to a spray transfer.

The **spray transfer** occurs when the welding current is above the current value that melts the electrode metal and projects fine particles through the arc to the liquid base metal. Large diameter wires are used (.045″ to 1/8″).

The high current density results in a high deposition rate and deep penetration that produces excellent welds in heavy steel sections, Fig. 9-37. The very high heat concentration limits the welding positions to horizontal or flat applications.

Shielding Gas Used with Steel. The shielding gas makes differences in the penetration, the wetting action of the weld, and the undercutting. Gases vary as to their ionization potential. The **ionization potential** is the electrical energy needed to remove an electron from the gas atom to produce an ion or charged atom. Charged atoms provide a path that will carry an electrical current. The ionization potential is a function of the welding gases' thermal conductivity and thus the arc density stream. A low thermal conductivity gas does not absorb as much heat from the arc stream, thus a dense arc with great penetration results. A gas such as helium has an ionization potential of 24.5 volts with a high thermal conductivity or the ability of the helium arc plasma to expand, using some of the heat energy in the expansion of the arc stream to reduce the density of the arc. This thermal loss will produce shallower penetration with a broader weld bead than another shielding gas such as argon, Fig. 9-38.

Carbon dioxide gas is a widely used shielding gas for the welding of steels, because it is inexpensive and produces excellent penetration. Carbon dioxide decomposes into carbon monoxide and oxygen in the arc stream. It produces enough oxygen to clean and wet out the weld area and reduces undercutting while still producing porosity-free welds. The oxygen in the arc may be further controlled by the addition of silicon to the welding wire. The silicon unites with any excess oxygen and forms a very light glass flux on the surface of the bead. Manganese is another deoxidizer used in welding wire.

A disadvantage of carbon dioxide as a G.M.A. welding gas is a harsh arc stream which produces spatter. The spatter may be controlled by holding the torch nozzle close to the work, thus confining the spatter to the weld area.

Flux-Cored Arc Welding. Flux-cored wire welding is a development of the gas metal arc process. The welding wire is made out of a sheet of metal rolled into a wire and a granulated flux is inside the tubular wire, Fig. 9-39. Flux-cored wire welding produces the advantages of the shielded metal arc but is in a form that can be handled by

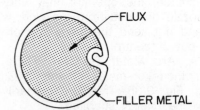

Fig. 9-39. Tubular wire welding has the flux on the inside of the filler rod tube. For heavy welds carbon dioxide is used as a shielding gas to produce deep penetrating quality welds.

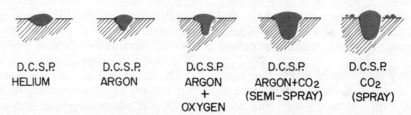

| D.C.S.P. HELIUM | D.C.S.P. ARGON | D.C.S.P. ARGON + OXYGEN | D.C.S.P. ARGON+CO₂ (SEMI-SPRAY) | D.C.S.P. CO₂ (SPRAY) |

Fig. 9-38. The penetration of a weld changes as different shielding gases, polarity, and energy level of current are applied.

semi- and fully automatic equipment, Fig. 9-40. The time of the welder changing the stick rod is eliminated. Also, an adjustment of the control of the arc length and machine need not be done by the operator. Instead, they are done by the wire feeder.

The flux-cored arc is of two types: flux-cored welding with a carbon dioxide gas shield and self-shielding flux-cored welding, Fig. 9-41. The flux-cored wire process with a carbon dioxide gas shield may be operated with energies that are in the arc spray condition, Fig. 9-42. The spray-type arc at high arc density produces deep penetration of welds. High production on heavy welds make some applications competitive with the submerged arc welding process. Semi-automatic or welder-operated wire equipment of this size will weld large fillets and groove welds in the flat or horizontal positions with 3/32" diameter electrode. With the development of small tubular wires and the use of globular transfer rather than spray, tubular wire welding has become an all-position welding process. Flux-cored wire is manufactured as small as 0.035" in

diameter, making vertical and overhead welding possible. Flux-cored arc welding meets the high quality standards of welding but does leave a flux slag on the bead which must be removed by wire brushing or with a deslagging hammer and wire brush.

Self-shielding wire process is a simplified process, because none of the gas shielding equipment is necessary. The electrode

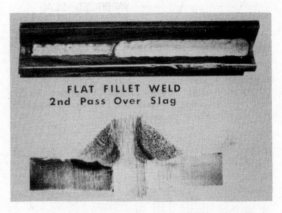

FLAT FILLET WELD
2nd Pass Over Slag

Fig. 9-41. A ⅝" cover pass weld made at 15 inches per minute and 500 amperes over a root pass with slag left in place. The welding material was ½" thick.

Fig. 9-40. Flux-cored wire welding of a barge bottom. High amperages and metal disposition rates are achieved with flux-cored wire welding.

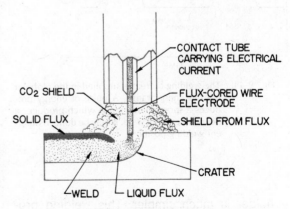

CONTACT TUBE CARRYING ELECTRICAL CURRENT

CO_2 SHIELD

FLUX-CORED WIRE ELECTRODE

SOLID FLUX

SHIELD FROM FLUX

CRATER

WELD   LIQUID FLUX

Fig. 9-42. Flux-cored wire processes with a carbon dioxide gas shield are operated with a high-arc density or spray transfer. These energy levels produce a penetration weld with high production rates.

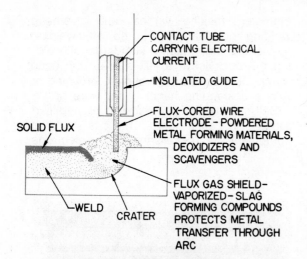

Fig. 9-43. Electrode wire with flux on the inside of the wire functions as a shielded stick electrode. All of the shielding, slagging, and deoxidizing materials are in a tubular wire.

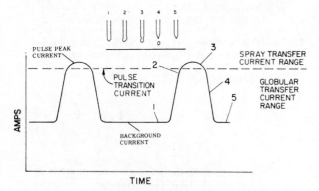

Fig. 9-44. The welding current is supplied in the globular transfer energy range. A pulse of energy produces a spray energy range which produces a deep-penetrating weld. It may be done in different welding positions.

holder is much simpler. This welding process does not have as much penetration, because the arc transfer is globular or short-circuiting. With the use of small tubular wires, this process may be used for out-of-position welding, Fig. 9-43. The quality of the welding is lower than that of welds produced with auxiliary gas. It is not recommended for steels having a yield strength above 42,000 psi. However, for much light fabrication, this strength is adequate.

Pulsed-Arc Transfer. The pulsed-arc transfer is produced by a power supply which is really two power supplies in one machine. There are two constant voltage supplies. One power supplies a steady D.C. current with enegry level in the globular transfer welding range and is referred to as the **background current,** Fig. 9-44. At the beginning of the globular transfer, but before the transfer occurs, the second power supply pulses the current into the amperage range of spray transfer. This pulse continues until the transfer of the metal takes place, and then the amperage returns to the globular transfer welding range.

Normally a spray transfer is not obtained below 190 amperes. With a pulsed power source, droplet to spray transfer will occur around 155 amperes, making the use of a larger diameter electrode possible. Weld porosity and wire feeding problems are reduced. The advantages of G.M.A. spray transfer welding out of the flat and horizontal position are maintained. The welding of light-gauge materials out of position can be done also by the pulse-arc transfer process.

Gas Tungsten Arc

Gas tungsten arc welding (G.T.A.) is a manufacturing process developed in the early 1940's to weld the difficult-to-weld metals, Fig. 9-45. The aircraft industry was concerned with high quality welds in aluminum, magnesium, chrome, and molybdenum steels.

The gas tungsten arc produces a high quality weld, because the inert shielding gas pushes the atmosphere away from the weld area and a tungsten rod provides an electrode that produces the arc between itself and the workpiece, Fig. 9-46. The tungsten does not melt, but it provides an arc which produces a clean source of heat. When the arc has produced a molten pool in the base metal while it is under the gas shield, the filler rod is melted into the forward edge of the molten pool. The filler metal fuses with the base metal. It cools and makes a continuous weld.

The G.T.A. process may be applied by manual operation or set up for automatic welding. All programs of welding such as continuous beads, skip welds, or intermittent and spot welds may be done by G.T.A.

The diameter of the tungsten electrode may be changed to suit the welding job. The smaller electrodes require lower amperages. Metal as thin as 0.005″ can be welded with these. A great deal of control may be gained with automatic welding, and difficult, thin parts may be successfully joined.

G.T.A. will weld most metals and alloys, including alloy steels, carbon steels, stainless steels, aluminum alloys, beryllium alloys, copper-based alloys, nickel-based alloys, titanium, and zirconium alloys, Fig. 9-47.

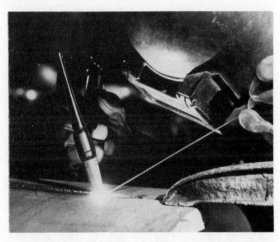

Fig. 9-45.　Gas tungsten arc welding is applied to difficult-to-weld metals. The inert shielding gas makes high quality welds possible on commercial metals.

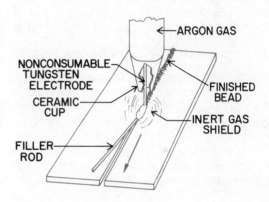

Fig. 9-46.　Gas tungsten arc produces the welding heat by holding an arc between the tungsten rod tip and the base metal. The high temperature area is protected by an inert gas.

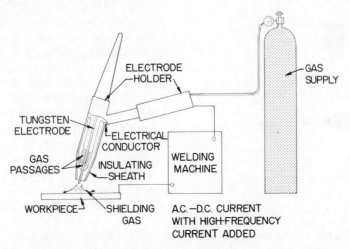

Fig. 9-47.　The basic equipment needed for gas metal arc welding.

G.T.A. Power Source. The gas tungsten arc welding uses the constant current type of welding power supply. This type of welder has been in use for many years and is very efficient. The constant current machine is known as a "drooping" machine. The amperage is controlled by the machine. The voltage is controlled partly by the design of the machine, but also by the length of arc the welder maintains. The voltage is important in G.T.A. welding because it is a main factor for determining the available current to put heat into the metal. This is related to Ohm's Law: $E = I \times R^*$. (Voltage is equal to the amperage times the resistance.) Amperage delivers the heat but is controlled by the resistance and the voltage. ($I = \dfrac{E}{R}$). In the arc stream, as the arc length gets longer, the resistance increases. As the resistance increases, the voltage drop is greater. When the arc length gets longer, the amperage is lower and the heat available for the weld is reduced. The welder can see the length of the arc and thus has an additional control over the width of the molten pool during welding.

* E = Voltage, I = Amperes,
 R = Resistance

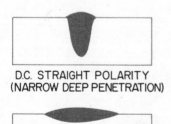

D.C. STRAIGHT POLARITY
(NARROW DEEP PENETRATION)

D.C. REVERSE POLARITY
(WIDE SHALLOW PENETRATION)

Fig. 9-48.   The change in welding polarity affects the depth of weld penetration and the size of the weld.

The welder moves the welding torch to make the correct bead and to control the metal's welding heat. This may disrupt or break the arc stream. To keep the arc from going out, G.T.A. welding supplies a high-frequency alternating current which is superimposed on the welding current. The advantage of this high-voltage, high-frequency current (R.F.) is that it provides an ionized gas path for the welding current to follow, thus greater arc stability is achieved. The arc may be started easily without the electrode touching the work, and the welder can see the workpiece. The high frequency aids in breaking up the oxides on the surface of the work, and a longer arc can be maintained while manipulating the rod and torch.

The currents used in G.T.A. welding are direct current straight polarity (electrode negative −), alternating current (electrode alternating +, −) and (infrequently) direct current reverse polarity (electrode positive +).

Most nonferrous materials are G.T.A. welded with alternating current. The A.C. provides good penetration with good surface oxide removal for clean, low-profile weld reinforcement. A.C. is highly recommended for welding aluminum, magnesium, and beryllium copper alloys.

Straight polarity current has the electrons moving to the weld area at a high velocity. As they strike the metal, additional heat is produced, Fig. 9-48. The workpiece is heated more than the tungsten electrode tip is. This action produces a narrow, deep heat penetration zone. A smaller diameter tungsten electrode for a given current can be used because of the lesser transfer of heat to the work from the electrode.

When a welding requirement of 125 amperes straight polarity is necessary, a 1/16″ diameter tungsten electrode is satisfactory. If reverse polarity is required, at 125 amperes, a tungsten electrode of 1/4″ diameter would be needed to control the amount of heat, Fig. 9-49. The reverse polarity transfers the electron heat flow to the tung-

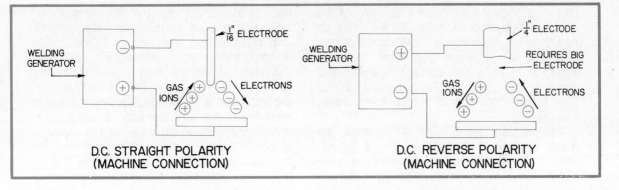

Fig. 9-49. The size of the electrode needed and the heat available for welding will be affected by the polarity applied.

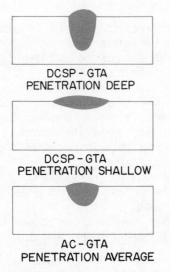

Fig. 9-50. The type of welding current applied will produce different patterns of penetration.

Fig. 9-51. A power-pulsed control added to a conventional power supply will aid in puddle control and welding speed pacing.

sten rather than to the work. A large tungsten water-cooled torch is required.

Reverse polarity produces an oxide film-cleaning action and a broad weld bead with a shallow penetration. Reverse polarity is not generally recommended for G.T.A. welding. This polarity is useful in the setting up of the G.T.A. welding equipment in that this current is used to melt a ball on the end of a new tungsten electrode prior to welding. The penetration of the A.C. is close to an average between the D.C.S.P and D.C.R.P. for G.T.A. welding, Fig. 9-50.

Pulsed gas tungsten welding currents may be added to conventional power sources. The extra pulse of welding energy is designed to aid the operator by pacing the welding speed, Fig. 9-51. This additional equipment makes difficult weldments easier by providing greater puddle control as the result of the pulsing action of the current.

Shielding Gas. The shielding gases used for G.T.A. welding are argon, helium, argon-helium mixtures, and argon-hydrogen mixtures. Argon is widely preferred for manufacturing because it makes a soft arc which is stable and smooth. Argon is 1.4 times as heavy as air, therefore it provides weld area coverage with low gas flow rates. Because of its weight, it is less affected by drafts and wind. Argon is an easy arc-starting shield. This ease in starting is helpful for welding thin metal.

Helium provides a high arc voltage, so it delivers a hotter arc. A hot arc is an advantage in the welding of materials over 3/16″ thick. In automatic welding in which speeds above 25″ per minute are reached, helium produces welds with less porosity and undercutting than argon. Since helium is lighter than air, a greater rate of flow of the gas to shield the weld is needed. Helium is more expensive and not as available as argon gas. In automatic welding, argon-helium mixtures are used to increase penetration and reduce porosity on the welds of heavy thicknesses. Argon-hydrogen mixtures are used for welding stainless steel, monel, and silver, because it controls porosity.

Gas Tungsten Arc Welding

Aluminum Welding Procedure

1. Examine the thickness of the metal to be welded and the equipment available for use. Aluminum plate of 1/8″ thickness will require an alternating current of 100 to 160 amperes, a 3/32 tungsten electrode, a 5/16 ceramic cup on the nozzle, and an argon flow meter that will deliver 12 cubic feet of argon shielding gas per hour.
2. Select the tungsten 3/32″ rod and grind the rod to a slender point, Fig. 9-52. Put on welding safety equipment and helmet. Place the ground tungsten in the collet and adjust it so that the tungsten projects 1/4″ beyond the gas cup of the torch. Turn on the argon gas and adjust to 12 cubic feet per hour.
3. The welding machine is **temporarily** put on reverse polarity, and the high frequency current and welding current are turned on.
4. The tungsten and torch are held vertically over a brass block, and the foot control is slowly pushed down. The face and eyes are covered with the welding hood before the foot control is touched.
5. The torch is zigzagged down until the high-frequency current establishes an arc path.
6. Lower the electrode down to within 3/32″ of the metal, and adjust the foot

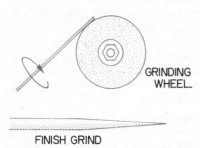

FINISH GRIND

Fig. 9-52. The welding tungsten is ground to a long, slender taper.

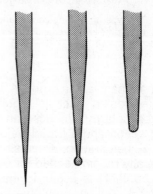

Fig. 9-53. Reverse polarity is used only for forming the ball on the tungsten electrode. The machine will next be changed to alternating current or straight polarity for welding.

control until a ball appears on the end of the tungsten. This is called balling the tungsten, Fig. 9-53.

7. Turn the machine off, and move the current selector to A.C. Adjust the tungsten so that it projects out of the gas cup about 1/8″ to 3/16″.

8. Prepare and clean the joint to be welded. Attach the ground cable to the part.

9. Obtain 1/8″ aluminum welding rod and turn the A.C. welding current and the high-frequency current on.

10. Do not touch the tungsten electrode to the metal. With the foot control down, zigzag down until the high frequency establishes the arc stream, Fig. 9-54.

11. Move the torch in a small circle until a small molten pool is established, Fig. 9-55.

12. When the liquid pool appears, tilt the torch off the vertical position about 20°, Fig. 9-56.

13. Move the torch slowly back and forth along the direction of travel until an elongated puddle is formed, Fig. 9-57.

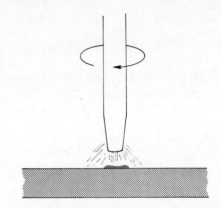

Fig. 9-55.   Move the torch in a small circle until a liquid pool appears.

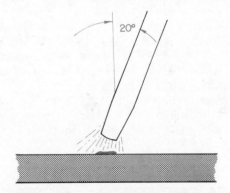

Fig. 9-56.   Tilt the torch off in a vertical direction about 20°.

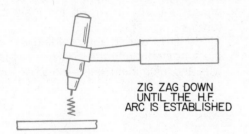

ZIG ZAG DOWN UNTIL THE H.F. ARC IS ESTABLISHED

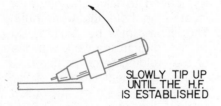

SLOWLY TIP UP UNTIL THE H.F. IS ESTABLISHED

Fig. 9-54.   Two methods of establishing the arc in gas tungsten arc welding.
    (1)   Zigzag down until the high-frequency current establishes the arc. Then the welding position may be located.
    (2)   Place the side of the ceramic cup on the work and slowly tip the torch up until a high-frequency arc is established. Then move to the welding position.

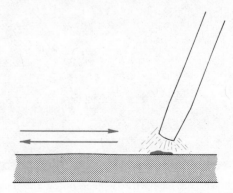

Fig. 9-57.   The established arc is moved back and forth along the direction of the weld.

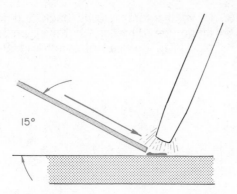

Fig. 9-58.   Move the torch back to the starting position and deposit filler metal on the leading edge of the liquid pool.

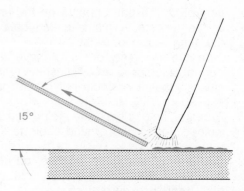

Fig. 9-59.   Remove the filler rod and pass the torch over the new filler material, flowing the filler metal and base metal together and creating the new molten pool ahead.

14. Move the torch back to the starting position and apply the welding rod to the front of the liquid pool. Deposit filler metal, Fig. 9-58.
15. Remove the filler rod and move the torch over the filler material, flowing the filler metal and base metal together while melting the metal ahead of the liquid pool, Fig. 9-59.
16. The rhythm of the torch and filler rod is repeated until the weld is completed with the proper bead height.

17. When the weld is completed, the crater is filled and the foot is removed from the foot control. The arc is out, but the torch should be held over the work while the argon continues to flow to protect the metal as it cools.
18. Shut down the equipment, turn off the argon tank, and hang up the tools and torch.
19. At no time during the welding should the tungsten electrode come in contact with the aluminum base metal.

Fig. 9-60.   Gas tungsten arc welding around a steel pipe. The operator is welding in the overhead position.

Automatic Welding. Automatic welding is planned to take advantage of fixtures to locate and hold the work during welding. The fixtures may be designed to clamp the work and align the work with the movement of the welding torch or head.

The equipment used is of various types. Longitudinal seamers are used for long, straight butt joints, Fig. 9-61. **Circumferential** (girth) **seamers** are used for large diameter pipes or pressure vessels and tanks, Fig. 9-62. A track and carriage to move the welding torch are mounted to one side. They provide movement of the head during welding. A welding filler wire feeder is supplied. Other equipment includes the

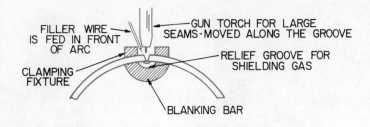

FILLER WIRE IS FED IN FRONT OF ARC

GUN TORCH FOR LARGE SEAMS-MOVED ALONG THE GROOVE

CLAMPING FIXTURE

RELIEF GROOVE FOR SHIELDING GAS

BLANKING BAR

LONGITUDINAL SEAMING FIXTURE

ROLLERS TO POSITION TANK WITH SEAM AT TOP OF CIRCLE

Fig. 9-61.   Longitudinal seaming fixture.

Fig. 9-62.   A circumferential or girth seamer welding an aluminum alloy spacecraft fuel tank.

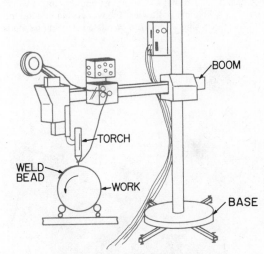

BOOM

TORCH

WELD BEAD

WORK

BASE

Fig. 9-63.   Pedestal, boom, and manipulator.

machine welding head, control panel, power source, shielding gas and flow meters, motorized carriages, and (in some cases) a work positioner.

Equipment should be designed so that the parts may be assembled and aligned quickly for welding. Equally important is the ability to remove the welded parts quickly and economically from the welding fixture.

Products of different shapes may require a pedestal boom manipulator, Fig. 9-63. Rotating positioners may be used in both manual and automatic welding. **Positioners** are circular or square tables that may be moved in a number of orientations as well as rotated at various speeds to meet the

Fig. 9-64. A pedestal and boom manipulator and a work-rotating positioner. The work is clamped to the positioner and may be rotated at welding speeds in the correct welding position.

Fig. 9-66. Skate welders are mounted on small wheels which allow the wire welder to follow the seam being welded. They are used on large workpieces such as ship plates or large tanks.

Fig. 9-65. Large workpieces are located and turned by a positioner with radiating steel arms that hold the work.

welding requirements, Fig. 9-64. The table has T-slots cut into its surface so that holding and clamping fixtures or work may be bolted to the positioner. Large positioners have steel arms which radiate out from a central hub and hold large pieces of work, Fig. 9-65.

Skate welders are equipment used for very large workpieces such as ships or very large tanks. Skate welders are lightweight wire feeders mounted on a carriage with the welding head, Fig. 9-66. The skate carriage synchronizes its movements down the seam to be welded with the required weld and wire feeder. The skate welder rolls on straight or curved tracks which are attached directly to the workpiece.

Spot Welding. Spot welding using G.T.A. is successful because of G.T.A.'s ability to join two pieces of metal together when only one surface is available for welding, Fig. 9-67. It is used to spot-weld automobile bodies, thin metal to thick metal, double-walled structures, or other assemblies where there is restricted access to the back of the work. The metallic cup is held firmly against

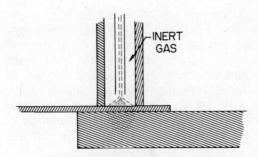

Fig. 9-67.   Spot welding using the gas tungsten arc process. The arc is delivered with in the nozzle where it is shielded. The penetration welds the two metal members together.

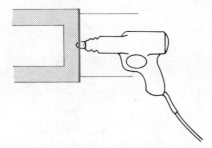

Fig. 9-68.   The gas metal arc spot-welding gun is used on double-walled structures, thick to thin metal, or areas that are difficult to weld because of space limitations.

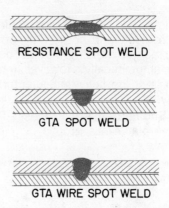

RESISTANCE SPOT WELD

GTA SPOT WELD

GTA WIRE SPOT WELD

Fig. 9-69.   A spot weld with filler wire added prevents crater cracking and concave welded surfaces.

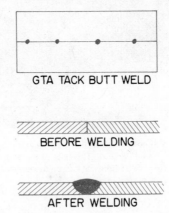

GTA TACK BUTT WELD

BEFORE WELDING

AFTER WELDING

Fig. 9-70.   Spot welding with filler rod is frequently done to tack welding of flat sheets and large, tubular sections.

the metal to be welded, and the spot-weld gun trigger is pulled, Fig. 9-68. The argon automatically begins to flow, the cooling water comes on, and the high frequency starts the arc. The welding current comes on, and the liquid pool forms under the tungsten electrode. The amperage and time of the arc duration determine the size and depth of the spot weld.

Ferrous materials may be spot-welded without the high frequency by having the electrode move forward about 1/16″ and contacting the base metal to establish the welding arc.

Spot welding may be made automatically or made according to a program of welds. To prevent crater cracking and concave weld surfaces, filler wire is added to the spot-weld gun. A programmed controller provides for the starting of the shield gas, welding current, wire feed, current decay for shutting down, and weld timing. The controller shuts the sequence down. With the addition of filler wire, a very strong tack weld or spot weld is possible.

This type of spot welding has the added advantage of being able to produce spot butt welds very rapidly on both tubular and flat material, Figs. 9-69 and 9-70.

Submerged Arc Welding

Submerged arc welding is a process that has been used in manufacturing for many years, Fig. 9-71. It uses a granular flux, a bare wire filler, and heavy-duty power supply. Submerged welding may be done manually or automatically. However, most of the welding is of the heavy manufacturing automatic type.

The Submerged Arc Process. The welding takes place in a groove or joint prepa-

ration with the arc under a mound of granular flux, Fig. 9-72. There is no flash, because the arc takes place deep under the flux. The submerged arc has very deep penetration and high production rates. It produces very high-quality welds on such items as butane tanks, heavy water tanks, heavy structural members, automotive starter motor frames, earth-moving machinery, and other heavy welding.

The welding process uses currents to nearly 1600 amperes which deliver deep penetration. Steel butt welds up to 5/8"

Fig. 9-71. Submerged arc welding uses a granular flux and a bare filler wire. This type of welding produces very deep-penetration, high-quality welds at production rates.

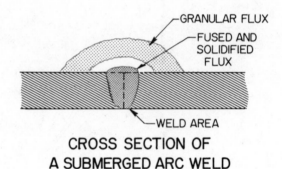

**CROSS SECTION OF
A SUBMERGED ARC WELD**

Fig. 9-73. Steel butt welds up to ⅝" thick may be made without any beveling or other edge preparation.

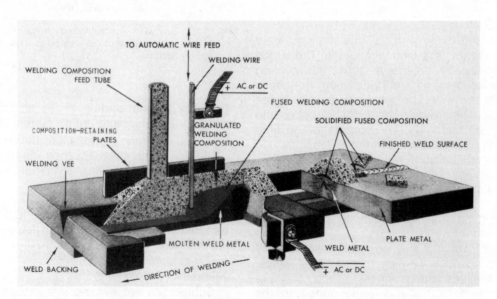

Fig. 9-72. The weld is produced under granulated flux. The extra flux is removed by a vacuum system and is later reused.

may be welded completely without any beveling or edge preparation, Fig. 9-73. The intense heat fuses the metal, filler wire, and the flux to produce a very sound weld.

After the weld is made, the unfused flux is vacuumed up and reused. The flux is made of oxides of silicon, manganese, calcium, aluminum, magnesium, titanium, zirconium, and other elements. In the high heat of the arc, the flux performs these functions: (1) cleans the weld surfaces, (2) provides a shield for the liquid metal, (3) helps form the bead, and (4) insulates the bead so that it does not cool too rapidly.

A large volume of metal manufacturing is done by the **submerged welding method**, Fig. 9-74. Most of the welding is done automatically with the use of positioners and is done in the flat position because of the granular flux and the amount of liquid metal involved, Fig. 9-75. Multiple electrode wires are fed into the weld zone when deeper, wider, and higher beads are desired at high welding speeds. The additional wire will be in front of or beside the original wire.

Electroslag and Electrogas Welding

Electroslag and electrogas welding are bonding processes which are very much the same except the electroslag process uses a molten flux over the weld to shield the weld, while electrogas welding is shielded by carbon dioxide or by a mixture of 80% argon and 20% carbon dioxide. Both welding processes weld vertically and fuse heavy sections of steel plate together.

Electroslag Welding. The electroslag welds are made in thick material usually from 1¼″ thick to as great as 36″. The weld is done vertically so the length of the weld can be almost any length, Fig. 9-76. The

Fig. 9-75. The welding of a nuclear reactor vessel head by the submerged arc process. Note the pre- and post-heating flames inside and outside the vessel. The man on top is setting up for the weld.

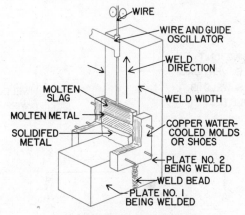

ELECTROSLAG WELDING

Fig. 9-76. The whole welding system moves vertically as the welding proceeds. Metal from 1¼″ thick to 36″ thick may be welded by this process. The length of the weld is dependent only on the width of the sheet being welded.

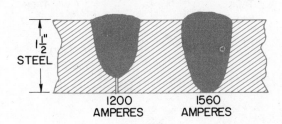

Fig. 9-74. Large volume welds and heavy welds are made with 40 volts and 1560 amperes at 12.5 inches per minute with 5/32″ steel wire. The joint is prepared with a 45° single V-groove butt joint.

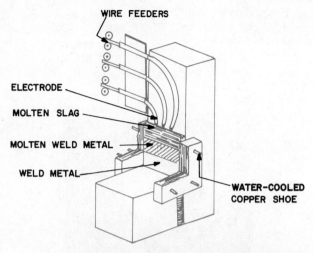

WIRE FEEDERS

ELECTRODE

MOLTEN SLAG

MOLTEN WELD METAL

WELD METAL

WATER-COOLED
COPPER SHOE

Fig. 9-77. Electroslag welding is carried out on large, thick work done vertically. Thick sections of bridges, dam equipment, and large tanks are welded with the electroslag welding process.

heat is so intense that no edge or joint preparation is needed.

On very long, vertical welds the filler wires come in from the side and the welding head contains an oscillating device to move the electrodes from side to side in the weld (for welds that are 6″ to 8″ thick), Fig. 9-77.

Electrogas Welding. Electrogas welding has been developed out of the principles of electroslag welding. Electrogas produces its heat by an electric arc, much as the G.M.A. process, under a shield of carbon dioxide or carbon dioxide-argon mixed gas. Only direct current is delivered to the arc. Electroslag uses the resistance of the slag to form the heat and either alternating or direct current may be used. The electrogas process uses either solid electrode wire or tubular wire with granulated flux rolled into the center of the tube.

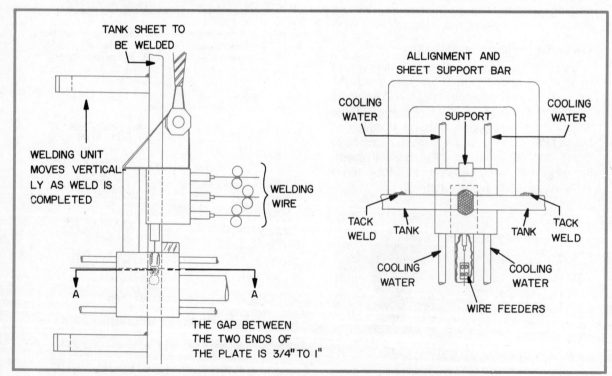

TANK SHEET TO
BE WELDED

WELDING UNIT
MOVES VERTICAL-
LY AS WELD IS
COMPLETED

WELDING
WIRE

A   A

THE GAP BETWEEN
THE TWO ENDS OF
THE PLATE IS 3/4″ TO 1″

ALLIGNMENT AND
SHEET SUPPORT BAR

COOLING
WATER

SUPPORT

COOLING
WATER

TACK
WELD

TANK

TANK

TACK
WELD

COOLING
WATER

COOLING
WATER

WIRE FEEDERS

Fig. 9-78. Electrogas welding makes long vertical welds on steel ½″ to 3″ thick. Large water or oil tanks may be welded at the site by using this process.

The electrogas process is used in the vertical position on steels from 1/2″ to 3″ thick. This is done with an automatic oscillator which moves the wire across the weld and dwells at the end of the weld to compensate for the heat that is lost to the copper-cooling dam.

Electrogas welding may be used for welding on heavy tanks, bridges, ship construction, heavy circular-pressure vessels, and nuclear reactor construction, Fig. 9-78. Heavy vertical welding in industrial construction is made more efficient by electrogas welding.

Plasma Arc Welding

Plasma arc welding is a process that uses an electric arc in a confined chamber. The ionized argon or helium gas is heated with so much energy that it changes its physical state into a plasma or a different form of matter. The plasma gas is in an unstable condition and has a great amount of energy. The plasma gas passes out of a restricted orifice to produce a plasma jet traveling at sonic speeds. When the jet strikes the workpiece, the plasma gas goes back to its natural gas state and in so doing releases the heat energy which was stored in the plasma. The temperatures produced by the plasma range between 500° F. to 60,000° F., far above the normal welding temperatures in the 6,500° F. range, Fig. 9-79.

The Plasma Arc. There are two types of plasma arc torches: the transferred type that is used for welding and cutting, and the non-

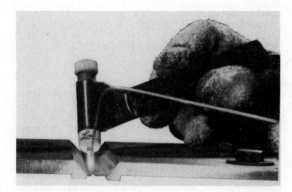

Fig. 9-79. Plasma arc welding performed manually. Temperatures up to 60,000° F. may be reached.

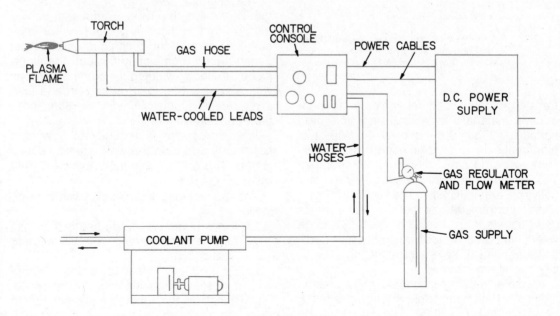

TORCH

CONTROL
CONSOLE

POWER CABLES

GAS HOSE

PLASMA
FLAME

WATER-COOLED LEADS

D.C. POWER
SUPPLY

WATER
HOSES

GAS REGULATOR
AND FLOW METER

COOLANT PUMP

GAS SUPPLY

Fig. 9-80. A plasma welding system requires the control of a large amount of electrical power.

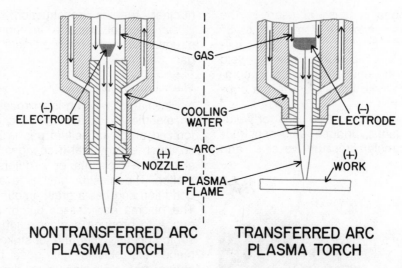

NONTRANSFERRED ARC
PLASMA TORCH

TRANSFERRED ARC
PLASMA TORCH

Fig. 9-81. The two types of plasma arcs are nontransferred arcs used for welding and heating nonconductive materials and the transferred arc for welding and cutting metals.

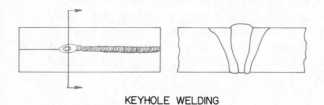

KEYHOLE WELDING

Fig. 9-82. Keyhole welding melts through the workpiece. As the hole is moved across the metal, the liquid steel edges flow together and cool as one piece.

transferred type used for welding nonconductive materials for metallizing and hard surfacing, Figs. 9-80 and 9-81. With this high heat source, welding speeds are greatly increased and new techniques are applied. The plasma torch may be operated manually or automatically. The welding operations performed may be the same as in gas tungsten arc welding such as edge welds and butt welds, but with the additional amount of heat energy, keyhole welding may be performed.

Keyhole welding is done by the plasma melting a hole completely through the metal being welded, Fig. 9-82. The edges of the hole are liquid and the metal flows around the edge of the hole and the plasma flame as the torch is moved forward. This liquid meets at the rear of the hole, and surface tension causes the liquid metal to flow together and produce a weld.

The keyhole weld may be carried across the work and the weld completed behind the keyhole without any addition of filler metal. The keyholing technique assures that penetration is completely through the workpiece. Keyhole welding is commonly done with a square butt joint on carbon and low-alloy steels, stainless steels, titanium, copper, and brass.

If reinforcement is required in a welded seam, it is added to the weld area by a wire feeder. The added filler metal builds a weld bead, Fig. 9-83.

The advantages of plasma welding are these:

1. It has higher welding speeds than gas tungsten arc welding in many welding applications.
2. It produces cleaner welds without tungsten contamination in the weld, because the operator cannot touch the tungsten to the molten metal. It requires a less skilled operator.

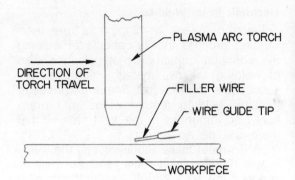

Fig. 9-83. A plasma welded seam may be reinforced by adding filler wire in the weld area, producing a strong joint rapidly without contamination.

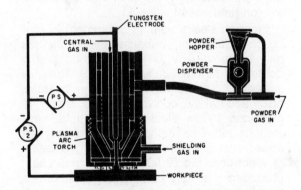

Fig. 9-84. Schematic drawing of plasma arc weld surface torch, powder connections, and powder dispenser.

Fig. 9-85. Close-up of a plasma arc surfacing. Plasma arc surfacing is especially suited to high-speed production of thin weld overlays with low dilution of the hardening materials on the base metal.

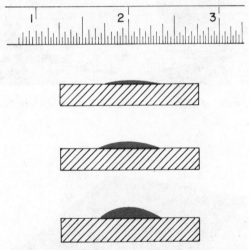

Fig. 9-86. Plasma arc overlays of cobalt surfacing applied in one pass, varying from 1/64" to 3/16" thick on carbon steel.

Plasma Arc Weld Surfacing. Thin coatings or hard facing overlays of heat-, wear-, and corrosion-resistant metals may be applied to other metals. The plasma arc used is the nontransfer type and metal or alloy powder is metered into the stream of argon gas. It is carried to a powder chamber and placed into the plasma stream beneath the upper arc-constructing orifice. The powder is partially melted, heated in the plasma, and then completely melted in the liquid pool, Fig. 9-84. The resulting hard deposit is welded to base metal.

Hard facings are frequently applied to ditch-digging teeth, plow points, bulldozer blade edges, extruder screw flights, turbine blades, or any surface that will be subjected to abrasive wear.

The advantages of plasma hard surfacing is that the deposit of expensive alloys may be welded to the base metal with a thin deposit, Fig. 9-85. The dilution or mixing of the hard surfacing alloy with the base metal may be precisely controlled, Fig. 9-86.

Speed, smoothness, and flatness of the deposit make for more efficient application of a hard coating on the hard surface of the metal, Fig. 9-87.

Plasma arc weld surfacing technology helps to produce better quality products at a competitive market price.

Fig. 9-87. A wear ring plasma arc surfaced with stellite alloy.

Electronic Beam Welding

Electron beam welding is a precision welding process that is capable of welding the common construction metals and also of welding reactive metals which are very susceptible to oxidation. Space and nuclear industry welding require welds with practically no impurities in the weld area so that there will be no defects in the weld, Fig. 9-88. To meet these requirements, the welding is done in a vacuum, and the heat is supplied by an electron beam. Welding in a high or hard vacuum makes welding of the highly reactive space age metals of titanium, zirconium, columbium, and hafnium possible with the quality and purity of the original metal.

Electron beam welding equipment is precision equipment, because the workpiece may be moved by .001″ in location and is controlled automatically, Fig. 9-89. The size and depth of the fusion from the beam is also accurately controlled. The beam is produced by an electron gun with a tungsten or tantalum cathode, a forming electrode, and an anode accelerating a beam of electrons, Fig. 9-90. The electron beam produced is a very narrow flow of electrons traveling about one tenth the speed of light.

Fig. 9-88. Electron beam welding produces a narrow, deep weld of high purity. This sample is of a 304 stainless steel, 1½″-thick electron beam welded with a gun-to-work distance of 15″.

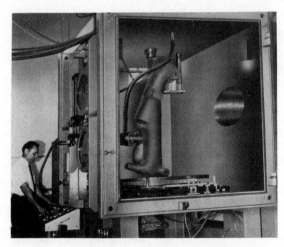

Fig. 9-89. Electron beam welder set up to weld an aircraft landing gear.

When this dense stream of high speed electrons strike a metal, the electrons release kinetic energy in the form of heat, Fig. 9-91. As each electron strikes the metal, the heat supplied melts a narrow path through the metal. The beams are focused so that the diameter is typically 0.10″ to 0.30″. At these speeds the metal is very quickly liquefied in a very narrow zone.

The penetration ratio may be controlled from 25:1 to a few thousandths of an inch, meaning a very narrow and deep weld may be obtained, Fig. 9-92. This weld characteristic allows joint designs to be welded which were impossible to do before electron beam welding, Fig. 9-93.

The welding is performed by the keyhole technique. The electron beam makes a hole in the metal and as the beam is moved along the work, the liquid metal moves behind the beam and the surface tension of the liquid metal causes the metal from **the**

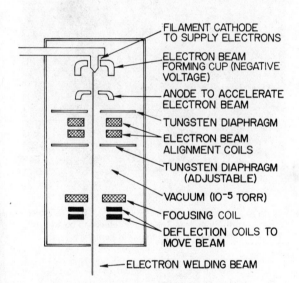

Fig. 9-90. Schematic cross section of an electron beam gun that produces the welding beam.

FILAMENT CATHODE TO SUPPLY ELECTRONS
ELECTRON BEAM FORMING CUP (NEGATIVE VOLTAGE)
ANODE TO ACCELERATE ELECTRON BEAM
TUNGSTEN DIAPHRAGM
ELECTRON BEAM ALIGNMENT COILS
TUNGSTEN DIAPHRAGM (ADJUSTABLE)
VACUUM (10⁻⁵ TORR)
FOCUSING COIL
DEFLECTION COILS TO MOVE BEAM
ELECTRON WELDING BEAM

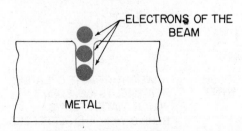

Fig. 9-91. The electron beam strikes the metal, converting the electron's kinetic energy into heat and melting a narrow hole into the metal.

ELECTRONS OF THE BEAM
METAL

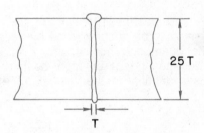

Fig. 9-92. Electron beam welding may produce a penetration ratio of 25:1. This means that deep, narrow welds may be made. With this deep weld, joint design may be changed.

25 T

T

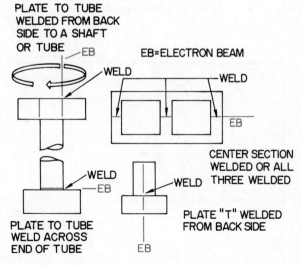

Fig. 9-93. Electron beam welding makes possible the redesigning of welded joints.

PLATE TO TUBE WELDED FROM BACK SIDE TO A SHAFT OR TUBE

EB

WELD

EB=ELECTRON BEAM

WELD

EB

CENTER SECTION WELDED OR ALL THREE WELDED

WELD
EB

PLATE TO TUBE WELD ACROSS END OF TUBE

WELD

EB

PLATE "T" WELDED FROM BACK SIDE

sides of the weld to intermingle and to be joined upon freezing, Figs. 9-94 and 9-95.

Electron beam welders may be programmed with the use of numerical control or direct computer control to accurately locate and weld the work. The work is usually accurately machined and fit before it is placed in the welding fixture or manipulator. The work may be given its final cleaning in the vacuum chamber by focusing the beam into a circle and traversing the weld area. This broad beam scrubs any loose scale or oxide films away from the weld area and eliminates these impurities from the weld.

Pre- and post-heat treating of the weld area may also be accomplished by defocusing the beam and heating the weld area. This post-heating of electron beam welds increases ductility in the weld, Fig. 9-96.

Electron beam welders will vary in size and shape, depending upon the customer's requirements, Figs. 9-97 and 9-98. In some cases the vacuum system will be larger than the welder and power supply. The weld may be observed by optical telescopes attached to the welding gun or by the use of a television-like camera. In any case, soft X-rays are generated by the electron beam. The operator must be shielded by leaded glass.

Fig. 9-96. A microsection of .480″ thick 4340 M-360 steel, showing how weld parameters can be adjusted to obtain a different weld geometry.

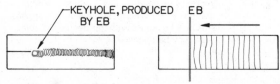

Fig. 9-94. Keyhole welding is performed by having the beam melt a hole through the metal welded.

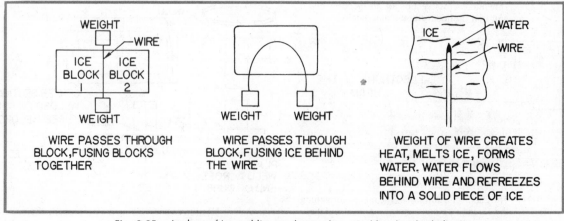

WIRE PASSES THROUGH BLOCK, FUSING BLOCKS TOGETHER

WIRE PASSES THROUGH BLOCK, FUSING ICE BEHIND THE WIRE

WEIGHT OF WIRE CREATES HEAT, MELTS ICE, FORMS WATER. WATER FLOWS BEHIND WIRE AND REFREEZES INTO A SOLID PIECE OF ICE

Fig. 9-95. Analogy of ice welding to electron beam welding by ''keyholing''.

Advantages of Electron Beam Welding

1. Weld thickness from 0.003″ to 1.000″ in stainless steel and as much as 2.000″ in aluminum may be joined.
2. The heat-affected zone of the weld is very narrow, making welding possible after heat treating with little loss of strength. The distortion of parts is minimal, and parts may be finished prior to welding.
3. The welds are deep, narrow, uniform, and free of inclusions and porosity.
4. Dissimilar materials may be joined, Fig. 9-99 (page 372).

Disadvantages of Electron Beam Welding

1. The high cost of equipment, joint preparation, and tooling for holding and manipulating the work.
2. The limitations of the vacuum chamber size which determines the largest piece that can be welded.
3. The cost of pumping down a vacuum necessary for welding.

Soft Vacuum Welding. The manufacturing of mass-produced electron beam welded products has been achieved by designing equipment that keeps the electron gun and its parts in a hard vacuum but allowing the welding area to be in a less hard or soft vacuum area. This was made possible by building a valve which may be opened when the beam is welding and closed the

Fig. 9-98. Clamshell welding machine designed to provide wide-open access for large parts or multiple parts of titanium and other metal aircraft parts.

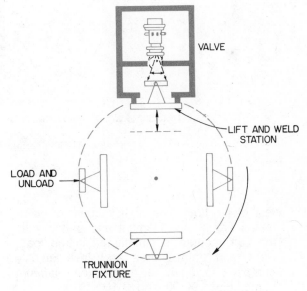

Fig. 9-100. A soft vacuum welding system combines all the standard advantages of electron beam welding with high production techniques and rapid pump down.

Fig. 9-97. Electron beam welders vary in size, depending upon the customer's requirements.

WELDABILITY OF DISSIMILAR METAL COMBINATIONS

	Ag	Al	Au	Be	Cd	Co	Cr	Cu	Fe	Mg	Mn	Mo	Nb	Ni	Pb	Pt	Re	Sn	Ta	Ti	V	W	Zr
Ag – SILVER	■	C	S	⊠	C	D	C	C	D	⊠	C	D	N	C	C	S	D	C	⊠	C	D	D	⊠
Al – ALUMINUM	C	■	⊠	C	⊠	⊠	C	⊠	C	⊠	⊠	⊠	⊠	C	N	C	⊠	N	C	⊠	⊠	⊠	⊠
Au – GOLD	S	⊠	■	⊠	C	D	S	C	⊠	⊠	C	N	S	⊠	S	N	⊠	N	⊠	N	⊠	D	N
Be – BERYLLIUM	⊠	C	⊠	■	N	⊠	⊠	⊠	⊠	⊠	⊠	N	⊠	N	⊠	D	D	⊠	⊠	⊠	⊠	⊠	⊠
Cd – CADMIUM	C	⊠	N	N	■	D	D	⊠	D	S	D	N	N	D	⊠	N	C	N	⊠	N	N	N	D
Co – COBALT	D	⊠	C	⊠	D	■	C	C	C	⊠	⊠	S	C	S	⊠	S	C	⊠	⊠	⊠	⊠	⊠	⊠
Cr – CHROMIUM	C	C	D	⊠	D	C	■	C	C	⊠	C	S	C	C	C	S	C	⊠	S	D	S	⊠	⊠
Cu – COPPER	C	⊠	S	⊠	⊠	C	C	■	C	⊠	S	D	C	S	C	S	D	C	D	⊠	D	D	⊠
Fe – IRON	D	⊠	C	⊠	D	C	C	C	■	⊠	D	C	C	⊠	C	C	S	⊠	⊠	⊠	S	⊠	⊠
Mg – MAGNESIUM	⊠	C	⊠	S	⊠	⊠	D	⊠	⊠	■	D	N	⊠	⊠	N	⊠	⊠	N	N	D	N	D	D
Mn – MANGANESE	C	⊠	C	⊠	D	C	C	S	C	⊠	■	D	⊠	C	⊠	N	C	N	⊠	⊠	⊠	D	⊠
Mo – MOLYBDENUM	D	⊠	C	N	⊠	S	D	C	D	D	⊠	■	⊠	D	C	⊠	D	S	S	S	S	⊠	⊠
Nb – NIOBIUM (COLUMBIUM)	N	⊠	N	⊠	N	⊠	⊠	C	⊠	N	⊠	S	■	⊠	N	⊠	⊠	⊠	S	S	S	S	S
Ni – NICKEL	C	⊠	S	⊠	D	S	C	S	C	⊠	C	⊠	⊠	■	C	S	D	⊠	⊠	⊠	⊠	⊠	⊠
Pb – LEAD	C	C	⊠	N	C	C	C	C	C	⊠	⊠	C	D	N	■	C	⊠	N	C	N	⊠	N	D
Pt – PLATINUM	S	⊠	S	⊠	S	C	S	S	⊠	⊠	C	S	⊠	C	S	■	C	⊠	S	⊠	⊠	⊠	S
Re – RHENIUM	D	N	N	⊠	N	S	S	D	N	N	⊠	D	N	C	⊠	D	■	⊠	D	⊠	D	⊠	⊠
Sn – TIN	C	C	⊠	D	C	⊠	C	C	⊠	⊠	⊠	D	⊠	C	D	⊠	⊠	■	⊠	⊠	D	⊠	⊠
Ta – TANTALUM	⊠	⊠	N	D	N	⊠	D	⊠	N	⊠	S	S	N	⊠	⊠	⊠	⊠	⊠	■	S	⊠	S	C
Ti – TITANIUM	C	⊠	⊠	⊠	S	⊠	D	⊠	S	S	⊠	S	S	⊠	⊠	⊠	S	⊠	S	■	S	C	S
V – VANADIUM	D	⊠	D	N	⊠	D	D	S	N	⊠	S	S	S	N	⊠	D	⊠	S	⊠	S	■	S	⊠
W – TUNGSTEN	D	⊠	N	⊠	N	S	D	⊠	D	D	S	S	⊠	D	S	⊠	D	S	C	S	⊠	■	⊠
Zr – ZIRCONIUM	⊠	⊠	D	⊠	⊠	⊠	⊠	D	⊠	⊠	⊠	S	⊠	⊠	⊠	⊠	⊠	⊠	C	S	⊠	⊠	■

⊠ INTERMETALLIC COMPOUNDS FORMED – UNDESIRABLE COMBINATION.

S SOLID SOLUBITY EXISTS IN ALL ALLOY COMBINATIONS – VERY DESIRABLE COMBINATION

C COMPLEX STRUCTURES MAY EXIST – PROBABLY ACCEPTABLE COMBINATION

D INSUFFICIENT DATA FOR PROPER EVALUATION – USE WITH CAUTION

N NO DATA AVAILABLE – USE WITH EXTREME CAUTION

COPYRIGHT JANUARY 1965, HAMILTON STANDARD DIVISION OF UNITED AIRCRAFT CORPORATION

Fig. 9-99.   Electron beam weldability of dissimilar metals.

rest of the time. The high vacuum is maintained around the filament and parts, while the soft vacuum is maintained in the load, weld, and unload area and then quickly pumped down by a mechanical vacuum pump, Figs. 9-100 and 9-101.

With the use of a soft vacuum, the electron-beam welding of parts may be automated to produce high quality and economical parts.

Laser Welding

Laser welding is done by a burst of light that is concentrated into a diameter of 0.010″ in diameter and is capable of melting metals. Welding is accomplished by a series of small spot welds. Lasers are used in the field of microelectronics to weld circuits together, joining metal foils and wires 0.015″ to 0.030″ in diameter. The output of most lasers is in milliwatts, but the energy

Fig. 9-101. A high-production electron beam welding machine.

is so concentrated in area that it will melt, join, and solidify in microseconds.

Laser beam welders weld dissimilar metals, but the materials welded are only those with light-absorbing surfaces, because the photon radiation must be absorbed and be converted to heat, Fig. 9-102. Because of this characteristic, welds may be made through transparent materials such as glass or plastic. The laser has the added advantage of fusing refractory metals, ceramics, and very thin materials without warpage or heat damage to adjacent parts. Lasers have developed rapidly in the fields of nuclear, space, medical, communications, and measurement technologies.

Activity

Design and build a project that will require a number of welding processes.

Related Occupations

These occupations are related to bonding:

Welders
Welding engineer
Shielded arc welder
Gas metal arc welder
Plasma welding operator
Gas tungsten arc welder
Electron beam welding operator
Laser beam welding operator
Submerged arc welding operator
Induction welding operator
Hard facing operator
Electroslag welding operator
Stud welding operator

Nondestructive testing technicians
Magnetic particle inspector
Dye penetrant inspector
Eddy current testing inspector
Radiographic inspector
Ultrasonic inspector
Hardness inspector
Corrosion tester
Welding cost estimator

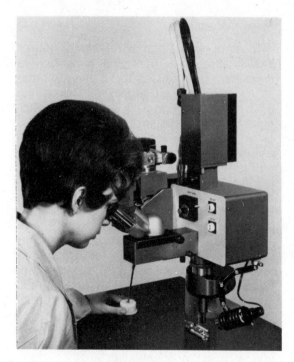

Fig. 9-102. Lasers can be used to join dissimilar metals, including common and space-age alloys.

Bonding by Heat and Pressure

Chapter 10

Words You Should Know

Weld nugget — The fused area in the metal of a spot weld.

Squeeze time — The time that pressure is held on the metal.

Weld time — The time the current is producing the welding heat.

Hold time — The time that the pressure is held on while the nugget solidifies.

Roll spot welding — A series of spot welds made by passing the metals between rolls.

Seam welding — A series of spot welds close enough that the nuggets overlap, making a liquid-and gas-tight seal.

Projection welding — Points or projections are embossed on the metal's surface. The points of the metal comes in contact with the base metal, and a spot weld takes place on each embossed contact.

Fig. 10-1. Several large stamped parts are positioned and joined automatically by spot welding.

Flash welding — The ends of metal are positioned together, and an electric arc occurs between the ends of the metal. The arc makes the metal molten, and the two ends are forced together, making the weld.

Spot Welding

In spot welding, an electrical current passes through a restricted area, generating heat. This causes the mating surfaces of the metal to melt, Fig. 10-1. The conversion of electrical energy to heat is brought about by the resistance of the metal. The arms (or horns) and electrodes are made of heavy copper which has a low electrical resistance. The metal being melted has a relatively high electrical resistance, so as the current passes through the metal, heat is produced. Ohms' law states that voltage = amperage × resistance. As the amperage flows through the resistance of the joints, the conversion of electrical energy to heat results. The metallic oxide and dirt add further resistance to the electrical circuit. Thus, the highest resistance is between the two metals, Fig. 10-2. The weld nugget forms at this point. The electrode pressure holds the two metals together while the nugget cools and solidifies.

Spot welding is used extensively in manufacturing to bond parts together very rapidly. There are no filler materials, fluxes, or warping with spot welding.

Spot Welding Cycle

Spot welding requires a cycle time for these four processes which make up the spot weld, Fig. 10-3.

1. A **closing and squeeze time** when the part is placed between the electrodes and the electrodes are closed.
2. A **weld time** when pressure is on the tips as the current flows through the workpiece, making the nugget.
3. A **holding time** which allows the nugget to solidify while the pressure is holding the metals together.
4. A **down or off time** when the parts to be welded are unloaded and the new parts to be welded are placed in the machine.

The power supplied to the electrode for the welding cycle comes from a large transformer which steps the voltage down but greatly increases the amperage.

Types of Spot Welding Work

Spot welding is frequently used on sheet metal products and metal stampings ranging from .005″ to .5″ in thickness. Most of the spot welding is done on material which is ⅛″ or less in thickness. Products that are not required to be liquid- or gas-tight may be economically joined by spot welding.

The capacities and complexities of welding machines vary greatly. Heavy water-cooled automatic welders using thousands of amperes weld automobile bodies to-

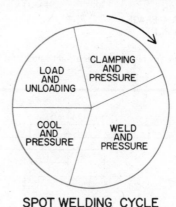

SPOT WELDING CYCLE

Fig. 10-3. There are four timed processes in the spot welding cycle: loading and unloading; clamping and pressure; weld and pressure; and cool and pressure.

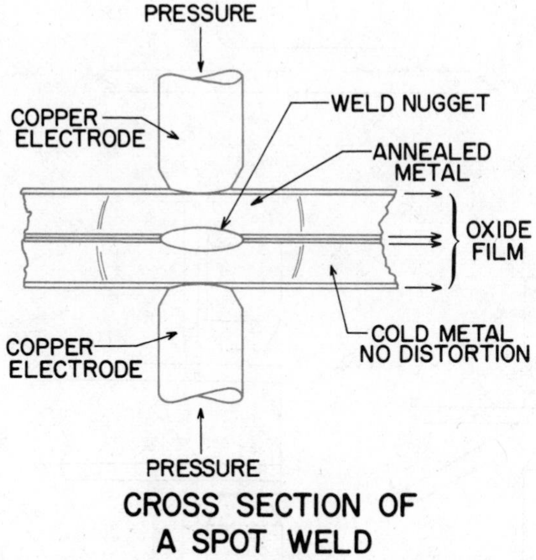

Fig. 10-4. A heavy-duty, air-clamped, water-cooled-tip spot welder.

CROSS SECTION OF A SPOT WELD

Fig. 10-2. A metal oxide film and impurities between the weld surfaces make for the highest resistance to the electrical current, thus the point where the heat is developed and the metal fused.

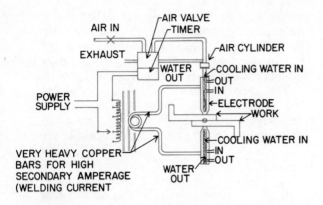

Fig. 10-5.   Schematic of an air-activated spot welder.

gether, Figs. 10-4 and 10-5. Floor-mounted spot welders are capable of welding two pieces of ⅛″ sheet metal. Small portable welders which may be moved around the work can weld sheet metal of 20-gage thickness.

Boxes, brackets, braces, and reinforcing members may be rapidly and easily fastened together by spot welding, Fig. 10-6. This method of welding may be made automatic, and countless articles such as sheet metal parts for automobiles, stoves, and refrigerators are spot-welded together, Fig. 10-7.

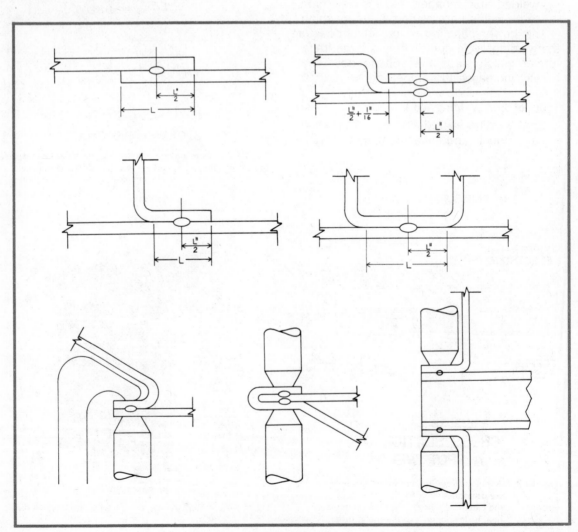

Fig. 10-6.   Typical spot weld applications.

Roll Spot Welding

Roll–spot welding is an adaption of spot welding. The electrodes are rolls or wheels. The metal to be fastened passes between the rolls, and the current is turned on at intervals to make the welds, Fig. 10-8.

Roll spot welding does not produce liquid- or gas-proof welds, but it does make strong, continuous seams, Figs. 10-9, 10-10,

Fig. 10-7. A multispot welder welding the stiffener supports to a floor pan.

Fig. 10-8. In a roll spot-welding machine the metal to be welded passes between the rolls. The current is periodically sent through the rolls and metal to make the weld.

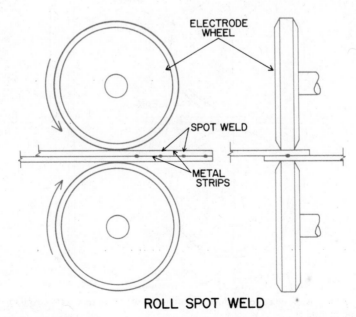

ROLL SPOT WELD

Fig. 10-9. Roll spot welding is performed on long or circular work, producing evenly spaced spot welds.

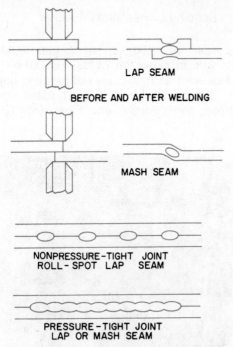

LAP SEAM

BEFORE AND AFTER WELDING

MASH SEAM

NONPRESSURE-TIGHT JOINT
ROLL-SPOT LAP SEAM

PRESSURE-TIGHT JOINT
LAP OR MASH SEAM

Fig. 10-10.   Types of resistance of roll seam welds.

and 10-11. Sheet metal that has been pressed, drawn, or bent may be fastened by roll spot welding to produce commercial structural shapes, components, and products.

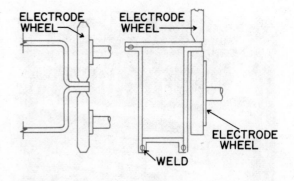

Fig. 10-11.   Typical flange-type, roll-spotting applications.

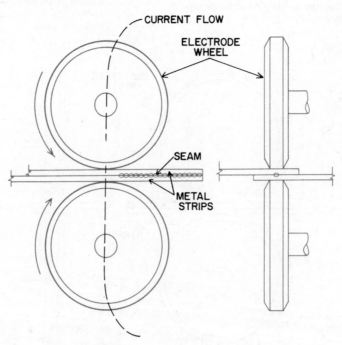

Fig. 10-12.   Seam welding is designed to produce a continuous seal. The spot welds overlap, making liquid-proof and gas-proof joints.

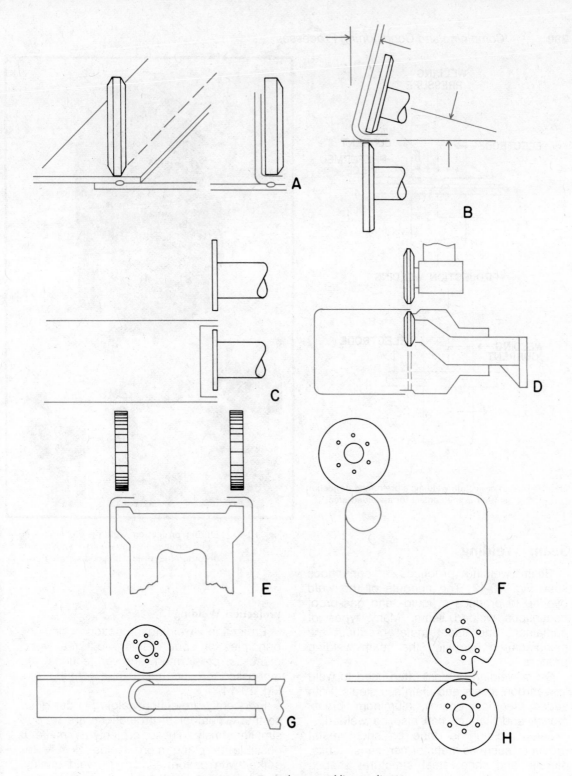

Fig. 10-13. Typical seam welding applications.

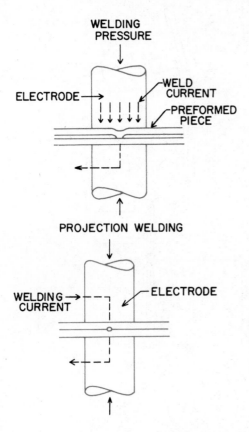

Fig. 10-14. In projection welding, a preformed projection is used as the contact for a welded area.

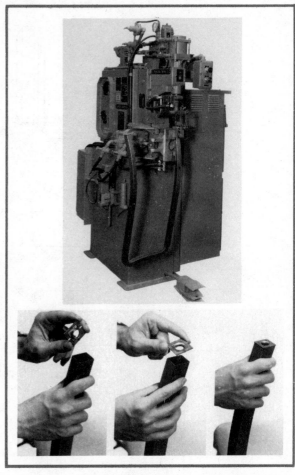

Fig. 10-15. During projection welding the embossed metal projections are melted and form a distortion-free weld with good surface finish.

Seam Welding

Seam welding produces a continuous seal, Fig. 10-12. The nuggets of the weld overlap to produce a liquid- and gas-proof continuous welded seam. Many types of containers and sheet metal products are manufactured by using the seam welding process.

Seam welding is most often used to weld low-carbon steels and stainless steels. With added electrical energy, aluminum, brass, monel, and titanium can also be welded.

Seam welding is done on sheet metal products such as transformer cans, tanks, barrels, and sheet steel structural shapes which are flat or cylindrical in shape, Fig. 10-13.

Projection Welding

Projection welding functions on the principles of spot welding, but the workpiece is pre-formed and projections are embossed on the metal in a forming die, Fig. 10-14.

Frequently projection welding is designed so that a number of projections are welded simultaneously. These projections make it possible to place more welds closely together with uniform and high weld quality. By using the projections on the metal which is being melted as the heat concentration, the welding electrode life is greatly lengthened. The projections are often used to locate the parts with respect to the other workpiece, thus holding fixtures are greatly simplified and the welded part is accurately positioned for welding. With this type of welding the spaced projections aid in the control of heat distortion of the metal, and a commercially good surface finish and appearance is obtained, Fig. 10-15.

Metal shapes such as fastener nuts, angle iron, and tubing that ordinarily would not be able to be spot welded can be welded by projection welding, Fig. 10-16. Normally

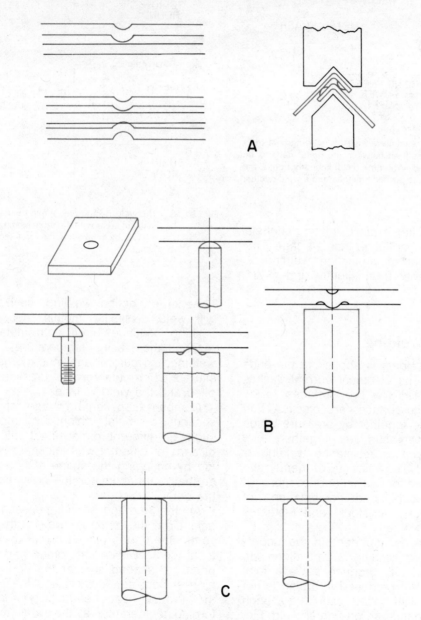

Fig. 10-16.   Applications of a few of many arrangements of projection welds.

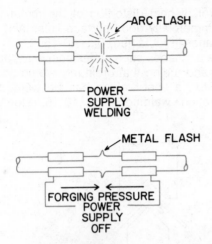

Fig. 10-17. Flash welding receives its welding heat from the arc flash between the metal parts to be welded. When the metal is nearly liquid, the parts are pushed together, and a weld occurs.

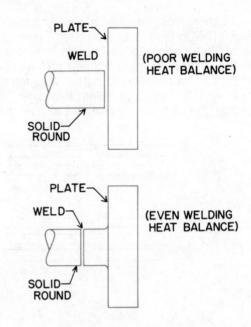

Fig. 10-18. The weld is improved when the flash weld is made in areas with the same heating and cooling rate by controlling the cross section of the metal.

projection welding is not used on materials greater than ¼″ or lighter than 24 gage. The greatest amount of projection welding is done on material from slightly under ⅛″ to ¼″ steel.

Flash-Butt Welding

Flash-butt welding is applied to the ends of the metal. The electrical current flashes or arcs between the two surfaces to be welded until the surfaces are plastic. Then they are forged together by pressure which pushes the two surfaces together, Fig. 10-17. This type of resistance welding is applied to metals which have nearly the same cross sections. The end sections of tubes, rods, bars, and combinations of heavy flat stock, bar stock, and fasteners are flash-butt welded.

A factor that is important in the understanding of flash welding is part alignment. Poor alignment will produce a weak joint because the heat produced in arcing is not uniform and will cause sliding by each part or will produce an off-center weld, Fig. 10-18.

As in projection welding, flash welding will weld dissimilar metals, because the flashing continues until both metals have reached their fusing temperatures. In flash welding, the current and thus the heating is started before the forging pressure is applied and the weld is made.

Compensation can be made for welding sections of metal which are unbalanced such as different dimensional thicknesses of parts or different wall thicknesses of tubing by changing the shape of one member so that it approximates the cross section of the other.

Another method of compensating is by balancing the temperature in the fusion zone. One part is placed closer to the electrical contact than the other part, or the parts are shaped so that the part with the highest heat conductivity and the part with the lowest heat conductivity reach the forging temperature at the same time, Fig. 10-19.

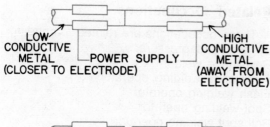

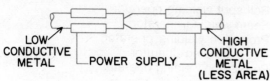

BALANCING TEMPERATURE

Fig. 10-19. When metals have different conductivity, their fusion zone temperatures are balanced by placing the high conductivity metal farther from the electrode holder. Shaping the weld part will frequently aid in balancing the welding temperatures between metals.

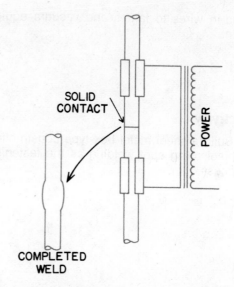

Fig. 10-20. In upset butt welding, the pressure is applied to the weld before the electrical current is turned on, thus a larger area is upset-forged in the weld area.

At the completion of flash welding, the flash may be removed by grinding or machining in order to improve the appearance of the part.

Upset Butt Welding

Upset welding is very similar to flash welding, but the process gathers or thickens the stock more than flash welding does. In upset welding, this thickening (upsetting) results because pressure is applied to the joint before the electrical current is turned on, Fig. 10-20. A larger length of metal is affected in the forging process while the metal is coming up to a welding heat. This method is used primarily for small or medium cross sections in ferrous materials because of the heavy current requirements.

Percussive Butt Welding

The percussive welding, also similar to flash welding, is performed by having the current flow and the parts forged together at the same time. This process is very rapid and free from heat distortion, because it receives its welding energy from a high-voltage condensor or the collapsing field of a transformer.

The parts are held in a slide which aligns and forges the parts. The parts are held about ¾" apart and the condenser is charged to 3000-5000 volts. A spring is tripped, and the two parts move together at high speed. When the parts are about 1/16" apart, the electrical flash occurs and heats the welding surface. The energy released at this time is about equal to 70,000 amperes and produces a thin welding surface on the ends of the parts. The welding takes place so rapidly that there is almost no heat flow in the parts welded.

This process is almost completely confined to butt welding of materials of ½ square inch in area or less, but it has the ability to weld threaded steel studs to aluminum extrusions, forgings or castings: or to weld silver electrical contact tips to copper conductor springs or stellite cutting tool tips to steel holders. The process is also used in electronic and electrical plants to weld lead-in wires to lamps and vacuum equipment.

Activity

Build a sheet metal box-type construction project using spot welding as the fastening process.

Related Occupations

These occupations are related to bonding by heat and pressure:
Resistance welder
Resistance welding operator
Flash welding operator
Spot-welding operator
Roll spot welding operator
Seam welding operator
Projection welding operator
Upset welding operator
Percussion welding operator

Bonding with Chemical Heat

Chapter 11

Words You Should Know

Acetylene — A colorless gas with an obnoxious sweet odor. It is a fuel for oxyacetylene welding.

Acetone — A volatile, flammable liquid.

Regulator — A device to reduce and control the amount of gas delivered to the welding torch.

Gas Welding

Gas welding is a fusion welding process in which metal is heated to the liquid condition by the burning of gases with oxygen or other oxidizers, Fig. 11-1. The fuels used commercially are acetylene, butane, hydrogen, napp, natural gas, and propane. These gases are mixed with oxygen in a welding torch to produce a very small, hot flame that can be manipulated to melt the surfaces of the metals to be welded. The two metal surfaces melt and flow together. Upon cooling, the two metals are joined in the fused area.

The gas-welding processes are much slower than the arc-welding processes, but because of this, metal temperature control is gained. This is important in the joining of alloys which have low melting points.

Oxyacetylene Welding

Oxyacetylene welding is performed by the burning of two gases: oxygen and acetylene.

These two gases burn to produce a very hot flame. When acetylene is burned with oxygen in the air, the acetylene gas (C_2H_2) unites with the oxygen (O_2) to produce two different chemical compounds, carbon dioxide (CO_2) and water (H_2O) at a temperature of 6300° F.

Fig. 11-1. Oxyacetylene welding requires a concentrated heat source and safety equipment.

Acetylene Gas. Acetylene gas is the most widely used hydrocarbon gas of the welding industry. The gas has more carbon available for burning in its molecule than other welding commercial gases. Acetylene gas is generated from calcium carbide (Ca C_2) and water (H_2O). Carbide is obtained from smelting of coke and lime together in an electric furnace. The coke (c) and the lime Ca $(OH)_2$ are fused together to make carbide (Ca C_2). The carbide is crushed to different sizes for convenience in handling and use. The acetylene gas is generated by adding measured amounts of carbide (Ca C_2) to water to generate the gas. Ca C_2 + 2 H_2O = Ca (OH_2) + C_2H_2. The product of this chemical reaction is acetylene gas and slacked lime. The acetylene gas is an obnoxiously sweet-smelling, colorless gas which is compressed into cylinders and distributed to the welding supply companies.

Acetylene requires special handling because above 27 psi and at some temperatures, acetylene gas will dissociate and cause an explosion. No air is needed for such an explosion. It may start by a shock or spark. Because of the danger, no torch or manifold should have an acetylene pressure above 15 psi for safety.

The gas stored in an acetylene cylinder has a pressure of 225 psi when it is full, but the tank has special techniques for acetylene storage built into it, Fig. 11-2. The tank is filled with a porous filler material (such as coke and asbestos) saturated with acetone. Acetone can dissolve large volumes of acetylene gas, thus the acetylene is in solution in the acetone and does not explode in the high-pressure cylinder. Because the acetone is in the cylinder, the tank should always be stored and used in the vertical position. It takes time for the acetylene gas to come out of the acetone when needed. A tank, therefore, should not be emptied at a rate greater than one-fifth the total weight of the gas in the tank per hour, Fig. 11-3. This is the reason that more

Acetylene Capacity
Approx. 275 Cu. Ft.
at 250 Lb. per Sq. In.
Pressure and 70° F.

REMOVABLE METAL CAP
STEEL VALVE
ASBESTOS CLOTH
SAFETY FUSE PLUG
LONG FIBRE ASBESTOS
25"
12" I.D.
40.5"
34.5"
MONOLITHIC FILLER OR BALSA WOOD
FINE ASBESTOS
SAFETY FUSE PLUGS

Typical Airco No. 5 Acetylene Cylinder

Fig. 11-2. The filler material of coke, asbestos, earth filler, and acetone provides the porosity of 80% for the acetone and acetylene. The acetone absorbs 25 times its own volume of acetylene for each atmosphere (of pressure) applied.

Fig. 11-3. Acetylene tanks must not be emptied faster than one-fifth the weight of gas per hour. Tanks are chained vertically for safety. Anti-flashback devices are on each cylinder.

than one tank is used to supply acetylene to three or more welders.

The acetylene cylinder has other safety features designed into its construction. It is strong, made of steel, and has fuse plugs at the top or bottom to release the pressure if the tank is in a fire or over-heated. The hose fittings are left-handed so that wrong hoses cannot be attached. This acetylene cylinder has a shutoff valve with a handle. At all times the shutoff valve is opened only one and one-half turns so that the tank could be closed quickly in an emergency.

Oxygen Gas. Oxygen gas is used to support combustion in gas welding to provide the high temperatures for metal fusion. Most oxygen is obtained by extraction from the air. The air is compressed to a pressure of about 3000 psi. During the compression, much heat is driven off from the gas. The air is passed through cooling coils and allowed to expand into a lower pressure. The temperature of the air greatly drops during the expansion, because the coils cool the incoming air. The temperature drops the air in the coils until liquefaction takes place. The gases are separated in a rectifying tower, where the liquid air is sprayed onto a series of trays and evaporation takes place. The nitrogen and other gases of the air have lower boiling temperatures than oxygen. The gases with lower boiling points gasify and are collected at the top of the tower. The highly pure liquid oxygen remains on the tray. This is collected and shipped as liquid oxygen, or it is expanded and pumped into oxygen cylinders.

Oxygen Cylinders. Oxygen is transported to the point of use in oxygen cylinders. Oxygen cylinders are under a pressure of 2200 psi when they are full and at room temperature (70°F.). There is over one ton of force being exerted on every square inch of surface inside the cylinder. Therefore, the cylinder must be very strong and must be given special care when it is used or moved.

The oxygen cylinders are forged from one piece of chrome molybdenum steel, Fig. 11-4. They are forged so they have a bottom

which is about ⅞″ at the thickest part where the cylinder is most likely to be damaged by dropping, sliding, or scuffing. The thickness on the walls tapers to about 5/16″.

The top of the cylinder is swaged down to receive the fitting of the valve, and it is also thick. The identification and inspection numbers are stamped where the metal is thick.

The oxygen cylinder has a special valve to hold these very high pressures of gas, Fig. 11-5. The valve is constructed so a nylon surface covers the top of the rounded hole when the tank is shut off. When the tank is open, a neophrene washer is crushed between the sides of the valve opening and stem of the valve by brass

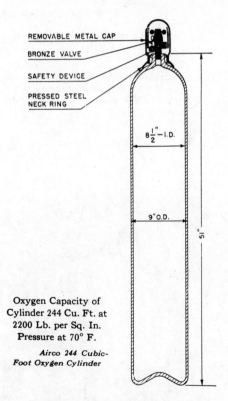

REMOVABLE METAL CAP
BRONZE VALVE
SAFETY DEVICE
PRESSED STEEL NECK RING

$8\frac{1}{2}″$ – I.D.

9″ O.D.

51″

Oxygen Capacity of Cylinder 244 Cu. Ft. at 2200 Lb. per Sq. In. Pressure at 70° F.

Airco 244 Cubic-Foot Oxygen Cylinder

Fig. 11-4. Oxygen cylinders are designed to transport the oxygen safely to the point of use. A very strong cylinder is needed for these pressures, and a forging is used to provide the strength for safety.

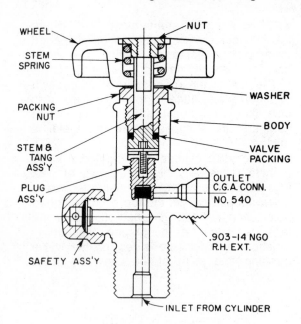

WHEEL — NUT

STEM SPRING

WASHER

PACKING NUT

BODY

STEM & TANG ASS'Y

VALVE PACKING

PLUG ASS'Y

OUTLET C.G.A. CONN. NO. 540

.903-14 NGO R.H. EXT.

SAFETY ASS'Y

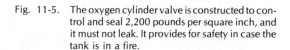

INLET FROM CYLINDER

Cut-away Drawing of Airco Oxygen Cylinder Valve

Fig. 11-5.   The oxygen cylinder valve is constructed to control and seal 2,200 pounds per square inch, and it must not leak. It provides for safety in case the tank is in a fire.

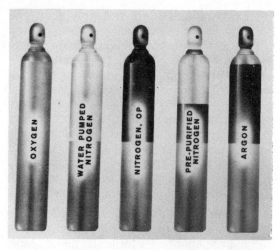

Fig. 11-7.   Other gases are supplied in the same type of cylinder as the oxygen cylinder. The gas inside is identified by the color of the tank.

Fig. 11-6.   A manifold system is used when a high volume of oxygen is needed.

washers. These washers compress the neophrene (a valve packaging), making the seal. For this reason, the oxygen cylinder must be open all the way or closed all the way in order to hold the high-pressure gas in the tank, Fig. 11-6. A bursting disc is placed in the valve to relieve the pressure in case the tank should be in a fire. This will rupture at 3200 psi and lower the pressure in the tank slowly, thus preventing an explosion. The oxygen-type tanks are used for other gases, but these gases are identified by a different color tank, Fig. 11-7.

Regulators. The high pressures of the oxygen and acetylene tanks have to be reduced to a low, even pressure while welding is in progress. Most welding is performed with from one to five pounds of pressure in the welding torch. The pressure used is related to the size of the tip being used, Fig. 11-8.

The regulator-adjusting screw is always backed out before the tank valve is opened. When the pressure is on the regulator, the regulator-adjusting screw is turned in until the required pressure is registered on the welding gage. Because of the gas, this pressure will remain the same as the pressure in the tank is dropped. This always gives the operator a smooth flow of gas.

Welding Torch. There are two types of welding torches. The **injector type** uses acetylene pressures from generators at less than 1 psi. The **positive-pressure type** receives acetylene above 1 psi and is mixed by turbulence, Fig. 11-9. Today the majority of torches which use cylinder gas supplies are positive-pressure torches with the same pressure of acetylene and oxygen.

Torches are manufactured so that the tip may be screwed off and quickly changed for another tip. This means that one welding torch butt can be used for many tips. Tips come in sizes from 000 to 10, Figs. 11-10

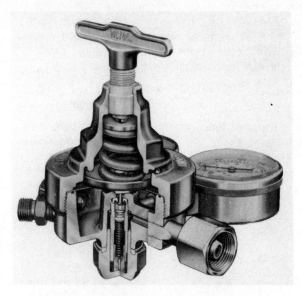

Fig. 11-8. The regulator reduces the tank pressure down to the pressure required by a specific tip.

Fig. 11-9. A welding torch is designed to mix acetylene and oxygen to provide a very hot welding flame (6300° F.).

Tip Size	Drill Size	Oxygen Pressure P.S.I.		Acetylene Pressure P.S.I.		Acetylene Consumption CFH°		Metal Thickness
		Min.	Max.	Min.	Max.	Min.	Max.	
000	75	½	2	½	2	½	3	up to ¹⁄₃₂″
00	70	1	2	1	2	1	4	¹⁄₆₄″ - ³⁄₆₄″
0	65	1	3	1	3	2	6	¹⁄₃₂″ - ⁵⁄₆₄″
1	60	1	4	1	4	4	8	³⁄₆₄″ - ³⁄₃₂″
2	56	2	5	2	5	7	13	¹⁄₁₆″ - ⅛″
3	53	3	7	3	7	8	36	⅛″ - ³⁄₁₆″
4	49	4	10	4	10	10	41	³⁄₁₆″ - ¼″
5	43	5	12	5	15	15	59	¼″ - ½″
6	36	6	14	6	15	55	127	½″ - ¾″
7	30	7	16	7	15	78	152	¾″ - 1¼″
8	29	9	19	8	15	81	160	1¼″ - 2″
9	28	10	20	9	15	90	166	2″ - 2½″
10	27	11	22	10	15	100	169	2½″ - 3″
11	26	13	24	11	15	106	175	3″ - 3½″
12	25	14	28	12	15	111	211	3½″ - 4″

Gas consumption data is merely for rough estimating purposes. It will vary greatly on the material being welded and the particular skill of the operator.

Pressures are approximate for hose length up to 25 ft. Increase for longer hose lengths about 1 psi per 25 feet.

°Oxygen consumption is 1.1 times the acetylene under neutral flame conditions.

Fig. 11-10. Welding tip data for oxyacetylene welding.

and 11-11. Each tip size produces a different volume of heat and is applied to welding different thicknesses of metal.

The Oxyacetylene Flame. To obtain the heat necessary to melt and weld, these gases must be mixed correctly and burned. When acetylene is burned completely in air, the oxygen and acetylene of the torch form one chemical reaction and these again combine with oxygen from the surrounding air to complete the chemical reaction or burning. The final products are heat, carbon dioxide, water, and heat. $(2C_2H_2 + 5O_2 - 4CO_2 + 2H_2)$ and heat.

Acetylene is actually burned in a two-step reaction. The first reaction is $2C_2H_2 + 2O_2$ which dissociates in the heat to become two gases which are combustible — carbon monoxide and hydrogen. Additional oxygen is taken from the surrounding air, and these two gases burn, forming heat, carbon dioxide and water.

$$4CO + 2O_2 \rightarrow 4CO_2 \text{ and heat}$$
$$2H_2 + O_2 \rightarrow 2H_2O \text{ and heat}$$

The acetylene gas is an excellent fuel, because there are carbon and hydrogen to burn to produce heat.

There are three types of flames that the acetylene torch delivers: neutral flame, carburizing flame, and oxidizing flame, Fig. 11-12.

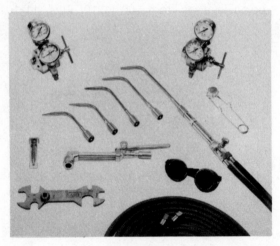

Fig. 11-11. Oxyacetylene welding and cutting equipment.

Manual Oxyacetylene Welding Technique

1. The correct pressures are set on the regulators.
2. The torch acetylene valve is opened, the gas is poured down into a friction lighter, and the gas is ignited.
3. The gas valve is opened until the flame is ready to jump off the end of the tip,

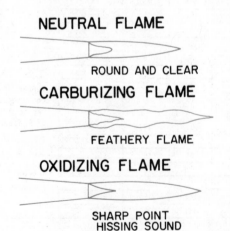

NEUTRAL FLAME

ROUND AND CLEAR

THIS FLAME WILL NOT REACT WITH THE BASE METAL AND IS USED IN WELDING

CARBURIZING FLAME

FEATHERY FLAME

THIS FLAME PRODUCES FREE OR EXTRA CARBON, WEAK AND DIRTY WELDS

OXIDIZING FLAME

SHARP POINT
HISSING SOUND

THIS FLAME IS SLIGHTLY OXIDIZING WHEN USED IN BRONZE WELDING AND BRONZE SURFACING, A VERY HOT FLAME

Fig. 11-12.   The three types of oxyacetylene flames.

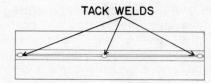

Fig. 11-13. When metal is being prepared for welding, it is positioned, squared, and tack-welded together. If the tacked metal is not square, the tacks are broken and the metal is realigned.

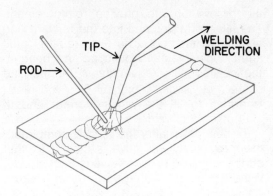

Fig. 11-15. Backhand welding is used on thick material.

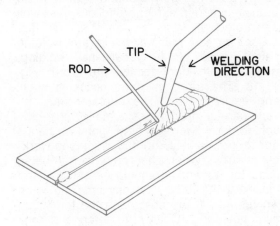

Fig. 11-14. Forehand welding is used on thin material.

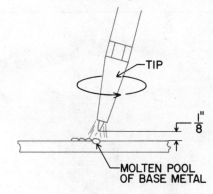

Fig. 11-16. The tip of the flame is held about ⅛" above the surface of the metal and is moved in a circle.

but the flame remains on the tip. This adjusts the volume of the acetylene gas required for that size of tip.

4. The oxygen valve is slowly opened until a neutral flame is obtained.

5. The metal is cleaned, and the joint is prepared for welding. A square butt joint is used for thin material, and a V-joint is used for heavy material.

6. The joint is positioned with a small root opening, and the parts are tack-welded together, Fig. 11-13.

7. If the metal to be welded is ⅛" thick or less, the metal is welded by the **forehand technique**, Fig. 11-14.

8. If the metal is thicker, it may be welded by the **backhand technique**, Fig. 11-15.

9. The inner tip of the flame is held about ⅛" from the metal and is moved in a circular motion until a liquid pool is established, Fig. 11-16.

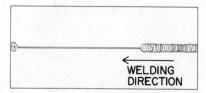

Fig. 11-17. The tip is moved in a crescent motion, carrying the molten pool of metal across the work to be welded. Filler rod is added in the forward edge of the molten pool to add reinforcement to the weld.

Torch Movement. The torch is moved in a crescent fashion across the workpiece, melting the base metal continuously, Fig. 11-17. The molten pool moves across the seam with the torch. Metal is added to fill the groove and to provide reinforcement. The filler rod is dipped into the forward end of the molten pool. The filler rod will melt and supply metal to build up the bead to the desired size requirements. At the end of the weld seam additional filler rod metal is added to fill the welding crater and to strengthen the weld.

To shut down after the weld is completed, the torch acetylene valve is closed to shut off the fuel. The oxygen valve is then closed. If the equipment is shut down for any length of time or if the operator will be leaving the immediate area, the oxygen and acetylene tank valves are closed. The pressure is drained from the hoses and regulator, one at a time, and each torch valve is drained. The regulator-adjusting screw handles are backed out.

Safety Considerations

The general safety procedure for fire and burns mentioned earlier are followed, as well as protection from the fire hazards discussed in the manual shield arc welding safety rules. In oxyacetylene welding, goggles with dark lenses are worn, Fig. 11-18. The torch in some cases is small and easily moved. The operator must be very careful not to accidentally move the lighted torch across the fingers or hands nor to turn quickly and burn someone working nearby. When the torch is not being used for welding, it should be shut off and placed in its rack. If at any time the torch whines like a jet engine and the torch butt gets hot very quickly, the torch and tank valves should be turned off immediately. A flashback may have occurred. The equipment should be checked by a welding technician to make sure it is safe.

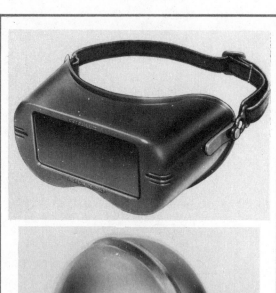

Fig. 11-18. Welding goggles with dark lenses are used during welding.

Activity

Design and build a project that requires cutting, fitting, and welding.

Related Occupations

These occupations are related to bonding with chemical heat:
Oxyacetylene welder
Aircraft welder
Alloy welder
Aluminum welder
Pipe welder
Railroad welder
Sheet metal welder
Welding layout worker
Welding setup technician
Welding inspector

Separating Bonds by Heat

Chapter 12

Words You Should Know

Oxidizing — The combining of materials with oxygen.

Cracking a valve — Slowly opening a valve momentarily so that dust may be blown from the opening of the valve, or pressure may build slowly in a regulator before the valve is opened fully.

Machine flame cutting — Cutting equipment is moved and controlled to cut a specific shape or size of cut. Cutting heads are frequently controlled by mechanical, electronic, or numerically controlled devices.

Plasma arc cutting — Melting process controlled by very high-temperatures.

Carbon dioxide laser cutting — A very narrow kerf is melted by a laser. Carbon dioxide is used in the process.

Cutting with Heat

As metals heat and melt, the molecules become free from their neighboring molecules and break the bonds holding them together as a rigid solid. A number of heat sources will produce the separation of metal by heat, including carbon arc, plasma, laser, oxyacetylene flame, or any intense heat source that can be controlled to keep its heat in a narrow area.

Flame Cutting

Oxygen has a strong chemical affinity for iron, as indicated by the rusting of steel and other iron-based alloys. This rusting or oxidation may be increased very rapidly by raising the heat of the metal and by supplying pure oxygen to do the rusting. Metals have a kindling temperature just as woods or other materials do. Metal will not burn as wood does, but it becomes very chemically active with oxygen when its temperature is between 1400° F. and 1600° F. When a narrow area of steel is heated to this temperature and a stream of oxygen is directed onto the metal, rapid oxidation takes place, and the oxygen stream cuts the metal along its direction of movement. This process is flame cutting. In the process of oxidizing the steel, a large amount of heat is produced, preheating the metal and bringing it to kindling temperature. The steel in the path of the oxygen stream is oxidized or melted and runs out of the cuts. About 30% to 40% of the metal in the cut is

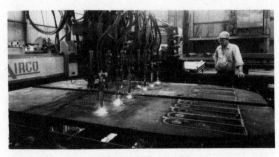

Fig. 12-1. A cutting machine producing small parts such as corner details and braces. The optical tracer automatically follows the outline (template) of the part to be cut.

washed out of the cut as liquid steel. Most of the steel is oxidized, Fig. 12-1.

Cutting Torch Construction. The oxyacetylene cutting torch is constructed differently from the welding torch, because it has a channel designed to by-pass oxygen through the torch to the cutting tip to perform the cutting, Figs. 12-2 and 12-3.

The cutting torch has three needle valves, two at the base of the torch and one valve midway on the torch. Most oxyacetylene cutting heads are designed to be an attachment to a standard welding torch. The tip

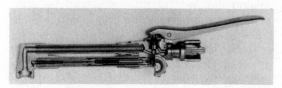

Fig. 12-2. The cutting attachment will screw on the end of most standard torches. The cutaway attachments illustrate the spiral mixer for the preheating flames.

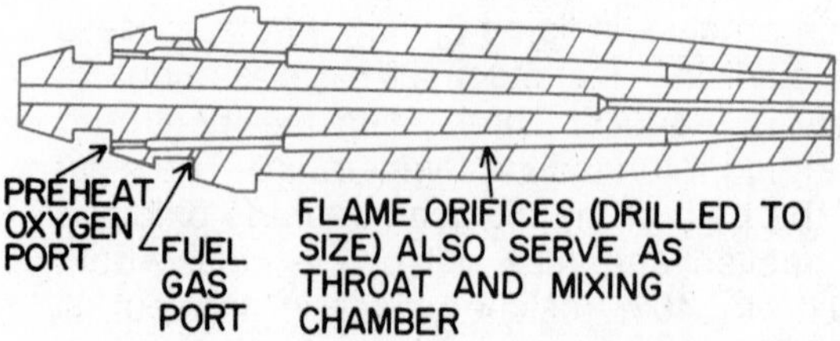

PREHEAT OXYGEN PORT
FUEL GAS PORT
FLAME ORIFICES (DRILLED TO SIZE) ALSO SERVE AS THROAT AND MIXING CHAMBER

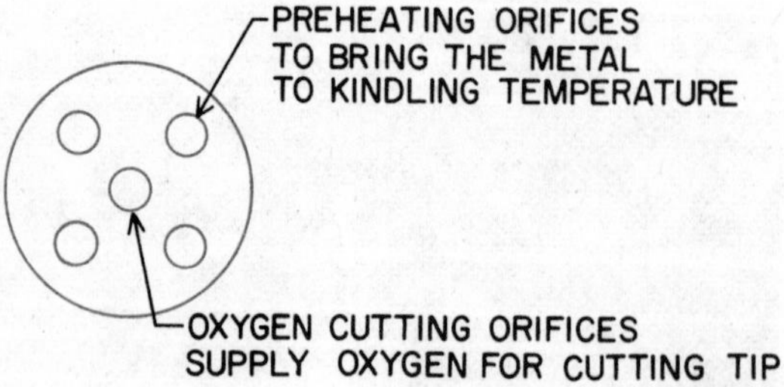

PREHEATING ORIFICES TO BRING THE METAL TO KINDLING TEMPERATURE

OXYGEN CUTTING ORIFICES SUPPLY OXYGEN FOR CUTTING TIP

Fig. 12-3. The end of an oxyacetylene cutting tip has a series of holes drilled into it. The center hole is the oxygen orifice. The smaller holes surrounding the center hole are preheating orifices.

screws off and the cutting head screws in its place.

Setting Up the Cutting Torch. The cutting operation requires different pressures from those necessary for welding. The welding manufacturers' recommendations for the pressures should be consulted. The size of the tip and the cutting pressures are related to the thickness of the steel being cut, Fig. 12-4.

The regulator adjustment screw handle is checked to see if the regulator screw is backed out and the regulators are closed on both tanks. The oxygen valve is "cracked" slowly so that the high pressure in the tank will build up slowly within the regulator housing. The acetylene valve is slowly opened so that the gas will fill the high-pressure side of the gage. The acetylene valve is opened on the torch briefly while the regulator is set for 4 psi. The torch valve is closed. This procedure sets the pressure and purges the hose of other gases. The same process is repeated for the oxygen, except the oxygen regulator is set for 35 psi. The pressures are now set for cutting ½″ steel plate with a number 1 sized cutting tip.

Lighting the Cutting Torch. The oxygen valve on the torch butt is opened all the way. The acetylene valve is opened a small amount, and the gas at the tip is lit with a friction lighter. The flame is held away from flammable materials. The valve midway up the torch is slowly opened, and oxygen is supplied to the preheat flames. The preheat flames are adjusted until they are neutral, between ¼″ to 5/16″ long. The flame is tested by pushing down on the oxygen lever on the top of the torch. The oxygen should flow through the center hole between the preheat flames without changing their shape. They should not jump or become reducing flames. If there is a change in the preheat flames, it should be readjusted slightly with the cutting oxygen lever open.

Cutting Procedure for Steel. The ½″ steel plate is laid out with a soapstone line and center-punched every 2″. The plate is laid

Acetylene and Oxygen Medium Pressure Cutting Tips

Metal Thick-ness	OXYGEN PRESSURE PSI			Cutting Speed	Kerf Width	DRILL SIZES		Recomm. No. of Cylinders (Sgl. or Manif.)
	Cutting Pressure		Fuel Gas Pressure			Cutting Jet	Pre-heat SC10 Series	
	At Regulator	At Torch						
⅛″	20	20	3	28	.035	72	71	1
3/16″	20	20	3	26	.050	68	71	1
¼″	30	30	4	22	.055	62	70	1
⅜″	35	35	4	20	.055	62	70	1
½″	35	35	4	19	.080	56	68	1
⅝″	40	40	4	17	.080	56	68	1
¾″	36	35	4	16	.095	54	65	1
1″	41	40	4	14	.095	54	65	1
1¼″	51	50	4	13	.095	54	65	1
1½″	42	40	5	12	.100	51	65	1
2″	47	45	5	10	.100	51	65	1
2½″	38	35	5	9	.125	45	60	1
3″	44	40	5	8	.125	45	60	2
4″	54	50	5	7	.125	45	60	2
5″	56	50	6	7	.150	41	60	2
6″	67	60	6	6	.150	41	60	2
8″	78	70	6	5.5	.150	41	60	2
10″	83	70	6	5	.203	32		2
12″	125	90	6	4.5	.230	32		2
14″	100	82	7	4	.250	28		3

Fig. 12-4. Acetylene and oxygen pressures for medium preheat — general hand-cutting attachments.

on the cutting table so that the cut will be over the edge or between the supports of the cutting table.

Goggles, gloves, and safety equipment are put on and adjusted. The torch is lit and the preheat flames adjusted. The hands and arms hold the torch in a comfortable position so that the torch can be moved through long arcs without having to move the body or upper arms so a straight cut can be made, Fig. 12-5.

The torch is held at a 45° angle to the plate over the edge of the plate until the edge is red hot. Then the torch is moved so that the torch is 90° from all surfaces,

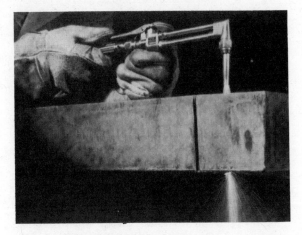

Fig. 12-5. When hand-held cutting tools are used, workers wear gloves to protect their hands.

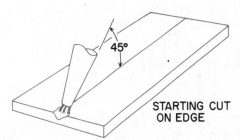

Fig. 12-6. Preheating and starting the cut on a steel sheet.

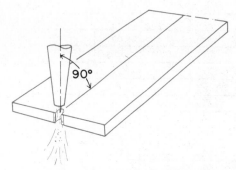

Fig. 12-7. The oxygen lever is pressed and the oxygen cuts through the steel. The cutting torch is raised to a 90° position.

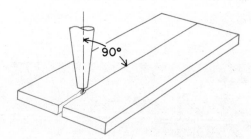

Fig. 12-8. The tip is held above the sheet and moved along the sheet at the number of inches per minute recommended by the cutting pressure chart. For example, ½″ steel would require 19 inches per minute.

Fig. 12-6. When the edge of the plate has become red-hot, the oxygen cutting lever is pressed down, and cutting will proceed, Fig. 12-7. Once the cutting has started, the cutting tip is held ⅜″ to ½″ above the metal through the rest of the cut, Fig. 12-8.

Fig. 12-9. Hand-cutting bevels on heavy pipe.

Cutting with an oxyacetylene torch has a number of applications in metalwork. Straight cuts, holes, bevels, arcs, and combinations of cuts are used to fabricate pipe, bar stock, and plate, Fig. 12-9. The cutting torch is also applied to special cutting work such as removing rivets, gouging metal to prepare for a welded joint, and cutting cast iron.

Machine Flame Cutting

In the manufacturing of heavy equipment and construction, structural members are used which are cut from heavy sections of steel, Fig. 12-10. These members are gen-

Fig. 12-10. Numerically controlled multitorch machines automatically cut complex shapes out of heavy steel plate.

Fig. 12-11. Cutting torch mounted on a tractor or skate can cut straight cuts, bevel cuts, and (with a radium rod) circles and rings.

erally cut to shape by machine flame cutting.

Machine flame cutting will cut structural steel and plate with considerable accuracy, speed, and economy. Flame cutters will make as long a straight cut as necessary. They cut circles, radii, and arcs as well as bevels, plate-edge preparation, and intricate shapes, Fig. 12-11. Flame cutters will cut through one thickness of metal or a series of thicknesses such as stacked plates.

Mechanical Shape Control. Shape cutting may be controlled by a number of methods. The oldest method is mechanical. Controlling the cutting torch by a tractor and a track or radius rod is excellent for long cuts and large arcs. Smaller shapes are accurately cut to shape by using a cam and template to guide the torch, Fig. 12-12.

Photoelectric Shape Control. Photoelectric tracers are burning machines that utilize a photoelectric cell (electric eye) to follow paper drawings of pencil or ink. With a motorized unit they move the cutting torch correctly to produce an exact size or a re-

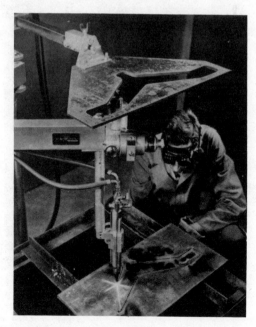

Fig. 12-12. Cams and templates are used to guide contoured cuts for small repetitive work.

duced size shape, Fig. 12-13. Accurate intricate shapes may be cut in this manner.

Numerical Shape Control. Numerically controlled flame-cutting machines are capable of very detailed and accurate cutting. They operate torches for multiple cuts simultaneously, Fig. 12-14. Numerical control produces any number of shapes that can be programmed on an X/Y axis. The 1″ tape may be shipped anywhere in the world. An identical part, in shape and in size, can then be cut on a like machine. In many cases a light tape may be shipped and parts cut on site. This procedure is much simpler than shipping heavy steel parts.

Numerically controlled cutting is used for machine cutting because of the additional information that can be programmed which offers more control over the cutting for very reliable, accurate work.

Plasma Arc Cutting

Plasma arc cutting is used to cut nonferrous metals, stainless steels, refractories, and carbon steels at high speeds, Figs. 12-15, 12-16, and 12-17.

Speed of Cutting. Carbon steel up to 2″ thick can be cut up to 10 times faster by plasma arc cutting than by oxyacetylene flame. A ¼″ carbon steel plate can be cut up to 200 inches per minute equal in quality to a cut by oxyacetylene flame, Fig. 12-18.

Fig. 12-13.   An optical tracer automatically follows the outline (a template) of the part to be cut.

Fig. 12-14.   Tape-O-Graph, a numerically controlled flame-cutting machine. Operating from a punched tape, it cuts circles and complex contoured shapes from steel ⅛″ to 12″ thick.

Fig. 12-15. Plasma arc cutting is designed to cut metals at very high speeds.

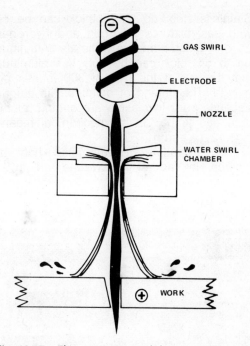

Fig. 12-17. The water-constricted plasma arc concentrates the plasma jet into a narrow cutting stream.

GAS SWIRL

ELECTRODE

NOZZLE

WATER SWIRL CHAMBER

WORK

Fig. 12-16. Plasma cutting of two sandwiched ⅛" sheets of steel. This is a very difficult operation to do by oxyacetylene cutting, because the second sheet will deflect the oxygen jet stream. The plasma's heat is so great that it cuts both sheets accurately.

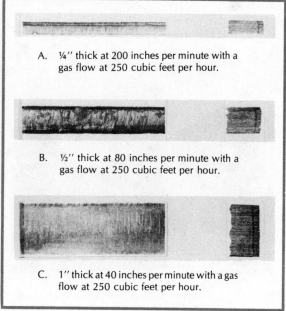

A. ¼" thick at 200 inches per minute with a gas flow at 250 cubic feet per hour.

B. ½" thick at 80 inches per minute with a gas flow at 250 cubic feet per hour.

C. 1" thick at 40 inches per minute with a gas flow at 250 cubic feet per hour.

Fig. 12-18. Plasma cutting of carbon steel.

Stainless steel up to 2″ thick can be cut with gas mixtures of argon and hydrogen, as can most nonferrous metals. Aluminum up to 5″ thick can be cut; ¼″ aluminum may be cut at the rate of 300 inches per minute. The cutting of aluminum is unique, because prior to plasma arc cutting, aluminum had to be cut by mechanical means, Figs. 12-19 and 12-20.

Plasma cutting recently has had an oxygen jet added below the electrode,

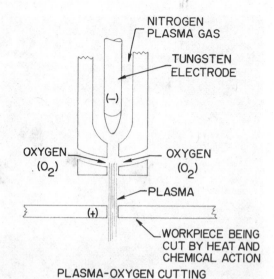

PLASMA-OXYGEN CUTTING

Fig. 12-21.   Plasma-oxygen cutting has increased the cutting speed for carbon steels above the original plasma jet cutting.

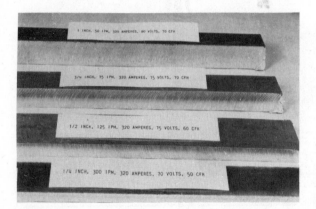

Fig. 12-19.   Aluminum cut with plasma arc.

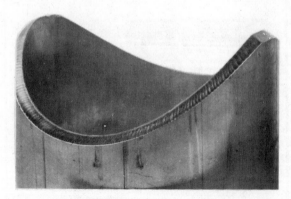

Fig. 12-20.   A contour cut made on schedule 40 aluminum pipe at 60 inches per minute with plasma equipment.

Fig. 12-22.   Compact, lightweight, movable, 500-watt laser cutting 3mm stainless steel at 1.75 metres per minute.

thus increasing the cutting action for steels without causing the erosion to the tungsten cathode. The addition of oxygen to the cutting system has brought more rapid cutting of steels, thus making cutting costs for large cutting contracts economical, Fig. 12-21. The investment for the equipment is high, but its production rate is such that large amounts of cutting are done and the savings ratio over that of oxyacetylene cutting is about 3:1.

Carbon Dioxide Laser Cutting

The carbon dioxide laser is designed to cut, slit, drill, and weld. The laser is used for precise cuts with very narrow kerfs (width of the cuts). Laser cutting offers great control over the cutting process, Fig. 12-22.

Activity

Lay out and cut out a project made of steel plate. Practice straight cuts, angular cuts, curved cuts, beveled cuts, and square and round holes.

Related Occupations

These occupations are related to separating by heat:

Oxyacetylene-cutting operator (burner)
Air carbon arc-cutting operator
Plasma arc-cutting operator
Flame-cutting machine operator
Radiograph cutting machine operator
Template-tracing cutting machine operator
Photoelectric control cutting machine operator
Numerically controlled cutting machine operator

Adhesive Bonding

Chapter
13

Words You Should Know

Bronze welding — The bonding of metals together with a bronze metal that melts above 800° F. but does not melt the base metal. Metal clearances and the bead are similar to oxyacetylene welding of steel. A large amount of filler metal is used.

Brazing (hard soldering) — As in bronze welding, metals are joined without melting the base metal. The difference between bronze welding and brazing is in the joint preparation. In brazing the fit between the joined parts is very close. The brazing alloy is carried through the joint by capillary action.

Soft soldering — The bonding of metals together with tin-lead alloys that melt below 800° F. The solder becomes a solvent and is diffused into the metal, making a new intermetallic alloy between the solder and the base metal.

Dip brazing — Cleaned, fluxed parts with alloy in position are dipped into a molten salt bath. The salt bath provides the heat to flow the alloy and join the parts.

Induction brazing — Prepared metal parts are surrounded with a high-frequency coil. When the current is on, an induced current provides heat that melts the alloy and completes the joint.

Adhesive bonding — Small molecules of adhesive produce attractive forces between the metal and the joint. With the addition of catalysts, long, chain-type molecules form and gradually solidify the material. A strong joint is formed.

Shear condition — An adhesive joint should be designed so that the center line of the load is on or parallel to the center line of the adhesive.

Thermoplastic adhesive — An adhesive that can be softened by heating and hardened by cooling.

Thermosetting adhesive — An adhesive that hardens by a chemical reaction. When the chemical reaction is completed, it cannot be reversed.

Elastomeric adhesive — An adhesive made of natural rubber or latex. These adhesives are vulcanized to increase their strength.

Joining with Intermediate Alloys

Bonding metals with filler metal by processes of diffusion below the melting point of the base metal but above 800° F. is known as **bronze welding** or **braze welding**, Fig. 13-1. The filler metal is applied to the metal as a puddle and is built into a bead. When the filler metal is melted and flows through a tightly fitted joint by capillary action, the process is called **brazing**. If the adhesion process is carried out by melting and flowing the alloys by capillary action through the joint below 800° F., the process is referred to as **soft soldering**.

Bronze-Welding Cast Iron

Steels and cast irons are the materials which are commonly bronze-welded. Bronze

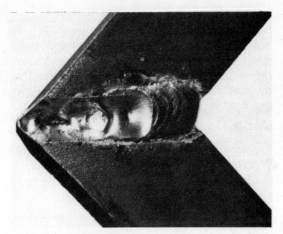

Fig. 13-1. Bonding metal by a diffusion process above 800° F. The metal is applied to a V-groove and built into a bead.

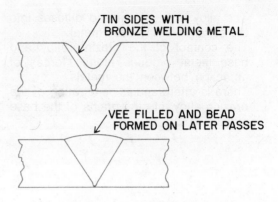

Fig. 13-2. The surfaces of the V-grooved weld are covered during the first pass to protect the base metal from oxidizing during the bronze-welding process.

welding is frequently used for repairing or rebuilding existing equipment or products.

The joint design prepared for bronze welding is the same as that made for fusion welding. The joint may be beveled by a cutting torch, ground, or machined to produce the V-groove for welding.

A broken casting may be repaired by bronze welding by grinding the cast iron broken area into a V. The surfaces are cleaned with a file to remove any graphite flakes that may have been spread over the face of the weld during the grinding. The graphite comes from within the cast iron. The scale should be ground away about ½″ back from each beveled edge, depending upon the size of the workpiece. A holding or clamping fixture is designed so that the surfaces are aligned and the root of the weld spaced about 1/16″ apart. A small portion of the break may be left at the bottom of the bevel to help align the workpiece. This alignment root section should be thin so that capillary action will carry the bronze through the joint.

The oxyacetylene flame should be adjusted so that it is moderately oxidizing for bronze welding. The point is preheated by holding the torch ½″ to ¾″ above the work and by warming the casting. The weld area is heated to a "black" heat, which means that no metal is hot enough to be seen as red in a darkened room.

The bronze rod is heated and dipped into the flux until the flux coats the rod. The flux and rod are heated and applied to the base metal. If the bronze "balls up," the base metal is not hot enough. If the bronze bubbles and runs around like water, the metal is too hot. When the metal is the proper temperature, it will spread out evenly over the beveled surface and "tin" (or "wet") the surface. The "tinning" should cover the total beveled surfaces as the weld is started. It will keep the base metal from oxidizing when the filler metal is being applied to the V-groove and bead, Fig. 13-2. The casting and bronze-welded area are placed in a lime box so that the temperature will drop slowly and allow the casting to stress-relieve itself.

Bonding by Brazing (Hard Soldering)

Brazing is the process of bonding in which a nonferrous filler metal flows through a closely fitted joint by capillary action. It must be heated above 800° F. but below the melting point of the base metal.

The forces which produce the joint strength by the brazing process are the result of these three things:

1. The alloy is a solvent and diffuses into the surfaces of the base metal.
2. The contact of the brazing alloy and base metal produce atomic forces of attraction between the metals.
3. There is intergranular penetration of the brazing alloy into the grains of the base metal, Fig. 13-3.

The metals which are commonly used for these brazing processes above 800° F. but below the melting points of the base metal are copper alloys, silver alloys, and other alloys, Fig. 13-4. Alloys may be formulated from a combination of these following metals for brazing:

> copper
> silver
> zinc
> gold
> cadmium
> phosphorous*
> tin
> manganese
> nickle

With the selection and composition of these listed metals, a series of liquid and freezing temperature ranges may be obtained.

The fit of the silver brazed alloy joint is important in order to gain the greatest strength from the alloy used. The best fit for silver brazing between the mating parts is about 0.0015″. With a larger clearance, the strength drops quickly. If the space is tighter than 0.0015″, there is not enough space for capillary action to pull the alloy through the joint, Fig. 13-5.

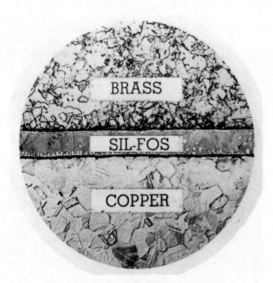

Fig. 13-3. Nonferrous joints are made possible by full penetration and alloying with metal surfaces. These joints are usually stronger than the base metals.

*phosphorous is used to improve the flowing of the alloys

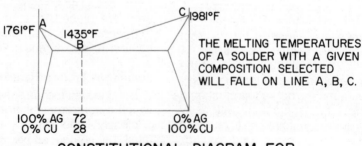

THE MELTING TEMPERATURES OF A SOLDER WITH A GIVEN COMPOSITION SELECTED WILL FALL ON LINE A, B, C.

CONSTITUTIONAL DIAGRAM FOR SILVER/COPPER

Fig. 13-4. The flowing temperature of a brazing alloy is determined by the composition of the alloy. Different flowing temperatures may be obtained by different percentages of metals in the brazing alloy.

As an illustration, stainless steel with a tensile strength of 160,000 psi was joined in a butt joint with a silver-based alloy which tested at 70,000 psi. When the stainless steel test pieces were brazed together with a clearance of 0.0015″ between the parts, the joint strength was tested at 130,000 psi. The strength of the joint dropped proportionately as the clearance was increased to 0.024″ (to about 40,000 psi).

Brazing Alloy Forms. Alloy forms are available in a number of shapes. The selection of the shape will be dependent upon the application or use. Manufacturing organizations frequently use pre-forms of

Fig. 13-6. Brazing alloy forms commonly come in wire, rod, sheet, rings, washers, and special shapes.

alloys which are specially shaped and engineered for the special joint to be brazed, Fig. 13-6. These parts may be formed out of wire alloy or stamped out of sheet or strip alloy. Brazing alloy is often used in the forms of sheets, wire, snippets, and filings. Because these alloys contain a relatively high content of silver, the alloy is used carefully and economically. Silver alloys should not be heated above 1500° F. Silver will be damaged by a temperature above 1700° F. Most silver-copper brazing will take place between 1175° F. and 1500° F.

Another consideration necessary in selection of a silver alloy is its color. A series of colors from yellow to silver-white may be obtained if it is necessary to match the color of the base metal to beautify the product.

Techniques for Brazing

Brazing procedure is the same, regardless of which brazing process (torch, furnace, dip, or induction) is utilized.

1. Shape or machine the part to be brazed to provide the proper fit and joint clearance for brazing.
2. Clean the parts **mechanically** by sanding or scraping, or **chemically** by dipping the part in an effective solvent to remove all dirt, grease, oil, and oxides.
3. Flux the surface to be brazed. In special cases a light wash coat will be applied to the whole part to protect it from oxidation during brazing.
4. Select the desired brazing alloy and form of application (pre-form, wire, sheet, etc.)

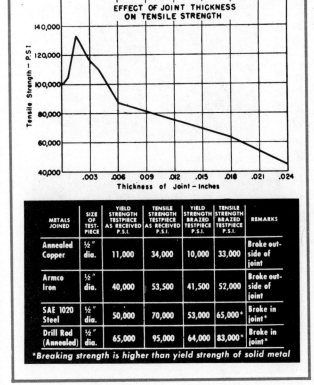

EFFECT OF JOINT THICKNESS ON TENSILE STRENGTH

METALS JOINED	SIZE OF TEST-PIECE	YIELD STRENGTH TESTPIECE AS RECEIVED P.S.I.	TENSILE STRENGTH TESTPIECE AS RECEIVED P.S.I.	YIELD STRENGTH BRAZED TESTPIECE P.S.I.	TENSILE STRENGTH BRAZED TESTPIECE P.S.I.	REMARKS
Annealed Copper	½″ dia.	11,000	34,000	10,000	33,000	Broke outside of joint
Armco Iron	½″ dia.	40,000	53,500	41,500	52,000	Broke outside of joint
SAE 1020 Steel	½″ dia.	50,000	70,000	53,000	65,000*	Broke in joint*
Drill Rod (Annealed)	½″ dia.	65,000	95,000	64,000	83,000*	Broke in joint*

*Breaking strength is higher than yield strength of solid metal

Fig. 13-5. The thickness of the joint or fit is an important factor in producing a strong joint. Silver solder joints with a fit of 0.0015″ to 0.002″ produce the highest tensile strengths.

5. Clamp, wire, or place in a brazing fixture so the parts have the correct relationship to each other for brazing.
6. Heat the part to be brazed, and allow the alloy to flow through the brazed area.
7. Clean the work to remove the flux by washing in water or other solvents.

To control the capillary action on art workpieces, a graphite pencil line may be drawn across the work, Fig. 13-7. The graphite will contaminate the workpiece, and silver brazing alloy will not cross the line.

Brazing Processes

Brazing may be carried out by a number of processes. The one selected will be decided by the production rate needed and the availability of equipment.

Torch Brazing. The most versatile brazing process is torch brazing. The heat is applied with an oxyacetylene torch and a tip that delivers a broad flame. The torch is moved around the work slowly and evenly to raise the work to the flow temperature of the brazing alloy. Fuels such as propane,

A. Miltiflame

B. Multitorch

C. Aligning fixtures

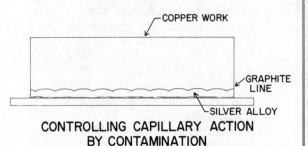

CONTROLLING CAPILLARY ACTION
BY CONTAMINATION

Fig. 13-7. Silver alloy will move by capillary action to a place where the metal is clean, fluxed, and heated to the proper temperature. Control may be obtained by contaminating the surface with a pencil line.

Fig. 13-8. Combinations of fixed gas-air burners providing brazing heat.

butane, hydrogen, or natural gas are used to reduce the manufacturing costs.

The hand-held torch is advantageous when the mass of the material is unequal or if the thermal conductivity varies. The operator can vary the heat input by moving the torch to compensate for the heating differences.

Large production rates are achieved by using multiflame tips and multitorch fixtures.

Machine brazing will utilize holding and aligning fixtures, rotating mechanisms, and multiple torches to speed up production, Fig. 13-8.

The work may also be placed on an automatic conveyor, where it is heated by a stationary multiflame burner for a predetermined interval of time and then passed on to a cooling position, Fig. 13-9. The heating torches may also be set so that the work is on a rotary table and is loaded, brazed, cooled, cleaned, and unloaded, Fig. 13-10.

Furnace Brazing. Furnace brazing is a high-production brazing process in which the work is continuously moving through the furnace, Fig. 13-11. The parts have a holding

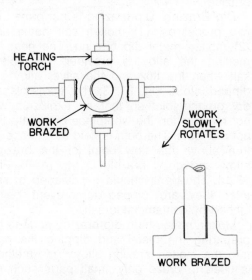

Fig. 13-9. Multitorch brazing provides an even brazing temperature by having the work slowly turn in the heating flames.

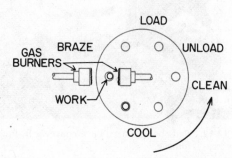

Fig. 13-10. Brazing with a rotary table makes possible the continuous production of parts.

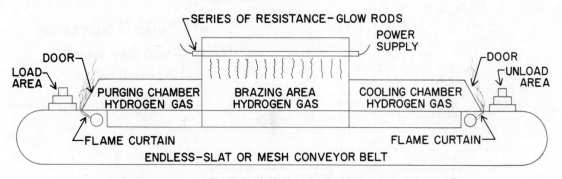

HYDROGEN ATMOSPHERE FURNACE SCHEMATIC

Fig. 13-11. Oxygen-free brazing is done in a hydrogen atmosphere furnace. The hydrogen protects the hot metal from atmospheric oxygen and provides a clean, bright, finished part.

and aligning fixture or are designed so that the joint is self-positioning and/or self-aligning. The workpieces are prepared for brazing (cleaned, fluxed, and pre-forms placed). The work is now placed on the endless belt which slowly carries the parts to be brazed through the furnace, Fig. 13-12. With a hydrogen atmosphere furnace, clean copper parts require no flux. The hydrogen removes any oxides and produces a beau-

Fig. 13-12.   Parts being brazed on a conveyor hydrogen furnace.

tiful, clean, bright part. The temperature of the furnace is very accurately controlled by the electrical resistance heating elements. The time is controlled by the speed of the belt and the length of the chamber. The openings to the furnace usually have doors and flame curtains with a small amount of hydrogen burning at each end. This aids in keeping oxygen out of the furnace.

Large amounts of electronic and vacuum system parts are manufactured by furnace brazing.

Dip Brazing. Dip brazing is performed by two processes: (1) molten salt baths and (2) molten metal dip brazing. The heat for melting the alloy is provided by molten salt when the flux and the silver alloy are dipped into it, Fig. 13-13. The flux is placed dry on the joint so that an explosion will not result when the workpiece goes into the hot salt bath. The bath is held at a temperature above the flow point of the brazing alloy, and good brazing is the result, Fig. 13-14. The parts should be cleaned of salt after they are dipped to prevent being etched by contaminated salt.

Molten metal, bath-dip brazing is done by preparing the metal and dipping the part into the molten brazing alloy. This method is used to braze only small parts such as

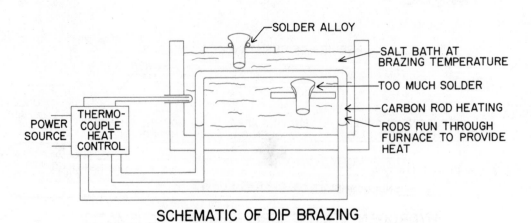

SCHEMATIC OF DIP BRAZING

Fig. 13-13.   Parts are dipped into a hot salt bath. The hot bath melts the brazing alloy and flows it throughout the joint. Many parts are aligned in a holding fixture during the dipping.

metal strips, wires, and tubes because of the large amount of expensive alloy which must be heated. To keep the alloy from oxidizing, the molten metal is kept covered with flux.

Induction Brazing. An electric coil is formed into a shape to surround the part to be brazed, and a high-frequency current is passed through the coil, Fig. 13-15. This coil and current induce eddy currents into the metal parts being brazed. The eddy currents flowing in the surface of the parts are converted into heat that melts the brazing alloy and completes the joint, Fig. 13-16. Pre-form brazing alloy melts and forms the brazed joint, Fig. 13-17.

Soft Soldering

Soft soldering is done wtih temperatures below 800° F., and it creates a joint by adhesion. Soft solders make a bond with other metal, because the hot solder acts

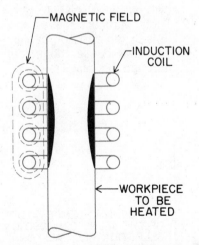

Fig. 13-14. Heat-exchange, unit-return bends and beaders are dip-brazed all at once in this salt bath furnace.

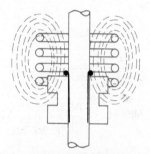

PRE-FORM BRAZING ALLOY MELTS AND FORMS A BRAZED JOINT

Fig. 13-16. The induced heat causes the brazing alloy to flow throughout the joint.

—MAGNETIC FIELD

—INDUCTION COIL

←WORKPIECE TO BE HEATED

SCHEMATIC OF HEAT-AFFECTED AREA IN INDUCTION HEATING

Fig. 13-15. The high-frequency electrical current induces current into the surface of the workpiece. The result of the shorted current flowing in the work is heat. The heat is used to flow the brazing alloy.

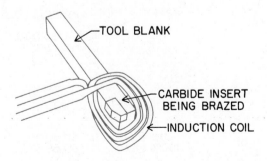

—TOOL BLANK

—CARBIDE INSERT BEING BRAZED

←INDUCTION COIL

Fig. 13-17. Induction brazing being used to braze a carbide cutting tool into a tool blank.

as a metal solvent. This makes diffusion of the solder into the base metal possible. This solvent action produces an intermetallic alloy between the solder and the base metal, Fig. 13-18.

Soft solders are made of metals with low melting points. The major metals used in soft solders are tin and lead. Small amounts of other metals which vary the properties of the alloy may be found in solders. Such metals are antimony, bismuth, cadmium, or silver. Antimony may be present in the alloy as an impurity.

There is need in manufacturing for solders with different melting points. These different melting points are obtained by changing the composition of the alloy by varying the ratio of tin to lead, Fig. 13-19.

The alloy 63% tin, 37% lead is of special interest to the electronics industry. The soldering of wires, components, and circuit boards is subject to high electrical resistance if the solder joint is vibrated or shaken as the solder is solidifying. This can be minimized by soldering with a 63% tin, 37% lead alloy. When it cools, the solder changes from a liquid to a solid metallic condition within a very small temperature range. This quick freezing greatly reduces the opportunity for high electrical-resistant joints to occur.

Techniques for Soft Soldering

In order to bond metals together with soft solder with the maximum strength, careful preparation of the metal and the joint must be completed.

1. The metal must be tinnable, and it must wet with solder, so the solder will act as a solvent upon it.
2. Metal parts are shaped so that when it is fluxed and hot, capillary attraction will cause the solder to flow through the joint.

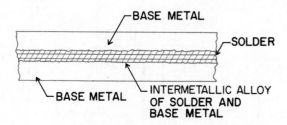

Fig. 13-18. Soft soldering bonds a joint by diffusing into the base metal. This causes an intermetallic alloy to form between the solder and base metal.

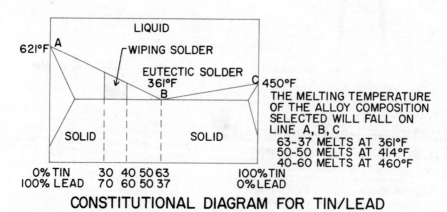

CONSTITUTIONAL DIAGRAM FOR TIN/LEAD

Fig. 13-19. Different melting points of solder are made by having different compositions of tin and lead in the alloy.

3. The areas to be soldered are cleaned by abrading, scraping, or etching before the parts are assembled.
4. The proper flux is applied to remove any remaining oxides.
5. The parts are assembled and held in position by clamps, fixtures, rivets, or spot welds for heating.
6. The workpiece is heated, and the solder is applied.
7. The solder is flowed through the joint by capillary action, and excess solder is removed.
8. The workpiece and solder are allowed to cool. While the work is held still, the acid flux or residue is washed or removed as required.

Soft Soldering Processes. Soft soldering may be performed by many processes — soldering coppers, electrically heated soldering irons, or soldering guns. It may be performed by flame heating (natural gas and air-aspirating), dip soldering, induction heating, resistance heating, oven heating, and ultrasonic soldering. The soldering process will be selected by considering the joint design, product size and shape, and the number of parts desired.

A method which has always been economically competitive is the dip process, because large objects such as automobile radiator cores have many joints soldering in one dipping. Electronic circuit boards may also have many joints soldered in one dipping, Fig. 13-20.

Polymer Adhesive Bonding

An adhesive bond fastens materials together by surface attachment at temperatures from below room temperature to about 375° F. (with some exceptions), Fig. 13-21. This bonding process has been referred to as "chemical fastening." During the fastening process, the molecules of the adhesive form large molecules from smaller ones. These long chain-type molecules produce attraction forces between the metal of the joint and the adhesive. As the large molecules form, polymerization takes place. The adhesive becomes a solid, producing a bond between the metal surfaces. The adhesives become a part of the structure and are influenced by the materials which they contact. The efficiency of the bond and thus the joint is affected by the environment, whether it is heat, chemical, or stress.

Adhesive substances are used in conditions where the adhesive and the mating surfaces are in continuous contact, thus the load is distributed over the whole area of the joint. Local stress concentrations are eliminated, because the load is over the large area, Fig. 13-22.

Fig. 13-20. Soldering electronic components with a wave-soldering machine.

Fig. 13-21. An adhesive is applied to two low-cost aluminum chainsaw castings and cured in an oven at 375° F. for 20 minutes. The bonded seam is the strongest part of the tank.

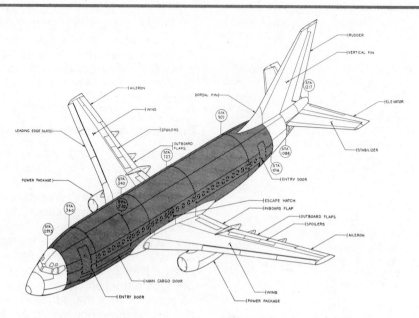

A. Boeing 737 short-range jetliner. Shaded area indicates laminated aluminum fuselage skin and doubler panels.

B. A total of 560 square yards of adhesive is used in bonding the 737 skin panels required for the cabin section of the 737 twin jetliner.

Fig. 13-22.   Adhesives are used to laminate the fuselage panels on Boeing aircraft.

C. Stretch press (foreground) is used for forming of 737 bonded skin and doubler panels prior to chemical milling. Autoclave (background) is used to bond the flat laminated sheets prior to stretch forming and chemical milling.

D. Masked skin and doubler panel prior to chemical milling. Dark areas will remain after milling.

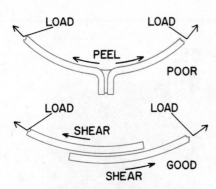

LOAD   LOAD

PEEL

POOR

LOAD   LOAD

SHEAR

GOOD

SHEAR

Fig. 13-23. Adhesive joints are designed so that the load placed on the adhesive will be a shear load rather than a peel load.

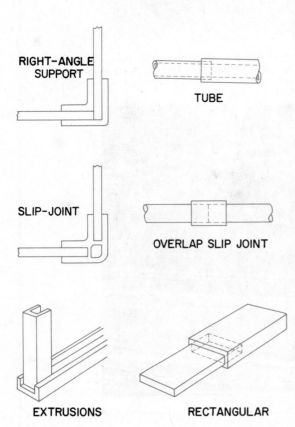

RIGHT-ANGLE SUPPORT

TUBE

SLIP-JOINT

OVERLAP SLIP JOINT

EXTRUSIONS

RECTANGULAR

Fig. 13-24. Typical adhesive joint design. The increasing of the contact area provides holding strength.

A. Applying structural adhesive to magnets by means of a spatula.

B. Clamping adhesive-coated magnets to inside of steel stator ring. Magnets are clamped so they will stay in alignment during the curing operation.

C. Assembled stator rings are conveyed through electric oven to cure adhesive bond between magnet and steel ring.

D. Magnet bonded to steel ring forms three stators for new DC motor.

Fig. 13-25. A structural adhesive used to bond ceramic permanent magnets to a 3-stator-ring steel housing.

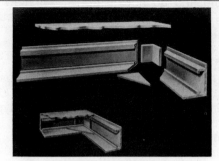

A.   Joint corner construction of frame.

B.   Adhesive-bonded power supply cabinet.

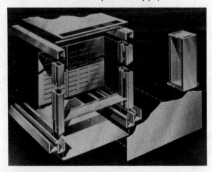

C.   Joint construction of equipment rack.

D.   Equipment rack group.

Fig. 13-26.   The application of adhesive assembly technique to production structures.

Joint Design

The design of the adhesive joint is very important. By increasing the contact area of the adhesive and the metal and by using the adhesive so that it will be in a shear condition rather than a peel condition, excellent joints will be produced, Fig. 13-23.

Metals fastened by adhesives will have their area increased by utilizing joints which are similar to those in woodworking, Fig. 13-24. Adhesives may be used to join a number of different materials: metals, woods, plastics, glass, ceramics, or a combination of materials, Figs. 13-25 and 13-26. A range of difficult materials to fasten (such as honeycomb core) may be bonded by adhesives, Fig. 13-27.

Adhesive Functions

Functions different from bonding may also be performed by adhesives. Sealing may be done by applying the film to metal parts before they are spot-welded. The welding current breaks through the sealant film and forms the weld, but the sealant fills the

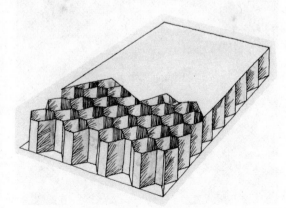

Fig. 13-27.   Honeycomb sandwich skins can be aluminum, stainless steel, titanium, hardboard, plywood, composition board, plastic, galvanized, cold-roll, etc. Adhesives make is possible to assemble these strong, light structural materials.

space between and around the welds to form an airtight and watertight joint. These sealants are used in automotive body and aircraft pressurized cabin sealing in combination with spot welding, Fig. 13-28.

Insulating is the function of some adhesives. The adhesive may be an electrical or a thermal insulator. Thermal insulation may be done by foaming the adhesive within the structure. Spaces provide an area between the working surfaces. This process is utilized in the manufacture of fiberglass boat hulls, as well as in metal panels.

Adhesives can also be used to prevent electrochemical corrosion between joints of dissimilar metals. By waterproofing and by keeping the two metal surfaces from contacting each other, the adhesive can control corrosion. Vibration may be dampened, thus the reduction of joint fatigue may be controlled with the application of the proper adhesive to the joint.

An additional function unique to adhesive bonding is that it produces an aerodynamic contour between surfaces when it is applied. The skin or surfaces of aircraft fuselages, wings, and control surfaces may be made

Fig. 13-28. Adhesive is applied to the surfaces of the 1C-foot propellant tank with this "glue gun." The adhesive and tank are spot-welded through, and hot air is used to cure the adhesive. The adhesive seals the space between the spot welds.

smooth and do not disrupt the air flow as rivets or other mechanical fasteners would, Fig. 13-29.

Types of Adhesives

A number of methods have been used to classify and organize adhesives. One system of classification is by chemical type. These are the major chemical types: (1) thermoplastic (2) thermosetting, and (3) elastomeric.

Thermoplastic Adhesives. Thermoplastic adhesive materials are those that can be softened by heating and hardened by cooling. The adhesive must be in the fluid stage when a joint is made. The heated adhesive must wet the point parts adequately to produce a bond upon cooling. Thermoplastics used are polyvinyl acetate, polyvinyl alcohol, acrylic, cellulose nitrate, asphalt, and oleoresin. These thermoplastic adhesives have good strength to 150° F. Above this temperature, the adhesive in the joint will creep or slip. To improve the bond, the joint will be reinforced with caps, overlaps, or stiffeners, Fig. 13-30.

Thermoplastic materials are used in the bonding of all materials, but the most frequent use is the bonding of metal to wood, leather, cork, paper, or other nonmetallic materials which frequently operate in a nonstress joint.

Thermosetting Adhesives. A thermosetting adhesive material has a chemical reaction which takes place during curing. This chemical reaction may be started by heat, catalysts, ultraviolet light, or pressure. When the chemical reaction has taken place, it cannot be reversed. When the adhesive has set, it can only be removed by destroying the bond permanently.

The thermosetting adhesive materials commonly used are phenolic, resorcinol, phenolresorcinal, epoxy, urea, melamine, and alkyd. For the bonding of metals, epoxy and alkyds are most commonly used. These adhesives function to temperatures up to 200° - 500° F.

The thermosets are applied primarily in three conditions — liquid, paste, or solid.

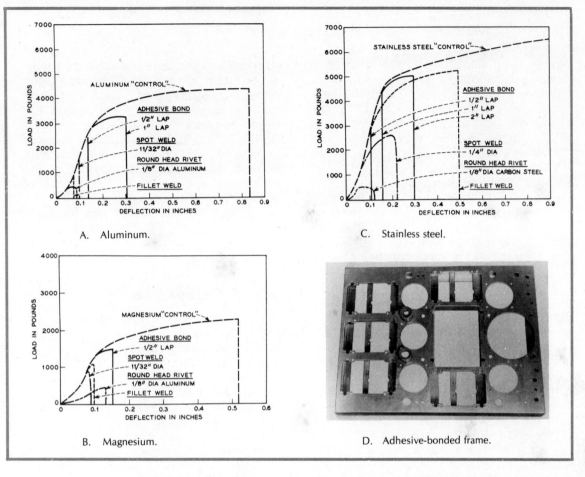

A. Aluminum.

C. Stainless steel.

B. Magnesium.

D. Adhesive-bonded frame.

Fig. 13-29. Comparison of mechanical-bonded joints and adhesive-bonded joints.

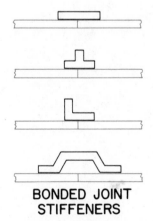

BONDED JOINT STIFFENERS

Fig. 13-30. Metal may be added to a joint to improve the bond. Caps, overlaps, or stiffeners reinforce the adhesive joint.

Liquids are free-flowing when they are applied. They are made in a two-part system to produce the chemical reaction. Thermosetting adhesives also come in other forms: powder, dry film, and tape. These materials are applied to the joint and subjected to heat and pressure to complete the bond.

Epoxies are recommended for bonding dissimilar materials. Phenolics will bond wood, metal, and glass into a structural unit. The alkyds are used as metal laminators.

Elastomeric Adhesives. Elastomers are materials which are capable of stretching at room temperature. The most commonly used thermosetting elastomers are natural rubbers known as **latex**. This material may have sulphur added to it, and when it is heated, vulcanization takes place. Other elastomers are butadiene-systrene, neoprene, acrylonitrile-butadiene, and silicone. These are used as adhesives primarily for rubber, paper, leather, wood, fabric, foil, and plastic films.

Cleaning for Adhesion. When adhesives are used, the surface to be bonded must first be treated. The surface must have the loose or weak boundary layer molecules removed so that the adhesive will have a chemical attraction to a clean and solid metal layer. The cleaning of the metals to be fastened is very important, thus the metal is carefully degreased. Scrubbing with a solvent or vapor degreasing will remove the surface grease, but the solvents must not evaporate on the metal, thus redepositing the grease. Metals with oxide scale or loose surface material should be sanded or ground to clean the surfaces down to the bare metal. Aluminum alloys may be further cleaned by dipping the surface to be bonded in an alkali bath for 1 to 10 minutes at 170° F. This cleans and slightly roughens the surface for the adhesive. The metal is washed in warm water to remove all traces of the alkali solution. The bond area is further prepared by being dipped in sulfochromate solution for 10 minutes at 155° F. This is followed by a warm, fresh rinse with running water. It is allowed to dry or it is blown with warm air to dry the bond surface.

The cleaning is carefully done and must not be touched with bare hands. Clean cotton gloves are worn, because grease from the fingers would prevent good adhesion. The adhesive is applied to the joint, and the joint is clamped together. In some applications such as an aircraft wing flap, the parts are placed in a plastic bag, the bag sealed, and the air is pumped out of the bag. The outside air pressure (14.7 psi) then pushes the plastic bag and the parts being joined together. The whole unit is placed into an autoclave, where the thermoset is heated to its bonding temperature.

Related Occupations

These occupations are related to adhesive bonding:
Oxyacetylene bronze welder
Oxyacetylene brazer
Silver brazer
Copper alloy brazer
Soldering machine operator
Soldering machine feeder
Dip solderer (radiator-core)
Torch solderer
Small parts solderer
Electronic solderer
Adhesive supervisor
Polymer chemist
Adhesive joint designer
Adhesive jig and fixture builder
Adhesive worker

Conditioning Metal

Chapter 14

Words You Should Know

Annealing — A softening process for steels. It is done by heating the steel above its upper critical temperature and then cooling it very slowly. The metal is cooled in the furnace or insulated by lime or vermiculite.

Normalizing — A method of controlling grain size after forging or working. The steel is heated above its upper critical temperature and cooled slowly in still air.

Spheroidizing — High-carbon steels are heated slightly below the lower critical temperature and held for a period of time. The carbon gathers into ball-like shapes.

Stress relieving — Removing the stresses in steel which are caused by cold-working or welding. It is done by heating the metal below the lower critical temperature between 800° F. - 1250° F.

Allotropic change — The same metal that exists in two forms or states. Above 1333° F. steels will start changing atomic structure and start forming a solid solution called austenite.

Austenite — A solid solution of carbon and other elements in iron which forms when the allotropic change take place. The metal has a face-centered, cubic space lattice.

Eutectoid — A eutectoid steel has the lowest possible melting point.

Solid solution — A solid solution exists when the carbon in the steel grains is dissolved and is redistributed throughout the metal. This is dependent upon temperature. When the carbon has been dissolved, the remaining material is called austenite.

Lower critical temperature — The lowest temperature that carbon steel may be heated and quenched and hardening will still occur. Transformation has started in the steel at this temperature.

Heat treatment — The heating of metals, quenching, and tempering to produce the desired physical and mechanical properties.

Upper critical temperature — The highest temperature that carbon steel may be heated and quenched to make the maximum hardness with the finest grain size. At this temperature the carbon should be dissolved into a solid solution.

Hypereutectoid — A steel with a carbon content above 0.83%.

Hypoeutectoid — A steel with a carbon content below 0.83%.

Martensite — A fully hardened steel which is very strong, but brittle unless further heat treated. The metal's structure is needle-like.

Cementite — Iron carbide material that is very hard, strong and brittle. Cementite adds hardness to steel.

Pearlite — Strong, tough grains made up of layered ferrite (iron) and cementite (iron carbide.)

Ferrite — Pure iron grains that exist in large amounts in low-carbon steel.

Quench hardening — Steel is heated above the lower critical temperature and rapidly cooled in oil or water.

Tempering or drawing — The reduction of hardness of steels by controlled heating between 300° F. and 750° F.

Recrystallization temperature — A temperature where old stressed grains are dissolved. Upon cooling, new annealed grains reform.

Solution heat treatment — The metal is heated until the metals are taken into solid solution (dissolved by another metal). At a later time upon heating at a lower temperature, small particles will precipitate out of solution. A new compound which hardens the metal will be formed.

Metals are conditioned to make them more workable or to improve their physical characteristics. They may be conditioned before, during, or after manufacture. Metals may be conditioned by heat treatment or by mechanical conditioning, Fig. 14-1. Grain size, hardness, toughness, elasticity, ductility, or size are some of the characteristics which may be controlled by conditioning.

Heat Treatment of Steels

By heating and cooling the metal through a temperature sequence, the material's characteristics (and thereby the product manufactured) may be improved or modified. Heat treatment alters the metal metallurgically and either softens it, hardens it, or controls the toughness and the metal grain size. The softening processes are annealing, normalizing, spheroidizing, and stress relieving. The hardening processes are quenching, hardening, surface hardening, and case hardening.

Understanding Changes in Steels

The heat treating of steel can best be understood by studying the changes that take place in steel as it is heated and cooled. The common steel is made up of carbon and iron. As the carbon content changes in the steel, the characteristics and strength are drastically changed. The phys-

Fig. 14-1. Metal conditioning: a weldment being heat treated in a car-type furnace.

Fig. 14-2. Steel bars are annealed in controlled atmosphere in a continuous annealing furnace. The physical changes in the steel are due to the allotropic changes in the steel.

ical changes in the metal relate to the atomic changes in the steel as the heat is applied or withdrawn. These changes are **allotropic** changes in steel, Fig. 14-2.

Allotropic Changes. These allotropic changes are reversible and will occur with rising or falling temperatures, but the allotropic change temperatures will be offset. A change indicated by A_R is reached by cooling the metal. The change indicated by A_c is reached by a rising temperature, Fig. 14-3.

A piece of steel is heated in a furnace where the temperature of the steel and the temperature of the furnace may be controlled and measured very accurately. This piece of steel will show that the furnace temperature and the metal temperature remain together. However, at a critical point, the temperature of the furnace will continue to rise as planned. The steel will remain at the same temperature for a period of time before it gains in temperature again. This phenomenon is referred to as an allotropic change. Metallurgists explain that there has been a change in the atomic structure. The additional heat absorbed is used in this structural change. The steel changed from a body-centered, cubic-space-lattice structure to a face-centered, cubic-space lattice structure, Fig. 14-4. The chemical composition of the material at this temperature remains the same, but there are changes in electrical resistance, loss of magnetism, and changes in atomic structure.

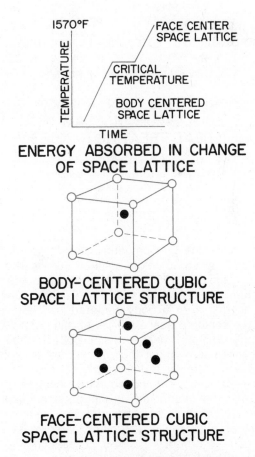

ENERGY ABSORBED IN CHANGE OF SPACE LATTICE

BODY-CENTERED CUBIC SPACE LATTICE STRUCTURE

FACE-CENTERED CUBIC SPACE LATTICE STRUCTURE

Fig. 14-4. Below the lower critical temperature steel is magnetic and has a body-centered cubic space lattice. Above the upper critical temperature steel is nonmagnetic and has a face-centered cubic space lattice. Iron in this nonmagnetic form has the power to dissolve carbon and alloying elements while in a solid form.

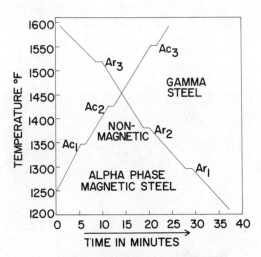

ALLOTROPIC CHANGES IN CARBON STEEL

Fig. 14-3. Allotropic changes are changes in the atomic structure of the metal. The metal has its internal physical characteristics modified at those critical points of temperature.

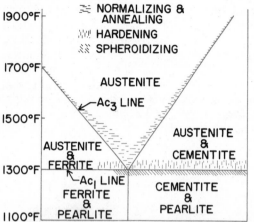

PARTIAL IRON–CARBON
EQUILIBRIUM DIAGRAM WITH
HEAT TREATING RANGES

Fig. 14-5. The physical properties of steel are changed by heating it above its critical temperature.

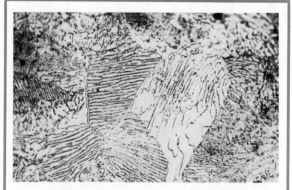

A. A.I.S.I. 1040 steel austenitized at 1600° F. (magnified 750 times). It is cooled in air to form ferrite and pearlite. The steel is tough and soft.

B. A.I.S.I. 1090 steel austenitized at 1600° F. (magnified 750 times). It is furnace-cooled to form pearlite. The steel is very strong, tough, and soft.

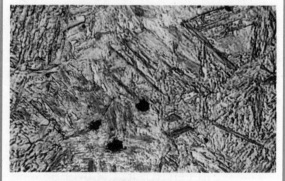

C. A.I.S.I. 1040 steel austenitized at 1600° F. (magnified 750 times). It is water-quenched to form martensite. The steel is very strong, hard, and brittle.

Fig. 14-6. Austenitized steel.

This same phenomenon occurs in carbon steels. When eutectoid steel of .83% carbon is heated to 1333° F., the critical temperature is reached, Fig. 14-5. The steel changes from a body-centered lattice to a face-centered space lattice. At this point the carbon in the steel is dissolved and redistributed throughout the metal. The steel is in solid form at a dull, red heat. The metal is said to be in **solid solution** because the carbon is distributed throughout the metal. However, only one material is visible — grains of austenite. The other materials have been absorbed, Fig. 14-6. These changes make heat treatment possible, because the carbon distribution can be controlled within the steel, thus the characteristics of the steel can be controlled, Fig. 14-7.

As the steel reaches critical temperature, solid solution occurs, and the carbon is

TEMPIL°
Basic Guide to Ferrous Metallurgy

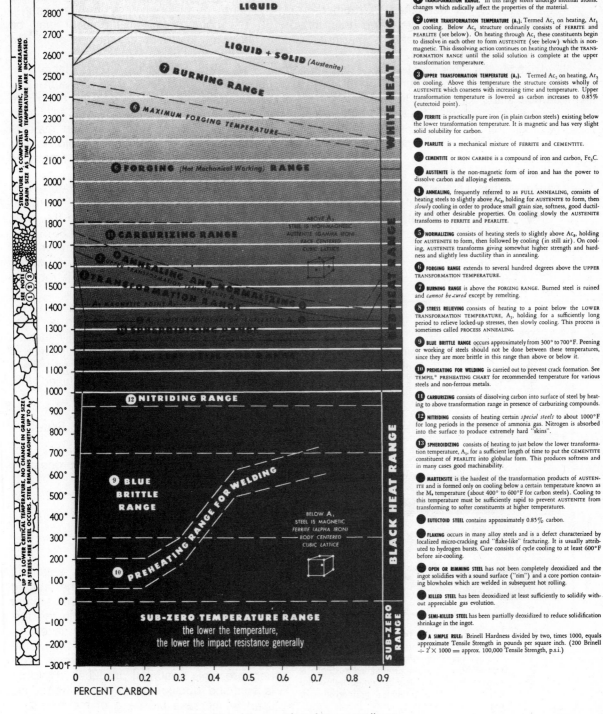

Fig. 14-7. Basic guide to ferrous metallurgy.

absorbed, creating **austenite,** Fig. 14-8. The grain size has been refined. If the metal is cooled rapidly or quenched to room temperature, a steel with a very fine grain size will result. The new material is **martensite,** and it is hard and strong.

To soften it, this steel is reheated at a low, controlled temperature. The dissolved carbon starts to migrate to produce a new crystalline structure. The new structure will have different properties, as determined by the amount of carbon and other alloys in the sample.

Heat treatment can be accomplished because of these factors:

1. The grain size of steel may be changed by placing its carbon in solid solution and hardening.

2. Hardening may be reduced and controlled by tempering (drawing out the hardness by applying controlled heat).
3. The composition of the steel may be selected to control the amount of carbon in the steel.

The rate that cooling takes place from above the upper critical temperature line for a certain steel determines the grain size and hardness of the steel. Steels with carbon content above 0.35% carbon may be hardened by a quenching or rapid cooling process. Their toughness may be controlled by later tempering.

A time, transformation, temperature diagram indicates the relationship of the cooling rapidity and the hardness and the grain size, Fig. 14-9. Eutectoid steel (0.83% carbon) produces a very hard, fine-grained martensite when it is quenched in water. It is very strong and brittle, so it requires further heat treatment to be useful.

Austenite is produced by dissolving carbon and/or alloying elements in a high-temperature iron or steel while all materials are in a solid solution. When austenite is rapidly cooled below 450° F., a very hard material — martensite — is produced, Fig. 14-10.

Heat Treating of Softening Steels

Heat treating of softening steels is accomplished by the following four processes:
1. Annealing
2. Normalizing
3. Stress relieving
4. Spheroidizing

Annealing

Annealing is the process of softening steels, Fig. 14-11. The steel is either heated 75° F. above its upper critical temperature and allowed to cool very slowly in the furnace, or it is insulated in lime, ashes, or vermiculite. The metal is cooled below the lower critical temperature or to room temperature. This treatment produces steel that is soft and ductile. Annealing is frequently used to increase the ease of machining.

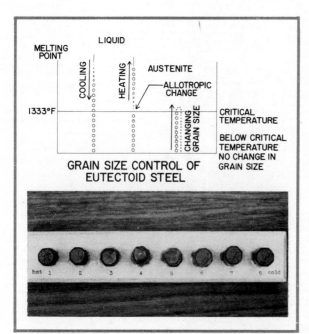

Fig. 14-8. When the critical temperature of steel is reached, it has the ability to dissolve carbon. The grain structure of the metal changes form, and a fine-grained material called austenite is formed. When the steel is cooled below the critical temperature, the fine-grained size of martensite is formed.

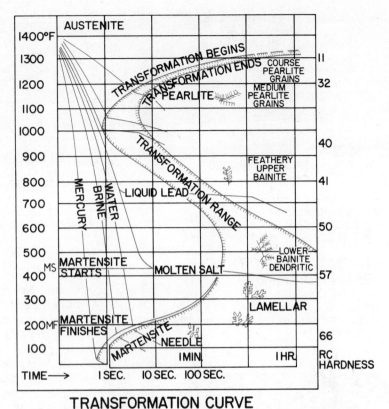

TRANSFORMATION CURVE
OF 0.83 STEEL

Fig. 14-9. The heat treatment of steel applies the information from a time, temperature, transformation diagram to obtain the desired characteristics in the steel product.

Fig. 14-10. Martensite is produced by the rapid cooling of austenitized steel.

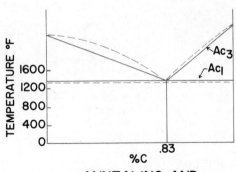

ANNEALING AND
NORMALIZING RANGES

Fig. 14-11. Annealing is a process for softening steel and making it more machinable.

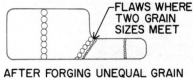

AFTER FORGING UNEQUAL GRAIN
SIZE THROUGHOUT WORK

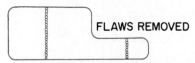

AFTER NORMALIZING NEARLY
EQUAL GRAIN SIZES
THROUGHOUT WORK

NORMALIZING FOR GRAIN REFINEMENT

Fig. 14-12.   Normalizing is used to control grain size. It improves the metal by producing a uniform fine-grained metal.

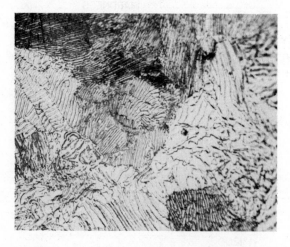

Fig. 14-13.   Normalizing produces a uniform and small grain size throughout the steel.

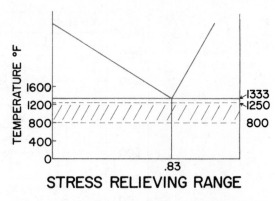

STRESS RELIEVING RANGE

Fig. 14-14.   Stress relieving removes the metal's locked up energy produced by welding or cold working.

critical temperature and cooling the part slowly in still air. Normalizing is used to produce a more nearly uniform grain size throughout forgings and castings, Fig. 14-12. In rough forging, the thin section cools first, producing fine grains. The thick section cools slowly, forming large grains. Where the large and small grains meet, an area of flaws may occur in the metal. After the normalizing process, the grain size will be more uniform throughout the castings or forgings, so there will be less opportunity for flaws, Fig. 14-13.

Stress Relieving

Work may become highly stressed by being cold-worked, Fig. 14-14. Rolling, bending, stretching, machining, or welding may cause a considerable amount of stress. These forces utilize much of the metal's strength and may cause cracking or failure of the metal if they are not removed. The stress is removed by heating the part below the lower transformation temperature for a period of time until the locked-up stresses are relieved. The temperature varies between 800° F. and 1250° F.

Spheroidizing

Medium- and high-carbon steels may be made more machineable by **spheroidizing**, Fig. 14-15. The high-carbon steels have

Normalizing

Normalizing is a heat treating process which controls the grain size throughout the workpiece. The process consists of heating the workpiece 100° F. above its upper

carbides dispersed throughout the metal, and they are hard to machine.

The workpieces are heated slightly below the lower critical temperature for a long period of time. This slow heating converts the hard lamellar or platelike carbides into large spherical particles. They are coalesced (fused) and gathered into large spherical shapes which are surrounded by ferrite or iron and thus become machineable.

In special cases, small high-carbon parts are quickly spheroidized by being heated rapidly above the lower critical temperature and allowed to cool. Then they are heated above the lower critical temperature again. This cycling across the lower critical temperature speeds up the spheroidizing process, Fig. 14-16. This conditioning makes the **hypereutectoid** (high-carbon) steel more easily machined, cold-drawn, and cold-

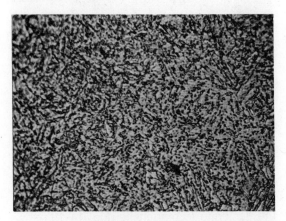

Fig. 14-15. Spheroidizing makes high-carbon steel more machinable than annealing does. The carbon compounds collect into tiny spherical-shaped particles. The remainder of the matrix is soft iron compounds.

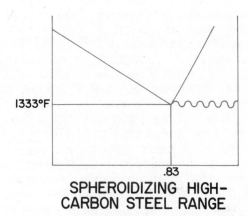

SPHEROIDIZING HIGH-CARBON STEEL RANGE

Fig. 14-16. Small, high-carbon steel parts are spheroidized by cycling the temperature of the parts across the lower critical temperature.

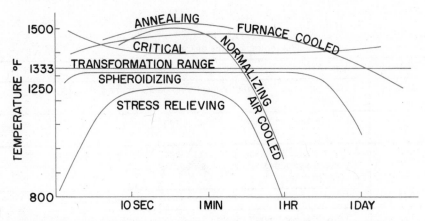

Fig. 14-17. Summary of thermal conditioning processes used to soften, improve the ductility, and to improve the machinability of steels.

formed because of the spheroid carbides instead of the lamellar carbides in the steel structure.

Heat Treating of Steel — Hardening Processes

Steels are hardened so their wearing qualities will be improved. The physical properties of fatigue, torque, or the response to different loads or stresses are largely determined by the heat treatment of steel before or after the manufacture of the part.

Austenitizing and quenching are used to harden steels with a carbon content above .30%. The carbon necessary for the hardening is contained throughout the workpiece. When the material is heated above the critical temperature, austenite is formed. It then is quenched in water or oil so that martensite is formed and hardness is produced in the workpiece, Fig. 14-17A.

If metal has below .30% carbon, carbon or nitrogen are added to its outside surface by the diffusion of the carbon or nitrogen into the surface. The diffused carbon or nitrogen makes it possible for the surface then to be hardened by austenitizing and quenching as in tool steel.

Quench Hardening

The hardening of steel is usually related to the production of martensite. The hard-

Fig. 14-18. Dropping a basket of small parts into the quenching oil hardens the steel.

ness is dependent largely upon two factors: (1) the amount of carbon in the steel that is available and contained in an interstitial (between the atoms) solid solution, and (2) the quenching rate or the rate that the heat can be removed from the metal to transform the austenite to martensite, Fig. 14-18.

Hardness is apparently produced by the distortion of the primary crystal lattice by straining or by invasion of different atoms which distort the metal lattice and block the movements by dislocation. These distortions produce resistance to penetration or move-

FIG. 14-17A.
Hardening Grades of Carbon Steels

AISI Type	Use
C 1030	Bolts and nuts, general flywheel, ring gears, connecting rods, camshafts
C 1040 C 1050	Pipe flanges, industrial shafting, tie rods
C 1060 C 1080	Cutting tools such as lawnmower blades, rock bits, plow points, bumper supports, torsion and sway bars, grind mill balls, and various coil springs
C 1095	Hammers, axes, wrenches, grinding mill balls and bars, springs
C 1137 C 1141 C 1144	Parts that are manufactured on screw machines and automatic lathes such as shafts, splines, gears, electrical tool ammatures, and bearing surfaces.

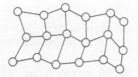 COMPRESSING
OR
STRETCHING

DISTORTION HARDNESS PRODUCED
BY STRAINING (COLD WORK–
COLD ROLLING, DRAWING, FORGING,
EXTRUDING)

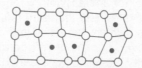 QUENCHING
OR
PRECIPITATION

DISTORTION HARDNESS PRODUCED
BY INTERSTITIAL ATOMS (FOREIGN
ATOMS–CARBIDES–OCCUPY SPACE
BETWEEN BASE METAL ATOMS AND
DISTORT THE CRYSTAL LATTICE)

DISTORTION HARDENS METALS

Fig. 14-19.   Distortion of the metal's crystal lattice produces hardness and strength as a result of blocked metal lattice movements.

ment in the metal, resulting in strength and hardness, Fig. 14-19.

The rapid quenching appears to lock these different atoms in the host metal space lattice in the distorted condition at room temperature. The hardness can be decreased by reheating at a low temperature when the carbon starts to coalesce into the formation of cementite, an ingredient of pearlite.

The quenching media largely determines the rate that the steel may be cooled. However, the size and the shape of the workpiece will determine the volume and flow of the media, as well as the kind of quenching media used. The most commonly used quenching medias are oil, water, and air. The cooling must be rapid enough to stop the transformation of austenite into spheroidite or pearlite. It must not be so drastic that it causes warping or cracking of the workpiece.

Oil. Oil is the most widely used quenching media. It produces deep hardening of the workpiece. The quenching media should not volatilize (pass off in vapor) or boil easily because small bubbles may form on the surface of the metal. These bubbles would keep the media from removing the heat at that point. They will retard the cooling rate of the workpiece and produce soft spots. The workpiece should be moved vigorously in the quench media so that

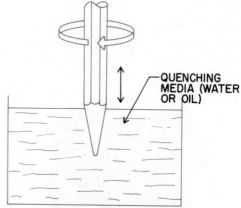

QUENCHING
MEDIA (WATER
OR OIL)

QUENCH HARDENING A COLD CHISEL

Fig. 14-20.   The circular movement of the work breaks up steam or gas bubbles on the surface of the quenched metal. The vertical movement produces a transition zone of hardness between the hot and cold parts of the work.

bubbles will be broken and cool-quenching media will come in contact with the metal. The workpiece should be plunged into the quenching media so that all of the surfaces will have as nearly the same cooling rate as possible. This will help to control warping and cracking, Fig. 14-20.

Water. Water and salts in waters are used as quenching medias and are used on water-hardening steels. Water will produce shallow hardening for steels which require a fast quenching rate. The quench rate is controlled by the water temperature. For very rapid quenching, cooled or refrigerated brine solutions may be used as quenching media. Water quenching must be done with great care, since many tool steels will crack if given a severe quench in water.

Air. Air quenching is possible when tungsten and other alloys are contained in the steel in sufficient amounts. The air temperature produces a cooling rate which causes hardness in the metal. An example of air quenching is the hardening of high-speed steel lathe tools. The air quenching maintains the hardness in the steel up to a red heat.

Surface Hardening

There are a number of processes that harden the surface of the steel and leave a soft, shock-resistant interior. These processes make it possible to use less expensive materials and still have performance of the workpiece at a very high level. The surface-hardening process may be accomplished by adding carbon to the surface of a low-carbon steel. If the steel has enough carbon in it to harden, the surface may be hardened quickly by applying heat to the surface, raising it above the critical temperature, and quenching. A hardened skin is formed on the surface of the workpiece.

Flame Hardening. In flame hardening, the heat is supplied by an oxyacetylene flame from a burner which is shaped to heat only the area to be hardened. The hardening is dependent upon the carbon content of the metal and the metal's critical temperature. The heating is followed by a jet of cold water which quenches the area to be hardened, Figs. 14-21 and 14-22. A hard surface up to about 1/8″ thick may be obtained by this process. The burners are designed in various shapes, ranging from rings to contoured or flat so that different shaped parts may be hardened. Typical work for flame hardening includes gear teeth, sprockets,

Fig. 14-21.   In flame hardening, the surface of the metal is heated above the critical temperature of the metal and is then water-quenched.

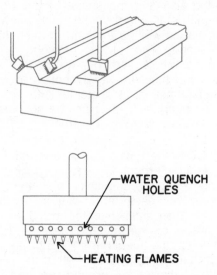

Fig. 14-22.   Burners may be designed in many shapes: into rings, flat, or contoured to fit the workpiece. They produce the flame-hardening heat.

sliding ways of machine tools, bearing surfaces, cams, rollers, pins, and shafts.

Pack Carburizing. Manufactured steel products (such as gears, firearm parts, roller bearings, and conveyor chains) which require a hard finished surface may be made from low-carbon steel. These products are heat treated to add carbon to the surface of the steel. In pack carburizing, the parts are packed into a container with a carbon material such as charred peach pits or coke. The steel box is closed, and the carbon, parts, and box are heated above the lower critical temperature and held for a period of time. The longer the soaking time, the deeper will be the hard case on the work, Fig. 14-23. At these temperatures, the carbon forms carbon monoxide and the gas transfers the carbon to the surface of the hot steel. Steel above the lower critical temperature has an affinity for carbon, and it absorbs the carbon from the carbon monoxide, thus converting it to carbon dioxide. The carbon dioxide in turn moves back to the carbon (charred peach pits) and picks up another carbon atom, converting the carbon dioxide to carbon monoxide, Fig. 14-23A.

In this manner the carbon is transferred and diffused into a thin case over the surface of the workpiece, Fig. 14-24. When the metal is quenched in oil or water, a fine-grained, hard case is produced around the surface of the work. A hardness thickness of 0.003" to 0.050" may be produced, depending on the time and the temperature, Fig. 14-25.

Fig. 14-24. Carburizing, or case hardening, produces a hard wearing surface with a soft, strong interior, as shown in this roller bearing.

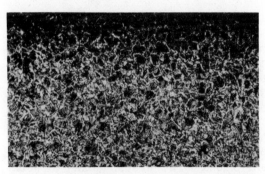

Fig. 14-23. Microstructure of case-hardened steel surface (magnification 100X).

FIG. 14-23A.
Carburizing Grades of Carbon Steel

AISI Type	Use/Carburized Parts
C 1015 C 1020	Case hardened parts for cam shafts, bearings, agricultural parts — wherever resistance is important.
C 1022	Workpieces similar to C 1015 and C 1020, but yielding higher strengths.
C 1117 C 1118	Used for parts that require machine-ability and case hardening. Screw machine manufactured parts, piston pins, gears, bearings.

Fig. 14-25. Typical case-hardened machine parts.

Fig. 14-26.   Gas carburizing is carried out by converting natural gas or other hydrocarbons into an atmosphere that will diffuse into the surface of the metal. It produces a hardened surface when quenched.

Fig. 14-28.   Liquid carburizing is performed by dipping the parts into a molten salt bath.

Fig. 14-27.   Heat treating and hardening furnace.

Gas Carburizing. Gas carburizing may be carried out by using other carburizing materials. Converted natural gas, alcohol, or other hydrocarbons may be placed in the chamber which holds the parts, and the carbon will be transferred and diffused into the metal. A hardened shell is produced upon quenching, Fig. 14-26.

Nitriding. Nitriding is a gaseous case-hardening process. Nitrogen is used instead of carbon to cause the hardening. The parts are placed in a furnace, heated to about 950° F., and exposed to ammonia, Fig. 14-27. The ammonia gas dissociates into nitrogen and hydrogen, then the nitrogen diffuses into the metal's surface to form iron nitrides. The nitride case is very hard and thin.

Nitriding has many advantages. Because of the low temperature involved and the absence of quenching, very little distortion appears. The treated surfaces are very hard, although special areas may be kept soft if necessary by masking sections. The gas is kept from reaching the steel, and soft spots will remain. This is important if drilling or machining must be performed after the parts are hardened. Hardened parts are frequently machined by being ground. Nitriding is an economical hardening process for large volumes of small parts such as gun parts, wrist pins, steel molds, gears, rollers, cams, and pins.

Liquid Carburizing. Liquid carburizing is a process in which the part to be hardened is submerged into a salt bath of cyanide compounds at a temperature above the lower critical temperature of the metal to be hardened, Fig. 14-28. Sodium cyanide salt or other salts are melted, and the workpieces are dipped into the liquid. The carbon from the cyanide molecule, as well as some of the nitrogen from the cyanide, is diffused into the steel. After a period of time, the work is quenched in oil, water, or brine. The nitrogen increases the hardness of the carburized case. This process is quick and produces a hard, thin case on small parts. Sodium cyanide can produce almost instant death, thus this process must be done by professional heat treaters.

Induction Hardening. Induction hardening is a process in which an alternating electrical current at high frequencies is used to generate heat in the workpiece surface by electromagnetic induction, Fig. 14-29. A heavy coil is shaped around the surface to be hardened, and a high-frequency current is applied. The workpiece has current induced in its surface. The resistance offered by the workpiece to the flow of the induced current produces heat. The heat is proportional to the electrical resistance of the workpiece and to the square of the current flowing. The work is heated above its critical temperature and is then water-quenched.

The high frequencies necessary for this heating vary from 1000 to 500,000 cycles per second. The lower frequencies will usually deliver more power and will produce a deeper heating, thus a thicker hard case is produced after quenching. The high frequencies are used for thin hard cases on the workpieces.

Induction hardening is used on steels with a 0.4% to 0.5% carbon content and on cast irons containing 4% to 8% carbon. Typical

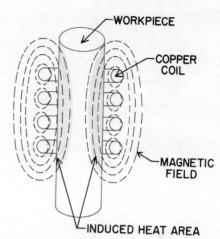

Fig. 14-29. Induction hardening produces the heat in the part by inducing a high-frequency electrical current into its surface. The resistance to the flow of the current causes the metal to heat.

Fig. 14-30. An induction heating coil is heating the outside surfaces of the castings. When the castings reach their critical temperature, they will be removed from the coil and quenched. A hard surface will be the result.

work for induction hardening are camshafts, gears, heavy crank shafts, hubs, bearing races, and many other parts, Fig. 14-30.

Precipitation Hardening

Ultrahigh-strength steels are produced by an age-hardening heat treatment. The metal

MARAGING STEEL
IN ANNEALED CODITION

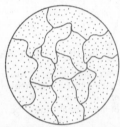

MARAGING STEEL AFTER
PRECIPITATION HARDENING

Fig. 14-31.   Heat treating maraging steels may cause the precipitation of intermetallic compounds or the growing of new compounds from the metals within the alloy.

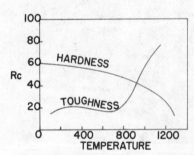

TEMPERATURE FOR HARDNESS
AND TOUGHNESS PLAIN-CARBON
AND LOW ALLOY STEELS (4140)

Fig. 14-32.   Workpieces that require hardness or wear resistance are tempered below 400°F. If toughness is required, the part is tempered above 800° F.

is heated to about 900° F. for one to four hours. Their yield strengths of 100,000 psi are increased to yield strengths of 200,000 to 350,000 psi, depending on the grade of steel used.

Maraging Steels. Maraging steels are high-nickel, low-carbon content steels which contain other alloys of cobalt, molybdenum, titanium, aluminum, and (in some cases) columbium. Strength is added during the heat treatment by the precipitation of intermetallic compounds (growth of new compounds). These new products form between dissimilar metals. The compounds are built out of the alloy metals. The finely dispersed materials form and distort the lattice structure of the parent alloy, thereby increasing its hardness, tensile strength, and yield strength.

Maraging steels are tough and ductile in the annealed condition and are capable of being formed, machined, and welded, Fig. 14-31. After the fabrication of the products, the precipitation hardening (aging treatment) may be carried out with a furnace or a heat treating torch. Because of the low temperatures involved, there is no distortion or dimensional change in the work.

Maraging steels are suited for strong, tough precision parts such as aircraft parts, rocket motors, torsion bars, extrusion rams and dies, and high-strength fasteners.

Tempering or Drawing

Tempering, or drawing, is a heat treating process that reduces the hardness and increases the ductility of carbon steels. When the hardened steel is in a martensitic condition, it is hard, brittle, and highly stressed. If the workpiece were left in this condition, it probably would develop stress cracks which would destroy it. This fully hardened steel is unsatisfactory, but it can be made tough, shock-resistant, and strong by the tempering process.

Tempering is performed by reheating the steel below its lower critical temperature, Fig. 14-32. The carbides of martensite will slowly be decomposed into a softer and more ductile material by coalescence and diffusion. The temperatures used for tempering will vary from 400° - 750° F. Products will receive different physical characteristics by receiving different tempering temper-

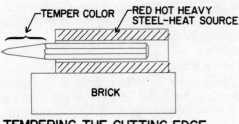

TEMPERING THE CUTTING EDGE OF A COLD CHISEL

Fig. 14-33. A cold chisel being tempered by a hot, heavy block of steel with a hole through it. The hot block supplies the heat for tempering, and the cutting edge is drawn until a purple color is reached.

FIG. 14-33A.
Oxide Color Temperatures

Light blue to gray	above 600°F. (soft)
Dark blue	600°F.
Bright blue	560°F.
Purple	530°F.
Brown dappled with purple	510°F.
Brown	490°F.
Dark straw	470°F.
Light straw	450°F.
Pale yellow	430°F. (very hard)

FIG. 14-33B.
Tempering Colors for Various Carbon Steel Tools

Bright blue (soft)	Screwdrivers
Purple	Center punches, cold chisels
Dark straw	Taps and dies
Light straw	Milling cutters, drills
Pale yellow (very hard)	Cutting tools

atures. Steels may be heated to 300° to 400° F. immediately after the quenching process to remove the snap from the steel or to remove the stress and peak hardness which may cause cracking. Following this, tempering will be done, using precision temperature-controlled drawing furnaces or molten salt baths.

Tempering Colors. When heated, polished carbon tool steel will pass through a series of colors ranging from a pale yellow or lemon color to a light blue. This color change is caused by the thickening of the oxide film on the steel. Small shops may use oxide colors as a tempering heat indicator, Figs. 14-33 and 14-33A.

Different products require different hardness and toughness. The use of the product and its requirements will determine the tempering temperature, Fig. 14-33B.

Austempering

In austempering, the steel is heated above the upper critical temperature and is then quenched in a molten salt bath between 400° - 800° F. Bainite, a hard but tough material, is produced by this method, Fig. 14-34. Austempering eliminates traditional

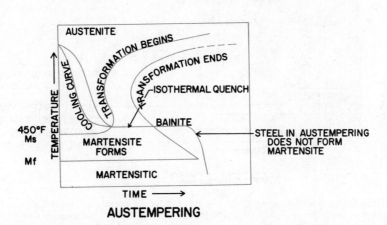

AUSTEMPERING

Fig. 14-34. Austempering produces a bainitic structure used for components which do not break under shock loads. Parts such as seat belt buckles and wrenches make use of these properties.

tempering, because the steel is not cooled below its drawing or tempering temperature during the process, Fig. 14-35. At no time is the steel fully hardened to the martensitic state, and the steel is not reheated after the process. Thus, the risk of quenching cracks and part warpage is greatly reduced. Aus-

tempering creates Rockwell C hardness of 45 to 55, while increasing the steel's ductility and toughness a great amount.

Martempering

Martempering differs from austempering, because the steel is quenched in a liquid salt bath which is held at a temperature slightly above where martensite begins to form. It is held at this temperature until the same heat is throughout the workpiece, Fig. 14-36. Before bainite can form, the steel is removed and air-cooled. Martensite forms without greater stress in the workpiece. The cooled work is later reheated to a temperature until the desired hardness is obtained. The major objective of martempering is to reduce distortion and temper cracks and still have the desired hardness, Fig. 14-37.

Alloy Conditioning Processes

The conditioning of nonferrous metals and alloys is frequently carried out by heat treating. Heat treatment is the application of temperature changes to cause the alloying ingredients to redistribute and produce hardness or softness in the metal.

Fig. 14-35. Austempering wrench bodies. The wrenches are loaded on racks that are carried through three salt baths.

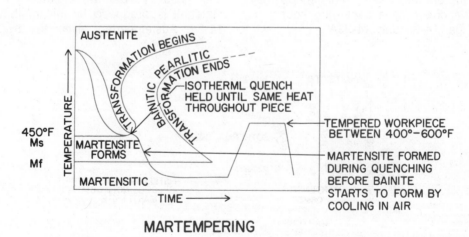

MARTEMPERING

Fig. 14-36.   The rapid cooling in martempering is interrupted just above the martensitic transformation temperature. The work is held in a constant-temperature bath until the temperature is equalized throughout the work. The product is cooled in air to room temperature and tempered to produce the desired hardness. No bainite is formed.

Aluminum-Based Alloys

Pure aluminum is a silver-colored, soft, ductile metal. It can be conditioned by a work-hardening process (rolling, drawing, stretching) to the desired hardness of metal. Distortion of the metal uses or locks the slip planes of the metal and thus produces strength. Aluminum alloys are hardened by the addition of copper, manganese, silicon, magnesium, and zinc. These metals are used in aluminum alloys in very small amounts, but they create great changes in the physical characteristics of the metal.

The aluminum and its alloy metal (for example, copper) is heated to at least 996° F. The copper is taken into solid solution or dissolved by the aluminum. The aluminum and copper make up a supersaturated solution. When the aluminum and copper are rapidly quenched, they stay in solution. This process is called **solution heat treatment.** This soft metal may be formed and worked in this condition. At room temperature a copper compound cannot stay in solution within the aluminum metal, and very small particles of aluminum-copper compounds form or precipitate as new compounds within the metal, Fig. 14-38. These aluminum-copper compounds distort and/or lock the slip planes within the metal, creating hardness and strength throughout the metal.

This precipitation phenomenon of copper in the metal means that aluminum may be conditioned by the following process: (1) heating the metal until the copper goes into solution, (2) quenching the metal, and (3) reheating with precise temperature control to form intermetallic compounds to produce the strength. In addition to the heat treating of the aluminum alloy, the metal may be strain-hardened or cold-worked to produce the desired properties.

Aluminum alloys must be annealed below 1018° F., because this is the melting point of the aluminum-copper eutectic alloy.

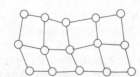

COPPER IN SOLID SOLUTION OF ALUMINUM (SOFT)

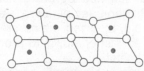

COPPER ALUMINUM COMPOUND PARTICLES PRECIPITATE OUT OF ALUMINUM, DISTORTING, BLOCKING, AND LOCKING GRAIN MOVEMENT (HARD)

Fig. 14-37.   Martempering lawnmower blades by isothermal quenching.

Fig. 14-38.   Aluminum may be strengthened and hardened by the precipitation of aluminum copper compounds within the metal at room temperature.

Copper-Based Alloys

Copper-based alloys are common metals of zinc; tin; nickel; aluminum; bronze; beryllium; and elements of silicon and phosphorus which are mixed with copper to produce many alloys with varying characteristics.

Pure copper and most copper alloys may be hardened by the work-hardening processes of cold-drawing, rolling or stretching. The temper or hardness of sheet brass (copper and zinc) are indicated as quarter hard (1); half hard (2); three-quarter hard (3); hard (4); extra hard (6); spring (8); and extra spring (10). The numbers following the temper — such as spring (8) — indicate that the brass was cold-rolled down 8 Brown and Sharp gage numbers in thickness in order to produce the hardness. The greater the reduction in thickness, the harder the brass product.

Copper alloys which are hardened by precipitation hardening are aluminum bronze, beryllium copper, chromium copper, copper-nickel-silicon, copper-nickel-phosphors, and zirconium copper. These copper alloys are hardened by the forming of intermetallic compounds, the distortion of the crystal structure, and the locking of potential slip planes. Aluminum bronzes with an aluminum content greater than 10% are treated differently. They are treated as steel, because they are hardened by quenching from a hot condition and tempered at a lower required temperature. Most copper alloys are annealed or softened by being heated to the **recrystallization** temperature. At this temperature old, stressed grains are dissolved, then cooled fast or slowly to produce a soft, formable metal.

Most copper alloys will anneal at a temperature of 1450° F. or below to produce a soft, formable workpiece.

Magnesium-Based Alloys

In its pure state, magnesium does not have enough strength for most commercial use. It is alloyed with other metals such as aluminum, manganese, zinc, zirconium, and thorium. The first three metals are most frequently used as magnesium alloys. The most important characteristic provided by magnesium alloys is their lightness. They have an average specific gravity of 1.80. The metal does not possess good corrosion resistance or resistance to shock.

With solution treatment, magnesium alloys are strengthened, and their toughness and shock resistance are improved. Precipitation heat treatment will further improve the alloy's hardness and yield strength.

The strength of most wrought magnesium alloys can be improved by strain hardening, solution heat treatment, and precipitation heat treatment. An alloy of 8.5% aluminum, 0.5% zinc, 0.15% manganese, and 90.85% magnesium which has been forged, rolled, or extruded may be solution heat treated at 750° F., cooled in air, and age-hardened at 350° F. for 16 to 24 hours. This process will produce an alloy part with a tensile strength of about 30,000 psi (about half the strength of mild steel).

Nickel-Based Alloys

Nickel-based alloys are strong, corrosion-resistant metals. They will withstand sea water, dilute sulfuric acid, caustic solutions, brine water, and food products. Nickel-based alloys fall into three main groups: monel, inconel, and hastelloy. These alloys are conditioned by work hardening, solution, and age hardening.

Monel. Monel is composed of 67% nickel, 30% copper, 1.4% iron, 1% manganese, 0.15% carbon, 0.1% silicon and 0.01% sulfur. Monel is primarily a copper-nickel alloy that can be fabricated with all the steel-working techniques such as rolling, drawing, spinning, pressing, and welding, Fig. 14-39.

Inconel. Inconel is an alloy that maintains its strength at high temperatures and resists corrosion. Its strength is developed by a combination of 80% nickel, 14% chromium, and 6% iron. Products which experience severe conditions of heat and corrosion (such as aircraft exhaust manifolds, turbine blades, and thrust reverser parts) are made

Fig. 14-39. Heat exchanger. Monel is used in place of steel where corrosion resistance is of prime importance.

Fig. 14-40. Inconel alloys combine the strength, toughness, and corrosion resistance of nickel with the resistance of high-temperature oxidation of chromium without becoming brittle.

of this rugged material, Fig. 14-40. Inconel is formed and welded by regular hot-working and cold-working processes. Hot-formed inconel has a strength of about 100,000 psi, making it an excellent fabrication material.

Hastelloy. Hastelloy, like most nickel alloys, is really a group of alloys. Hastelloy A is composed of 57% nickel, 20% molybdenum, and 20% iron. This alloy will harden when it is cold-worked to produce the ductility and strength of alloy steel. Hastelloy A is used for pipes, tanks, and transportation containers for hydrochloric, phosphoric, and other acids. It is used extensively in the chemical industry where corrosion is a serious problem.

Nickel-based alloys may be hardened by work hardening, rolling, deep drawing, stretching, solution treating, and precipitation hardening, Fig. 14-41. When magnesium, aluminum, silicon, or titanium is added to alloys, they are capable of precipitation hardening. A small amount of titanium, thoria, or other metals will cause submicroscopic particles to form throughout the

Fig. 14-41. When dispersion-strengthened nickel, titanium, or other metals are added to nickel, submicroscopic particles will form throughout the metal. As a result, its strength will be greatly increased.

crystal structure, thereby greatly increasing the metal's strength and hardness.

The solution treatment for monel typically is to heat it to 1725° F. for 1/2 to 1 hour and then quench it in water. The part will be soft. Age hardening is followed by heating it to 1100° F. and holding it in a furnace for 16 hours. The furnace is cooled to 1000° F. for 6 hours, 900° F. for 8 hours. Then the monel is air-cooled. The hardening usually takes place in sealed boxes inside the furnace to provide the metal protection from sulfur and from other elements in the fuel gases which would make the alloy brittle.

Chromium-Iron-Based Alloys

Chromium-iron alloys are known as stainless steels. These alloys were developed to improve the corrosion resistance of steel alloys. Presently there are more than 30 grades of wrought stainless. Stainless steels are made from chromium, carbon, manganese, silicon, and iron. Nickel, molybdenum, or columbium may be added to produce special characteristics.

Fig. 14-42.   Stainless steel jackets cover the insulation, pipes, and vessels of this anhydrous ammonia plant.

Stainless Steels. Stainless steels fall into three groups: (1) austenitic, (2) ferritic, and (3) martensitic.

Austenitic stainless steels (A.I.S.I. 201, 202, 301, 302, 309, 316) are made up of chromium-nickel and manganese and are assigned the 200 and 300 series of numbers for identification by the American Iron and Steel Institute. These alloys contain more than 6% nickel, but not more than 24% chromium and nickel combined. The famous alloy 302 (or 18-8 stainless steel) is generally used for cookware, food-handling equipment, jewelry, furniture, truck trailers, aircraft hose clamps, and high-strength applications in general, Fig. 14-42. The alloy 18-8 is made of 18% chromium, 8% nickel, and the rest steel. The austenitic stainless steels can only be hardened by cold work; they will not harden by heat treatment. The stainless steels may be formed hot or cold, machined, and welded. The material is non-magnetic.

Ferritic stainless steels (A.I.S.I. 405, 430, 446) contain chromium but no nickel. These alloys are designated as the 400 and 500 numbered series. The 400 group contains 11.5% to 30% chromium. Ferritic stainless steels may be hardened by cold work. They cannot be heat treated, because when they are quenched, a low-hardness ferrite is formed, and the hardness is not greatly improved. Ferritic grades of stainless steel are more machineable than austenitic grades and are both hot- and cold-worked. This grade is magnetic. It is used for annealing baskets, nitric acid tanks, decorative trim on cars, many stainless fasteners, and household items.

Martensitic stainless steels (A.I.S.I. 403, 410, 416, 420, 440A, 501, 502) contain 11.5% to 18% chromium and, with two exceptions, no nickel. For general purpose these alloys can be heat treated. These alloys will form martensite when they are quenched and thus will harden. These steels are used for cutlery, surgical instruments, valves, or products for which resistance to heat corrosion and wear are difficult problems. Some compositions of martensitic stainless steels can be hardened by precipitation. They are difficult to weld and tend to increase in brittleness in the weld area. This stainless is used for steam turbine blades and other highly stressed parts. The 440 C alloy will produce the hardest heat-treatable stainless and is used for balls and races of bearings.

Stainless steels are expensive — roughly twice the cost of carbon steels. Therefore, they should only be used where corrosive and special conditions require the material.

Titanium-Based Alloys

Titanium alloys are important to the space and aircraft construction industries because of their strength-to-weight ratio. These alloys are about one half the weight of stainless steel and can deliver the strengths of medium-carbon steel. The heat-treatable types of titanium can produce yield strengths as high as 150,000 psi.

Titanium processing and shaping is expensive because of titanium's attraction for oxygen and nitrogen. Titanium in the molten state requires a vacuum or inert atmosphere furnace to keep the oxygen and nitrogen from making the metal brittle.

Conditioning of titanium alloys includes stress relieving, annealing, solution treating, and aging.

Stress Relieving. Stress relieving is carried out at 1000° F. for two hours to relieve titanium parts after rough machining or forming. Heavy forming is done at 800° F. to 1000° F., while common bending is done cold with double the springback of stainless steels. Stress relieving returns the ductility to the metal.

Annealing. The annealing of titanium alloys is necessary to develop toughness, machineability, and ductility at room temperatures. The process of annealing is done by heating the alloy above the transformation point, slowly cooling it to 1000° F., and then air-cooling it to room temperature.

Annealed titanium and forgings frequently build up a scale of carbide and nitrates that are very abrasive to machining tools. The scale is removed by dipping the part into a weak solution of hydrofluoric acid and nitric acid. The parts can then be machined without damage to the cutting tools.

There are a number of titanium alloys referred to as **alpha weldable, alpha-beta weldable,** and **alpha-beta nonweldable** alloys. The alpha-beta weldable alloys are largely used alloys which can be heat treated by solution treating and aging.

The alloy Ti-6A1-4V is made up of 4% vanadium, 6% aluminum, and the rest titanium. It has tested at 155,000 psi tensile strength in a one-inch square section when it was solution-treated and aged. The alloy is heated to 1400° F. to 1850° F., and the intermetallic compounds in the metal dissolve. The heated Ti-6A1-4V alloy is quenched in ice brine at 30° F. The metal is reheated to 900° F. to 1000° F. for three to five hours for aging. The aging greatly increases the tensile strength and helps to control ductility of the metal.

The alloy Ti-13V-11Cr-3A1 is used in titanium forgings and heavy rolled parts. When it is aged at 900° F. for 8 to 10 hours, strengths of up to 200,000 psi are achieved.

Titanium alloys are used extensively in large modern aircraft for the fabrication of parts, from wing ribs, aircraft skin, and water tanks to jet compressor parts, Fig. 14-43.

Fig. 14-43.   A titanium water tank and parts used in the Boeing 747.

The metal is forged, welded, machined, and sized to produce these strong, lightweight parts.

Mechanical Conditioning

Metalwork that is done cold or below the recrystallization temperature of the metal may be conditioned by elongating the grain structure. This is done by introducing stress into the metal. Metalworking processes that condition metal during manufacture are the cold-working operations of drawing and squeezing. The best examples of these are the drawing of wire, sheets, tubes, and bars, and the squeezing operations of stamping, rolling, coining, and extruding. In all cases, the conditioning deforms the metal structure by elongating the grains. It increases the tensile strength, and improves the surface finish.

Cold Working

Low-carbon steel may be mechanically conditioned by being cold-worked until its physical properties are nearly equal to that of heat-treated high-carbon steel or low-alloy steel. The new strength is caused by the distortion of the grains of metal into fiber flow lines. These fibers are developed by the type and direction of cold work done on the metal, Fig. 14-44. Both ferrous and nonferrous metals respond to this treatment.

Deformation by Slip Planes. Slip takes place in metals along certain crystal planes which are determined by the space lattice of the crystal upon forming. The type of crystal determines the number of slip planes that are available to use in the direction of the load used to form the metal. A crystal which has many slip planes available will deform and take on the new shape rather easily, Fig. 14-45. Crystals with a few planes or planes in the wrong direction are metals which are strong and difficult to form.

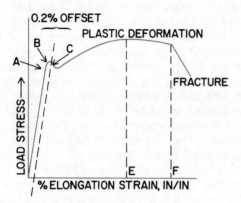

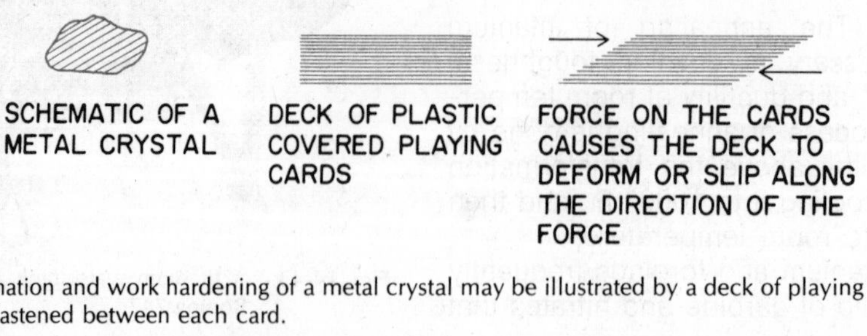

A = ELASTIC LIMIT
B = PROPORTIONAL LIMIT
C = YIELD LIMIT
AF = MAGNITUDE OF THE
      PLASTIC RANGE
E = THE ULTIMATE STRENGTH
    IS THE GREATEST TENSILE
    STRESS THAT A METAL
    CAN BEAR WITHOUT RUPTURE

STRESS, STRAIN CURVE
TO ILLUSTRATE FORMING
AND WORK HARDENING

Fig. 14-44.   Cold working causes new strength in the metal by distorting the grains of the metal fiber flow lines in the metal's slip planes.

SCHEMATIC OF A METAL CRYSTAL

DECK OF PLASTIC COVERED PLAYING CARDS

FORCE ON THE CARDS CAUSES THE DECK TO DEFORM OR SLIP ALONG THE DIRECTION OF THE FORCE

Fig. 14-45.   Deformation and work hardening of a metal crystal may be illustrated by a deck of playing cards with a stamp hinge fastened between each card.

If force is applied to the surfaces of a crystal and the stress is great enough, the crystal will slide along its slip planes. The movement will be one to a few atomic space lattice lengths along the slip plane. The movement is caused by many slip planes sliding because of the stress. When the available slip planes have moved and no more planes are capable of slipping, the deformation decreases and stops. The metal is now strain-hardened or work-hardened, and the tensile strength of the work has increased sharply, Figs. 14-46 through 14-52.

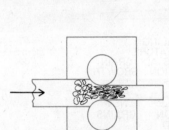

COLD ROLLING

Fig. 14-46. Cold rolling rotates and elongates the grain perpendicular to the forces. This distortion produces hardness and increases the metal's tensile strength.

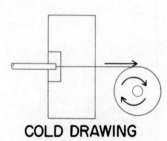

COLD DRAWING

Fig. 14-47. Cold drawing wire or bars elongates the grains. The cross section is reduced, and hardness and strength are increased.

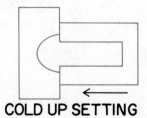

COLD UP SETTING

Fig. 14-48. Cold upsetting thickens the metal, distorting grains and producing strength and hardness.

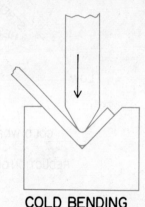

COLD BENDING

Fig. 14-49. Cold bending stretches the metal, distorting the grains and producing hardness and strength as in drawing.

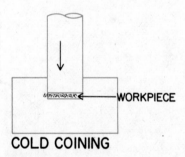

WORKPIECE

COLD COINING

Fig. 14-50. Cold-coined metal takes on the shapes of the punch and the die surfaces. The metal moves very little, but the grains are distorted.

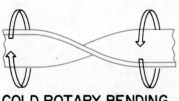

COLD ROTARY BENDING, TWISTING

Fig. 14-51. Cold rotary bending or twisting distorts the grains and causes hardness and strength.

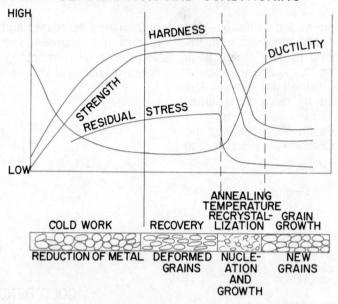

DEFORMATION AND CONDITIONING

HIGH

HARDNESS

DUCTILITY

STRENGTH

RESIDUAL STRESS

LOW

ANNEALING
TEMPERATURE
RECRYSTAL-   GRAIN
COLD WORK   RECOVERY   LIZATION   GROWTH

REDUCTION OF METAL   DEFORMED   NUCLE-   NEW
GRAINS   ATION   GRAINS
AND
GROWTH

Fig. 14-52.   Deformation and conditioning relate to work-hardening strength and residual stress in the metal. Conditioning by annealing provides ductility through the growth of new grains within the metal.

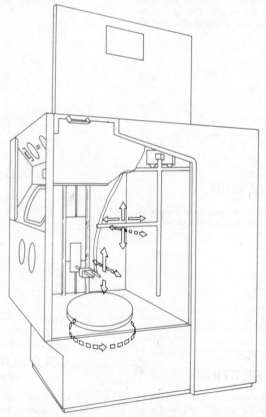

Fig. 14-53.   Shot peening stretches the surface of the metal by placing the surface metal under compression.

Shot Peening

Shot peening is a metal conditioning process which prevents fatigue crack from starting in a surface, Fig. 14-53. Steel shot, cast steel, cut wire, glass beads, or plastic is hurled from a wheel or impeller that rains the shot down on the moving workpieces. The shot strikes the surface of the metal. This stretches the metal, thus producing a thin layer that is slightly larger than that beneath it. This layer is under compression when the peening stops. The center metal resists this compression and is in tension. An example is the keystone and stone around the arch which is in compression. If the side walls were not there, the arch would have to be held up by tension rods across the base of the arch.

Fatigue failures in metal result from tension stress, but compression stress has been introduced into the surface. Thus, shot peening reduces the tension and the possibility of part failure by fatigue cracking. It also reduces the damaging stress concentrations found around fillets, notches, and other irregularities found in forged parts.

Shot peening is concerned with the size and type of shot as well as the speed at which it is thrown. Fatigue starts from sharp scratches or sharp notches, so the shot should not be sharp, nor should the work be rough or damaged. Too much shot peening may damage and weaken the steel part.

Activity

Forge a simple project and heat treat the steel. Use steel with at least 60 points of carbon.

Related Occupations

These occupations are related to metal conditioning:

Metallurgist
Heaters furnace tender
Heat treaters
Furnace / oven operator
Annealer
Normalizer
Hardener
Temperer
Trimmer
Chipper
Shotblaster
Pickler
Tumbler operator
Inspector

Making
Finished Products

Section
4

Principles and practices of assembly are concerned with completing the product in the most efficient manner. To do this, components are examined to determine how they may be simplified, combined with other components, or eliminated. Fastening and assembly are studied to determine which procedure or method will be most efficient. The same assembly operations of inserting, fastening, splicing, coupling, and adjusting assembly fasteners are performed whether the assembly is done manually or by automation.

Metal finishing processes provide beauty; corrosion resistance; a base for other finishes; a lubrication for cold forming and for precision work; and an accurate size of the part. Metal must be cleaned before it is finished. Finishes are applied to products by mechanical finishing, chemical and coat finishing, and metallic finishing.

Principles and Practices of Assembly

Chapter
15

Words You Should Know

Piece part — A one-piece part used to build up other components (washers, levers, frames, boxes, spacers, bushings, etc.).

Standard parts — Hardware such as bolts, screws, studs, pins, rivets, clips, etc.

Components — A group of assembled piece parts made into a functional unit (motors, gear boxes, electrical switches, transformers, etc.).

Subassembly — A group of components assembled into a large unit that may make up a section of the finished functioning unit.

Individual assembly — The product assembled at a bench by a craftsman who may make some of the parts, as well as adjust and test the product.

Lot size assembly — The product is assembled by a group of unskilled people. All add their part until the required number or the lot is completed.

Continuous assembly — The product is assembled continuously with unskilled workers adding components and assemblies as the product moves by.

Assembly Principles

Manufactured products are almost always built up from piece parts, components, and subassemblies into production units such as refrigerators, sports cars, or aircraft, Fig. 15-1. The unit is the end product of the assembly.

Assembly may be the largest single item in the total production cost because of the skills and the amount of labor involved, Fig.

Fig. 15-1.   Boeing 747 final assembly line. Four 747's are in the last stages of completion. One airplane is completed every 4½ working days.

15-2. Because of this, engineers are always looking for a better way to assemble the product and to improve its quality and sales appeal. New methods of fastening have been developed to simplify the products and assembly process. These methods may involve fastening, joining, or bonding processes.

A primary principle utilized in assembly practices is to simplify the unit by redesigning and combining a number of parts, Fig. 15-3. Parts and components may be redesigned to make use of crimped or rolled parts of the component frame for locking, holding, or adjusting end play, Fig. 15-4. Another principle that underlies assembly practices is the **interchangeability of parts.**

Piece parts that are accurately manufactured will fit with any of the other parts to build a component. The accuracy and consistency of the size, location, and finish of holes, tabs, bosses, channels, slots, and surfaces speed up assembly. Any part in the storage bin will go together with any other part to make up a satisfactory component. The interchangeability of parts is very important in assembly, but it is determined by earlier manufacturing processes, Figs. 15-5 and 15-6.

Product Elements

Assemblies made from basic parts are called piece parts. These parts are used to build up the product. The coil winder, punch press, welding fixture, machine tool, or the

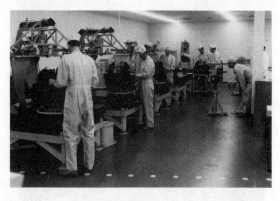

Fig. 15-2. The assembly cost is the largest single item in the total production cost.

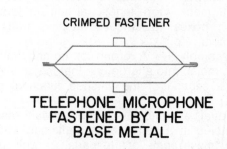

CRIMPED FASTENER

TELEPHONE MICROPHONE FASTENED BY THE BASE METAL

Fig. 15-4. Components are designed with the metal of the parts forming a crimped or rolled fastener. This solves a combination fastening and sealing problem.

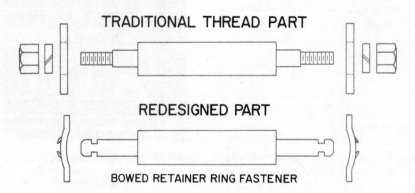

TRADITIONAL THREAD PART

REDESIGNED PART

BOWED RETAINER RING FASTENER

Fig. 15-3. To make assembly more efficient, engineers are redesigning and combining component parts.

foundry produce piece parts for the production system. Many piece parts are gathered together and organized into boxes, carts, bins, or material-handling belts or pallets. They are brought to the workers so they may assemble the parts into a component and/or a **subassembly**, Fig. 15-7. Components are made up of piece parts. Subassemblies are made up of piece parts and/or components. These workers assemble the parts by using fastening components — standard stock hardware such as screws, bolts, studs, pins, rivets, and clips. They may bond the materials together with adhesives or join them by welding, soldering, or brazing.

The piece parts may be made of many different materials. For instance, insulating washers may be made from ceramic, fiber, paper, or plastics. They are combined with parts from many different metals, Fig. 15-8.

Fig. 15-7. Piece parts from the coil winder are gathered together on a rolling cart and brought to the alternator stator. There they are assembled for a 3600-horsepower railroad locomotive.

Fig. 15-5. The principle of interchangeability of parts is made possible by accurate manufacture.

Fig. 15-6. Assembling electric stoves from standard components.

Fig. 15-8. Components are assembled or built up from simple subassemblies and are riveted into an electrical component.

Fig. 15-9. Components are used to assemble subassemblies. These condensor subassemblies will be used to assemble residential central air-conditioning systems.

Fig. 15-11. Tail section subassembly is assembled with air-driven tools and is inspected before shipment to a final assembly plant.

Fig. 15-10. A subassembly — a 747 fuselage panel — is being readied for shipment to the final assembly plant.

Components may be items such as motors, gear boxes, electrical contact points, transformers, switches, chains, clutches, brake wheels, or hydraulic cylinders. Countless items are used in the manufacture of an assembly; examples are stoves, air conditioners, boats, and cars, Fig. 15-9.

When large equipment (airplanes, cars, etc.) is built, the work is divided into many subassemblies. The fuselage is assembled and tested on a subassembly line, while the tail section and wing section are assembled on other subassemblies which may be in another factory, Fig. 15-10. The frame and wheel assemblies are produced for still other subassemblies. The principle of subassembly involves many people and machines in widely separated areas to make components which will fit together accurately to pass the final test and inspection, Fig. 15-11.

Assembly

Individual Assembly. When the production is small and the product is small in size, a craftsman will assemble, adjust, and test the parts at the workbench, Fig. 15-12. In an especially small shop, the craftsman

Fig. 15-12. Individual assembly is completed by a craftsman who assembles, adjusts, and tests. Each instrument is built up from standard parts and components.

Fig. 15-14. Long benches are used for sliding electrical assemblies to the next station, where additional parts are added.

Fig. 15-13. Lot size assembly of submersible water system motors. The components are brought to the assembly area and stored in tote trays or bins.

Fig. 15-15. Continuous subassembly in which the work and parts are brought to the worker and are assembled. An axle assembly is fitted with brake drums, brakes, and cables.

will **make** a number of the parts being assembled. As the number of products to be assembled increase, the craftsman will organize the work bench so that there are special holders for the tools which will be needed, and the small parts will be organized into boxes or bins.

Lot Size Assembly. As the assembly operation increases, the craftsman will place the benches in a line or circle and train unskilled people to assemble one part and place that component into the assembly, Fig. 15-13. The parts will move from bench to bench, Fig. 15-14. They may be moved by hand, by a tote tray, or on a pallet, depending upon the size. Assembly work done by unskilled workers is adjusted and tested to determine if the parts are correctly located, wired, and completed so their performance is satisfactory. An inspector is added to the group to check to see that the assemblies test up to standard.

Continuous Assembly. As the volume of assembly work increases, it is a more efficient process to have the work move past the workers, parts, and tools. Thus, the assembly line developed. Henry Ford was the first to use the concept of the assembly line on a large scale. The assembly line is the final assembly after a great number of subassemblies have been done, Fig. 15-15.

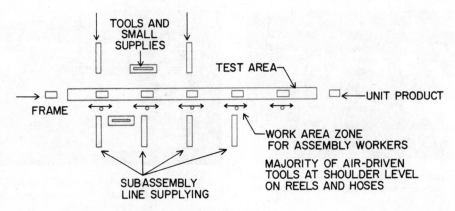

Fig. 15-16. Subassembly lines supply the final assembly line. Components may also be delivered by belts, monorails, pallets, or slides to the workstation.

Fig. 15-17.   Overhead monorails and hoists carry and position the heavy parts on the assembly line. Air-driven tools and adhesives are used.

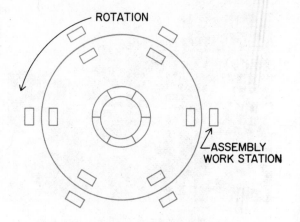

DIAL-TYPE OR CAROUSEL ASSEMBLY

Fig. 15-18.   Assembly which must be performed in a limited space is done on a dial, carousel, or trunnion type of system.

Assembly Line. The belt, slide, monorail, pallets, or wheeled parts are moved or slid down the assembly line while the workers fasten or install the parts to the assembly. To speed up the work, overhead hoists and locating fixtures are used to put the heavy parts into place, Fig. 15-16. Parts are fastened with air-driven tools which tighten one or many fasteners at one time. Adhesives are piped in lines under pressure so that the adhesive may be jetted into place on the product, Fig. 15-17. At the end of the assembly line, a run and test is performed to make sure the work is satisfactory.

The production line uses a large amount of space. Smaller assemblies or subassemblies may be organized around a dial assembly system, Fig. 15-18. The dial organization is like a telephone dial. The assembly table is built on a rotary base and rotates or is indexed to the various work stations. When more than 10 to 15 stations are needed for assembly, the "racetrack" shape is used, Fig. 15-19. This assembly system increases the number of work stations and is

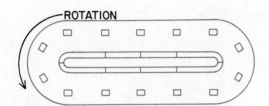

RACE TRACK TYPE ASSEMBLY

Fig. 15-19. When a larger number of assembly stations are needed, a race track configuration may be used.

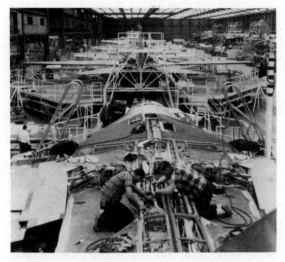

Fig. 15-20. Large, heavy, or complicated assemblies are frequently rolled along an assembly line from station to station, where the tools and parts are located.

Fig. 15-21. The basic assembly operations are inserting, fastening, splicing, coupling, and adjusting. Large numbers of people are employed in this type of work.

advantageous for small components and assemblies.

When the assembly is large, heavy, or complicated, the straight line assembly line is most frequently used. Airplanes are usually assembled by moving the aircraft from station to station down a long assembly line, Fig. 15-20.

Assembly Operations

The operations performed during assembly are the same for hand assembly as for automated assembly: inserting, fastening, splicing, coupling, and adjusting. These operations are used to put together piece parts or components for a fabricated product. Great numbers of people are employed in the assembly operations, Fig. 15-21. Hundreds of thousands of products, from minute electronic products to huge aircraft, are assembled on production lines.

Assembly Fastening Methods

Parts, components, or subassemblies are inserted into frames, cases, or blocks to position and locate the functioning part. Once the part or component is in place, the part is held in place by fasteners. The fasteners come from families of fasteners such as threaded or screw fasteners, rivets, welding, soldering and brazing, adhesives, friction-held interlocking, staking, and gripping fasteners.

In addition, assembly components frequently have to be "connected up." Such operations as splicing or connecting wires, cables, control rods; or coupling of shafts, pipes or tubing, and hoses are done this way.

Fig. 15-21A.
Assembly Fastening Methods

Threaded Fasteners		Rivet Fasteners
Machine bolt and nut		Solid
Machine screws		Tubular
Studs		One-side (blind fasteners)
Threaded pipe, bar		Two-part
Set screws		
Thread-forming tapping screws		Adhesive Fasteners
Thread-cutting tapping screws		Thermoplastic
Sheet metal nuts		Thermosetting
Clinching nuts		Elastomere
Inserts (internally and externally threaded)		
Thumb and wing screws		Interlocking Fasteners
Eye and V bolts		Formed sheet metal
Clevis		Seams and locks
Locking fasteners		Twisted tabs
		Crimping
Welded Fasteners		Snap-on caps and rings
Welded screws and studs		Wire twisting
Resistance-welded		
Arc-welded		Miscellaneous
Capacitor discharge		Staking
Ultrasonic-welded		Staples
		Wiring
Friction-Held Fasteners		Coupling
Nails	Press and shrink fits	
Dowels	Taper pins	
Clamps	Roll pins	
Straps	Spring clips	
Keys	Cotter keys	

Fig. 15-22. Screw-threaded fasteners are used to clamp two parts together. To speed up assembly, air tools are supplied to the workers.

Assembly Fasteners

The three most frequently used methods of fastening assemblies are the use of **threaded fasteners** (screws, bolts, and studs), **riveting,** and **welding.** Other methods of fastening such as the use of adhe-

sives is a fast-growing method of fastening and will continue to be so because of its ease of automation and its ability to bond, seal, and insulate the assembly.

Threaded Fasteners. Screw-threaded fasteners are widely used to clamp two parts together by using a helical inclined plane to provide the pressure, Fig. 15-22. Different thread shapes and sizes are used. The two most commonly used threads are the **national coarse** and the **national fine** series of threads. The national course threads are most often used on metal with large metal grain size or large diameter stock. National fine series of threads are used on strong, fine-grained materials and on smaller sized stock where the depth of thread would weaken the cross-sectional strength of the bolt or fastener.

The shapes of the bolt heads are designed to give the best results on the materials they are clamping or for the function that the head will perform.

Threaded fasteners are provided in many qualities, sizes, and shapes. **Machine bolts** are used to assemble and fasten parts which do not require close tolerances. **Cap screws** are used in machine manufacture where closer tolerances are needed, Fig. 15-23. The clamping surfaces are finished and

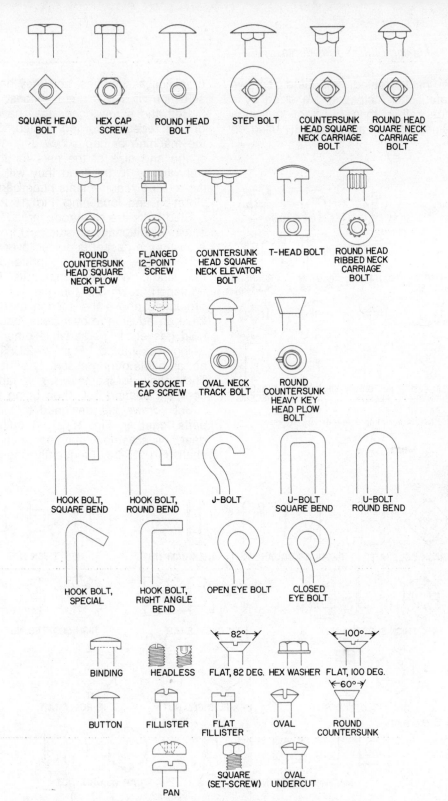

Fig. 15-23. Threaded fastener-head shapes. Threaded fasteners fall into large groups of bolts, cap screws and machine screws. Machine screws are used when small precision assemblies are required.

may be heat treated. **Machine screws** (which are small fasteners) are sized by a wire gage rather than by fractional inch measurements. They are frequently used on precision assemblies, and they have a wide selection of screw head shapes.

Bolt and screw heads may have different driving recesses forged, drilled, or cut into the machine or cap screw, Fig. 15-24. The shape and size of the nuts for the screws will vary as to the load they will carry and the locking requirements necessary to resist vibration and loosening, Fig. 15-25.

Stud bolts are steel rods threaded on both ends. Traditionally a stud is threaded into a casting, a gasket is placed over the studs, and another casting is slipped down over the studs. Washers and nuts would be tightened down on the part of the studs projecting above the second casting. Studs have an advantage over bolts, because they hold gaskets in place while other parts are being assembled, Fig. 15-26. They also act as pilots during assembly of heavy work. They allow for some error in tapping and act as a floating nut when being tightened.

Set screws are designed to lock mating parts together, Fig. 15-27. A pulley, collar, gear, or assembly may be anchored to a shaft, Fig. 15-28. Frequently one set screw

HEX SOCKET	FLUTED SOCKET	SLOTTED	CLUTCH RECESS	FREARSON RECESS
SLOTTED SPANNER	DRILLED SPANNER	POZIDRIV RECESS	ONE-WAY	PHILLIPS RECESS

Fig. 15-24.   Machine or cap screw driving recesses.

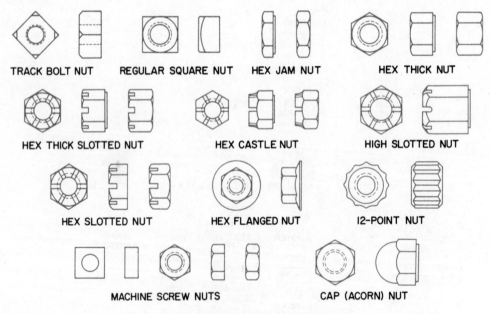

TRACK BOLT NUT   REGULAR SQUARE NUT   HEX JAM NUT   HEX THICK NUT

HEX THICK SLOTTED NUT   HEX CASTLE NUT   HIGH SLOTTED NUT

HEX SLOTTED NUT   HEX FLANGED NUT   12-POINT NUT

MACHINE SCREW NUTS   CAP (ACORN) NUT

Fig. 15-25.   Threated fastener nut shapes.

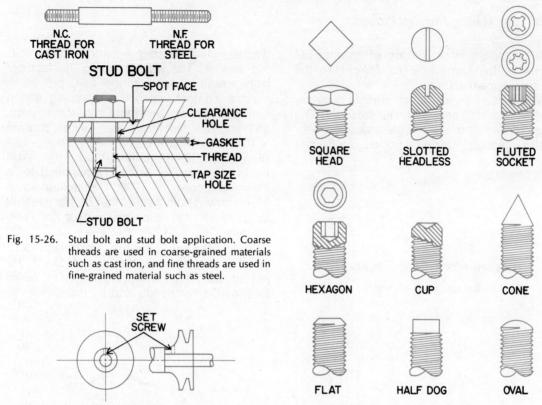

N.C. THREAD FOR CAST IRON

N.F. THREAD FOR STEEL

STUD BOLT

SPOT FACE

CLEARANCE HOLE

GASKET

THREAD

TAP SIZE HOLE

STUD BOLT

Fig. 15-26. Stud bolt and stud bolt application. Coarse threads are used in coarse-grained materials such as cast iron, and fine threads are used in fine-grained material such as steel.

SQUARE HEAD

SLOTTED HEADLESS

FLUTED SOCKET

HEXAGON

CUP

CONE

FLAT

HALF DOG

OVAL

SET SCREW

Fig. 15-28. Application of a set screw anchoring a pulley to a shaft.

Fig. 15-27. Set screw heads and points.

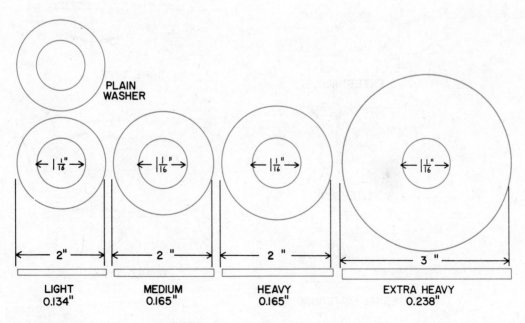

PLAIN WASHER

$1\frac{1}{16}"$

$1\frac{1}{16}"$

$1\frac{1}{16}"$

$1\frac{1}{16}"$

2"

2"

2"

3"

LIGHT 0.134"

MEDIUM 0.165"

HEAVY 0.165"

EXTRA HEAVY 0.238"

Fig. 15-29. Plain washers are available in four series for common sized bolts. These are the sizes for a 1'' bolt.

will be locked down on top of another set screw in the same hole to make sure the screws remain tight.

Washers. For the most reliable service, screw fasteners are used with washers. The common washers are the plain washer (flat), helical spring washer (lock), tooth-lock washer (star), finisher washer, and the fairing washer. The washers keep the fastener from working loose during use, provide a surface against which the clamping action can take place, distribute the thrust load, and improve the finish area of the screws.

Plain washers are supplied in four series of washers: light, medium, heavy, and extra heavy. The washer sizes are supplied for a machine screw of 0.060″ diameter to a 3″-diameter bolt. When the proper washer is used, it will slip freely over the bolt threads.

Helical spring washers are designed to keep a pressure on the underside of the nut or bolt. Helical spring washers resist loosening by vibration and by the shrinkage

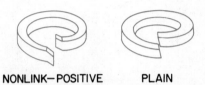

NONLINK−POSITIVE   PLAIN

Fig. 15-30. Helical spring lock washers.

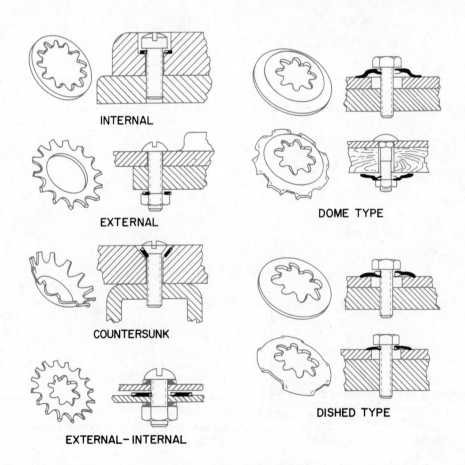

INTERNAL

EXTERNAL

COUNTERSUNK

EXTERNAL−INTERNAL

DOME TYPE

DISHED TYPE

Fig. 15-31. Tooth lock washers.

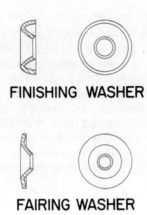

FINISHING WASHER

FAIRING WASHER

Fig. 15-32.   Finishing and fairing washers.

in parts and bolts due to slight temperature changes. Under normal conditions, tight bolts may work loose when the temperature drops.

Helical spring washers are also supplied in a series of four sizes: light, medium, heavy and extra heavy. Left-handed helical spring washers as well as right-handed washers are available, Fig. 15-30. The left-handed lock washer has a split angle which is the reverse of the right-handed washer.

Tooth-lock washers supply pressure and hard-edged teeth that cut into the metal and into the bolt to lock the fastener, Fig. 15-31.

The **finish washer** provides a countersunk hole for the fastener so machining of the base material is not necessary. It provides for clamping pressure away from the sides of the hole. With the use of oval screw heads and finish washers, a very decorative fastener is obtained. Finish washers are frequently used to attach fabric, plastic, and light metals to a base structure, Fig. 15-32.

A **fairing washer** is used to improve the aerodynamic flow of air over a fastener. This fastener is used on aircraft to spread the holding pressure over a large area and to keep the stress away from the screw hole on thin aluminum aircraft skins.

Self-Tapping Screws. Self-tapping screws are used to fasten sheet metal, plastics, and casting, Fig. 15-33. These screws are sup-

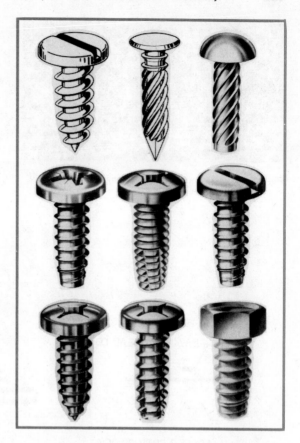

Fig. 15-33. Types of self-tapping screws widely used in industry to fasten sheet metal, plastics, and castings.

plied in a number of head and body styles. They are a quick fastener to install, and they save assembly time.

Single-Thread Engaging Fasteners. Single-thread fasteners take many forms and are designed for rapid or automatic assembly, Figs. 15-34 and 15-35 (page 462). They are formed by stamping and may be shaped or bent to hold and place tension in almost any position.

Caged and Clinched Nuts. A caged nut is a standard square nut that has a spring steel retainer bent in such a fashion that the nut is held from turning, but is allowed to float within the cage so that any misalign-

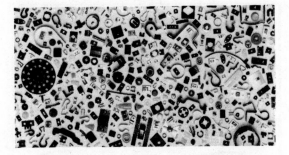

Fig. 15-34. Clip and single-thread engineered fasteners.

ment may be adjusted before the bolt tightens the fastener together, Fig. 15-36.

Clinched nuts are applied to sheet metal to provide a permanently fastened multiple thread fastener, Fig. 15-37. The nut is inserted into a punched or drilled hole, and the nut is clinched by peening, staking, or expanding. Clinched nuts are available in many forms and types and are very useful as blind fasteners or fasteners which may be reached from only one side of the workpiece.

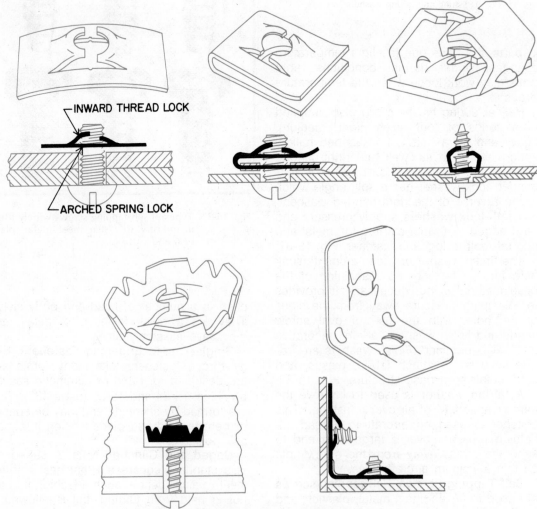

Fig. 15-35. Single-thread fasteners designed for rapid assembly.

Rivets. Riveting has many advantages in the assembly process. The primary one is the low cost for a permanent fastener that is rapidly put into place, Fig. 15-38. Rivets can fasten different kinds of materials together. Materials such as fabric, wood, metal, and plastic of various thicknesses may be fas-

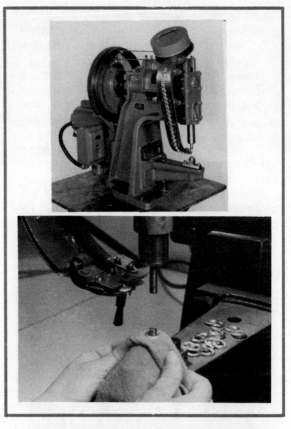

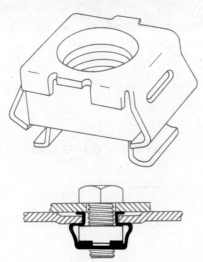

Fig. 15-36. Caged-nut retainer. This type of fastener is used when sheet metal panels are assembled with other components.

Fig. 15-38. Eyeletting machine performs a riveting process that fastens different materials and also provides a reinforced hole in the material for later fastening.

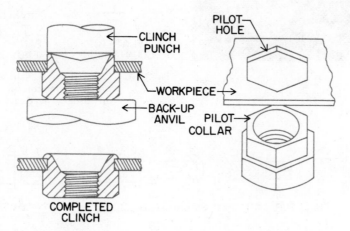

Fig. 15-37. Clinch nuts provide a permanent fastener for sheet metal products. The nut is clinched by expanding the metal back over the metal riveting nut.

tened together permanently with rivets. The rivet may function as an electrical contact, stop, spacer, insert, and pivot shaft, as well as functioning as a fastener.

Rivets may be used to assemble parts in machines at a rate of 1000 parts per hour. Rapid production and clinching are easy with hopper-fed, high-speed riveting machines, Fig. 15-39.

Small rivets are produced in four types: **semitubular, full tubular, bifurcated** (or split), and **compression,** Fig. 15-40.

Small solid rivets are used in large volumes to assemble products which require more strength than the tubular rivets can

Fig. 15-41. Small, solid rivets are driven or headed for greater strength in the assembly of small motors.

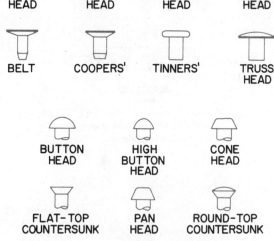

Fig. 15-42. Large rivets ½″ in diameter or larger provide strength for heavy loads. They are usually headed hot.

Fig. 15-39. Assemblies are fastened rapidly by hopper-feed, rivet-clinching machines.

SEMITUBULAR FULL TUBULAR BIFURCATED (SPLIT) COMPRESSION

Fig. 15-40. Rivets may fasten different kinds of materials together.

give, Fig. 15-41. Solid rivets are driven or headed instead of being clinched or squeezed.

Large rivets have heads which supply additional material to provide the strength for very heavy loads, Fig. 15-42. Large rivets are driven or headed hot, and they shrink up so they are very tight upon cooling. The length of the rivet is selected to fit the work and head shape. To supply the material for heading, a rough estimate of the length of the rivet may be obtained by having the rivet project through the metal 1½ times the diameter of the rivet.

Blind rivets are used when it is not possible to head the rivet from the back side, Fig. 15-43. The rivets are expanded from the finish side by a mandrel pull, by a thread, by a drive pin, or by a small charge explosion.

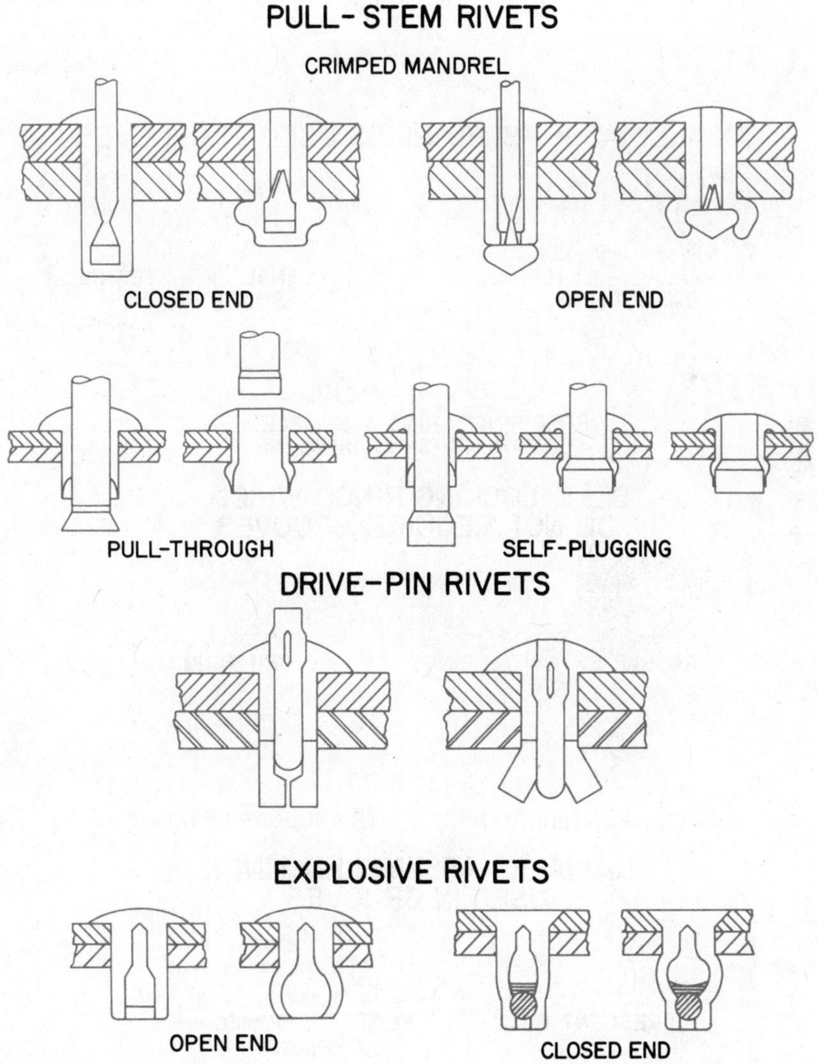

PULL-STEM RIVETS

CRIMPED MANDREL

CLOSED END   OPEN END

PULL-THROUGH   SELF-PLUGGING

DRIVE-PIN RIVETS

EXPLOSIVE RIVETS

OPEN END   CLOSED END

Fig. 15-43.   Blind rivets are assembled from one side of the work. The rivets are upset by pull-stems, explosives, or drive pins.

Retaining Rings. Retaining rings are spring steel stampings that fasten, locate parts, take up end play in an assembly, and provide a shoulder for a spring or other attachment, Fig. 15-44. Retaining rings snap into a groove on a circular piece of work or clinch down on the work without a groove.

Friction-Held and Interlocking Fasteners. Friction-held and interlocking methods of fastening assemblies are widely used. The friction-held fasteners that are in very common use are nails, dowels, clamps, straps, staples, keys, press-and-shrink fits, taper pins, roll pins, and spring clips.

Nails, dowels, clamps, staples, and straps are frequently used on soft materials such as wood, fiber, or paper. The material must be able to yield to receive a nail or other fastener. Materials which have relatively low strength are fastened by increasing the area of contact with the product and by holding the product with a steel strap or band. Materials with more strength and hardness are friction-held together with keys, press-and-

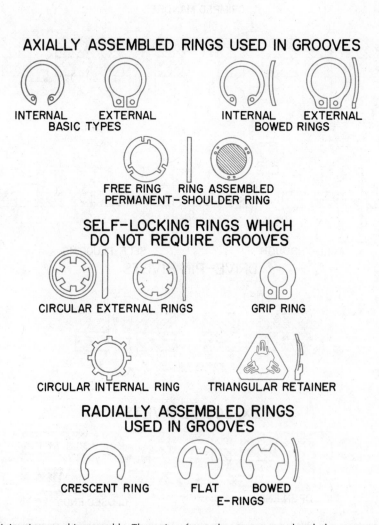

Fig. 15-44. Retaining rings used in assembly. These rings fasten, locate, or control end play on round shafts or parts.

shrink fits, taper pins, roll pins, and spring clips.

Keys are used to prevent movement between rotating parts. A gear may be "keyed" to its shaft so that the motion of the gear may be transmitted to the shaft. Heavy torque loads are transmitted by keys. They come in a number of types: **sunken key, dovetail key, beveled key, round key, saddle key,** and **woodruff key,** Fig. 15-45.

When very heavy loads are transferred from one member to another, splines are used, Fig. 15-46. Splines are multiple keyways and keys, and they are machined out of the mating parts. Besides transmitting torque, splines move on their shaft to compensate for change in shaft lengths or position of gears.

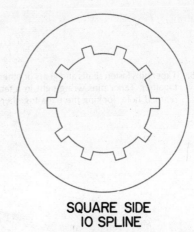

SQUARE SIDE
IO SPLINE

Fig. 15-46.   A spline is used to fasten shafts together and to provide for heavy torque, while still allowing length adjustment.

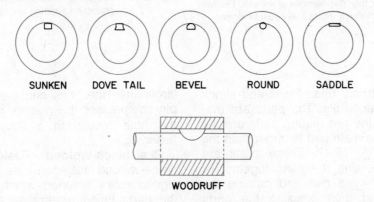

SUNKEN   DOVE TAIL   BEVEL   ROUND   SADDLE

WOODRUFF

Fig. 15-45.   Key fasteners used to transmit torque loads from one part to another. The key is interlocked between the parts.

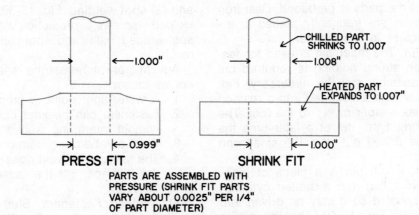

1.000"

0.999"

PRESS FIT

CHILLED PART
SHRINKS TO 1.007

1.008"

HEATED PART
EXPANDS TO 1.007"

1.000"

SHRINK FIT

PARTS ARE ASSEMBLED WITH
PRESSURE (SHRINK FIT PARTS
VARY ABOUT 0.0025" PER 1/4"
OF PART DIAMETER)

Fig. 15-47.   Shrink fits are used to permanently fasten two machined parts. The sizes of the parts are too large to fit together. The outside member is heated and expanded, then the parts are pressed together. When the parts cool, they are permanently fastened.

Fig. 15-48.   Taper pins fasten shafts and gears or other parts together. Taper pins wring tight in a tapering reamed hole, locking the parts together.

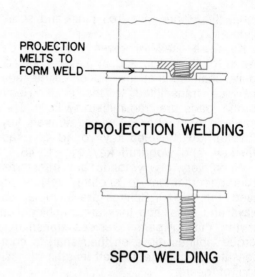

PROJECTION WELDING

SPOT WELDING

Fig. 15-50.   Resistance-welded fasteners are attached by projection welding and spot welding.

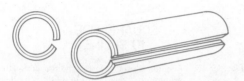

Fig. 15-49.   A roll pin is made of high-carbon steel that may be compressed when driven into a hole smaller than the diameter of the pin. The roll pin provides a strong shake-free fastener.

Press-and-Shrink Fits. Press-and-shrink fits are interference fits. The parts are machined so that the two mating parts are the same size or the male part is larger than the female part, Fig. 15-47. These parts are fastened by pressing the parts together or by cooling the male part and heating the female part and then pressing the parts together. As the one part cools and shrinks, the other heats and expands. A permanent fastening of the parts is produced. Cast iron piston sleeves are frequently placed in industrial engines in this manner.

Taper Pin. The taper pin is used for fastening when strong holding is required but when disassembly may be necessary, Fig. 15-48. Taper pins are rapped into a reamed hole that has a taper of ¼″ to the foot. The pin will wring tight, but at a later time the pin may be driven out from the small end of the pin.

Roll Pin. A roll pin is a piece of high-carbon steel rolled into a slotted cylinder. Its end is beveled so it may be driven into a drilled hole, Fig. 15-49. The roll pin is

larger than the hole, and the driving of the pin compresses it. Pressure on the walls of the hole results in a strong, shake-free fastening.

Resistance-Welded Fasteners. Resistance-welded fasteners are specially designed screw fasteners which are welded to the parts being assembled and become a permanent part of the assembly. The welding is of two types: (1) **projection welding** and (2) **spot welding,** Fig. 15-50. Projection-welded screws, projection-welded nuts, spot-welded nuts, and spot-welded screws result.

Advantages of resistance welded fasteners are these:

1. A water-tight point is produced.
2. Fasteners can be kept close together without interfering with one another.
3. Work can be done from only one side.
4. The fastener or weld does not damage the surface of the assembly, Fig. 15-51.

Arc-Welded Fasteners. Stud welding is used to weld heavy fasteners. The processes

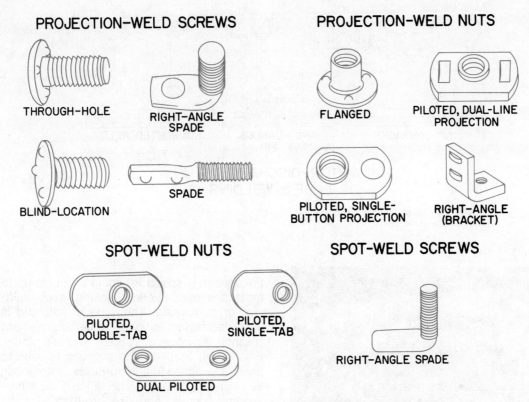

PROJECTION-WELD SCREWS

THROUGH-HOLE

RIGHT-ANGLE SPADE

BLIND-LOCATION

SPADE

PROJECTION-WELD NUTS

FLANGED

PILOTED, DUAL-LINE PROJECTION

PILOTED, SINGLE-BUTTON PROJECTION

RIGHT-ANGLE (BRACKET)

SPOT-WELD NUTS

PILOTED, DOUBLE-TAB

PILOTED, SINGLE-TAB

DUAL PILOTED

SPOT-WELD SCREWS

RIGHT-ANGLE SPADE

Fig. 15-51. Resistance-welded fasteners are manufactured in many forms. Fasteners may be welded in place as the product is being manufactured, thus reducing assembly time.

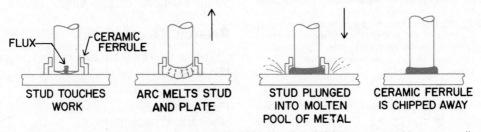

FLUX

CERAMIC FERRULE

STUD TOUCHES WORK

ARC MELTS STUD AND PLATE

STUD PLUNGED INTO MOLTEN POOL OF METAL

CERAMIC FERRULE IS CHIPPED AWAY

Fig. 15-52. Electric-arc stud welding is a method used to weld heavy threaded fasteners to steel beams or walls.

used for stud welding are the **electric-arc** and the **capacitor-discharge** processes.

The electric-arc stud welding passes a current through the stud to the work for a predetermined length of time, Fig. 15-52. While the metal is molten, the stud is plunged into the molten pool and held there until the metal solidifies. A ceramic ferrule holds the flux and liquid metal in place thus producing an excellent welded stud in any position. Capacitor-discharge welding obtains its welding heat from a bank of capacitors. This

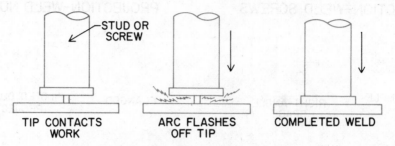

STUD OR SCREW

| TIP CONTACTS WORK | ARC FLASHES OFF TIP | COMPLETED WELD |

CAPACITOR-DISCHARGE STUD OR SCREW WELDING

Fig. 15-53. Capacitor-discharge fastener welding provides a process for welding to thin materials without distortion or discoloration of the metal.

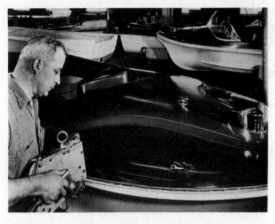

Fig. 15-54. Stapling and stitching is a process in which a metal staple is driven through the metal and is clinched on the reverse side. It permanently fastens the sheet materials together.

causes flashing at the stud tip and produces the welding heat, Fig. 15-53. The advantage of capacitor-discharge welding is that it does not burn through, distort, or discolor thin metals.

Stapling and Stitching. The stapling and stitching process is the driving of a stiff wire clip through the product and then clinching the wire tightly on the opposite side. Heavy-duty stitching tools will increase the assembly speed for some products up to ten times over riveting, bolting, spot welding, or cementing. This type of stitching is used for fastening steel, sheet metal, hard plastics, fiberboard, rubber, and aluminum, Fig. 15-54. Stapling and stitching is used to assemble air conditioning ducts, automobile sun visors, aircraft insulation, aluminum window frames, and like products.

Activity

Design a project, using a number of different methods of fastening.

Related Occupations

These occupations are related to assembly work:

Bench assembler
Floor assembler
Mechanical assembler
Electronic assembler
Hydraulic assembler
Pneumatic assembler
Power plant assembler
Subassembler
Assembly inspector
Final assembler
Fabrication inspector

Metal Finishing Processes

Chapter
16

Words You Should Know

Abrasive blast — A jet of sand and air, grit, or shot used for cleaning metals and other industrial materials.

Soil — Any industrial dirt, grease, or foreign material on the product.

Emulsion cleaner — Oil in a water solvent used for soaking and jetting away soil.

Vapor cleaning — A chlorinated solvent brought to a gaseous state that condenses on cool metal. The vapor and liquid act as a solvent.

Ultrasonic cleaning — High-frequency sound waves in a solvent that create bubbles and scrub the work clean.

Alkaline cleaners — Basic chemical compounds that dissolve oils, waxes, and greases.

Acid cleaners — Sulfuric acid, phosphoric acid, or dihydrogen phosphate used for cleaning light rust off steel.

Salt bath descaling — Molten salts that remove heavy oxides from iron-, nickel-, or copper-based products.

Electrolytic cleaning — An electric current is applied between the part and cleaning solution. The bubbles aid in the breaking up of scale and oxides.

Microinch — A measurement of surface quality which is one-millionth of an inch.

Hone — A bonded abrasive stone used to finish metal surfaces.

Lapping — The use of a free-abrasive flour and a cast iron plate or sleeve to cut the product. The abrasive imbeds in the cast iron and cuts the product when it is moved.

Endless belt grinding — A belt 1″ to 8″ wide with abrasive glued to the cutting surface.

Polishing — Cutting across previous scratches with a fine abrasive grit glued to a canvas wheel.

Buffing — Using a cotton buff charged with compound to put a glossy luster on the metal.

Power brushing — Use of a steel bristle brush to remove burrs, slags, or scale.

Diffusion coatings — Chemicals that enter the surface of the metal and produce a finish.

Conversion coatings — The finish produced by chemically altering the metal's surface.

Oil paint — A paint that dries by solvent evaporating and the oils oxidizing.

Varnish — Made of natural vegetation resins which dry by the evaporation of their solvents.

Alkyd resins — Frequently made from phthalic anhydride and glycerin and other organic compounds.

Acrylic coatings — Coating made from a group of thermoplastic resins.

Epoxy — A class of substances derived by polymerization of certain chemicals.

Vinyl — Any resin formed by polymerization of compounds containing the vinyl group of plastics.

Porcelain enamel — A glass coating applied to a metal surface.

Ceramic coating — Coating materials made from clay or similar materials.

Hot dip coating — Cleaned and fluxed metals are dipped into molten aluminum, zinc, lead, tin, or terne metals.

Vapor coating — Vacuum metallizing where metals are vaporized and moved to the surface of the product.

Electroless plating — A solution of metal that plates onto the metal surface by ion exchange.

Electroplating — An electrical supply moves the metal ions from bulk metal to the product being plated.

Hot spraying — The paint is heated to 160° F. and sprayed. It dries quickly, because little solvent is used.

Metal Finishing Principles

The purpose of the entire manufacturing process is to produce a salable product. The product's finish and its attractiveness to the customer is one of the most important factors which determine whether or not the product will sell, Fig. 16-1. The finish on an article frequently reflects the quality of the work throughout. When the product is competing with other like products and the function or use could be performed equally well by either, the beauty, attractiveness, and quality of the finish may be pivotal in making the sale.

Besides making the product attractive, finishing also protects the surface from rust and corrosion. It aids in cold forming of metal by providing lubrication, and it brings the part to size in precision finishing.

Metal finishing relies on a number of different manufacturing processes to create various surface finishes. Mechanical finishes are produced by abrasion or cutting. Non-precision finishing is done with abrasive wheels, discs or belts, barrel finishing, power brushing, and buffing, Fig. 16-2. Precision finishing is done by grinding, honing, superfinishing, and lapping. Chemical finishes and coatings are made by diffusing materials into the metals on the surface, Fig. 16-3. Chemical conversion finishes are made on the surface of metals by adding other metal compounds to the surface. High-temperature coatings are made when ceramic materials are blown on the surface of the metal, Fig. 16-4. Organic painting finishes are achieved by applying a broad range of paints.

Metal which is to be finished must first be cleaned. Dirt, scale, grease, and chem-

Fig. 16-1.   Metal finishing provides customer appeal and indicates the quality of work throughout the product.

Fig. 16-2.   A swing frame grinder removes large amounts of metal as an early process in metal finishing.

icals gather on the product from the raw materials during the manufacturing processes and storage. A finish that is satisfactory must be attached to solid metal; all other materials must be removed before the finish is applied.

Fig. 16-3. The chemical finish and coating on these sprockets are provided by electroplating.

Fig. 16-4. Measured quantities of oxygen, acetylene, and particles of coating materials are metered into the firing chamber. A spark detonates the mixture. A hot, high-speed gas stream melts the particles to a plastic state, hurls them down the gun barrel, and welds the particles to the workpiece to deposit a metal or ceramic coating.

Mechanical Cleaning

Mechanical cleaning is done by an abrasive blast of sand, shot, or grit which strikes the metal surfaces at high velocity and removes dry dirt, scale, rust, old paint, or oils. The metal surface is cut or pounded until all loose or foreign materials are removed. The blasting has the added advantage of roughening the surface so that paints have a clean and matte surface on which to adhere. Abrasive blasting also removes molding sands from castings and deburrs flash and wire edges from cast or machined parts, Fig. 16-5. When parts are removed from the blasting chamber, they have a matte surface that is clean except for dust.

Chemical Cleaning

Chemical cleaning involves cleaning with solvents, emulsions, alkaline, acid, or salt bath descaling. The cleaning of industrial

Fig. 16-5. Abrasive blasts clean and roughen the metal so that other metal finishing processes may be applied.

products with solvents has been done since products were first manufactured. The cleaning solvent used most is water. Water is inexpensive and it will not burn. It will not damage most products if it is not left to stand, and it is available frequently in large supply. Water is used to clean by dipping, rinsing, spraying, and high-pressure jetting. The high-velocity washing provides a hydraulic action as well as the solvent action to remove soil.

Organic Solvents. Organic solvents such as naptha, kerosene, and other petroleum products dissolve and remove oils, tars, and waxes, but they are expensive and also a fire hazard. They may be used in dipping and spray-cleaning tanks. For mass production, cleaning solvents are frequently used to loosen the soil; other less expensive processes of emulsion cleaning then are used to finish the cleaning.

Emulsion cleaning is an oil-in-water solvent in which a hydrocarbon solvent is dispersed in the water. The two solvents produce fast cleaning. Heavy deposits of oil and dirt are loosened from the metal by the emulsion solvent, which soaks or swabs the soil. The emulsion flushes it away to leave the cleaned part. Emulsion solutions are made of petroleum solvents (kerosene and mineral oil), soaps, and other compounds.

Vapor Cleaning. This is a special modification of a solvent cleaning process. A series of chlorinated solvents are heated until they vaporize in the cleaning tank, Fig. 16-6. Near the top of the tank there are cooling coils. When the hot vapor reaches the cooling zone, a thermal balance is created with hot vapor remaining below the cooling coils. When a cool workpiece is placed in the hot vapor zone, the vapor condenses on the part and dissolves the oil, grease, or wax. In addition, degreasing units may have a hot-liquid hand spray lance to aid in cleaning, or there may be a hot liquid bath into which the workpiece is submerged before it is placed in the vapor cleaning zone.

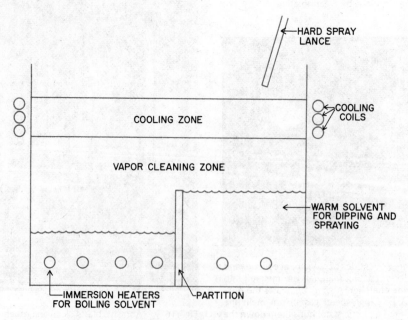

Fig. 16-6. Vapor cleaning is accomplished by placing the part to be cleaned into the vapor zone. The vapor condenses on the part and dissolves the oils or waxes.

Ultrasonic Cleaning. In this process, high-frequency sound waves are used in a solvent to shake soil loose from parts being cleaned, Fig. 16-7. A sonic transducer is mounted in the cleaning solvents, and the vibrations scrub the workpiece by causing millions of tiny vapor bubbles to form and collapse on the surface, in holes, and in corners, thus loosening soil. This method is used on small precision workpieces to degrease and prepare them for further coating processes.

Alkaline Cleaning. Alkaline cleaners such as sodium hydroxide or trisodium phosphate solutions will degrease parts quickly and will also include some inhibitors to keep the chemicals from attaching the metal. Alkaline cleaners are frequently used to remove buffing compounds and light oils and greases on metals that are to be electroplated, conversion-coated, or finished in other like finishing processes. The alkaline cleaning may involve soaking, spraying, or electrolytic methods of cleaning parts. Massive cleaning is automated with conveyors and tanks, and the cleaning is done with hot solutions. The cleaning is followed by a drain stage, a rinse with hot water, and hot air-drying. Alkaline cleaned parts are prepared enough for commercial phosphate priming for painting.

Acid Cleaning. These cleaners are used to remove light oils of machining, buffing compounds, and rust. These cleaners will remove soil and rust from steel parts that have been in transportation between plants or stored under high humidity conditions. The acids used are sulfuric acid, phosphoric acid, or ammonium dihydrogen phosphate. One of these acids is selected and used with a wetting agent and water on a cloth or sponge for a typical wiping cleaning of bar steel.

Large volumes of products are cleaned by dipping, spraying, or wet barrel tumbling, followed by rinsing and drying. The result is clean metal with a light phosphate coating that is desirable as a base for painting.

For cleaning heavy scale on steel forgings, castings, billets, sheet, and tubing, strong

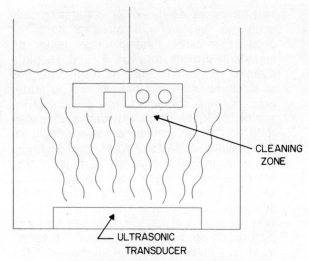

Fig. 16-7.   In ultrasonic cleaning, sonic vibrations are used to scrub the surface of the metal with small vapor bubbles.

Fig. 16-8.   High-temperature salt bath is used to remove the oxide scale from sheet, strip, or wire products before the metal is finished.

pickling solutions of sulfuric and hydrochloric acid are used. Nitric acid is used for stainless steels. After they are treated with the pickling solution, the parts are rinsed with hot water, and the acid is neutralized with a weak solution of sodium hydroxide. The parts are usually air-dried.

Salt Bath Descaling. This process is used for the removal of heavy oxide scale on mass-produced ferrous-, nickel-, or copper-based metals. Other metals may require alternative methods of descaling.

A molten salt bath dip tank is used for batch cleaning, Fig. 16-8. Special equipment and tanks are designed for continuous descaling of sheet, strip, or wire products. The process requires heating large tanks of sodium hydroxide to 700° F. and keeping them at this temperature as large volumes of steel are immersed into the tank. In addition, the bath contains from 1.5% to 2.0% sodium hydride, which provides a chemical reduction of the metallic oxides or scale in the metallic condition. When the oxide is removed, the reaction stops and the parts are removed. They are water-quenched and are thus rinsed. The work may be given an acid dip to remove the final loose scale. This process is expensive, because for long periods of production, the descaling tanks must be kept at temperatures between 700° F. and 1000° F., depending upon which descaling process is used.

Products which are heat treated and tempered below the heat of the descaling bath cannot be processed in this manner, because the heat would draw the hardness from the product.

Electrolytic Cleaning. An electrical current is applied between the cleaning solution and the part in electrolytic cleaning. This process is added to the alkaline, acid, pickling, and salt bath descaling processes described earlier. With the application of current, gas bubbles are released on the surface of the work. These bubbles loosen and remove the foreign material from the work. This process may generate hydrogen gas, so precaution against fire and explosion must be taken.

Mechanical Finishing

Mechanical finishing produces two general groups of finishes: (1) **precision finishes** and (2) **nonprecision finishes.** Precision finishes involve the size of the workpiece as well as the surface finish. Such processes are grinding, honing, superfinishing, and lapping. Nonprecision finishes produce an attractive surface finish, but accuracy is not a prime requirement. Offhand grinding, endless belt grinding, barrel finishing, polishing, buffing, and power brushing are processes used to enrich the beauty of the surface finish.

Precision Finishes. Precision grinding machines produce their accuracy and high-quality finish by applying abrasives as a cutting tool. The abrasives are controlled very accurately by machines which deliver very smooth finishes on hard steels. Each abrasive grain removes many very fine chips. The very fine chips removed ensure that any scratches remaining on the metal are very slight and are all lying in a plane.

The cutting action is related to the type of abrasive used. Either aluminum oxide or silicon carbide is used. On very hard and on very soft materials, silicon carbide is used. Common iron and steel require the use of aluminum oxide. The size of the abrasive grain used in manufacturing the grinding wheel will be a function of the finish desired, Fig. 16-9. The grade of the wheel or the strength of the bond which holds the abrasive grains into the wheel affect the work finish. The softer the grade, the easier the grains will break away from the grinding wheel and expose the sharp new grains. The harder the grade, the stronger the bond, and the longer the grain will remain in the grinding wheel, Fig. 16-10. To produce a mirror finish on a product, a harder wheel is selected so that the grains dull slightly and produce a light polishing cut without leaving the grinding wheel. The structure of the wheel is the "openness" or

Fig. 16-9. Grinding wheels are manufactured in many shapes and materials to provide the proper finish on the workpiece.

analysis of a typical

NORTON GRINDING WHEEL MARKING

ASD100-R100B56⅛✳

ABRASIVE				DEPTH OF DIAMOND SECTION	
DIAMOND				1/16 1/8 1/4	

D = Natural
SD = Manufactured
ASD = Armored (Wet)
ASDC = Armored (Dry)

GRIT SIZE			GRADE	CONCENTRATION	BOND TYPE	BOND MODIFICATION

Numeral to designate special bond modification. Example: Resinoid—56 and 69. This symbol may be sometimes omitted.

NOTE: No grade is shown for Hand Hones.
©1944 and 1969 by Norton Company   ✳ Manufacturer's Identification Symbol

Fig. 16-10. The standard marking system for grinding wheels.

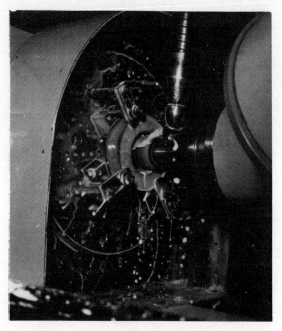

Fig. 16-11.   The finish grinding of carbide with a diamond wheel.

the space between the abrasive grains, and bonding material, thus the number of cutting edges per unit area. The openness provides room for chip clearance and coolant liquids. Fine-quality finishes will be obtained with a close spacing of the abrasive grains. The bond type or the material holding the grains together also affect the quality of the finish. A very high-quality surface finish is obtained with a grinding wheel with a rubber bond.

The surfaces developed by the grinding wheel are directly related to the type of grinding machines. Finish grinding produces finishes from 40 microinches to 0.75 microinches, a finish very close to a polished surface, Fig. 16-11.

Honing is an abrasive process for finishing round holes and other surfaces by the means of a bonded abrasive stone. The stones are mounted in holders so that the abrasives are moved in and out as well as around. A random cross-scratched surface is thus produced. Usually the workpiece is honed from either end so a very round, straight, smooth surface finish is obtained. Cutting fluids are used to remove waste metal and to provide cooling. The abrasive grit is anchored in the stone and thus does not become imbedded in the workpiece. Soft materials such as brass and bronzes, as well as hard steels, can be honed because of this abrasive condition.

Industrial internal hones produce surface finishes from 1.0 to 25 microinches. Very high quality finishes and fits are achieved.

Superfinishing is a precision method of finishing a surface. Superfinishing is similar to honing, because a solid bonded stone is used for the cutting in both. Honing is usually done to bring holes to size with two motions of the cutting grit. Superfinishing is used most extensively for exterior surfaces. Three, five, or more motions of the cutting grit are used. Superfinishing does little to change the size of the work, but it does produce very refined surface finishes.

Lapping is a surfacing process in which a free abrasive flour is rubbed between a lap and the workpiece, Fig. 16-12. The lap and work are moved through two different orbits so that a random movement of abrasive flour is obtained. The lapping surfaces usually have cuts in the surface 90° to one another so that extra abrasive, removed metal, and cutting fluid may be carried away from the lap. The abrasive flour becomes im-

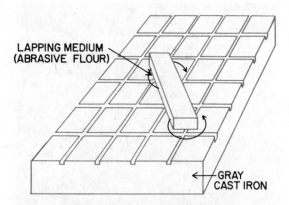

LAPPING MEDIUM
(ABRASIVE FLOUR)

GRAY
CAST IRON

HAND LAPPING CONCEPT:
WORK IS MOVED IN A FIGURE EIGHT MOTION
AND ROTATED AROUND THE PLATE TO
PRODUCE A RANDOM MOVEMENT OVER THE
ENTIRE PLATE'S SURFACE

MACHINE LAPPING CONCEPT:
WORK IS PLACED WITHIN
THE RINGS FOR LAPPING

Fig. 16-12. Lapping produces a random cutting pattern that results in a very accurate matte surface.

Fig. 16-13. Nonprecision grinding done on a stand grinder. Rough castings and forgings are ground to remove excess metal and imperfections in the metal.

Fig. 16-14. Portable grinders are powered by air or electricity and will grind welds or burrs or prepare metal for painting.

bedded in the lap and thus becomes the cutting tool. Laps take an abrasive charge, so they are made of cast iron, brass, copper, or lead. The laps become precision-cutting tools that continuously lap themselves flat and thus produce an accurate, flat surface on the workpiece. Lapping is a slow, expensive process, but it produces the flattest or roundest and most accurate surface finish. A lapped surface will have finishes from 0.2 microinches to 22 microinches.

Nonprecision Finishes. This type of finish makes the product attractive, but it does not produce an accurate surface. **Offhand grinding** is a finishing process done on a stand grinder or a portable grinder to remove burrs, excess metal, and imperfections, Figs. 16-13 and 16-14. This type of grinding improves the surface finish. Most frequently the size of the part is not a function of the grinding. Most welding processes are finished by offhand grinding before they are painted.

Endless belt and disk grinding is used to remove scale, adjust angles, and finish complex contours of metal. Abrasive belts

DETAIL

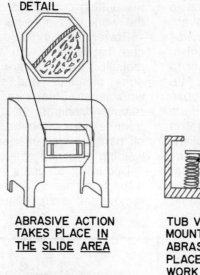

ABRASIVE ACTION
TAKES PLACE <u>IN</u>
<u>THE SLIDE</u> AREA

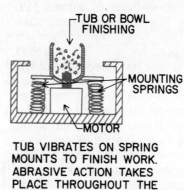

TUB OR BOWL
FINISHING

MOUNTING
SPRINGS

MOTOR

TUB VIBRATES ON SPRING
MOUNTS TO FINISH WORK.
ABRASIVE ACTION TAKES
PLACE THROUGHOUT THE
WORK

Fig. 16-15.   Barrel finishing is caused by the abrasive material rubbing the work.

Fig. 16-16.   Artificial preformed abrasive shapes used in tumbling and vibratory finishers.

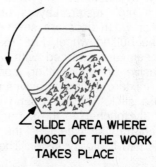

SLIDE AREA WHERE
MOST OF THE WORK
TAKES PLACE

Fig. 16-17.   The parts and abrasive shapes slide down the front of the mass. The rubbing smoothes the workpieces.

may be formed to many shapes by the use of a formed contact wheel that provides the correct curve to the belt. Endless belt grinding is often used to cut down the rough surfaces on complex curved products prior to polishing and buffing for a gloss finish.

Barrel finishing or tumbling is a process in which small parts may be economically deburred, and a high luster produced on the work, Fig. 16-15. The finishing is either done in a rotating barrel-like container with angu-

lar sides or in a vibratory tub filled with water and abrasive. Tumbling may be applied to many materials — metals, plastic, rubber, stone, and others.

The finishing is performed by the sliding of the work and abrasive in the mass to produce high-quality surfaces, Figs. 16-16 and 16-17. Tumbling takes time but only loading and unloading requires workers. A number of tumblers can be maintained at one time.

Polishing is a surface finish operation in which small scratches and tool marks are removed. Polishing does not change the size of the workpiece, but it improves the finish. Polishing is done by gluing abrasive grit to the surface of a wheel made of canvas, felt, cotton, or leather. The abrasive is kept loose in a long box. A freshly glued wheel surface is rolled in the abrasive, and the wheel is charged. After the glue has dried, the polishing wheel is mounted on an arbor. The surface of the wheel is struck with a steel bar so the glue will crack and the surface of the wheel will flex to the shape of the work. Polishing will be done from different positions on the workpiece so the abrasive will cross the old scratches and marks to cut them out. Polishing may also be done by hand or with fine abrasive on endless belts.

Buffing follows the polishing operation in which all the scratches should have been removed. Buffing should impart the highest gloss finish available on the work. Buffing produces a high luster. Layers of cotton, linen, felt, or wool discs are sewn together to make the buffing wheel, Fig. 16-18. The buffing wheels are charged with very fine loose abrasive or with buffing compound, which is a mixture of fine abrasives and wax made into a stick. The compound is lightly but frequently touched to the surface of the buff to recharge the wheel as the work is being buffed to a glossy luster. The wax of the buffing compound is later cleaned from the work with solvents.

Power brushing is done with a wheel or cup that is filled with steel, brass, or non-metallic fibers that are used for deburring, edge blending, and surface finishing on workpieces. Steel wire bristles are used to remove flashings, sand, and sharp edges from castings. Steel brushes will clean paint off parts as well as cleaning scale, rust, or welding slag and spatter.

Deburring of gears and splines is done with short, densely filled wire brushes that remove the sharp edges on the end of the

ADHESIVE

SEWN BUFF

GRIT IN BOX

Fig. 16-18. A canvas or sewn cloth buff with wet adhesive is rolled in a trough of free abrasive to charge the edge of the wheel with abrasive grit.

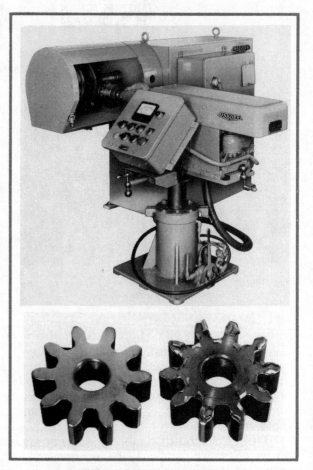

Fig. 16-19. Power brushing will deburr and blend sharp corners on metal parts to reduce stress concentration.

gear tooth and blend in the wired radius without damage to the machined surfaces, Fig. 16-19. Flat metal surfaces of steel, stainless, and brass may be wire-brushed to blend scratches and to produce a uniform textured surface of great beauty.

Chemical Finishes

Chemical finishes and coatings are produced by a chemical reaction between the metal being coated and the materials being applied. Heat may be used to cause one material to diffuse into another. Also, a strong chemical reaction of an acid or base may form a new material on the surface of the metal.

DIFFUSION COATINGS. In a number of hardening processes, elements are diffused into the surface of the metal to create new compounds for hardness and also to provide a surface finish.

Carburizing is a process in which carbon is introduced into the surface of the metal by heat. When the metal is quenched, a hard finish of lustrous black with a play of colors is given to the steel. The hard case provides an excellent finish if the surface is kept oiled and free from moisture.

Chromizing is a diffusion-hardening process in which the work is packed in a powdered chromium compound and heated. A thin case of chrome alloy on the surface results. This improves the work's resistance to corrosion and wear.

Nitriding produces a hard, thin shell on the work that improves the metal's wear resistance. Nitriding is a diffusion finish in which nitrogen combines with the compounds in the metal. It is used for many small parts.

Siliconizing is the diffusion of silicon into steel, white, malleable, and cast irons. Increasing the silicon content on the surface of the metal produces a nongalling bearing surface as well as a surface that is resistant to corrosion and wear. It is useful for dry bearings of earth-moving and agricultural machinery sliding parts.

CONVERSION COATINGS. Conversion coatings are formed by a chemical reaction with the metal on the surface of the product. A very thin metallic film that is bonded very tightly is produced. This type of coating effects practically no change in the size of the workpiece, but it does have other functions. The phosphate and chromate coating provide an excellent base for paints on metal surfaces. All the conversion coatings improve the corrosion resistance of the metals. Phosphate and oxide coatings also aid in the cold forming of metal. The phosphate coatings provide lubrication and prevent metal-to-metal contact with the metal forming dies or rolls. Phosphate and anodic films on metal surfaces prevent wear between surfaces and pressure welding when metal surfaces are under great load.

Phosphate coating may be sprayed, dipped, or painted on metals in an acid phosphate solution, Fig. 16-20. They result in a uniform crystalline structure which

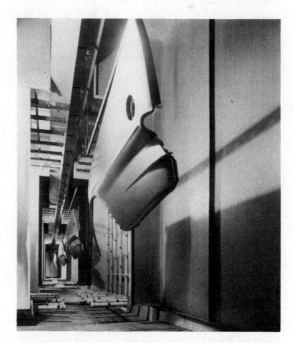

Fig. 16-20. Interior view of stage power washing and phosphatizing machine.

covers the surface of the metal. This chemical reaction with the metal provides a dull surface that is a phosphate salt, and the process is known under trade names of Parkerizing, Granodising, and Bonderizing. Dull black hand tools are finished with this process.

Chromate coatings are frequently applied to aluminum, zinc, cadmium, and magnesium workpieces. The chrome salts combine with the surface metal, and a nonporous surface is the result. This tough, tight coating provides excellent protection from corrosion. The color is natural and nonfading; often the green color is seen in the interior of aircraft structures.

Black oxide coatings are applied to exposed metal parts such as guns, wrenches, metal brackets, and small parts. The parts are boiled in a concentrated solution of sodium hydroxide and different nitrates and nitrites. These are chemically very active and are used on ferrous alloys. The chemical reaction produces a black oxide that has the same luster as the metal before it was oxidized, but it is deep black in color. This finish is washed and oiled to stop the reaction and aid in corrosion resistance.

Anodic coatings are a reverse electroplating process. Aluminum or magnesium alloys are placed in a sulfuric acid solution, and an electrical current is passed through the solution and parts. A small amount of aluminum is removed, leaving an oxidized and porous surface. This oxidized surface provides corrosion resistance and also makes it possible for color to penetrate the metal.

Organic Coatings

Organic coatings cover the surface of metals and adhere to a primer coat or to the roughness of the metal itself. Some materials go onto the workpiece and "dry" by the oxidizing of oils to produce a paint film. Other coatings will set up as the result of the evaporation of their solvents. Lacquer and shellac are this type of coating cure.

The resin coatings have many different substances used to produce their protective coatings. They are cured by a catalyst that causes cross-linking by heat that polymerizes the resin. Alkyd, acrylic, epoxy, vinyl, and other materials are used to produce these polymerization types of curing coatings.

Oil Paint. An oil paint is used on metal or wood for indoor or outdoor use. It is made by mixing pigments with drying oils and a solvent or thinner. As the solvent evaporates, the paint dries. It continues to dry over a longer period, depending on the oxidation of the oil in the paint. Linseed oil and soybean oils dry very slowly.

Harder paint finishes are produced by using tung or castor oils. These oils oxidize rapidly. Oil-based paints provide a flexible finish that will not chip as rapidly as other finishes. Oil paints do not resist chemical attack as well as other coatings and will not produce as high a gloss as other materials. Most paint of this type is applied by a paint brush, but it may be sprayed on wood or metal. Using cheap paint on exterior surfaces is poor economy, because it will peel, chalk, or flake in a short time. The saving of the money for paint is offset by the cost of labor for repainting. Oil paint is not used as frequently as it was in the past, but it is still applied to large areas such as houses, bridges, storage tanks, water towers, industrial facilities, ships, and marine equipment.

Varnish and Enamels. Varnish and enamels are made of natural resins and are products of vegetation, copals, bater, dammar, and shellac. These resins are first dissolved in a solvent and are then applied. The solvents evaporate, and the resins slowly oxidize to produce a glossy, tough finish. Enamels are color-pigmented varnishes which are tougher than clear varnish but less flexible. Varnishes and enamels are slow-drying, so large industrial finish drying rooms are required. This slow drying makes production schedules difficult. For this reason, varnish, enamels, and paints are rapidly being replaced by synthetic plastic resins and elastomers as coatings.

Alkyd Coatings. Alkyd resin coatings are widely used on metal products. These coatings may be air-dried or baked to produce a quick-drying resistant coating. Most **alkyd resins** are made from phthalic anhydride and glycerin or other similar materials. The amount of and type of drying oils used in alkyd coatings produce a dramatic effect on the coating capabilities. The amount of oil is referred to in the terms: short, medium, or long oil alkyds.

Short oil alkyds used with urea or melamine formaldehyde produce a colorfast and quick-baking coating. These coatings are resistant to salt spray and moisture; and they produce a high gloss. Products on which these materials are used are automobile bodies, exterior metal signs, kitchen cabinets, refrigerators, and hospital equipment.

Medium oil alkyds are modified with soya or linseed oil for brushing and spraying, Fig. 16-21. These oxidizing alkyds are fast air-drying coatings that are tough and flexible and have a high gloss. This material resists gasoline, oils, water, salt spray, and alkalies. Medium oil alkyd is used for industrial machinery enamels, automotive painting, and exterior metal painting.

Long oil alkyds are air-drying coatings which are low in cost and provide good protection. These oxidizing alkyds are used in maintenance because of their resistance to water, salt spray, and cleaning materials. Long oil alkyds are excellent for marine use, and they perform better than oil paints and varnishes when they are air-dried.

Alkyd coatings may be modified further by being combined with phenolic, vinyl, silicone, and styrene to improve their water and chemical resistance, flexibility, gloss, color, heat resistance and air-drying time, Fig. 16-22. Alkyd emulsions are of great help in fire hazard areas because they are water-thinned and are safe. The emulsion coatings give protection from outdoor wear and are used in steel and castings in hot areas for rust protection.

Acrylic Coatings. Acrylic coatings are made from a material in the plexiglass and

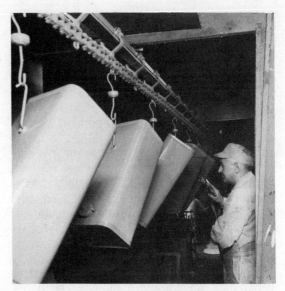

Fig. 16-21.   Alkyd resin coatings with medium oil will resist exposure to gasoline.

Fig. 16-22.   Silicone-alkyd enamels have been designed for bridges and other exterior metal structures. The silicone coating normally lasts 50 percent longer than conventional paint.

lucite family. It is used often on metal as a clear coating or as a pigmented enamel. **Acrylic coatings** produce a clear or metal powder filled coat that is tough and durable indoors and outdoors. Acrylic coatings provide protection for metals such as copper, brass, and aluminum that react quickly to atmospheric fumes and moisture. They seal the surface from chemical reactions.

Acrylic enamels are applied to machinery and equipment that is exposed to fruit juices or fumes from chemical processing. Modified acrylics may be applied to sheet metal in sheet form that later is sheared and rolled into its finished shape, all prepainted. Water-thinned acrylic primers are applied to metal products to reduce the fire hazard in flammable work areas.

Epoxy Coatings. Epoxy coatings have outstanding adhesion, toughness, flexibility, and chemical, electrical, and heat resistance. There are a number of blends with other resins to produce special characteristics.

The epoxy-phenolic coating is an outstanding covering if color is not important. It has high solvent and chemical resistance, as well as flexibility. Thus this coating becomes important for wire coating and varnishes for cans, drums, and tanks. Epoxy-urea has good color retention and is used to coat hospital and laboratory furniture when chemical protection is a problem. The coating may be applied colored or clear for the protection of copper and brass hardware or jewelry.

Vinyl Coatings. Vinyl coatings are weather-resistant, tough, and strong. This resin is odorless, tasteless, and nonpoisonous. Vinyl solution coatings are used in food packing, interiors of beer and soft drink cans, toothpaste tubes, bottle caps, and dairy equipment. Vinyl is applied in other forms to cover metals and to provide color and abrasion resistance.

There are many more organic coating systems not mentioned in this section because the coating industry is a very broad, complex, and changing technology. Just a few of the major resins have been discussed from the hundreds of possible resin combinations available for special coatings.

Nonmetallic Finishes

Porcelain Enamels. Porcelain enamels are glass coatings on metal. The glass is frequently an alkali borosilicate that is heated in a furnace. It melts and covers the metal with a glass coating. This process is used on road signs and industrial products because the surface is smooth, hard, and resistant to weather, acids, bases, and mechanical abrasion.

Porcelain enamels are complex glass systems ground from premetered frits (materials from which glass is made) composed of from 5 to 15 or more chemical compounds, Fig. 16-23. Pigmented and opaque glass coats may be added to the enamel to give it different colors or to cover colors already on the metal.

The metals used are enameling iron, a low-carbon rimmed steel, cold-rolled steel A.I.S.I. 1010, hot-rolled steel plate, and a special enameling steel of low carbon and titanium. Gray cast irons of a specific composition and a dense structure are used for enameled castings. Stainless steel and precious metals may be enameled for decorative products and artwork. Aluminum alloys can be enameled, but the maturing or flowing temperature of the enamel is usually above the annealing temperature of the aluminum alloy and thus may soften the aluminum and reduce the product's strength.

The standard practice for steel enameling is to first apply a **ground coat,** usually a dark enamel which vitrifies (becomes a glass) at

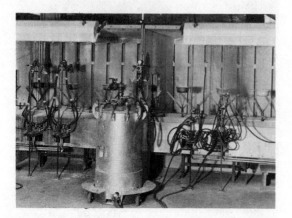

Fig. 16-23. Continuous coating of two-color cookware with porcelain enamel.

a higher temperature than the cover coat. This is followed with cover coats of white, blue, or another color.

The enamel is handled in a water solution called slip. The enamel is applied to the metal by painting, spraying, dipping, or flow coating. Porcelain enamels are fired at different temperatures dependent on the chemical makeup of the enamel, but most enamels applied to steel are fired between 1350° F. to 1600° F. The majority of the firing is done at 1450° F. to 1550° F. with the ground coat fired 25° to 50° higher than the cover coats.

Cast iron and heavy sections of steel are coated by a dry covering process. The casting or steel is heated to about 1700° F. and removed from the furnace. The enamel frit or powder is sprinkled on the hot metal with a sieve. When the enamel is thick enough to cover, the metal is put back into the furnace so the enamel will fire-polish and fuse.

The products produced by enameling include bathtubs, washing machine tops, tubs, sinks, water heater liners, cookware, road signs, exterior signs, architectural panelling, and architectural finishes. Enameling provides metals with one of the most weather-resistant coatings available with little maintenance.

Ceramic Coatings. Ceramic coatings are applied to metals for protection from oxidation and corrosion at high temperatures. Ceramics have higher fusion temperature materials than enamels. They are made from oxides and silicates, but in addition special coatings of borides, carbides, nitrides and cermets are used. These coatings protect the metals in rocket motor nozzles, jet engines, combustion chambers, exhaust manifolds, and heat–treatment heat shields, Fig. 16-24.

Ceramic coatings are used to protect metals which are exposed to high temperatures over long periods of time, Fig. 16-25. These coatings also produce resistance to wear and electrical discharges. They also provide a highly reflective surface for heat. They are used and have been greatly improved by the aerospace and research industries. They have aided in the development of surfaces that can withstand gas flows beyond Mach 10 and the resulting temperatures, Fig. 16-26 (page 486).

Ceramic coatings are applied to metal surfaces in a number of ways. The largest amount of the application is done by spraying or dipping the surfaces with a slurry or slip made of water and the ceramic formula. The parts are first sand-blasted with a chemically clean abrasive. The work is later chemically cleaned with an acid or alkaline cleaning solution and then carefully washed. The ceramic coating is ground and mixed in mills which are lined with a noncontaminating

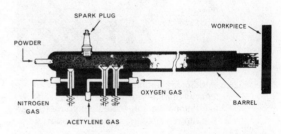

Fig. 16-24. The detonation gun process makes possible the application of extremely hard, wear-resistant, heat-resistant materials.

Fig. 16-25. Remote control operation of the detonation gun applying a ceramic coating.

Fig. 16-26.   A titanium compressor blade used in an aircraft gas turbine engine. The midspan stiffeners are detonation gun-coated with tungsten carbide to resist wear from sliding impact.

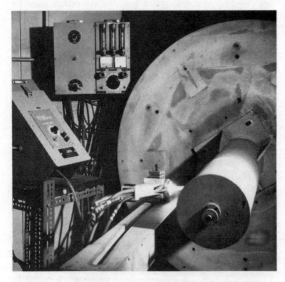

Fig. 16-27.   Plasma spraying is done by heating inert or semi-inert gas passing through the arc. The gas reaches temperatures of 15,000° F. or higher, well above the melting temperature of known materials.

coating, and water is added or taken out of the slurry until the desired specific gravity of the ceramic coating is reached. This slip is applied to the work with a commercial spray gun designed for spraying heavy fluids. Special care is taken at all times to eliminate grease, rust, or dirt from the workpiece and the mixing equipment.

Ceramic coatings are applied by dipping the cleaned product into the slip and placing it on a rack for the excess ceramic coating to drip back into the tank. The parts are carefully dried at a determined speed and temperature that removes all moisture from the surface.

The coated pieces are fired in a contamination-free furnace. The temperature and time of firing will be related to the kind and thickness of the metal necessary to develop good adhesion of the ceramic coating and the substance or metal.

Other methods of applying coatings are troweling, flow coating, flame spraying, plasma arc spraying, and detonation gun flame spraying, Fig. 16-27.

Metallic Finishes

Metal finishes are made by coating metal with other metals by dipping the part in liquid metal or by transferring metal atoms to the surface.

Hot Dip Coatings. Hot dipping of metals into a molten bath of other metals coats the base metal to give it protection from corrosion and heat and to improve its beauty. The metals that are frequently applied as coats are aluminum, zinc, lead, tin, and an alloy of tin and lead called **terne.**

Aluminum is coated on steels and cast iron to produce a surface that can withstand high temperatures and resist corrosion. The aluminum coating resists oxidation from the atmosphere up to 1000° F. The coatings are used on hot piping, furnace parts and fasteners, oil refinery, plumbing, and industrial equipment which is subject to heat.

The dipping is accomplished by either batch or continuous dipping. In continuous dipping, products such as steel strip or wire or wire products are cleaned by oxidizing the surface and removing any soil. The sheet is passed through a reducing atmosphere that removes the oxide coating. The reducing atmosphere also acts as a flux as the sheet or wire products are dipped into the molten aluminum to provide the protective cover.

Batch dipping of steel parts to be coated in aluminum is similar. The parts are cleaned by hot alkaline cleaning followed by a water rinse. They are then abrasive-blasted to remove any scale or loosened soil which in turn is followed by an acid pickling and rinse. Malleable iron parts may be given an additional cleaning in a molten salt bath to remove any carbon dirt. The baskets of parts are dipped into the aluminum to coat all surfaces. The baskets are in constant motion to assure contact of the aluminum with all surfaces. The baskets of parts are then removed, vibrated, shaken, or spun in a centrifuge to remove the excess aluminum. The parts are later removed from the baskets, quenched in water, dried, and inspected. The cleaning of the metal parts is a very important part of successful dip coating with aluminum.

Applying molten **zinc** to the surface of the metal is called galvanizing. The zinc is coated on the steel products to protect the base metal from corrosion. Freeway signs, supports, guards, steel power poles and towers, fencing, hardware, garbage cans, and most steel products that will be exposed to the atmosphere, water, or soil are galvanized.

Galvanizing begins with cleaning the sheet, wire, or workpiece. Steel is pickled with acid to remove scale from the steel. The work is rinsed and scrubbed to remove any remaining particles of metal. It is then fluxed with ammonium chloride, dried, and dipped into the molten zinc. Small work such as bolts, nuts, washers, and other such parts are moved while they are in the molten zinc to produce an even coat of zinc on them. The thickness of the zinc coating is controlled by the speed that the work is drawn from the zinc bath. The excess zinc

runs off the work as it is being withdrawn from the molten zinc. The method of cooling determines the amount and type of iron zinc alloy that is developed on the metal surface. Heavy work will be cooled to prevent the total coat from converting to an iron zinc alloy. When the metal is air-cooled, a common spangled zinc crystal surface on sheet steel is produced. A slow or elevated heat produces a surface in which the zinc has diffused into the metal, creating a zinc alloy surface that is matte and is excellent for painting.

Tin coatings are applied to metal by a dipping process. A tontoxic, nonstaining finish is produced on metals. The metal may be formed by drawing, rolling, and bending without damage to the coating. Tin coatings are used on dairy and other food-handling equipment. Tin coating prepares the surface for later soldering of electronic components, or for improving bonds to other lead- or tin-based alloyed fasteners.

Work to be tin-coated is first cleaned by degreasing to remove any forming oils or waxes and then by pickling to remove any scale or rust. The work is fluxed to aid in the reaction between the steel and the tin so that the tin will hold tightly and evenly to the metal's surface.

Tin coating is a very old process and has been used for all types of food containers and corrosion-resistant surfaces. Today tin is used extensively for solder in the electronic industries and for coating heavy food-handling products such as milk cans, food grinders, and industrial food-processing equipment. Tin is applied electrically in a very thin coat to the outside of tin cans.

Terne coating for steels and copper contains about 80% lead and 20% tin. The coating provides resistance to corrosion for chemicals and gives a surface that is easy to join by soldering. The chemical inactivity of the lead makes terne coating useful for roof gutters, oil filters, gasoline tanks, and cabinet linings that are subject to vapors and liquids of chemicals. Terne coatings with additional tin are referred to as **solder coatings** and are used in the electronics industry for printed circuit boards, coatings of condensers, and capacitor containers.

Vapor Deposit Coating. Vapor coating is also known as **vacuum metallizing.** This process produces a very thin film of metal on the object coated. The film may be as thin as 3 to 16 millionths of an inch. Metal coating may be applied to plastic, glass, metal, paper, and other materials.

The work to be coated is placed in a vacuum chamber with a filament or a resistance heater to evaporate the metal to be deposited. A vacuum is pumped in the chamber. The degree of vacuum needed will be related to the metal being deposited, and the distance the work will be from the vapor source. Quality vacuum coating will require pressures which are quite low. Thus expensive vacuum equipment and pumps are needed.

When the chamber has pumped down to the required pressure, the metal that is to be vaporized is heated by electrical resistance, by an electron beam, or by a laser to evaporate the metal. At these pressures, evaporation of the metal takes place more easily than at atmospheric pressures. The atoms leave the metal and travel in straight lines. Thus when the atoms strike a cool object, they will coat it. Because of the straight travel of the atoms of metal, products have to be carefully positioned or rotated to assure that the coating will reach all surfaces that need the metal deposits.

The process is about one-fourth as costly as plating. Therefore it has grown rapidly as a decorative coating such as on plastic toy cars, horns, and pistols which have the look of highly plated and polished metal. The thin metal coats will not withstand wear, so they are protected by a coat of clear lacquer or plastic. Automotive plastic trim may have metal deposited on the back side so the metal finish shows through and still has the plastic to protect the thin coat.

Other applications of vacuum coatings are in electrical coatings of gold, platinum, copper, silver, and aluminum, Fig. 16-28. These deposits are used to make electrical circuits in miniaturized electronics. The circuit may be masked onto the work during the deposit or etched with acid through the coating later. Metallic compounds of aluminum oxide, cadmium sulfide, silicon oxide, and tantalum oxide and other materials are coated and etched in combination with layers of metal coatings to produce miniature integrated electronic circuits.

Vacuum deposits are applied to another field of technology — optical coatings. Examples are precision mirrors for instruments with the coatings on the front of the mirror for reflection with no distortion of the image and mirrors in the reflectors of automobile headlights. Metal deposits may be controlled so accurately that an engineered percentage of light falling on the surface will pass through the deposit and the remainder of the light will be reflected. This principle is applied in the dividing of one image into two images for optical measuring instruments. Vacuum technology has added new instruments and products to manufacturing processes.

Chemical Immersion Coating. Chemical immersion coating is done by applying metals to the surface of other metals without the use of an electrical current. Metals such

Fig. 16-28.   The evaporated metal travels through the space until it contacts a surface. At that point the metal coats all materials with a thin deposit of the metal.

as nickel, copper, tin, gold, silver, and platinum may be deposited by this process.

Nickel is deposited on large areas that would be difficult to coat with other methods. Depositing a protective coating on the inside of a railroad tank car or on the interior surfaces of missile fuel systems is done with nickel. The nickel is deposited from a solution onto the metal surface by ion exchange, and a very uniform metal coating is produced. This process plates into corners and recesses that electroplating does not always reach. However, this process is more expensive than electroplating. **Electroless plating** produces a thin coating, but it is a denser coating. Thus it has as great corrosive protection as does heavier electrocoatings.

Nickel is used for corrosion protection on the interior of fuel-handling systems, tanks, and pumps. Tin is plated to make bearing surfaces (such as pistons and bearings) and to provide low friction during an engine break-in mileage.

Copper is plated to improve the electrical conductivity of the workpiece. It is used in the manufacture of printed circuits or in the forming of metals when the copper functions as a lubricant — such as the shaping of steel in a forming die.

Gold is plated for costume jewelry because of its beauty. It is also used for instruments because of its corrosion resistance, electrical conductivity, and ability to join steel with other metals for electrical connections.

Silver applications are similar to gold and copper. Silver provides an inexpensive decoration to jewelry and trophies, as well as improving the electrical conductivity of electronic parts. Silver immersion coatings are used to service components that are in production, because they are useful for maintenance and repair plating. Platinum metals are applied to immersion films used on special printed circuit board developments.

Electroplating. Electroplating is a widely used coating process in manufacturing, Fig. 16-29. It improves the decorative finish,

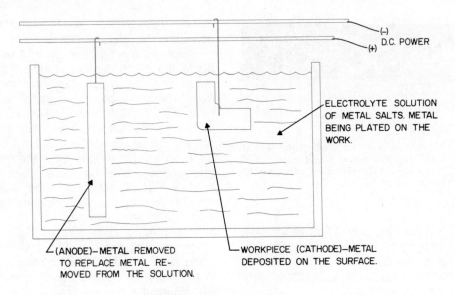

ELECTROLYTE SOLUTION OF METAL SALTS. METAL BEING PLATED ON THE WORK.

(ANODE)—METAL REMOVED TO REPLACE METAL REMOVED FROM THE SOLUTION.

WORKPIECE (CATHODE)—METAL DEPOSITED ON THE SURFACE.

(−) D.C. POWER
(+)

ELECTROPLATING PROCESS

Fig. 16-29. Electroplating may be done on all common metals as well as on nonmetals if the surface is treated so that it will conduct an electrical current.

Fig. 16-30. A plated part is removed from the plating tank. A low-voltage, high-amperage current transfers the metal ions to the surface of the workpiece.

Fig. 16-32. Giant bumper plating machine that is carefully watched by engineers to make certain that lustrous, corrosion-resistant bumpers are produced.

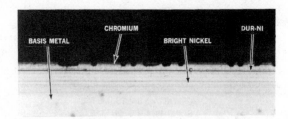

Fig. 16-31. A cross section of a steel panel plated with only 0.6 mil of Bright Nickel and 5 minutes of Dur-Ni after exposure to 144 hours of testing.

electrical conductivity, corrosion resistance, and wear resistance of the product. Electroplating can be used with all common metals and nonmetals if the surface has been prepared and made conductive.

The metals most commonly used for commercial electroplating are nickel, chromium, cadmium, zinc, tin, silver, and gold. These metals may also be alloyed, and the alloy may then be plated.

Electroplating takes place in a tank which holds the electroplating solution, the workpiece and the metal slab, and the source from which the plating material will come, Figs. 16-30, 16-31, and 16-32. The workpiece is hung on the work rod, and direct current power is applied. Electrical energy removes the metal ions from the metal slab (the anode), and the metal goes into the solution. The metal is then transferred through the solution and redeposited on the workpiece (the cathode). A low-voltage, high-amperage current is used.

The amount of current, the shape of the workpiece, and the distance from the anode to the cathode are important factors in plating. When the current flow is high, the deposits of metal build up rapidly. When the current becomes too high, the plating becomes rough and spongy rather than smooth.

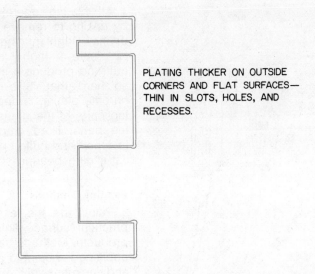

PLATING THICKER ON OUTSIDE
CORNERS AND FLAT SURFACES—
THIN IN SLOTS, HOLES, AND
RECESSES.

Fig. 16-33. Electroplating will deposit a thinner coating on edges, in corners, and down holes and recesses. Heavy deposits will occur where a larger volume of electrolyte is available, such as outside corners and flat surfaces.

The temperature, agitation of the parts, acidity of the plating solution, and the amount of current per area of part must be carefully controlled by the plater.

The shape of the work becomes a design problem, because sharp corners or thin, deep recesses of cuts will have a very thin coating of plated metal, Fig. 16-33. Flat surfaces and projections will have thicker coatings. If possible, the designer should reduce the slots and recesses, and the plater should move and rotate the parts while they are being plated. Plating is also aided by shaping the anode to the curvature of the workpiece and getting it as close to the cathode as is practical.

Small parts such as nuts, bolts, washers, screws, machine screw parts, or metal stampings are plated in a nonconductive drum or barrel. This is called **barrel plating**, Fig. 16-34. The barrel has a hole in it so the solutions will enter the barrel when it is filled with parts and they will then be rotated in the plating solution. The work-

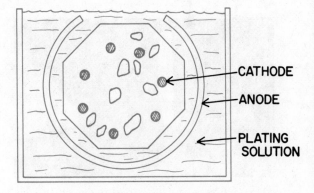

CATHODE
ANODE
PLATING SOLUTION

Fig. 16-34. Barrel plating applies a metal coating to small parts. The tumbling deburrs and burnishes the parts while they are plated.

pieces are connected to the cathode by a series of bars running through the barrel, and the work is against the rods. The anodes on the outside of the barrel are curved to

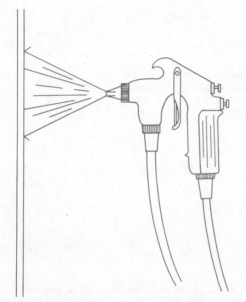

THE GUN SHOULD BE HELD PERPENDICULAR
TO THE SURFACE, 6" TO 8" AWAY.

Fig. 16-35. Paint is broken up into a mist by air pressure to produce a fan of paint. The paint may be applied by hand or applied by an automated painting system.

Fig. 16-36. Sheet metal stampings are being loaded and inspected as they enter an automated spray production system in a large factory.

fit around the slowly rotating barrel. The small hardware in the barrel will have a smooth, bright plated surface which is free of burrs because of the burnishing action during the tumbling.

All plating operations are finished first by rinsing to remove chemicals and then by drying. Bright plating will usually be buffed to create a high luster. Nickel plates will buff and produce a fine polish. Chromium is so hard that it is difficult to buff. Hard chrome plating is applied to dies and wearing parts of machines to lengthen the life of the parts. Hard chromium plating is used to repair and build up worn shafts, dies, and flat-sliding bearing surfaces.

Painting Methods

Paints are the least expensive and most practical surface finish for many industrial products. Methods have been developed for applying the paint to various products rapidly and inexpensively.

Spray Painting. This method is frequently used for applying paint to the metal surfaces of products, whether in high volume or in limited production.

The paint is broken up into a mist by air pressure and into a fan shape by the spray gun nozzle, Fig. 16-35. The gun is either passed by the work or the work is passed by the paint fan. The spray method of applying paint is very flexible and may be applied in almost any position. It is widely used in industry, because it produces a fast and dependable quality-coated surface. Spray equipment may range from an inexpensive hand sprayer to a completely automated spray production in a large factory, Fig. 16-36.

In **hot spraying,** the paint is heated to about 160° F. and is used for heavy-volume industrial painting. Since the paint is at a high temperature, little paint solvent is used, and thicker coating films are obtained. The number of coats of paint needed is reduced, resulting in greater production with less time and labor.

Large areas are quickly covered by **air-less spraying.** The paint is pumped at high pressure into the paint nozzle, where the paint is broken up into a fan. A large volume of paint flows through the nozzle so that less spray time is needed on the product. Be-

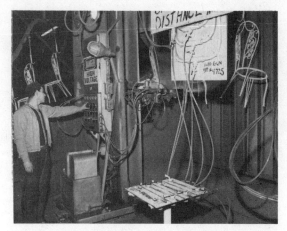

Fig. 16-37. In electrostatic spraying a high voltage is placed on the metal chairs. The opposite polarity is induced into the paint. The paint thus is attracted to all metal surfaced with little wasted paint.

Fig. 16-38. Dip coating provides coverage to combat corrosion in metal areas that would otherwise be very difficult to protect. This body shell is being submerged into a 65,000-gallon tank of paint.

cause there is no air involved, less paint is lost as overspray or mist off the work. High-pressure pumping and paint temperature control reduce the solvents needed to produce a thick coat on the work without runs or sags.

Electrostatic spraying is done by developing an electrostatic charge between the paint and the workpiece, Fig. 16-37. The charge is about 100,000 volts. It causes the paint to be drawn to the work surface.

The paint is broken into small droplets by spraying or by spinning the paint off a disc or a bell-shaped atomizer. The work is grounded with the opposite electrical polarity to that of the paint, and the paint is drawn by the charge to the nearest surface of the product. The process does not waste paint, and it quickly covers large areas evenly. The paint does not cover as well in corners or in recesses as do other coating processes, so electrostatic coating is most frequently adapted to automated mass production of simple surfaces.

Dip Coating. Dip coating is used on large amounts of relatively small workpieces (such as wire frames) and on recesses or surfaces which are not easily reached. The paint is in a tank, and the cleaned parts are dipped into the paint and moved. The paint flows into holes, around rivets, and in the back side of flanges and bent edges. The parts are removed from the paint and allowed to drip or are spun to throw off the excess paint. Small parts are placed in baskets and dipped into the paint. Dip painting is not a finish operation, because the paint will be thicker on the bottom edges of the work, and there may be paint bumps on the edges of the work. Dip coating provides paint coverage to combat corrosion, Fig. 16-38. Automobile frames and bodies are dip-coated to cover exposed metal surfaces that would be difficult to cover by other methods.

Flow Coating. As products increase in size, dip coating becomes impractical because of the size of the tank needed and the amount of paint needed for the tank. Flow coating is applied to larger workpieces.

The paint is pumped to a group of nozzles that wet the surface of the work, and the excess paint is collected in a drain and recycled. The paint is not sprayed; instead, the liquid flows over the workpieces. Flow coating is automated for high production. Flat work with notches and recesses — such as panels for dishwashers, freezers, and

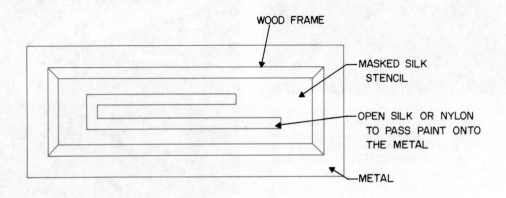

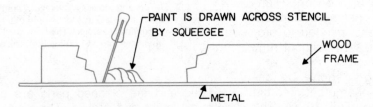

Fig. 16-39. In silk screening the paint is drawn across the nylon screen and the paint flows through the unmasked areas onto the metal to produce the design.

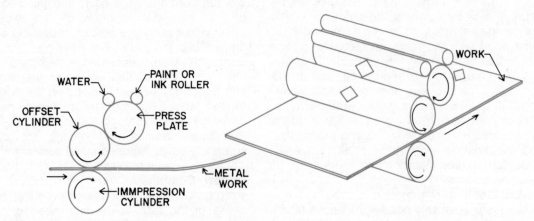

Fig. 16-40. Roll coating and lithography are used to place detailed designs and instructions on metal products. The process is an outgrowth of the printing industry.

refrigerators — is frequently finished by this process.

Silk Screening. Silk screening is the coating process which is specialized for covering controlled areas such as labels, instructions, and directions. They are often silk-screened over a previously coated surface. A stencil is placed on a fine screen of silk, nylon, or other fabric. The paint is passed through the screen onto the product, making a painted area the shape of the stencil. This type of area coating is utilized for low-volume production such as warning signs and instructions or directions on flat metal or cylindrical surfaces, Fig. 16-39.

Roll Coating and Lithography. Organic coatings are transferred from a roller to sheet or strip metal at high speeds in roll coating. Coating may be done with one pass with excellent coverage and control of the coat thickness. Labels or instructions on a mass-produced product such as soft drink cans, aerosol spray cans, toys, and other products receive their finish by a more precise process of lithography, Fig. 16-40. A metal plate is prepared photographically with the design or instructions to be printed on the work. The surface is prepared by chemicals so that only the areas of the plate that are to print will pick up ink from the ink roller. The other areas of the plate will pick up moisture from a water roller. The ink is transferred to the plate only in the shape that is to print and is again transferred to the offset cylinder. This in turn transfers the ink coating and the design to the metal. Lithography may be done on sheet strip material before round cylinders such as aerosol spray cans are formed.

Fluidized Bed Coating (Powder). Small products are frequently coated with epoxy, nylon, polyethylene, and other thermoplastic powders. These metal parts are dipped into a tank containing the powder which has been fluidized by vibrating the powder and passing air through it, Fig. 16-41. The air and vibration separate the powder so that all the particles are loose or fluidized. Warm metal parts may be dipped into the dry

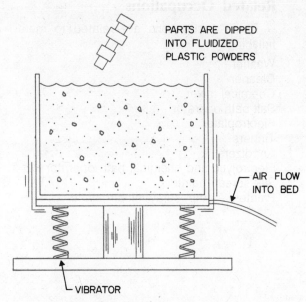

Fig. 16-41. Warmed parts are dipped into the fluidized polymer powder to receive a coating. The coating is later cured in a low-temperature oven.

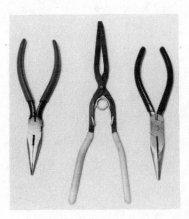

Fig. 16-42. Heated parts are immersed into dry fluidized powder to coat the parts.

powder as if it were water. The plastic coats the warm metal parts which are then cured by a lower temperature by baking for a short period. Pliers, springs, and numerous other tools may be coated in this fashion, Fig. 16-42.

Related Occupations

These occupations are related to metal finishing:

Washer
Cleaner
Chemical mixer
Salt bath operator
Electroplater
Tinners
Anodizers
Electrolytic and chemical bath operator
Silk screen painter
Painters — dip, spray, electrostatic

Etching equipment operator
Commercial artist
Sealer
Tester
Inspector
Metal finisher
Sprayer
Cutter
Trimmer
Filer
Grinder
Buffer
Polisher

Automated Manufacturing

Section 5

Automation principles are based upon the control of a production machine or process by a logic device. Control is gained by receiving information from products and comparing it with information which is stored within the logic device. The discrepancy between "what is" and "what is wanted" is the correction given to the automated system. Automation production is powered by all types of basic machines to provide mechanical power and transmission systems to start, couple, and brake the power.

Automation technology is rapidly moving toward multiple machine automation by the use of modular block tool units that are designed to move the product in a circular path or straight line configuration.

Numerical Control

Numerical control is a method of communicating to a machine. As a result, maximum flexibility of accurate machine control is achieved. Numerically controlled machines fall into groups related to the machine's movements: straight or angular movements, point to point, and continuous path or contouring machines. A manufacturing complex that utilizes a computer to completely control the functions of a machine is referred to as direct numerical control. The application of the computer makes possible adaptive control, whereby the action of the machine is sensed. Thus, the cutting action may be modified by the computer while cutting is in progress to achieve the greatest production efficiency.

With the addition of automatic transfer of pallatized workpieces, a computer-controlled transfer machine may be assembled out of modular units.

Manufacturing processes are moving in the direction of combining machining, inspection, and assembly into one transfer machine. Factories will bid on producing only products that can be shuttled around the factory on their special equipment pallet sizes. The future will bring the development of mobile factories in which the factory will manufacture the product while it is moving over the site. A pipeline factory would build and lay the pipe as it traveled across the country.

Automation Principles

Chapter
17

Words You Should Know

Logic device — A decision-making module that provides a function signal.

Automation — Combining of electronic sensing controls, power supplies, movement of materials, the function of the machine, and the control of the machines, all in one large manufacturing system (combining of many machines controlled automatically).

Sensor — A device that obtains information about a process.

Analog data — Continuous data; such as data concerning a specific pressure.

Digital data — Pulses that may be counted.

Closed loop — Feeding back information from the process to control the process.

Set point — What is desired in the process.

Actuator — A device such as a fuel valve that may be controlled and may make a change in the system.

Transducer — A device (such as a thermocouple) that converts one form of energy to another and will provide a proportional signal to the variable being measured.

Controller — Determines whether the sensed variables are within tolerance of the set point.

Reaction curve — The time required to change a system from one set point value to a new value.

Wall attachment — A flow held against a surface by differences of speed and/or pressures.

Boolean logic — Propositions or conditions are defined as "true or false," or "on or off."

Gates — Gates perform any desired logic function.

Fluidics — The use of flowing gases or liquids to sense, transmit, and control equipment.

Actuator — These devices carry out the controller's signal and make a change in the manufacturing system. They may be valves, motors, hydraulic cylinders, switches, stepping motors, air diaphragms, or cylinders.

Servomechanism — A self-correcting device (such as a governor) with the feedback or closed loop sensing information frequently built into the actuator.

Mechanism — An elementary machine — a motion converter such as levers, gears, cams, etc.

Cam — A mechanism that transmits a predetermined irregular motion to a cam follower.

Inversion — Anchoring or locking different members of a mechanism and producing a totally new movement or machine.

Electromagnetic induction — An electrical charge-producing magnetism in a neighboring body without contact (or vice versa).

Pneumatic power — Power generated from pressurized gases.

Fluid multiplication — Increasing the area of the actuator increases the power that is delivered. However, a greater volume of fluid must be pumped at a given pressure.

Serpentine drive — A belt wrapped around a pulley like a snake to give the belt more contact area with the pulley.

Eddy current — A current inducted into another body at variance with the main current.

Coupling — A device for joining two rotating shafts.

Automation Is Control

Automation is the control of the production machine or process by a logic device. It is the result of a growing technology that has passed through a series of stages of mechanical development. Decisions concerning size, location, length, depth, form, or sequence are applied to a machine by limit switches, timers, cams, electronic circuits, numerical control tape, or by computers, Fig. 17-1.

Early machine tools produced single parts and were controlled by a worker. The decisions were made by the craftsmen with the aid of blueprints. **Mechanization** was the process of aiding the craftsmen by making the machines automatic by the addition of stock feeders, tooling devices, and automatic tool turrets. Workers were given devices to speed up and improve their work, but they still made the decisions in controlling the machine.

Completely automatic machines such as automatic screw machines and lathes have been in existence for some time, but these machines are not automation, Fig. 17-2. They are single machines that produce single parts. When a number of machines are linked together by material-handling machines or are built into one large machine with many production stations, this machine is called a transfer machine. The machine

Fig. 17-1. The control of a production process makes automation possible. Decisions as to where to move the part, the size and depth of the various holes and surfaces, and the sequence of operations are pre-engineered in an automated machine.

Fig. 17-2. An automatic screw machine is an example of high development of mechanization. The craftsman sets up the machine and maintains the decision of size control.

becomes automated with the combining of electronic controls for sensing and controlling, the hydraulics to supply power for movement or action, and the motors and mechanisms to support and multiply the working forces, Fig. 17-3. Numerical control

Fig. 17-3. This six-way boring and facing dial-type machine uses electronic sensing to control power supplies and hydraulic pumps. These pumps provide energy for the precise movement to the location needed to produce the workpieces.

or computers may be added to supply program data of size, location, and facility, for feed back, and correction of machine control.

Automation results when these technical devices are added to a transfer machine. Automation, then, is a creation of the developing technologies, which supply the essential parts of mechanics, electronic control, hydraulics, and pneumatics to make a product. When these technologies are applied to one system, an automated manufacturing system is built. It is the combination of many machines which are all controlled automatically that creates automation.

Automation is based upon the concept of a self-correcting system. Controlling an automated system is based upon the use of the "closed loop," Fig. 17-4. The closed loop, in turn, makes use of the **feedback** principle to provide information for control purposes. The closed loop receives its information from the process, Fig. 17-5. This information is gathered by a **sensor** and is fed back to a decision-making device called a **controller**. The controller receives the values or information and compares it with a **set point**. The set point is "what is desired," and the sensor tells "what is" in the

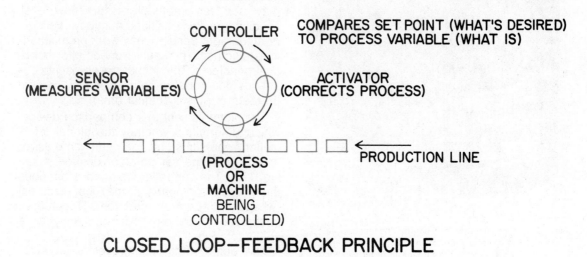

CLOSED LOOP—FEEDBACK PRINCIPLE

Fig. 17-4. The closed loop is a self-correcting system.

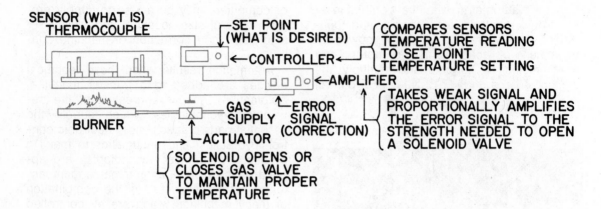

SENSOR (WHAT IS)
THERMOCOUPLE

SET POINT
(WHAT IS DESIRED)

CONTROLLER

COMPARES SENSORS
TEMPERATURE READING
TO SET POINT
TEMPERATURE SETTING

AMPLIFIER

BURNER

GAS
SUPPLY

ERROR
SIGNAL
(CORRECTION)

TAKES WEAK SIGNAL AND
PROPORTIONALLY AMPLIFIES
THE ERROR SIGNAL TO THE
STRENGTH NEEDED TO OPEN
A SOLENOID VALVE

ACTUATOR

SOLENOID OPENS OR
CLOSES GAS VALVE
TO MAINTAIN PROPER
TEMPERATURE

Fig. 17-5.   Here a drawing furnace illustrates the closed loop concept. The idea of closed loop control may be applied to any machine or production system.

Fig. 17-6.   The controller may control one function such as a furnace with continuous data (analog data) coming from the furnace in order to keep the furnace on the set point.

process. The controller compares the two values and an **error** signal is the controller's output. The error signal is transmitted to an **actuator,** a device used to make the desired change in the system. The actuator may be a valve, relay, air cylinder, hydraulic cylinder, servomechanism, or a stepping motor.

The sensor is the device that obtains the information about the process. A sensor may "see" temperature, pressure, flow, level, specific gravity, position, humidity, voltage, or amperage. Sensors may work on mechanical, electrical, hydraulic, pneumatic or optical principles. The sensor may present continuous data (analog data), or it may present the data in pulses (digital data).

The signal is sent to a comparison device, which determines whether the signal information represents less than, the same as, or greater than the **set point** or desired value, Fig. 17-6. The controller instrument that does the process comparison may be electrical (wheatstone bridge), electronic (counters), pneumatic (force balance), or it may be a computer, Fig. 17-7.

The output of the controller is an **error signal.** The error signal is amplified so that

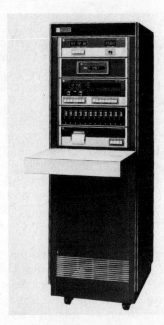

Fig. 17-8. Building block for a computer memory. A total of 4608 bits of information are held on tiny rods.

Fig. 17-7. A digital data surveillance facility makes possible the monitoring of many points in the manufacturing processes throughout the plant. The sensors are scanned by the controller for one-tenth of a second per point. The surveillance facility will scan 576 process points.

enough energy is available to move an actuator. The actuator may be a valve, a switch, stepping motors on machine tools, or an automated process.

Information Supplied to Controllers

Set Point Information. The set point is the reference point of the system and may employ different devices. The set point may be a position of a screw, a lever, a pressure, or a voltage. In each case, the set point will represent the desired value needed. The system will **adjust itself** to match the set point. An example of a set point is the dial of a household thermostat set at 68°F. The furnace or cooler will start and will function until the temperature of the air around the thermostat is 68°F., and then the system will shut itself down.

Tape Information. What is desired (or a set point) may be established by instruments other than a controller. The set point may be placed in permanent form by putting the information on a punched tape or a magnetic tape. When it has been coded onto a tape, the information may be returned to a system, and the system will try to equal the information supplied by the tape. In this case, when the set point is changed, the machine will try to produce the new position or condition. This same concept applied to numerical control means that each reading of a block of tape will produce a new machine position.

Card Information. Information for controlling may be punched into cards. Each punched card will contain the coded information necessary to program a computer or to process data that may be recorded, classified, calculated, and summarized. The information frequently required for controlling manufacturing is summarized information, and the output is frequently printed out on a typewriter or fed to an actuator to change the system. The card information is a convenient method of getting information into a computer so that a vast number of positions throughout a manufacturing plant may be controlled.

Memory Core, Disc, or Drum Information. Memory cores, discs, or drums are parts of a computer that store volumes of information, Fig. 17-8. The bits of information are stored as a magnetic charge on a ferrite core or on the surface of a disc or drum. Many discs may be applied to a system and increase its capacity. Information is coded onto these discs and is later used by the computer.

Computer Application. The computer is used for many applications. In automation manufacturing, it is frequently used for controlling all phases of manufacturing, including machine control.

Like a set point, the computer can operate its program in the memory disc with what is being put into it by the card reader, tape reader, or sensor. It can follow a logic, perform a mathematic calculation, and produce an output. The output is like an error signal — it may change the system. If the computer is attached to a numerically controlled machine tool or an automated system, the desired position change will be made.

Sensing to Gain Information

Transducers. These are sensors that convert a sample of the variable being measured into a proportional signal and also convert one form of energy to another. The signal is frequently electrical and is concerned with a temperature, pressure, level, flow, or a position. Some examples of these transducers are thermocouples, differential transformers, bourdon tubes, strain gages, photoelectric cells, and synchros.

Probes. Probes are fingers or feelers that contact and provide "yes or no," "on or off," or "full or empty" information. They indicate whether something is or is not there, such as a broken tap in a hole of a casting. A probe or finger is pushed into a tapped hole. If no contact is made, it is assumed that a tap has not been broken in the hole. Operations such as float level, position of switches, or position of a piston in a cylinder are determined by probes. These probes give information, but they do not measure the amount. They merely show that the function has or has not been completed.

Pickups. "Pickup" is an electrical term used for gathering information that converts the information to a signal. It may be a dimension, an amount, or a process factor that can be used for process or machine regulation.

Switches. Switches complete or break an electrical circuit or turn power on or off. They are widely used as sensors. The limit switch is a positive and simple method of transmitting position information. A microswitch is a device which is either on or off. It is operated by a number of actions: mechanical, pressure, photoelectric, or proximity.

Common Sensors. Information for an automated system comes from primary sensors. The sensor measures the product which is in the process of manufacture, whether it is in an oil refinery, power generating plant, plastic manufacturing plant, transfer machine, or numerically controlled machine, Fig. 17-9.

Decision

The automatic controller is an instrument that compares what is desired to make the process work successfully with what is actually happening in the process, Fig. 17-10 (page 506). The controller determines whether the process condition is within the tolerance of the set point or whether it is moving out of tolerance from the set point. This decision may be made by a group of different instruments. The major controllers are pneumatic, fluidic, electronic, and mechanical.

Pneumatic Controller. The pneumatic controller was developed early in the manufacturing process control field. The use of air bellows and air relays have proven to be very reliable and trouble-free over years of service. The force-balance type of decision-making device has been applied to the control of temperature, pressure, flow, level, and density measurements. Force-balance controllers are produced in different shapes, but they perform the same controlling function. The basic concept of pneumatic control is

FACTORS BEING SENSORED	SENSING ELEMENT	DIAGRAM OF SENSORS
TEMPERATURE	THERMOCOUPLE THERMOPILE RESISTANCE THERMOMETER	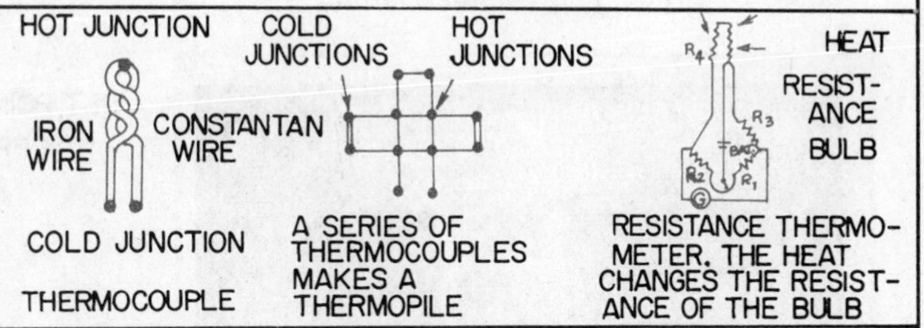 HOT JUNCTION COLD JUNCTIONS HOT JUNCTIONS HEAT RESISTANCE BULB IRON WIRE CONSTANTAN WIRE COLD JUNCTION THERMOCOUPLE A SERIES OF THERMOCOUPLES MAKES A THERMOPILE RESISTANCE THERMOMETER, THE HEAT CHANGES THE RESISTANCE OF THE BULB
PRESSURE	BOURDON TUBE DIAPHRAGM BELLOWS	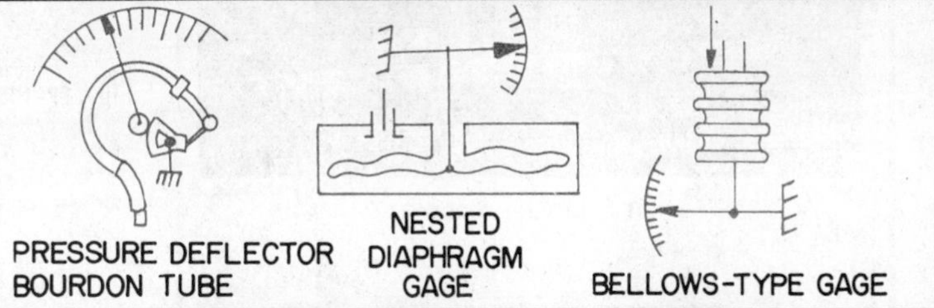 PRESSURE DEFLECTOR BOURDON TUBE NESTED DIAPHRAGM GAGE BELLOWS-TYPE GAGE
FLOW	ORIFICE ROTAMETER	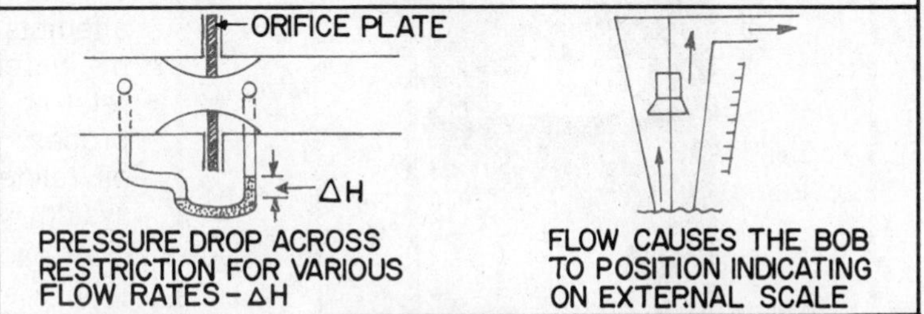 ORIFICE PLATE ΔH PRESSURE DROP ACROSS RESTRICTION FOR VARIOUS FLOW RATES – ΔH FLOW CAUSES THE BOB TO POSITION INDICATING ON EXTERNAL SCALE
DENSITY	NUCLEAR RADIATION FLOAT DENSITY	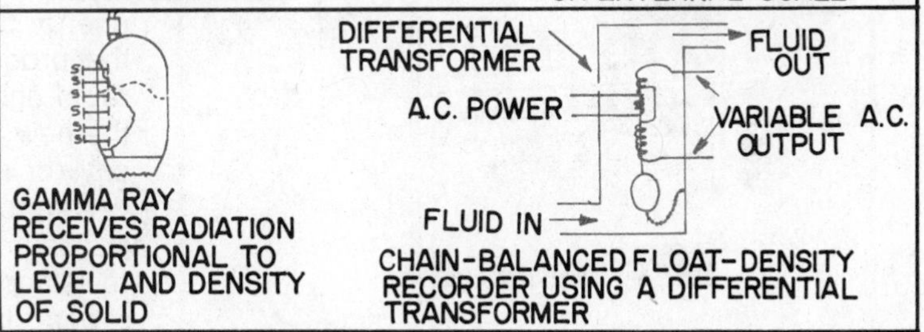 DIFFERENTIAL TRANSFORMER FLUID OUT A.C. POWER VARIABLE A.C. OUTPUT GAMMA RAY RECEIVES RADIATION PROPORTIONAL TO LEVEL AND DENSITY OF SOLID FLUID IN CHAIN-BALANCED FLOAT-DENSITY RECORDER USING A DIFFERENTIAL TRANSFORMER
POSITION LINEAR	DIFFERENTIAL TRANSFORMER	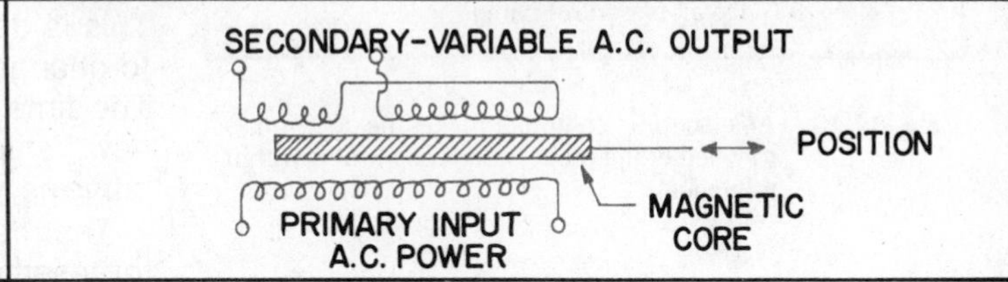 SECONDARY-VARIABLE A.C. OUTPUT POSITION PRIMARY INPUT A.C. POWER MAGNETIC CORE
POSITION	STEPPING MOTOR	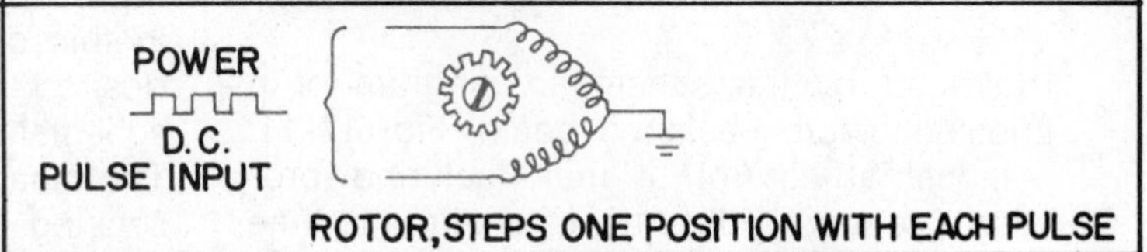 POWER D.C. PULSE INPUT ROTOR, STEPS ONE POSITION WITH EACH PULSE
ROTATION	TACKOMETER	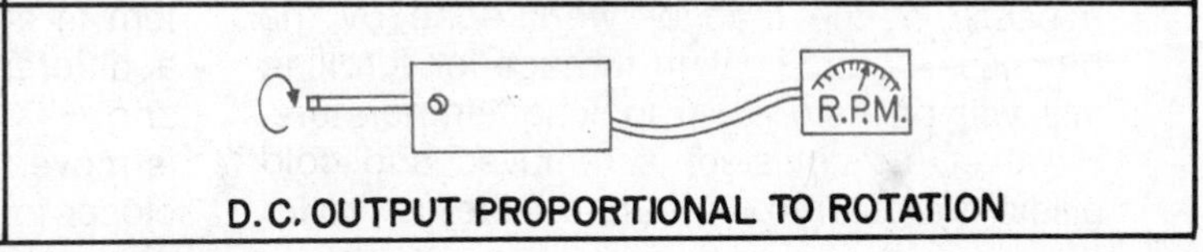 R.P.M. D.C. OUTPUT PROPORTIONAL TO ROTATION

Fig. 17-9. Common Sensors.

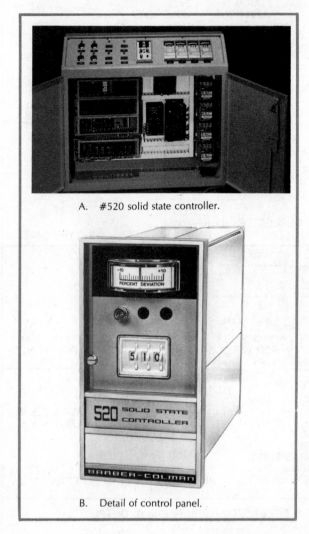

A. #520 solid state controller.

B. Detail of control panel.

Fig. 17-10. An electronic controller makes the decision as to whether the process under control is within tolerance.

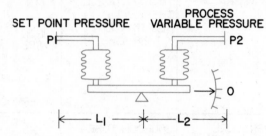

Fig. 17-11. The principle of pneumatic force balance beam is based upon the idea that when the pressures are equal in like bellows and the lengths (L_1 and L_2) are equal, the beam will balance.

attempts to keep the furnace at temperature by increasing the fuel. These changes are referred to as load. When the manufacturing process becomes complex (such as in an oil refinery), the reaction to changes in the system is called a **process reaction curve.**

A reaction curve such as "A" in Fig. 17-12 is in a system that has little resistance. The energy change in the process starts to affect the process at once. The rate of change is rapid and then slows down as it approaches the new value. This system has a high capacity (or rate) of change.

A reaction curve such as "B" in Fig. 17-12 is a complicated process with considerable resistance. Energy is put into the system for a period of time before any change occurs. This is **dead time.** The system slowly begins to change and finally reaches the new value. The time it takes to go from one value to a new value is called **process lag.** A reaction curve is produced.

The desired condition of the system is represented by the set point. The set point in this case is presented as pressure. The desired condition is represented in the schematic as a value of 10 psi, Fig. 17-13. The measured pressure is received from the sensing element and indicates that the system is actually at a value of 8 psi. There is a difference of 2 psi as to what is wanted and what actually exists. The balance beam is moved down, causing the flapper to move closer to the air nozzle. More air is restricted

illustrated by the schematic diagram of a pneumatic force balance beam, Fig. 17-11.

Industrial control of manufacturing processes requires adjustment from time to time because of the load or work done by the process. A steel heating furnace for a rolling mill will be held at a soaking temperature. However, as hot steel is removed and cold billets are put into the furnace, the controller

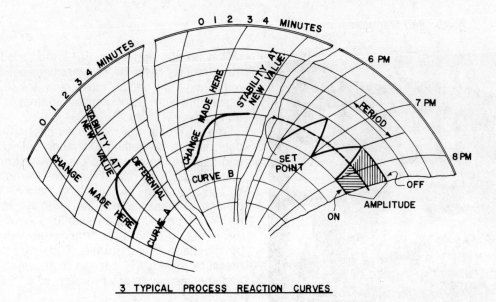

3 TYPICAL PROCESS REACTION CURVES

Fig. 17-12. A reaction curve gives a visual description of what is happening in a system which is being controlled.

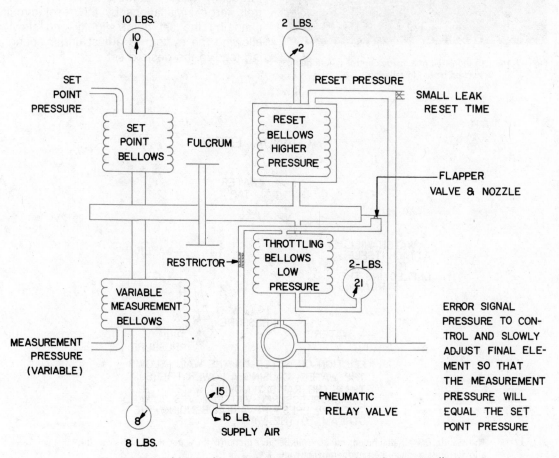

Fig. 17-13. Schematic diagram of a pneumatic force balance beam controller.

Fig. 17-14. An industrial pneumatic control system being checked before installation.

at the nozzle, and pressure is increased in the throttling bellows. The throttling bellows repositions the relay valve, and the error signal pressure is increased to make the correction.

The controller must not overshoot the set point setting by getting too much deflection and causing the system to cycle. A reset circuit adds a pressure that dampens the action of the throttling bellows and allows the change to take place at the rate of a small leak. The pressure soon stabilizes between the throttling bellows and the reset bellows by the flapper and nozzle. The correction pressure makes the change in the system, and the measurement pressure is equal to the set point.

The proportional reset control is the type frequently used in industrial work. It controls large installations by the reset circuit changing the set point slightly and slowly allowing the system to adjust under changing load to the desired set point, Fig. 17-14.

COANDA EFFECT

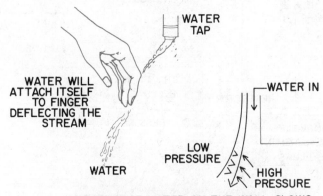

WATER TAP

WATER WILL ATTACH ITSELF TO FINGER DEFLECTING THE STREAM

WATER

WATER IN

LOW PRESSURE

HIGH PRESSURE

FRICTION OF WATER ON THE WALL SLOWS THE WATER, CAUSING A LOWER PRESSURE THAN THE RAPIDLY MOVING WATER. THE HIGHER PRESSURED WATER MOVES TOWARD THE LOW PRESSURE AREA, BENDING THE STREAM TO THE WALL.

Fig. 17-15. The coanda, or wall-attachment, effect results because friction of the water on the wall slows the water and causes a lower pressure than the rapidly moving water.

Fluidic Controller. A fluidic controller is a device in which flow gases or liquids are used to sense, transmit energy, and to control. Fluidics bring the control sophistication of electronics to a nonelectrical system that can perform the same control functions of pneumatics, hydraulics, or electronics.

Fluidic controllers are based upon two fluid dynamics principles: (1) the Coanda Effect and (2) the Beam Deflection Principle, Fig. 17-15. Most devices function upon the Coanda or "wall attachment" effect, Fig. 17-16. When a jet of fluid is introduced near a curved or flat plate, it will attach and follow the plate, even when the jet is moved, so the flow is away from the direction of the jet. The flow of the jet may be shifted from one output channel to another by a pulse in the control jet. The stream will remain attached to the new wall until a pulse from the opposite control jet is given.

When a control jet pulse is given, the jet stream shifts, so the basic information that is required for binary numbers is obtained. A different kind of switch may be used to control or to compute. With the application of Boolean algebra where the logic or propositions are defined as either true or false (or on or off), a series of devices called **gates** are designed, Fig. 17-17 (page 510). These gates are capable of making decisions and are called "or," "and," "nor," and "nand," as in computer and design logic.

The advantages of fluidic control are that the systems have high reliability since there are no moving parts to wear, bend, or stick. Fluidic devices are not affected by electrical static, vibration, radiation, magnetism, or temperature changes so they may be used as a nonjammable electronic controlling system, Fig. 17-18.

Fluidics can be used for sensing, digital switching, counting and timing, amplification, binary logic, computing, information handling, and controlling.

Electronic Controllers. By applying electronics to the controllers, the size of the controller has been decreased and the speed of response increased. Electronic controllers are designed to utilize the basic

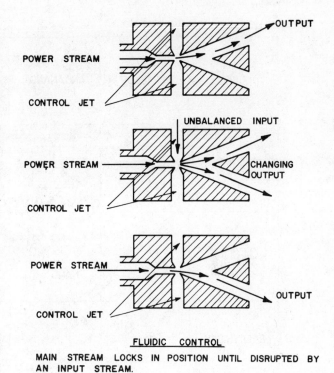

FLUIDIC CONTROL

MAIN STREAM LOCKS IN POSITION UNTIL DISRUPTED BY AN INPUT STREAM.

Fig. 17-16. Fluidic control using the principle of the wall effect.

Fig. 17-18. Pneumatic gates being tested.

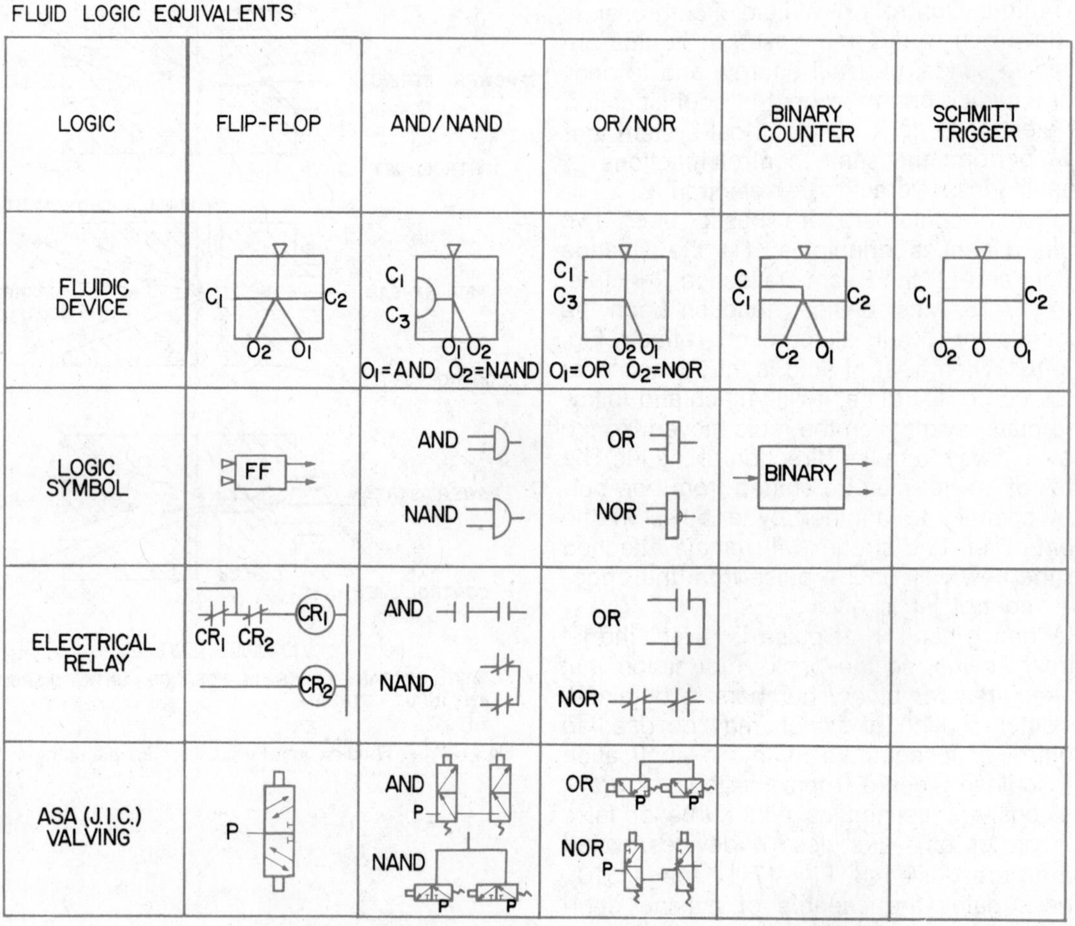

FLUID LOGIC EQUIVALENTS

LOGIC	FLIP-FLOP	AND/NAND	OR/NOR	BINARY COUNTER	SCHMITT TRIGGER
FLUIDIC DEVICE	C_1 ⊽ C_2 O_2 O_1	C_1 C_3 O_1 O_2 O_1=AND O_2=NAND	C_1 C_3 O_2 O_1 O_1=OR O_2=NOR	C C_1 C_2 C_2 O_1	C_1 C_2 O_2 O O_1
LOGIC SYMBOL	FF	AND NAND	OR NOR	BINARY	
ELECTRICAL RELAY	CR_1 CR_2 CR_1 CR_2	AND NAND	OR NOR		
ASA (J.I.C.) VALVING	P	AND P NAND P P	OR P P NOR P		

Fig. 17-17. Fluidic logic equivalents.

electronic functions of voltage, amperage, resistance, and induction.

A concept of an electronic controller is to oppose a regulated voltage (the set point) against a variable voltage (the sensor's input from the process), Fig. 17-19. The algebraic result becomes the error signal voltage. Error signals are low and will need an electronic amplifier to increase the signal to increase the power available to actuate valves or solenoids. These actuators in turn control the energy going into the industrial process. Alternating currents of from 60 to 400 cycles are used to release the power and reduce the sizes of the components.

Another electronic circuit is modified to make use of electromagnetic induction in the form of linear variable differential transformers, Fig. 17-20.

The two transformers and power supplies are identical, and the set point is regulated by the position of the core in the instrument. The two voltages are the algebraic sum of these voltages and produce a strong voltage or error signal.

Computer Controllers. Computer controllers are the result of development in computer technology. The control systems are rapidly turning to digital information handling. Use of the minicomputer is a con-

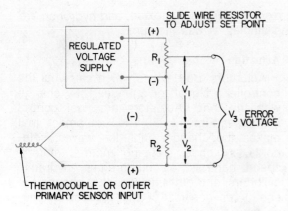

SLIDE WIRE RESISTOR
TO ADJUST SET POINT

OPPOSED VOLTAGE ELECTRONIC CONTROLLER

Fig. 17-19.   The output of the regulated voltage supply is seen across V_1 and is the set point voltage. The output of the process sensor is seen across V_2. The voltage available after adding, subtracting, or equalling voltages produces an error voltage V_3 used to change the actuator.

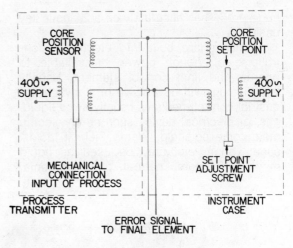

Fig. 17-20.   Electronic controller using linear variable differential transformers.

Fig. 17-21.   Computer controllers are a result of computer technology. Minicomputers monitor and control up to 100 control loop circuits.

venient method of analyzing information from a large number of sensors. Direct digital control (D.D.C.) is the application of a computer to control industrial manufacturing processes. A sensor's signal is transmitted in the form of a 0-5 voltage output to the computer. These voltages are scanned by the computer and compared. The computer compares its signal (set point) with the sampled signal of the sensor and produces a control signal proportional to the error or difference. The control signal is a digital quantity and is produced by counting away from the set point value. The countdown error signal is sent to adjust the final control element such as a valve or switch.

As new manufacturing plants are built, the process control functions will be grouped in one area. Minicomputers will be located throughout the plant. These in turn will be applied to many sensors. To be economically feasible, minicomputers will need to monitor and control about 100 control loop circuits, Fig. 17-21.

Electronic controls may be applied to nearly any industrial manufacturing process, Fig. 17-22. They are capable of controlling

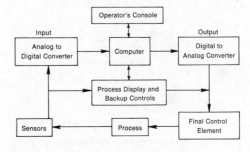

Fig. 17-22.   Computer control flowchart.

huge processes accurately on a 24-hour day basis.

Amplifiers are built into electronic controllers so enough power is available to operate the final control element. Usually this element will be a valve or a switch. Industrial control mediums usually are a liquid, gas, or electrical device such as an electric motor or a solenoid. Water, fuel oil, acids, sugar, brine, etc. may be used as liquid mediums. Steam, natural gas, ammonia, oxygen, carbon monoxide, and hydrogen are examples of gas mediums.

Actuation

Actuators are devices which carry out the controller's signal for change. They change with the load. Actuators are final control elements, Fig. 17-23. The equipment that makes the changes includes valves, solenoids, switches, stepping motors, motors, speed controls, air diaphragms, hydraulic cylinders, and other devices that change the energy or position of the process or machine.

Servomechanisms. A servomechanism is a device designed to self-correct itself. An automatic pilot in an airplane is an example of a servomechanism. When the heading on the compass is changed, the servomechanism changes the airplane to the new compass heading and holds it on that course. Servomechanisms are designed by using the principles of mechanical differential, synchros, resolvers, potentiometers, and hydraulics. The servomechanisms are closed loop devices and frequently are incorporated into the actuator with the change of control at another point.

Fig. 17-23. Actuators are final control elements. This value converts an air pressure signal into a position of the valve. The valve position will change the energy a machine or process receives.

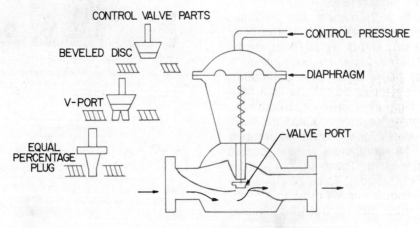

Fig. 17-24. Control valve ports are designed to control gases or liquids and are designed around a beveled disc, V-port, or an equal percentage plug.

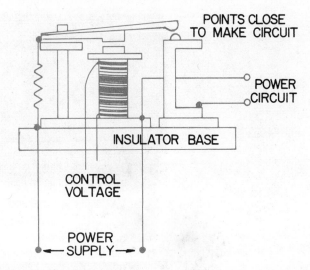

POINTS CLOSE
TO MAKE CIRCUIT

POWER
CIRCUIT

INSULATOR BASE

CONTROL
VOLTAGE

POWER
←— SUPPLY —→

Fig. 17-25. Electrical relays make possible remote control by closing power circuits distant from the control panel.

Control Valves. The most common control valves are those which control liquids and steam. Control valves are of many designs but they are generally designed around a beveled disc, a V-port, or an equal percentage plug, Fig. 17-24. These control valves may be operated by air pressure or by electrical means. The diaphragm top construction energized by air pressure is the most frequently used control valve. All control the amount of fluid which passes a given point.

Relays. Relays provide remote control, Fig. 17-25. They may be activated by vacuum, pneumatic, or hydraulic pressure; electricity; light; or heat. A small amount of energy is supplied to a solenoid coil which closes a set of contact points. These points in turn close a power circuit to high voltage and amperage to start an electric motor for a pump or other device, Fig. 17-26. With the use of a relay, large volumes of energy can be handled safely. Relays are widely used to turn on or off countless numbers of indirect circuits, Fig. 17-27.

Energy to Carry Out Control Change

The energy to operate actuators can come from any source, but the final control ele-

Fig. 17-26. Relays will handle heavy currents to operate machinery and keep these dangerous voltages and amperages away from the operator.

ments are usually powered by pneumatics, hydraulic, electrical, or mechanical equipment.

Fig. 17-27. Multipole relays are designed for precision instruments, business machines, copiers, and computer peripherals.

Fig. 17-29. Compressed air is controlled and provides the feed movement for this 5-way drilling machine.

Fig. 17-28. A multi-purpose ⅛" solenoid air valve that will precisely supply air to an activator such as a cylinder or motor. The valve functions as a 4-way, 5-port, 2-position directional flow control valve.

Fig. 17-30. Control switches are used to start or stop the electrical motors or circuits needed to keep a factory operating.

Pneumatic Power for Control. Pneumatic valves are frequently controlled remotely with electrical current. A multi-purpose 1/8" solenoid air valve provides control to air operated machines, Fig. 17-28. Machines that use air cylinders to clamp, move, locate, or rotate work include die casting machines, molding machines, and transfer devices, Fig. 17-29. They are very reliable and work under all weather conditions. Pneumatic systems are safe if a fire or explosion hazard exists because they produce no sparks or flashes.

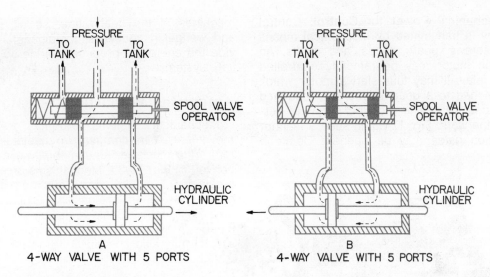

Fig. 17-31. Hydraulic directional control valves direct the movement of hydraulic cylinders so that very heavy loads may be accurately moved.

Electric Power for Control. Electric switches provide an off-on control over electrical power. Those frequently used are manual switches, electromagnetic switches, transistors, and circuit breakers. They provide the means to control the energy to countless types of motors and devices which use electrical energy, Fig. 17-30.

Switches are designed to complete or interrupt electric circuits from a few milliamperes to a thousand amperes momentarily. This surge is necessary to start a 200-horsepower motor.

Hydraulic Power for Control. Industrial processes which require great amounts of energy may be powered with a hydraulic system. Directional control valves control where the fluid pressure will be applied, Fig. 17-31. The sliding spool valve is a most important device for controlling fluids and directing the fluid power to the correct hydraulic cylinder. A powerful pump supplies the oil under great pressure to the large cylinder to do the work. Hydraulic circuits are designed to deliver precise amounts of movement with great power, Fig. 17-32.

Fig. 17-32. Hydraulic power is widely used throughout industry. Its chief contribution is almost limitless power with infinite control.

Mechanical Power tor Control. Control energy is transmitted by mechanical means with screws, gears, slides, and shafts. An electric motor may be remotely controlled by a relay. It may thus start a motor which is attached to a gear reducer and attached by a gear or belt to a driving screw to perform the work, Fig. 17-33. In such a manner irrigation gates may be opened. Common machines of the gear, slide, screw, lever, and wedge in countless combinations provide the energy to produce change-controlling systems.

Automated Production Actuators

Mechanical systems produce physical movement. A **mechanism** is an elementary form of a machine, fundamentally a motion converter, Fig. 17-34. Mechanisms are combinations of moving parts such as gears, cams, links, belts, chains, and springs held in a rigid frame. Besides its function as a means of converting motion, a mechanical system may also contain and transmit information.

Types of Individual Machines

Automation makes use of all types of basic machines: mechanical, electrical, pneumatic, hydraulic, and power transmission systems. Mechanisms are a troublesome part of automation, because they wear out, stick, get jammed, and sometimes respond too slowly. However, they are essential, because they supply the physical motion of the equipment.

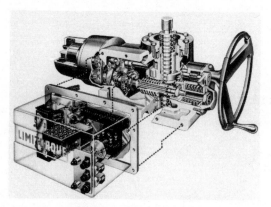

Fig. 17-33. An electric motor with remotely controlled powers activates mechanical valve operator. This mechanical control device has a lifting force of up to 500,000 pounds — enough to control dam or irrigation gates.

Fig. 17-34. Cams that provide physical movement on an automatic screw machine. These cams are motion converters that provide the machines timing, rate of movement, and the amount of movement for the various machining operations.

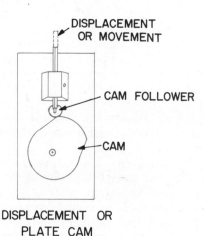

DISPLACEMENT OR MOVEMENT

CAM FOLLOWER

CAM

DISPLACEMENT OR PLATE CAM

Fig. 17-35. A cam may be designed to convert continuous rotation into a predetermined irregular motion to perform many manufacturing machine motions.

Cam. A cam made to rotate will cause the cam follower to move with relationship to the surface of the cam. A continuous movement of the cam will cause a displacement of the cam follower shaft. A cam may be in continuous rotation and transmit a predetermined irregular motion to the cam follower as is found in the valve assembly of a gasoline engine, Fig. 17-35. The cam is also one of the mechanisms that can produce a rotating, oscillating, or reciprocating motion. Cams are designed that can give very complex motions which are unobtainable by any other form of mechanism. The surface of the cam may be shaped so that analog information may be removed. When the cam is rotated a certain number of degrees from a reference point, a known amount of deflection is transmitted from the contact point of the follower on the cam. This information is read by the amount of displacement of the cam follower, Fig. 17-36. The positioning of the cam may be used to set another device to a precision setting very rapidly. Cams may also be designed to increase information by using three dimensions. The cam follower may remove information from the cam by having the cam slide along the follower as well as around the cam.

Three-dimensional cams are used for analog information that is instantaneous and accurate, Fig. 17-37. They are used in military ballistics coupled with range finders.

Ratchet. Ratchets provide power in one direction and are intermittent. They have a notched wheel and a pawl, Fig. 17-38. By regulating the distance the connecting arm is moved, the amount of rotation may be controlled. This mechanism is frequently used for feed devices. If the pawl is stationary but spring loaded, it will be used as a lock against rotation in one direction as on a wrench.

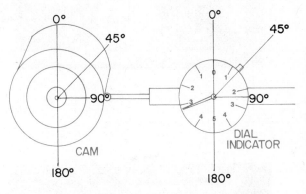

Fig. 17-36. A graphic analysis of the degrees of rotation of the cam and the distance of displacement may be made with the aid of a protractor and a dial indicator.

Fig. 17-37. A three-dimensional cam may have information removed from the surface of the cam by a cam follower. The follower slides along the cam while the cam rotates under the follower.

Fig. 17-38. This mechanism is frequently used in machine feeding devices.

Fig. 17-39. The silent rachet or overrunning clutch will transmit power in one direction like the rachet, but the work is free to move faster than the rachet is turning.

Fig. 17-40. A four-bar link system will transmit a number of motions. It is found in the basic designs of many machines such as engines, steering mechanisms, steam locomotive cranks, and connecting rods.

The overrunning clutch provides for the transfer of power in one direction. Thus, if the input of power is transferred, the clutch will run free and will not place energy back into the system. This action is done without a pawl, but the locking force is produced by a wedging action of the balls or rollers. These devices may be used to brake or stop inclined-loaded conveyor belts from rolling backwards. They may also be used as feed mechanisms or as free-wheeling devices in an automobile automatic transmission, Fig. 17-39.

Levers (Four-Bar Linkage). Levers are used throughout mechanisms. They may be moving or stationary members of the mechanism, Fig. 17-40. Levers will provide mechanical advantage to a mechanism or will change the direction of forces applied to the lever, its displacement path, velocity, and acceleration, Fig. 17-41.

Levers are usually rigid members which will follow restricted paths controlled by the ratio of their length to each other and the design of their joints to which the lever is anchored. A four-bar system may be used to analyze many machines, Fig. 17-42. The anchoring of the system's members in turn makes it possible to design these four motions: (1) rotary-to-rotary motion, (2) rotary-

Fig. 17-41. A linear/linear mechanism used as assembly and press-loading machine feeders.

to-oscillatory motion, (3) oscillatory-to-oscillatory motion, and (4) oscillatory-to-reciprocating motion.

By **inversion,** or locking different members of the bar system so that they become the mechanism's frame, totally new machines may be developed, Fig. 17-43.

Electrical Machines

Electrical systems used in machines for automation are really electromechanical machines. The most frequently used electrical machines are the electric (D.C. and A.C.) motors, switches, relays, solenoids, generators, and alternators. Often the electrical machines apply electromagnetic induction to convert the electrical energy into mechanical movement.

Electrical Motor. Electrical power sources for industrial motors are of 60 cycles per second and use nominal voltages of 120, 120/208, 240, 480, 600, 2400. Voltages of 120 are used on motors from 1 to 15 horsepower; 208, 240, 480, and 600 volts are used on motors from 1 to 200 horsepower, and 2400 volts are used on those that are 444 horsepower and larger. The types of motors used in manufacturing fall into groups based upon the needs of the equipment, Fig. 17-44.

Fractional horsepower induction motors are used in large numbers to power small

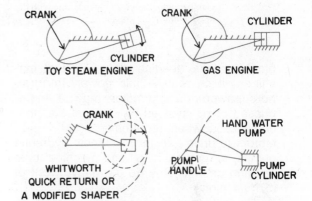

Fig. 17-43. With different bars held rigid, entirely new machines and motions are designed.

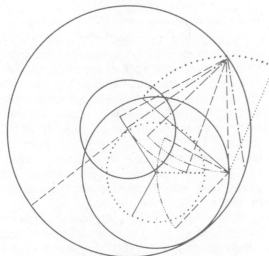

Fig. 17-42. Four-bar linkage system.
  A. An analysis of a four-bar link system for inversions of motions that may be applied to many different machines.
  B. By combining levers (bar linkage) and cams, part handlers and manipulators are designed to lift, transfer, and place parts in a machine.

Fig. 17-44. Electric motors are the prime source of mechanical energy in manufacturing. This is an exploded view of a 5000-horsepower industrial electric motor.

Fig. 17-45. The workpiece is moved to position and clamped into the broaching machine with the application of air pressure and cylinders.

blowers, fans, feeding devices, and machines which require little power. **Integral-horsepower** induction motors may be single-phased or polyphased motors that are generally used when requirements are larger. Motors to drive transfer machines, hydraulic pumps, air compressors, and machine tools needing great energy do the heavy work of manufacturing.

Pneumatic Activators

Automated pneumatic or compressed air machines shift control levers, load and unload machines, transfer work from one position to another, clamp the work during machining, and in some cases drive the spindles of machines, Fig. 17-45. **Pneumatic power** is frequently obtained from an air cylinder. The pressure on the piston provides the energy to the rod which performs the work.

Pneumatic devices are controlled by valves which are operated by levers, solenoids, microswitches, or by cams mounted on the parts of the machine to synchronize the movement of the machine. Collet chucking, vises, and other holding devices are frequently air-activated, because they are quick and sure holding tools.

When large amounts of energy are needed and accurate control is not needed, air is frequently used. The equipment is generally less expensive than hydraulic equipment. A small leak in a seal or connection does not contaminate the area as oil would, but it does cause the equipment to work less efficiently. Air is available at no initial cost and can be stored under pressure for instant use. It will operate devices at higher speeds than hydraulic equipment. The pipe circuits are less expensive, because the exhaust air may be released back into the air after use and need not be returned to the reservoir as in the case of hydraulics.

Compressed air has the disadvantages of temperature change that causes condensation of water in the air lines. This produces rust and scale that may damage equipment. Air systems have large contact areas such as between the piston and a cylinder wall. To reduce cylinder wear, a lubricator is installed in the system so that all the points of friction may be lubricated by an oil mist. Compressed air is a reliable source of energy when power, speed, and simple controls are needed to perform the work.

Hydraulic Activators

Hydraulically operated machines supply power and control for automated manufacturing. Heavy work done in extrusion presses, stamping presses, elevator rams, and machine tools is accurately and powerfully controlled. Hydraulic power is used in automated transfer machines for moving, lifting, rotating, and dumping out chips from the work between stations, Fig. 17-46. Hydraulics may be used to raise, position, and locate work and to operate chucks, clamps, or vises.

Fluid Multiplication. Hydraulic systems are designed to create great power with infinite control over the movement of the work or tool. This power and control is the result of the properties of liquids. For all practical purposes, liquids are not compressible. Thus, if a volume of liquid is forced into a hydraulic system, a pressure builds. The pressure in the hydraulic system will be equal in all branches of the circuit to which

Fig. 17-46. Hydraulic and electrical power do the heavy work of moving, lifting, rotating, and machining the parts in this automated transfer machine.

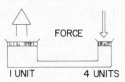

Fig. 17-47. Principle of fluid multiplication and transmission.

the pressure is applied. The liquid is applied to a device such as a hydraulic cylinder or a fluid motor which is capable of changing its volume, so the energy can be put to work. **Fluidic multiplication** is brought about by changing the area that the hydraulic fluid has to act upon. At a given pressure, a larger cylinder will deliver a greater amount of force. The volume of fluid is moved, and the force is increased by the difference of area and the pressure applied, Fig. 17-47.

Force = Pressure × Area

The advantages of hydraulic systems are control, energy, and cost. Industrial hydraulic systems operate in a range of 500 to 3000 psi (with special applications, up to 5000 psi). Thus, when great pressure is applied to different sized pistons, they will deliver vast amounts of energy over the designed length of the cylinder. A pressure of 2000 psi will deliver one ton of force for every square inch of piston surface. By changing the size of the cylinder and piston, the desired force can be controlled. Direction, speed, and amount of movement of a cylinder can be very accurately controlled by using volume and directional control valves.

Hydraulic Circuit Components. Hydraulic circuits to produce power and movement may be assembled out of components, Fig. 17-48.

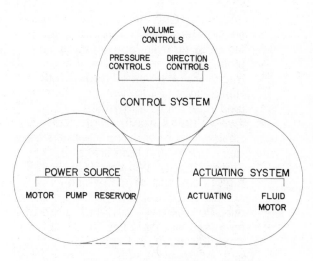

Fig. 17-48. Hydraulic power systems are made up of hydraulic circuits which provide a power source, control system, and an actuating system.

Hydraulic Circuits

1. **Direction Control Cricuit:** Direction of actuator by use of 2-, 3-, 4-, and 5-way valves.
2. **Flow Control Circuit:** Volume of fluid going to actuators, thus its speed of movement.
3. **Metering Out Circuit:** Speed control by controlling discharge rate of actuator.

4. **Metering-In Circuit:** Speed control by controlling the input rate into an actuator.

5. **Bleed-Off Circuits:** Speed control by returning to the reservoir part of the hydraulic pump's output.

6. **Reciprocating Circuits:** Single cycle of the cylinder or continuous cycles brought about by cams, limit switches, or sequence valves to pilot the directional change of a 4-way control valve.

7. **Sequence Circuits:** Second hydraulic cylinder does not begin to operate until first cylinder's action is complete.

8. **Rapid-Traverse and Feed Circuits:** Rapid advance to work, adjustable feed rate, and rapid return.

9. **Two-Volume Circuits:** High and low. Rapid advance of cylinder with a slow rate of travel and high pressure.

10. **Servo Circuits:** Will precisely control the velocity and/or the position of an actuator. Control of material, handling, location, velocity, and speed of machine tool spindles and tables. Control tension and speed of coiling and rolling machinery, control pressure and speed of forging press rams, control position and index of tables on punch presses.

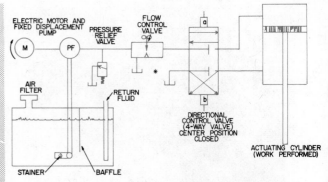

* THE SYMBOL ⊥⊥ MEANS THE LINE RETURNS TO THE RESERVOIR

Fig. 17-49. An example of a complete hydraulic circuit as an engineer would design it. This circuit will move the actuating cylinder forward or backward or lock it in any position.

Hydraulic Advantages. Hydraulic systems have many advantages over other control systems for automation, because they will withstand considerable physical abuse, they are self-lubricating, and they may also be stalled without destroying themselves. They are frequently more easily understood by operators and maintenance workers, they produce great power with precise control, and they are industrially safe to work around, Fig. 17-49.

Hydraulic systems require maintenance of screens, filters, pumps, hydraulic fluid levels, seals, or general leaks. Hydraulic systems are under high pressure and thus require a high quality of parts, fitting, and tubing. If a leak occurs, it may quickly be found, but a large amount of hydraulic oil spilled is messy and dangerous. The cost of a production run by a hydraulic system is about one-fifth that of a pneumatic system.

Mechanical Power Transmission Systems

Mechanical power transmission systems include chains, belts, gears, adjustable-speed drives, speed reducers, clutches, brakes, couplings, universal joints, and flexible shafts.

Chains. Mechanical power can be provided to the place of work by a chain drive. A chain is a positive (nonslipping) drive unit that locks the sprockets together as a power transmitting unit. Chains have an advantage in that they can transmit power between two shafts without serious concern for their center distances. Chains are easily assembled or replaced and may operate in high-temperature areas. Chains may be used for mechanical advantage by changing the ratio between the sprockets.

Fig. 17-50. Roller chains are made in a single or multiple strand, depending upon the horsepower to be transmitted. The chain may vary in pitch between ⅜" to 3".

Fig. 17-51. Inverted-tooth silent chains travel at high speeds. In heavy industrial power drives they will transmit up to 1200 horsepower.

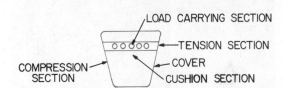

Fig. 17-52. A cross section of an industrial V-belt. When it is in use, the belt is wedged against the side walls of the pulley grooves.

There are a number of power transmission chain types. The most frequently used high-speed chains are the transmission roller chain and the inverted-tooth silent chain.

Roller chains may be single-stranded like a bicycle chain or multiple-stranded for high horsepower transmission. Roller chains are developed to operate at speeds up to 10,000 r.p.m. on small pitch sprockets. Chains with heavy multiple strands will operate up to 1000 to 1200 horsepower, Fig. 17-50.

Inverted-tooth silent chains are high-speed power drives, Fig. 17-51. These chains are made up of alternately assembled tooth links. They are joined with pins or joint components that make them powerful but flexible. This type of chain is found on drives of machine tools, pumps, power shovels, and power take-off drives up to 1200 horsepower.

Belts. Belts are used to transmit power from one pulley and shaft to another. Industrial power transmission with belts is largely performed by two types of belt designs: V-belts and flat. These belts are used in many different cross-sectional shapes and materials and are about 95% efficient in regards to slippage.

The standard industrial V-belt is constructed of a number of sections, Fig. 17-52. The load-carrying section may have cords which are made of nylon, rayon, dacron, glass, fiber, or steel. A cushion section is around the tensile carrying cords, and it absorbs the shock when quick load changes occur. The top section helps to carry the tensile load, but it retains the position of the tensile member and aids in equalizing the cords under load. The compression section holds the cords from the bottom and provides traction by the wedge pressure against the side walls of the pulley groove. The cover of the belt is made up of rubberized fabrics and provides the running surfaces and binds the belt into a unit.

Flat belts are an economical method of transmitting power from one pulley and shaft

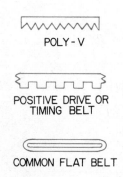

POLY-V

POSITIVE DRIVE OR
TIMING BELT

COMMON FLAT BELT

Fig. 17-53.   Types of flat belts used by industry for power transmission.

Fig. 17-55.   A 7000-horsepower drive designed to power a cement-grinding ball mill. The gear train reduces the speed of the shaft and increases the torque output. All bearings have high-pressure oil pumped to them for lubrication.

Fig. 17-54.   The groove belt transmits power efficiently, smoothly, and quietly.

to another pulley and shaft, Fig. 17-53. Flat-belting is efficient at high speeds and will work under adverse environmental conditions of dust, heat, and chemically active materials. Flat belts require tension for efficient power transmission or else slippage will occur at moderate and low speeds. To counteract this, **serpentine drives** are sometimes used, because they engage more surface of the driving pulleys.

Flat belts are made of rubber, plastic, rubberized fabric, cord, fabric, leather, reinforced rubber, plastic, or leather. These are the types of flat belts used by industry: (1) the **common flat belt,** (2) the **poly-V or groove,** and (3) the **positive-drive belt** (timing belt).

These belts are variations of the common flat belts. The grooved belt provides for additional "traction," and the power transmission efficiency is raised without seriously increasing bearing loads, Fig. 17-54. The positive-drive belt equals the efficiency of gears with no slippage or speed variation between the driving and driven pulleys.

The chief advantage of belts for power transmission is that it is an economical means of obtaining power when the center distance of the two shafts cannot be controlled or are a long distance apart.

Gears. Automation requires the use of gears to transmit power, time, movements, and give information to many parts of an automated machine. Gears are efficient, because there is a rolling contact between the gear teeth. The center distance between the two shafts of the gears must be very carefully designed and controlled. The distance is dependent upon the ratio required, the diametrical pitch, and circular pitch of the gears, Fig. 17-55.

Since gears are basically levers, if they are of a different diameter, the ratio between the shaft rotation is changed. With the speed change, a torque change results, Fig. 17-56, 17-57, and 17-58.

The rotation of gear trains can be changed by adding or removing an idler gear to the gear system, Fig. 17-59. This is done by a

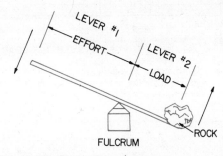

Fig. 17-57.   Gears running together function as a first class lever system.

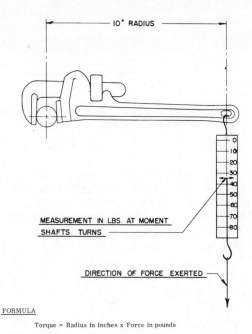

FORMULA

Torque = Radius in inches x Force in pounds

Torque = 10" x 35 # = 350"#

Fig. 17-56.   Gears are basically levers.

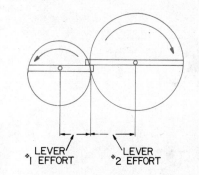

Fig. 17-58.   Torque multiplication is achieved by running gears with different diameters together.

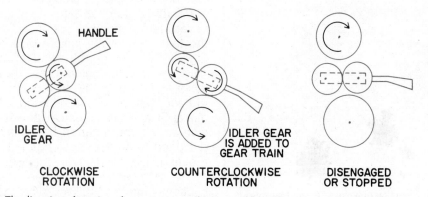

CLOCKWISE ROTATION

COUNTERCLOCKWISE ROTATION

DISENGAGED OR STOPPED

Fig. 17-59.   The direction of rotation of a gear train may be reversed by adding or subtracting an idler gear to the gear train.

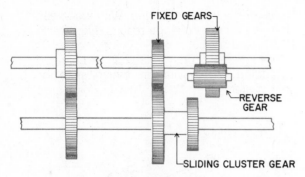

FIXED GEARS

REVERSE GEAR

SLIDING CLUSTER GEAR

Fig. 17-60. The direction of rotation may be changed by adding a sliding gear and a reverse gear to transmission.

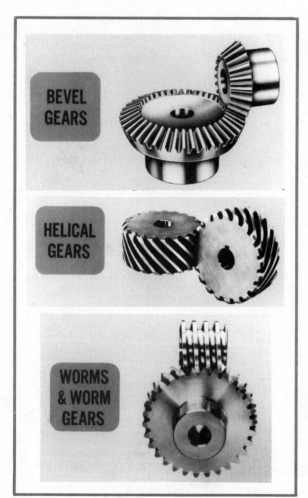

BEVEL GEARS

HELICAL GEARS

WORMS & WORM GEARS

Fig. 17-61. The direction or angle of power transmission may be changed by the use of bevel gears, miter gears, helical gears, and worm gears.

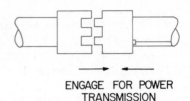

ENGAGE FOR POWER TRANSMISSION

Fig. 17-62. Positive clutch: power is transmitted in both directions of rotation.

Fig. 17-63. Spiral-jawed clutch power-transmitted in one direction of rotation.

movable gear or by a sliding gear, Fig. 17-60. A normal clockwise rotation may be reversed by adding a gear to the gear train. The power transmission may be stopped by disengaging a gear in the gear train.

The direction of power train may be turned 90° by bevel gears, miter gears, worm gears, and helical gears, Fig. 17-61. These gears may have intersecting shafts that may be skewed between 0° and 90°.

Clutches. A clutch is a device that is designed to engage, start, or disengage rotating machines without stopping the source of power. Clutches are activated by mechanical, electrical, or hydraulic systems. Mechanical clutches may be positive-action clutches or friction clutches.

Positive clutches are constructed with square-jawed contact surfaces which carry the total load at the instant of engagement, Fig. 17-62. The **spiral-jawed clutches** also have instant engagement, but they may release if or when the driven unit speed is greater than the driven unit, Fig. 17-63. This makes it a single-direction power transmission unit. A **tooth clutch** has many specially designed teeth that are engaged, thus distributing the load over many teeth, Fig. 17-64. This type of clutch has a high torque capacity and the ability to be engaged at different speeds.

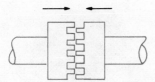

Fig. 17-64. Tooth clutch: power is transmitted in both directions of rotation for high torque loads.

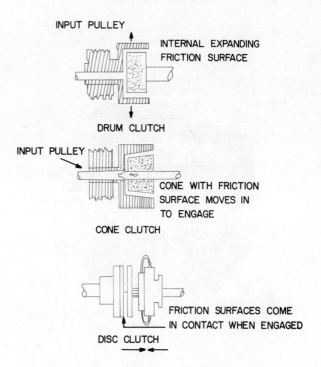

Fig. 17-65. Friction clutches are designed to slowly engage a load on a power transmission shaft.

Friction clutches are designed to slip as the load is being engaged in order to reduce destructive shock which would be transmitted to the power train, Fig. 17-65. Rim (or drum), cone, and disc clutches either expand or contract, depending upon whether the design is an internal or an external friction shoe. In each case a friction surface similar to brake lining comes into contact with a mating surface. It is powered by cams, levers, electricity, hydraulic, or pneumatic drives. The mating surfaces pick up the torque load gradually so that a minimum of shock is transmitted to the power transmission system.

Brakes. From time to time power trains need to be stopped or braked as a part of the operation or in case of an emergency. A brake is really a redesigned clutch. One member is stationary, and when the friction material is engaged with the rotating member, the energy of the rotating system is absorbed and converted to heat. Brakes are operated by mechanical levers and cams, pneumatic or hydraulic cylinders, or by electricity. They are available up to 250 horsepower, Fig. 17-66. Mechanical and hydraulic brakes are made with the same concepts as automobile brakes, while electric brakes use electrical principles of eddy currents, magnetic particles, or hysteresis, Fig. 17-67 (page 528). They operate very smoothly and do not wear. They may be controlled remotely and easily.

Couplings. Couplings are designed to join two power shafts. They may be rigid or flexible. Industrial shafts may be fastened rigidly with sleeve couplings when the shafts are perfectly aligned and are the same size.

Fig. 17-66. Electric brakes and clutches are engineered so they are usable on standard power transmission components, motors, reducers, shafts, and couplings.

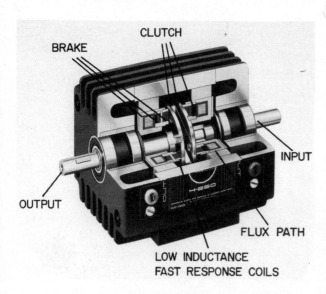

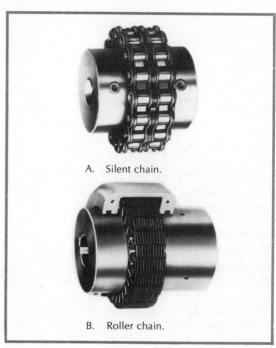

Fig. 17-67. A high-performance module of a clutch and brake unit designed for computer peripheral equipment. These units use electromagnetic forces to both start and stop the load.

A. Silent chain.

B. Roller chain.

Fig. 17-69. Silent and roller chains are used to make flexible couplings for low horsepower transmission shafts.

CR-TYPE STEEL

FC-TYPE THREE JAW

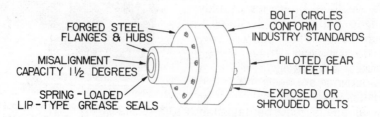

FORGED STEEL FLANGES & HUBS

BOLT CIRCLES CONFORM TO INDUSTRY STANDARDS

MISALIGNMENT CAPACITY 1½ DEGREES

PILOTED GEAR TEETH

SPRING-LOADED LIP-TYPE GREASE SEALS

EXPOSED OR SHROUDED BOLTS

Fig. 17-68. Rigid couplings are of two types: sleeve and flange design. A rigid coupling requires the shafts to be running in alignment.

Fig. 17-70. A dihedral gear-flexible coupling is used when torque and misalignment are severe.

Flanged, ribbed, or compression couplings are used when higher horsepower is transmitted or when shafts of two different diameters are fastened together, Fig. 17-68.

Flexible couplings are designed by a number of different methods, depending upon the horsepower that must be transmitted. Chain couplings will accommodate the same or different shaft diameters, and flexibility allows for slight misalignment. Chain coupling may be made from silent chain or roller chain design, Fig. 17-69.

Couplings for heavy power loads such as those required by cranes, dredge pumps, and mill drives will have dihedral gear couplings or a universal joint type of coupling if adjustment or severe misalignment is needed or occurs in a power train, Figs. 17-70 and 17-71.

Fig. 17-71. Universal flexible couplings are used to drive this automobile tailpipe in a tube mill. All shafts and rolls must revolve at the same speed and still be adjustable.

Related Occupations

These occupations are related to automation principles:

Control engineer
Mechanical engineer
Systems analyst
Computer systems technician
Instrument mechanic
Electronic technician
Electromechanical technician
Electrical engineer
Electrician
Mechanical technician
Millwright
Maintenance mechanic
Fluid power mechanic

Automation Technology

Chapter
18

Words You Should Know

Interlocked — All action is synchronized to ensure that a mechanism will operate in proper sequence with another mechanism.

Pallets — A platform or surface upon which a holding fixture and workpiece may be mounted as it is moved through an automated machine.

Modular block tool units — Standardized units or sections for easy construction or flexible arrangement.

Transfer bar mechanism — A shuttle with a reciprocating motion with a finger which engages the work or pallet and slides it to the new work station.

Numerical control (N/C) — The control of precise movements of machines by coded number data.

Real time — Information received quickly enough so that a part being machined that is going into an error can be corrected before it goes out of the tolerance range for the part.

Binary number — A number system based on two digits 0 to 1.

Position value — The number 2 raised to the 0 power yields a position value of 1.

Row — A series of holes 90° to the edge of an N/C tape.

Character — A collection of holes on a row that represents a number, letter, or symbol.

Word — A combination of characters using a number of rows to make up a command.

Block — A group of words together to make up a command for a specific machine movement or function.

Channel — A series of holes running the length of the tape.

Parity check — A safety check. A hole is added to an even number of holes in a word so that the total number of holes in the row will be an odd number.

End of block — The end of information and the command for machine to respond.

Miscellaneous functions — Machine control functions which are on or off.

Machine axis — Movements along and around the planes of a machine.

Point to point — The N/C movement of the work in a machine in a straight line from one location to another for a moving or cutting operation.

Continuous path — The N/C movement of the work in a machine in a contouring or sculpturing fashion.

Computer manufacturing control — A computer center that controls management, engineering, and manufacturing activities of production.

Computer memory — Coded information stored by an electromagnetic charge.

Programming — Coded instructions that describe the shape, size, locations of surface features, and holes on the part.

Monitor — A device to observe and record the operations of a machine or system.

Interface — A connecting unit between two pieces of electronic equipment.

Monorail — A single overhead rail that supports a part's moving system.

Dial assembly — Rotating tables with a number of assembly stations around the table's edge. The work moves from stations around the table.

Double tool plates — Tools mounted on circular plates above or below the assembly table.

In-line assembly — Parts are assembled as they move past work stations in a straight line.

Part feeder — A device that orients and positions parts and/or fasteners and transports them to the assembly point.

Surveillance — A system that watches over manufacturing so that information is available for management and controlling.

Multiple Machine Automation

Machine automation involves a process in which a machine or machines perform a sequence of production operations which are guided by other machines. The timing information, position, and power of the machine operations must be interlocked so that machine control is achieved. Transfer machines receive the coordinating information through microswitches, proximity switches, pressure switches, and sequence valves. Numerically controlled machines need greater information or communication, because a position and sequence of control must be defined. Numerical control (N/C) requires a machine language which communicates and defines the operations.

Transfer Machines

Transfer machines consolidate many manufacturing machines and operations into one machine. Material processing, material handling, and gaining information about the condition of the tools and work while the manufacture is in progress — all of these processes can be automatically performed within the transfer machine. The transfer machine moves the part to a point where it can

Fig. 18-1. Transfer machines are many machines interlocked into one machine. Large workpieces are moved and positioned from work station to station automatically. This 15-station transfer machine machines Cadillac cylinder blocks.

be positioned and clamped while the work is being performed. Then the part is moved to the next work station, where additional work is performed.

Transfer machines will accept work mounted on **pallets** or fixtures upon which one or more piece parts are mounted. Larger workpieces with reference surfaces or holes may be moved directly into the machine, located, and clamped at each processing station. Workpieces must be supplied constantly so that each time the machine cycles, a part is available for the machine to process at the next station, Fig. 18-1. Stations have additional individual control so that the tools may be corrected for tolerance drift during the manufacturing run. The total transfer machine with all its operations and stations must be electrically interlocked so that the transfer machine maintains its timing and sequence.

In the process of manufacturing in the transfer machine, castings or forgings will be moved, oriented, clamped, machined, turned over, cleaned, rotated, lifted, lowered,

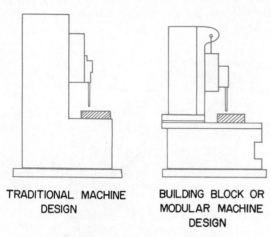

TRADITIONAL MACHINE DESIGN

BUILDING BLOCK OR MODULAR MACHINE DESIGN

Fig. 18-2. A machine tool of traditional design has the position and angle of sliding members cast into machine tool castings. The modular block machine tools are built up of a series of units that may be reassembled into new configurations on a new job.

Fig. 18-4. Angular wing base units, ways, and head units are combined to build this transfer machine. It completes 107 different operations on a V-8 cylinder head.

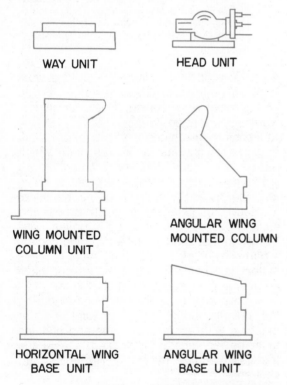

WAY UNIT

HEAD UNIT

WING MOUNTED COLUMN UNIT

ANGULAR WING MOUNTED COLUMN

HORIZONTAL WING BASE UNIT

ANGULAR WING BASE UNIT

Fig. 18-3. The basic units for modular machine tool building are base, column, way, and head. Other units such as angular units and wing units may be added.

gaged for broken tools and size, unclamped, and moved to the next process of manufacture. All of this occurs with automatic precision.

Modular Block Tool Units. Standardized units are designed so that they may be assembled into different machines. Traditionally machines have been built with a specific machining operation designed into the tool. The position and angle of the sliding tool member was designed into the casting, Fig. 18-2. Modular tools are built into different configurations or positions. The units may thus be replaced for repair or may be disassembled and reassembled into a new configuration for a different product or machine, Fig. 18-3. This concept of machine automation may be applied to build up any shape of production machine, Figs. 18-4 and 18-5.

Production Line Movement. The type of production line movement for automation depends upon the number of pieces to be made; the specifications and dimensions of the part; the methods of locating and clamping the parts; and the working heights required to load, machine, and unload the parts.

In automation, a number of types of machine tool organizations are used: single-

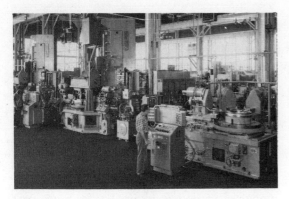

Fig. 18-5. Modular blocks of machine tools built into different shaped production machines.

Fig. 18-7. A circular indexing machine's chief advantage is the small amount of factory floor space needed for high production.

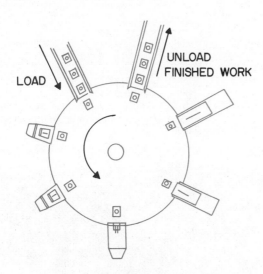

Fig. 18-6. Circular indexing machines move the work in a circular path between work stations.

Fig. 18-8. Circular indexing machines compact the electric, hydraulic, air circuits, and power supply into a small area.

station machines, rotary dial or circular indexing machines, vertical dial or truniontype machines (for medium production lines), indexing machines, and straight line machines (for mass or high production).

In **circular indexing automation,** the work is moved through a circular path, Fig. 18-6. The work is indexed or moved to a station and is clamped. The cutting, forming, or welding processes are done at that station. When the work is complete, the machine indexes to the next station and a new group of tools perform the next operations.

The chief advantage of the circular indexing machine is that it does not require a lot of production floor space, Fig. 18-7. The power, electric sources, and hydraulic sources are close at hand to supply and control the machine, Fig. 18-8. Time is saved because work-holding fixtures are returned to the area, where they will again be needed. Problems found in circular indexing deal with planning the work time at each station so that each operation requires the same amount of time, Fig. 18-9. The work can only index as fast as the time required for the slowest station of the machine. Also, vibration created at one station may interfere with work being done at another. With proper design and maintenance, circular indexing machines are efficient and productive, Fig. 18-10.

Straight line automation has intermittent movement of the part from one work station to another, Fig. 18-11. The parts are moved through the machine by **transfer bar mechanisms.** The work is moved station-to-station between horizontal, angular, or vertical machining units, Fig. 18-12. If the work is small or irregular, it may slide through the machine mounted on pallets, Fig. 18-13. If the work is large and has machined surfaces such as an engine block or head, the workpiece itself may slide through the machine.

There is no limit to the number of part faces that can be machined in a transfer machine, because hydraulic-powered part

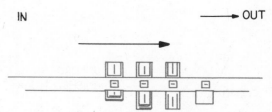

Fig. 18-11. A straight line transfer machine moves the work intermittently from work station to work station.

Fig. 18-9. In a circular transfer machine the work is loaded in and out of the machine at or near the same station, increasing the efficiency of part handling.

Fig. 18-10. Circular indexing machines are also turned so that the work path is in a vertical plane. This 12-station machine head consists of nine drills, six boring spindles, and two accelerated reaming clusters used to produce oil pump bodies.

Fig. 18-12. A straight line transfer machine that will machine V-type diesel engine blocks. The blocks may be made of aluminum or cast iron and of 6-, 8-, or 12-cylinder construction.

Fig. 18-13. An 18-station palletized transfer machine producing differential carrier covers at the rate of 55 parts per hour.

Fig. 18-15. When a tool or group of tools is placed in a machine, the toolmeter is set to permit this tool to machine a certain number of parts. When the toolmeter cycle indicator reaches zero, the machine is automatically stopped and the dull tools are replaced by sharp tools stored in the board.

Fig. 18-14. Many part faces may be machined by including part turning and orienting devices between work stations.

turnover and orienting devices may be placed at any point in the transfer machines, Figs. 18-14 and 18-15. A transfer machine's movements are interlocked electrically within and between work stations. The machine automatically performs the following functions:

Transfer Machine Movements

1. Transfer the pallet or workpiece to the station.
2. Locate the pallet or work accurately under the tool and lock or clamp the work.
3. Start the tool down to the work at high-speed feed (rapid traverse).
4. Slow the tool to predetermined feed rate, cutting speed, and depth of cut to perform the machining operation.
5. Automatically return the cutting tool to starting position at high speed (rapid traverse).
6. Release the clamps on the workpiece so the transfer bar may move the work to the next station.

Intermittent line transferring mechanisms have various methods of transferring the work:

1. The pallet or workpiece is supported by round or flat bar ways and slides along by a mechanism.
2. The pallet or workpiece is supported and moved on rollers by gravity or by a mechanism.
3. The pallet or work is moved by a shuttle, making use of finger, pawl, or walking

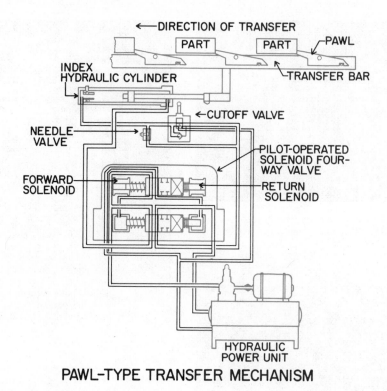

←DIRECTION OF TRANSFER

PART   PART   PAWL

INDEX
HYDRAULIC CYLINDER   TRANSFER BAR

←CUTOFF VALVE

NEEDLE
VALVE

PILOT-OPERATED
SOLENOID FOUR-
WAY VALVE

FORWARD
SOLENOID   RETURN
SOLENOID

HYDRAULIC
POWER UNIT

PAWL-TYPE TRANSFER MECHANISM

Fig. 18-16. Pawl-type transfer bar mechanism is an inexpensive and reliable method of moving pallets or large-sized workpieces through a transfer machine.

Fig. 18-17. The pawl in the transfer bar engages the work-piece and slides it intermittently down round or flat bar ways.

beam transfer bar mechanism, Fig. 18-16.

A shuttle transfer unit may be powered by air, hydraulics, or by a chain. In each case the shuttle uses a reciprocating motion. A pawl or finger engages the work and moves it to the new station. The shuttle moves back to the first position without moving the part and is ready for the next index, Fig. 18-17.

A Different Manufacturing Concept — Numerical Control

In mass production transfer machines use a separate tool and machine within the transfer machine for each manufacturing operation. This results in many specialized holding and clamping fixtures, jigs, and specialized tools. Mass-produced items are manufactured by using conventional tooling, jigs, and fixtures to produce accurate parts quickly.

Numerical control is a fundamentally different concept of manufacturing. It has only a simple holding fixture for the work and the accuracy of the workpiece is not given

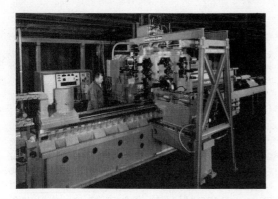

Fig. 18-18. "The Parts Maker" is a numerically controlled machine that will make a series of identical finished parts from a single piece of bar stock, tubing, or extrusion. Its accuracy is dependent upon the N/C tape-controlled precise movement of the machines.

Fig. 18-19. The ability of the machine to be programmed for many difficult part shapes gives the numerically controlled machine great flexibility for production.

Fig. 18-20. The machine's instructions are coded into a punched tape that the machine tool tape reader can use to precisely control the machine.

by a jig. Instead, the accuracy results from the accuracy of the machine, its precise movements, and by the use of a number of accurately preset length tools. With accuracy built throughout the machine, tools, and positioning system, great machine flexibility is achieved, Fig. 18-18. Work which has changing dimensions on two or more axes may be accurately manufactured. Before the work would have been machined, repositioned, and finished by the use of different machine tools and setups.

Numerical control is not for mass production. However, because of its flexibility and its ability to be programmed for many different and difficult jobs, this new concept of manufacturing is most successful, Fig. 18-19.

Communication to a Machine. An accurate method of providing information to machines is essential for numerical control. A reliable way to communicate between thinking people and a working machine is the use of the binary code system that an electronically controlled machine and a person can both understand. The machine's instructions are coded into a punched tape that is later read by a machine tape reader, Fig. 18-20. The machine receives these precise instruc-

tions, which may be repeated endless times to produce duplicate manufactured parts.

A more recent method of communicating with a machine is to program these instructions into a computer by a thinking person via punched cards, a punched tap, magnetic tape, digitizer, and keyboard terminal, thus to the machine. The computer will direct all

Fig. 18-21. The contour milling of a star-shaped cam slot on the face of the part in a numerically controlled lathe. Many points of location per inch are computed, and the information is punched into the control tape. The tape controls the movement of the end mill and the rotation of the work.

the machine's movements and will also control them precisely.

Information Processing Requirement. When numerical control is used, a great volume of information must be accepted and processed quickly so that a logical and accurate decision may be made. A great deal of information is needed to cause a machine to cut a curve on a workpiece. As many as a thousand points per inch may need to be calculated and located for the machine. The computer has the speed and the ability to accept massive volumes of data, rapidly reduce the data according to a mathematical equation, and command the machine control unit to position the workpiece so that the cut is accurately made. The cut may be made immediately when the machine is activated, or the information may be punched into 1″ tape and the cut made at a later time, Fig. 18-21.

Economic Requirements. Since the machines in numerical control can be reprogrammed for new products readily, small lot size work can be economically manufactured.

Through the use of these rapid, efficient manufacturing systems, production may be increased with the same number of workers. Scrap or wasted parts are reduced because of the accuracy of the machine. Greater running time of machines is achieved, thus higher production results. Work is performed at a higher rate of speed, thus additional productivity is possible.

Planning for Numerical Control. Manufacturing by numerical control requires detailed planning. How many parts are needed? How complicated is the part's shape? How costly will the part be if it is manufactured by numerical control? These are questions that will be asked before the decision will be made for numerical control production. Another manufacturing decision will be made concerning the metal from which the parts will be made. The principle manufacturing tools today are designed for ferrous metal cutting. Ferrous materials have serious limitations because of their low metal removal rates.

The machineability of metals and alloys varies greatly, thus this factor must be considered when numerical control is used for manufacturing. The light metals have a high metal removal rate that may be used advantageously. For example, gray cast iron may be milled at a cutting speed of 125 to 200 surface feet per minute, while aluminum may be milled at a cutting speed of 1000 to 2000 surface feet per minute. The choice of material and design can greatly reduce the cost of production, even when a more costly metal is used.

Another planning decision that is important to manufacturing is the detailed scheduling of machines. The traditional craftsman-operated machine tools are actually cutting or working about 20% of the total manufacturing machine work time. The operator reads a blueprint, follows an operation sheet,

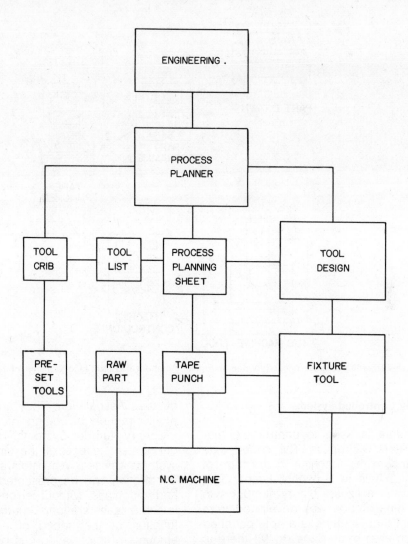

NUMERICAL CONTROL PROCESS

Fig. 18-22. Planning for numerical control permits the manufacturing engineer to direct the complete production of the part. Thus, greater productivity is achieved.

changes tools, measures, adjusts the machines, or plans the next operations. When machines are programmed, all movements are preplanned, tooling is preset, and the cutting speeds and feeds are precoded into the tape. The tape readers read all the necessary information so fast that the machine "down time" is greatly reduced, and the machine operates at near capacity.

The decision to manufacture with N/C starts with the planning procedure of the design, drawing, and specifications for the part. Time is needed to develop the process planning needed and to program the part. Schedules, operations, bill of materials, and tooling sheets will need to be completed. These planning efforts will result in a program on tape, the fixture tooling, the necessary preset cutting tools, the availability of the machine and raw parts at the scheduled place and time ready for work, Figs. 18-22 and 18-23.

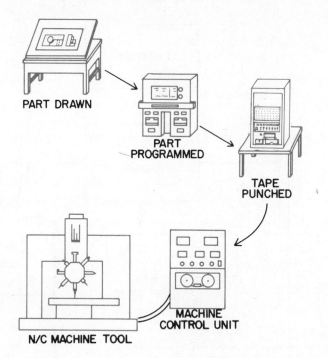

Fig. 18-23. The procedure needed to plan and make a numerical control tape to manufacture a part.

Numerically Controlled Systems

Coded data is used to control the machines in a N/C process. The coded data is presented to the machine in the form of a positioned hole in a punched tape. This coding system utilizes a binary number system. The binary code was developed from the concept that (1) there **is a hole** punched in a certain area of the tape or, (2) there **is no hole** punched in a certain position. These two alternatives mean that an electrical contact is or is not made. The patterns of holes in a tape will be located so that a code is built, and information may be transmitted.

The holes in the tape are used to construct a code similar to the Morse telegraphic code except that the information is presented by a position and pattern of holes and no holes. Having eight positions and combinations of patterns plus the increased information about the hole provides a code in which vast volumes of information may be quickly stored or read as instructions. The tapes used in the metal manufacturing industry were standardized by the Electronic In-

dustries Association. They are 1″ tape with eight channels of coding information.

Binary Number Systems. The binary number system is one of the simplest number systems devised. When this system is used, information can be presented to computers and machines without error. The symbols in some cases are long, but the machine can function at the speed of electricity. Vast amounts of information can be presented or read in a short period of time.

The binary numbers are based on two digits: 0 and 1. The system may be thought of in terms of an electrical light switch. If a switch is on and the light is burning, the position of the switch indicates a number 1 electrically. If the light switch is off, the light is not burning, and this indicates a number 0 electrically.

Each symbol of 1 or 0 is called a bit or binary digit and is represented in a numerical control tape by the absence or presence of a hole in the tape or a magnetic charge in a computer.

In binary coding the power value of the number 2 is important. Thus, 2 raised to the

Decimal Number	Power Value (Decimally Added)	Position Value 32 16 8 4 2 1	Binary Number
5	4+1	1 0 1	101
8	8	1 0 0 0	1000
9	8+1	1 0 0 1	1001
12	8+4	1 1 0 0	1100
27	16+8+2+1	1 1 0 1 1	11011

Fig. 18-24. Power values in a common number.

Power Value (of 2)	2^8	2^7	2^6	2^5	2^4	2^3	2^2	2^1	2^0
Position Value (Common Number)	256	128	64	32	16	8	4	2	1

Fig. 18-25. Coding a binary number.

0 power yields the position value or our common number value of 1, Fig. 18-24. The number 2 raised to the first power gives a position value in regular, common numbers of 2. An example of a binary number 5 would be coded into a tape or computer memory core as 101. In the binary code, in the number 5, there is 1 number 4 position value, 0 number 2 position value, and 1 number 1 position value. This adds up to 4 + 0 + 1 = 5. The binary number code is the system that is applied to computers and the numerical control tape, Fig. 18-25. These coded instructions are punched into a paper or metal-plastic tape. When they are read, they give precise instructions to the machine being controlled.

Binary Coding on Control Tape. A hole in the tape indicates a binary one (1) and the absence of a hole indicates a binary zero (0). A **row** is a series of holes 90° to the edge of the tape. A **character** is a collection of holes on a row and represents a number letter or symbol. A **word** is a combination of characters using a number of rows to make up a **command,** such as the four rows used to produce the command "1976," Fig. 18-26. A **block** is a group of words which together give complete infor-

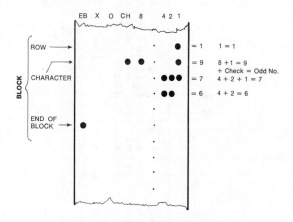

Fig. 18-26. Binary code on tape — 1976.

mation for a specific movement of the workpiece. The block gives all information of movement including the direction of movements, as well as the distance to be moved such as movement on the X axis of a machine + 68750 which would move the machine table to the left 6.8750″.

Coding Control Tape. One-inch control tape is made up of eight channels, levels,

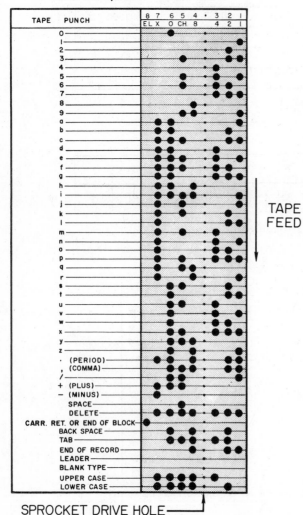

E.I.A. STANDARD CODING
FOR
1 INCH WIDE, EIGHT TRACK TAPE

TAPE FEED

SPROCKET DRIVE HOLE

Fig. 18-27.   E.I.A. standard coding for 1″ wide 8-track tape.

and a hole punched in channel 3 represents the number 4. The next series of holes down the tape is the sprocket drive holes. They are used to move the tape and are not a channel. A hole punched in channel 4 represents a number 8, and a hole punched in channel 5 represents a parity check or safety check. When two holes are punched to make a number, a check hole will be added to the character to produce a self-check in the code. An odd number of holes must appear in each row, or the machine will stop. This parity check indicates whether mechanical failures or errors have been made in the making of the tape.

Channels 6, 7, and 8 are used for information other than numerical values. Channel 6 is always represented as 0. Channel 7 is the code for X and is not used for dimensions. It also indicates the various machine functions represented by letters and special characters which are necessary to perform the machine operations. Preparatory functions required to drill, bore, ream, mill, or turn will be indicated here. An example of this code number would be G 81 for the movement of feed in which the machine would be set for a drilling or spot-facing operation.

Channel 8 is labelled EL and indicates the ending of a line of information. More frequently it is referred to as **End of Block.** It completes the command for the information to move the machine to the desired position.

Miscellaneous functions are of the "on" or "off" variety and control the function specified until an opposite command is read, Fig. 18-28.

M 00	Stop Program
M 03	Start Spindle Rotation Clockwise
M 06	Tool Change
M 08	Coolant On
M 09	Coolant Off
M 25	Skip Feed
M 27	Partial Retraction of Spindle

Fig. 18-28.   Partial list of miscellaneous functions.

or tracks of holes on the tape, Fig. 18-27. The channels or tracks run down the length of the tape and are coded so that a hole punched in channel 1 represents number 1, a hole punched in channel 2 represents 2,

Macnine Axis. Numerically controlled machines must have the axis of the **machine** identified so that the programmer who is writing the process planning sheet will have the work move the correct direction. To locate the work, the Cartesian Coordinate System is used, Fig. 18-29.

The axis of the machine movements may be thought of as the corner of a box. The X axis is the longest working travel of the machine and is usually horizontal. The Y axis is in the same plane but 90° to the X axis. It is the lateral movement or in-and-out work movement. The Z direction is vertical or at 90° to the XY plane. The Z direction is illustrated by the up-and-down motion of the quill of a drill press.

There are movements around each axis which are designated as a, b, or c, Fig.

18-30. The drill press illustrated above has a, − z direction and a, + C movement, because the drill is turning clockwise. The primary movements of the workpiece on the axes are + or − X, + or − Y, + or − z, + or − a, + or − b, + or − c.

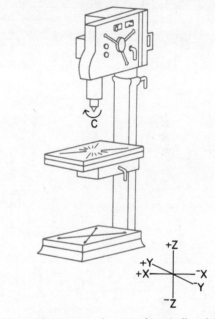

Fig. 18-31. The axes of a single-spindle drilling machine.

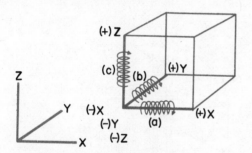

Fig. 18-29. Cartesian coordinate system.

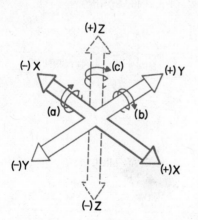

Fig. 18-30. Machine axes: the rotation around a machine's axes are a, b, c. The direction of rotation is + or − in each case.

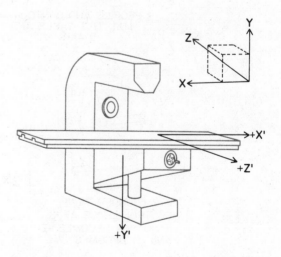

Fig. 18-32. The axes of a horizontal knee mill.

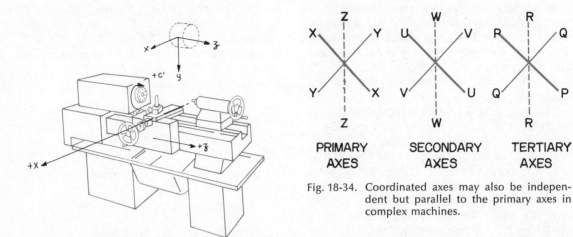

Fig. 18-33. The axes of an engine lathe.

PRIMARY AXES

SECONDARY AXES

TERTIARY AXES

Fig. 18-34. Coordinated axes may also be independent but parallel to the primary axes in complex machines.

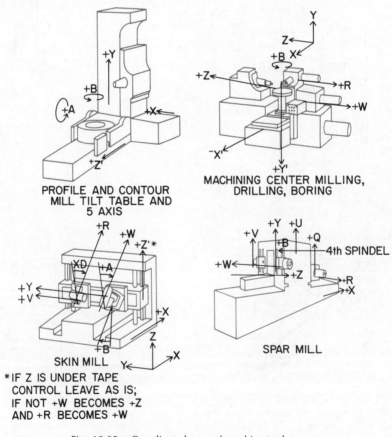

PROFILE AND CONTOUR MILL TILT TABLE AND 5 AXIS

MACHINING CENTER MILLING, DRILLING, BORING

SKIN MILL

*IF Z IS UNDER TAPE CONTROL LEAVE AS IS; IF NOT +W BECOMES +Z AND +R BECOMES +W

SPAR MILL

Fig. 18-35. Coordinated axes of machine tools.

Other motion-coordinated axes which are independent of but parallel to the primary x, y, z axes are called secondary axes and tertiary axes.

Bases of Dimensioning for Numerical Control

The Cartesian Coordinate System is the measuring system used for numerical control, Fig. 18-36. The system is the same as that used in mathematics with the central point located as 0,0.

In the first quadrant, all the numbers of points located are positive numbers, Fig. 18-37. In order to simplify coding and eliminate as many errors as possible, numerically controlled parts are programmed in the first quadrant of the Cartesian System.

The machine tool is set up so that the workpiece will be in the first quadrant and the cutting or drilling positions will always be positive. However, the movement of the tool may be in a negative direction on an axis.

Point-to-Point Machine Movements. In the point-to-point system, numerical control is performed by moving the workpiece from one point to another in a straight line to locate and perform the cutting or forming operation, Fig. 18-38. This system is the

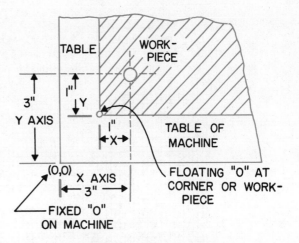

WORK IN IST QUADRANT

Fig. 18-37.   All the numbers of points in the first quadrant are positive.

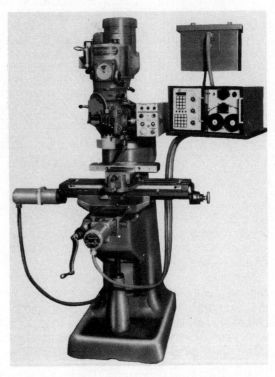

Fig. 18-38.   A point-to-point numerically controlled machine.

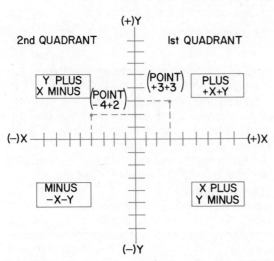

Fig. 18-36.   The cartesian coordinate measuring system.

ONE MOVEMENT   TWO MOVEMENTS   EQUAL RATE
MOVEMENT

Fig. 18-39.   Cutter paths for point-to-point machine movements.

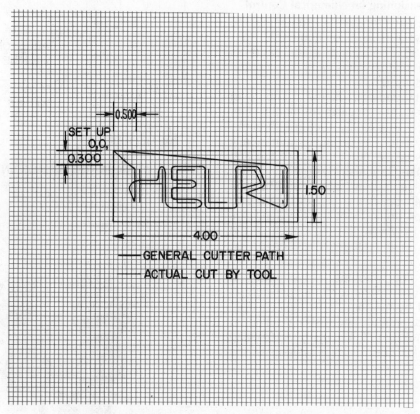

Fig. 18-40.   The program and layout for a desk paperweight — HELP.

simplest form of numerical control and is used to do drilling, straight-line milling, spot welding, riveting, electrical wiring, component assembly, and bending. It can be used on turret-type punch presses.

In drilling and milling operations the work will be controlled along the X/Y axis and the program can be written so that the machine will move along only one axis at a time, along both axes at the same time, or along both axes at equal rates, but not at differential rates, Figs. 18-39 and 18-40.

Continuous Path or Contouring Machine Movements. A continuous path machine is one in which the movements' cutters are constantly under the control of the tape, Fig. 18-41. It involves vast amounts of information to control the tools. This type of machine control provides a continuous path contouring or sculpturing in the x, y, and z planes. The contouring machine movements were originally designed for producing aircraft structural members such as wings and parts from solid slabs of aluminum to a very close dimensional tolerance, Fig. 18-42. They produced such a technological innovation that they overcame the near limit of forming by forging.

The tool can generate a complicated finished shape of the part, because the machine is able to be moved on from 1 to 5 axes continuously, whatever is required to produce the shape and desired tolerance. Only a computer is able to do all the calculations necessary to plot all the positions that a three-dimensional contouring cut requires, because vast quantities of informa-

Contouring machines are costly and very sophisticated. They are used only when other machining methods are not available or cannot perform the work, Fig. 18-43. tion are required to keep the tool in constant position.

Computer Manufacturing Control

Complete manufacturing control is achieved by the detailed planning and scheduling by the computer. Its capability to make a decision in "real time" and to adjust fast enough while the process is being carried out has

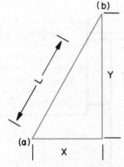

Fig. 18-41. A continuous path machine is constantly moving under the control of the tape. The length of a straight line which makes up a curve may be as small as .001″. Numerous points between (a) and (b) must be calculated and positioned on the X and Y axis of the machine generating the curved surface.

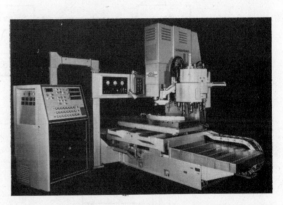

Fig. 18-43. Contouring machines will machine-curve workpieces.

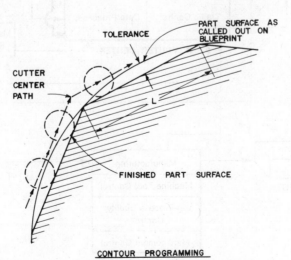

CONTOUR PROGRAMMING

Fig. 18-42. The cuts are calculated so that the distance between the straight line and arc remains within tolerance.

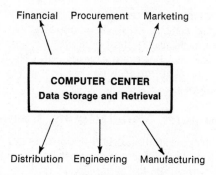

Fig. 18-44. Computer center for management and manufacturing control data.

made total manufacturing control possible, Fig. 18-44.

Machine control systems may include the total management, engineering, and manufacturing activities of production, Fig. 18-45. In this case the computer center becomes the bookkeeper and decision-maker for the marketing, financing, engineering, procurement, manufacturing, warehousing, and shipping departments.

The computer center has many functions. Control of inventory, material, tools, machines, and materials is a major function.

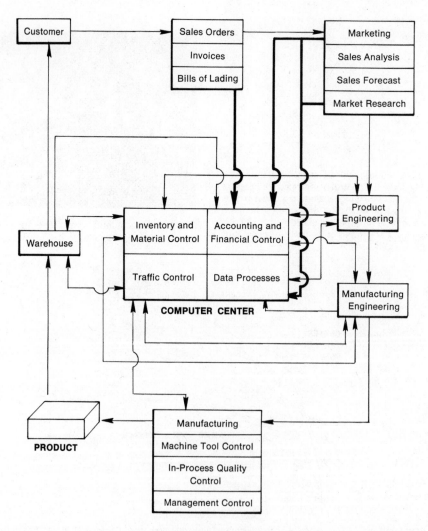

Fig. 18-45. Information inputs to the computer center for management and manufacturing data.

The entering of all accounts and financial control is another. The development of schedules to meet all production requirements and to keep all departments operating efficiently is still another.

The computer is also used for data processing for engineering and manufacturing. These functions for designing and machine control may be carried out by a second computer or by sharing time of another computer outside the production plant.

Product Engineering Complex

A product engineering complex uses the computer as a resource and information source for product design. The computer analyzes this information from sales and marketing data, industrial designer's anal-ysis, inventors' data, prior corporation design file, and management policy.

Analysis. With authorization from management to go ahead with a product design, the computer center becomes a retrieval system for prior design files, standard component parts file, computer data for calculations and analysis, design standards and programs to activate computer graphic consoles, and tape-driven drafting machines.

Design. The engineering group produces engineering drawings, bills of materials, and operation sheets. This data and information is expanded and analyzed by the computer center into materials on hand versus new bills of materials, operation sheets, shop prints, tool designs, and parts programming, Fig. 18-46.

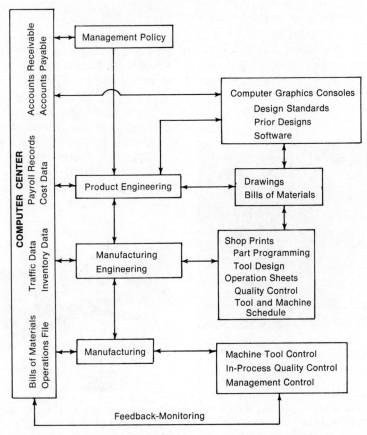

Fig. 18-46. Production engineering complex monitors the manufacturing processes of production, quality control, assembly-line scheduling, material control, attendance reporting, and costs.

Release Control. The manufacturing engineering group develops schedules of men, machines, and materials which control production, establish in-process quality control, and personnel management, Fig. 18-47.

The final stage of planning or production is the product, which is packaged and delivered to the warehouse for shipment to the customer.

Computer Control Center Inputs

The computer center receives information from many sources. The material is programmed and placed into the memory core of the computer; held in tape, disc, or drum storage; or placed in solid state circuits.

One method of entering information into a computer memory core is to punch the information in a coded form into a paper card by a tool similar to a typewriter called a card punch. The position of the hole punched in the card records a unit of information. Decks of these punched cards may be read by a card reader by passing electricity through the holes in the card and polarizing a portion of the memory core with that information. The input may also be in-

Fig. 18-47. Production reporting updates production records and compares actual performance with planned schedules. Items not following schedules are singled out for immediate attention and corrective action. A complete report is ready for the next day's activities to streamline the flow of materials and the assignment of job priorities.

troduced into the computer from coded magnetic tape or from a coded punched tape as that frequently used with numerically controlled tools. Information may be placed into the computer directly with a typewriter keyboard or terminal or by another computer, Fig. 18-48.

Computation. The decisions produced by the computer are based on the information put into the computer and the operations that are programmed into the computer (its instructions). The computer program controls how the computer will respond to the data or information that is fed into the computer. It will determine where the coded information will be stored in the memory of the computer. The memory is the storage of electromagnetic impulses which at a later time will be compared to the new input data. The input information may be directed to be added, subtracted, or compared to the program in the memory core. The program will also provide what will be done or the logic that will be carried out on the input data, Fig. 18-49. By the sequence, that input data is either added, subtracted, or compared. If a complicated algebraic formula is programmed into the computer, the data processing will be carried out at the speed of electricity into that algebraic expression.

After the coded mathematical sequence has been carried out, the result is printed out on paper, or a tape-punched or a control signal will be sent to a monitoring cen-

Fig. 18-48. Information may be placed into or received from a computer by a keyboard terminal and visual display.

ter, to the correction device or actuator to control a machine, Fig. 18-50.

When data, designs, schedules, bills of materials, tool design, and prints are completed and checked, the new product is released to the manufacturing engineering group.

Manufacturing Engineering and Control Center

Part Programming. The programming of parts is the description of the shape, size, and locations of surfaces and holes of a

Fig. 18-49. Computation is carried out in a computer center, where the data is processed at the speed of electricity.

Fig. 18-50. A large computer display in a monitoring center used to control the melting process in a casting center.

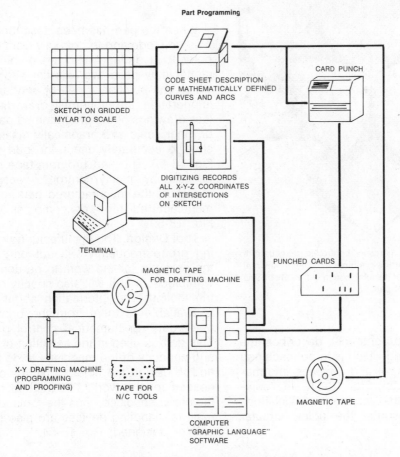

Fig. 18-51. Part programming flowchart.

Fig. 18-52. An automatic digitizing and drafting complex. The machine will draw the workpiece from computer data and proof numerical control programming.

Fig. 18-54. The selection of the machine tools, the redesigning of existing machines, and the designing of both holding fixtures and cutting tools are responsibilities of the manufacturing engineering group. They will use prints, computer data, and numerical control programs to solve their problems.

Fig. 18-53. A test numerically controlled tape used to check drafting machines or plotters. Any malfunction in the total system will easily be seen on the test pattern.

workpiece. The programmer describes the part from a drawing which is to be machined by numerical control and supplies information for a computer, Fig. 18-51. The computer performs the laborious calculations necessary to describe the point, circular arcs, and curves.

When the part has been described, a tape may be made and fed directly from the computer to a drafting machine or plotter. The plotter may be a two-axis flat bed machine which will position a pen or scribe. When it is operating, the plotter will draw dimensionless drawings of the programmed part to full- or other scale and graphically makes it possible to accurately check the part program, Fig. 18-52. The part program tape may also be used by quality control to verify the accuracy of the manufactured part against the manufactured tape or computer program, Fig. 18-53.

Tool Design. The manufacturing engineering group frequently will not only develop the program for the work to be done on the workpieces, but it will also specify or design the tooling. The information about the part is available to them from the prints, computer data, or numerical control programs. This data is used in the selection of the type and capacity of the machines to do the work and to design the holding fixtures or pallets needed for the work. The method of manufacture is specified, thus the tools, dies, and material-handling devices are also specified and/or designed, Fig. 18-54.

Fig. 18-55. Inspection machine is computer-controlled to measure, process, and record data without an operator. On part shown, location of 25 holes and surfaces on three axes is checked in less than six minutes. Dimensions or deviation from normal are recorded.

With the aid of part programming, computer analysis, and drafting machines or plotters, important and accurate decisions concerning manufacturing can be made. In this phase of manufacturing engineering, all the tools, machines, operation sheets, workers, and materials for the manufacturing of the product are supplied.

Quality Control. Reliability in parts and components is necessary to build a successful product. The economic success of the manufacturing organization will depend upon the performance of the product that is used by the consumer. Quality control is a procedure used in manufacturing to assure that the stated performance of the product will in fact be realized by the buyer.

The responsibilities of quality control begin with materials and components testing as they enter the manufacturing process. The checking of specifications to determine whether the incoming products are equal to the engineering and purchase order standards must be done.

Quality control has in-process inspection to make sure that the manufacturing system is producing work that is up to specifica-

tions. Reliability of parts is increased by the reduction of defects found on the parts. Quality control conducts a final inspection, operation, and test of the product.

Automation provides for monitoring of manufacturing processes during the production. With the applications of the highly efficient process of the computers, process control is achieved.

The auditing of production performance on a 24-hour basis is done by computers checking for conformance of the process against program sheets and standards. Deviations are thus revealed, and necessary corrections are shown. This procedure of quality control is readily adapted to such factors as temperature control of furnaces or baths, heat treatment of products, weighing and mixing of chemicals, the flow of liquids for chemical reactions, and the size of machined parts.

By using quality control and by auditing the production process, it is possible to eliminate defects before they have an opportunity to occur.

Size and location quality control of parts has been improved through the development of the three-axis coordinate measuring machine. The machine is a space digitizer. Many difficult measurements may be made in a short period of time, Fig. 18-55. The data can be statistically analyzed quickly enough so that the information of errors or defects is used to correct the production.

With the vast production achieved by automated production systems, quality control management requires fast accurate information that can be quickly passed to the manufacturing engineering group, analyzed, and diagnosed to correct the manufacturing problems, Fig. 18-56 (page 554).

Manufacturing Complex

The application of direct numerical control to manufacturing provides machine tools with the flexibility of adaptive control for machines. With constant monitoring for in-process quality control, the newest state of art in manufacturing is achieved.

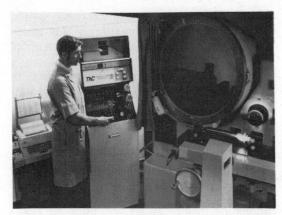

Fig. 18-56. The optical comparator and coordinated measuring machines define "points in space" controlled by a tape. The operator makes a visual analysis and then pushes a button. The tape control moves the table to the new position. The variation from the blueprint or tape and the part may be measured and recorded on a printed readout.

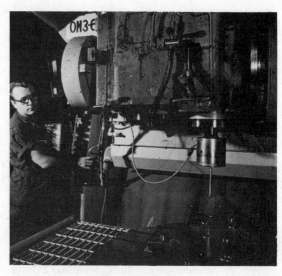

Fig. 18-58. In-process quality control measures in two minutes instead of the seven hours required by manual methods. A pencil-like measuring probe mounted in the machine tool's spindle electronically and precisely measures the dimensions of the part.

Fig. 18-57. The operator's console is used to give the command to the computer to start the machining operation. The computer will control the machine unless the operator stops the program.

Machine Tool Control. Machine tools that are direct numerically controlled do not have numerically controlled tapes or tape readers. The instructions are stored in digital form on a disc or magnetic tape or in solid state circuits provided with the computer. The machine tool has an operator's console from which the operator can command that a part program be sent from the computer storage disc into the computer and onto the machine, Fig. 18-57. The machine tool carries out the part program as well as the speeds, feeds, and depth of cut calculated by the computer.

In-process Quality Control. In-process quality control is the surveillance of the workpiece while it is in the process of manufacture. Computer-controlled machine tools can be checked by a comparison of the path of the tool in manufacturing against the geometry of the part and design standards. This is part of the information in the computer which was used by the product engineer to develop the manufactured part. It

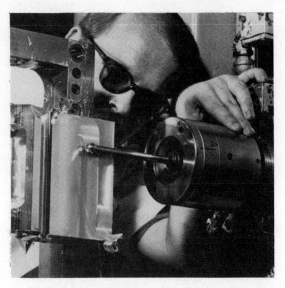

Fig. 18-59. A device which can check automatically the accuracy of the part while it is in position for machining.

Fig. 18-60. Adaptive control applies the correct feeds and speeds that the machine can deliver. Hard spots or air spaces are sensed, and the computer will make the change in the machine to keep it working efficiently.

is passed on to the manufacturing machine's consoles and is compared with the position probe mounted in the machine. The difference between the computer cutter position and the actual cutter position provides a check for any error in signal transmission or mechanical movement of the machine or position error, Fig. 18-58. The difference between the two positions is the amount the machine is moving out of tolerance. If the difference is not corrected, the machine is shut down.

Other in-process control systems monitor the machine tool cutter path before the cutter contacts the work. It checks to see that the cutter is following the theoretical programmed path, that the cutter is the correct size, that the correct tool length is preset into the holder assembly, and that the proper position is in the spindle, Fig. 18-59. The monitor further checks to discover if the assembly is running out of round.

If the cutters are functioning correctly, the system will run. If the system develops trouble or is moving out of tolerance, the machine is shut down before damage is done to the part. A diagnosis is given as to what has caused the error. With the high reliability of the manufacturing system inspection, scrap workpieces are vastly reduced.

Adaptive Control. The direct numerical control system increases the effectiveness of the machine tool by including adaptive control. Adaptive control is the sensing of the cutting action of cutter. Through a computer, the speed of the cutter and the feed of the cutter are changed to produce a part constant with specifications and efficiency, Fig. 18-60.

Sensors are placed on the machine tool's spindle and its deflection, vibration, and torque are monitored by the computer. If a hard spot, a boss, or a web is contacted in the metal, a gap of air appears in the cutter's path. The computer will maximize the cutting action by increasing or reducing the spindle speed and/or the work's feed. This is done without stalling the machine and with no damage to the cutters or work. A smooth surface finish is still produced on the workpiece.

The computer calculates the best machining conditions by comparing the metal's machineability, cutter size, capacity of the cutter, the surface finish required, the rigidity of the work, and the part-holding fixture. The machine's speeds and feeds are automatically changed to give the best machining conditions.

Fig. 18-61. Modular units being tested on a numerically controlled turning center in a slant bed lathe.

Fig. 18-62. Modular direct numerical control and computer numerical control machine control unit used to replace or update older machine control units.

Adaptive control of machine tools makes it possible to achieve the maximum efficiency output from a machine system, because the tools operate at capacity consistent with the material and tool cutting speeds.

Modular Numerically Controlled Machine Tools

Numerical manufacturing tools have been built by a number of machine tool builders. Many develop the tool, the machine control unit, and the tape readers. These units are designed and hard-wired for a peculiar machine control unit and numerical control machine. With technology of direct numerical controls, a device has been developed to convert machine control units to direct computer-controlled or direct numerical-controlled units, Figs. 18-61 and 18-62. This device is called a direct controller interface, Fig. 18-63. The interface converts the older machine control units to direct numerical control. This, too, will eventually be replaced. Added flexibility and capacity will be obtained by the ability to add standard machine control units and modular numerically controlled tools to a production system.

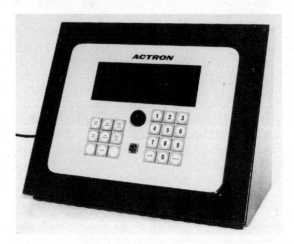

Fig. 18-63. Direct controller interfaces an advanced behind-the-tape reader for converting newer machine control units to direct numerical control.

The flexibility of the computer and multiplexor brings product flexibility to the manufacturing plant. Modular units are interchangeable standard units that are engineered so that they can be plugged in to produce a manufacturing system for new parts and products by direct numerical control.

Modular units will include soft wear or pre-engineered part programs, programs for standard geometry (curves, arcs, and shapes), and hardware such as digitizers, minicomputers, terminals, multiplexors, machine control units, machines, and material-handling devices. These standardized units may be assembled in a number of combinations to build a production system.

Programmable Controllers

Routine operation and control of a machine or process is being done by a computerlike controller called a programmable controller, Fig. 18-64. These controllers function as minicomputers and control the functions of a machine but do not supply the management data or monitoring.

The programmable controllers do not require an engineer for the programming. Their function is to store information, compare input sensor value against the memory value, and to control and produce an output. The input channels sense voltages from limit switches, pressure switches, push buttons, etc. They compare this data with the memory value and make a logical decision. The output drive channels actuate solenoids, starters, motors, indicator lights, and other like devices.

These programmable controllers are solid state plug-in modules that may also be plugged into minicomputers if needed or desired. If a tool, machine, or process is obsolete or replaced, the controller need only be programmed for a new job. Flexibility is gained with a controller that can withstand dust, heat, noise, and vibration of a manufacturing environment.

Automatic Transfer

Workpieces transferred from machine station-to-station are carried by transfer bars, walking beams, finger, and pawl. Smaller parts such as truck differential cases are transferred from stations by a pallet transfer method. Other products use a free transfer to move the part by having the part sit in a nest or cradle. It is transported to the next station, where it is lifted out of the nest and positioned for the machining operation, Fig. 18-65. The advantage of free transfer is that

Fig. 18-64. A programmable controller is an easy plug-in component. The control program for the process is entered into the controller memory from an elementary diagram via a programming panel.

Fig. 18-65. A palletized transfer machine. These differential carrier covers pass through an 18-station machine which machines the complete carrier.

Fig. 18-66. This die cast aluminum camshaft housing takes advantage of free transfer to be moved, rotated, and positioned for machining. This method of movement makes surfaces that frequently would be covered by the holding pallet now free for machining.

Fig. 18-68. A tool changer for a 5-axis, 3-spindle machining center with 24-position automatic tool changes. The tool manufactures rock bit arms.

Fig. 18-67. The length of the cutting tool must be considered when a machine is programmed. The tools are mounted into a standard quick-change tool holder and set for the tool's precise tool length before it is placed in a machine's magazine.

the part may be turned or rotated and surfaces machined. Normally the part would be covered by clamps or bolts or by the pallet holding the fixture itself. This increases the number of parts of a family which may be machined by the transfer machine, Fig. 18-66.

Automatic Tool Setting

Cutting tools have standard shanks so they can be mounted into standard quick-change tool holders. The lengths of tools are carefully set from a datum surface of the tool holder, Fig. 18-67. Micrometers and optical measuring instruments are used to accurately place the tool in the tool holder. At a later time the tool and holder are placed into the machine's tool magazine. The tool holders are all coded so that the magazine of the tool may rotate. The numerically controlled machine control unit may then select the proper tool for the correct operation.

Fig. 18-70. Automated assembly machine for truck door assembly lines. The machine is in the final stages of construction.

Fig. 18-69. Spray nozzle assembly machine feeds body, seats steel ball, assembles strainer screen (operator manually places compression spring), machines feeds, drives cap, inspects, and assembles automobile eject-reject at 35 assemblies per minute.

The machine's tool changer transfers the tool from the magazine to the machine spindle, where it is set and locked. The depth of cutting demands that the tools be accurately preset or the parts will be machined out of tolerance, Fig. 18-68.

Automation of Assembly

The assembly of products frequently is the most expensive item in the cost of manufacturing because of the labor involved. Automatic assembly is suited for products that are produced in large volumes and with products whose design is stable, Figs. 18-69 and 18-70. Many small subassemblies such as electric motor parts, solenoids, frames, wheel bearings, typewriter parts, valves, and nozzles can be automated easily. The designs in those parts change very slowly, and they can be produced in vast numbers by automation. Also assembly of parts which are hazardous from explosions, acids, electrical shock, or poisons are automated in order to reduce exposing people to danger.

Fig. 18-71. Single-station hand assembly of an absolute pressure transmitter.

Automated assembly is classified by the movement the part takes through the assembly machine: single-station, dial or rotary, and in-line assembly.

Types of Assembly

Single-Station Assembly Machine. This machine is the natural growth out of hand assembly, Fig. 18-71. The parts are positioned and fastened at a rapid rate. They are automatically fed, assembled, and ejected on a larger assembly line. Power screwdriver, nutrunning, or riveting machines are examples of single-station operations, Fig. 18-72.

Dial or Rotary Assembly Machines. These machines index the work from station-to-station around a table, Fig. 18-73. Dial assembly machines are available in standard tables and drives, making it possible to tool an assembly center. Due to the simple rotary motion of dial machines, they are usually accurate and adapted to small-to medium-sized parts. These machines may have index rates of up to 1,800 cycles per hour, Fig. 18-74. The dial assembly machines may have single or double tool plates above and below the assembly table. Numerous small parts are manufactured on popular dial systems of assembly.

In-Line Assembly. The assembly work of **in-line assembly** machines is done as the parts are transported through the machine on a straight line, Fig. 18-75. In-line assembly machines handle large-sized parts and perform at indexing rates of 500 to 1800 per hour. This type of assembly machine has from five to fifteen automatic stations. The capacity of the machine can be increased by the addition of new tooling stations or by the addition of double tooling to the assembly stations.

Fig. 18-72. Power screwdriving machine that will rapidly increase the production of a single-station assembly.

Fig. 18-74. A dial or rotary assembly machine that inserts and fastens parts in the assembly when the worktable indexes.

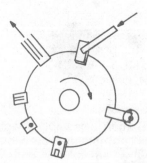

Fig. 18-73. A dial assembly machine takes little floor space and assembles small products efficiently.

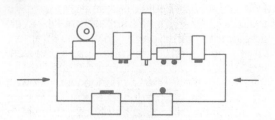

Fig. 18-75. In-line assembly machine may be used for large or small assemblies. Large products are nearly always done on in-line assembly lines.

In-line assembly machines can assemble large objects such as automobile bodies by spot welding. They can also assemble small products, Fig. 18-76.

In order to prevent a backlog of parts or reduced production, a manufacturing system which has obtained automated production also needs an automated assembly system. Automated assembly begins by getting the parts together into a position in which they may be handled. Large parts frequently come to the assembly point on an overhead monorail or other conveyor, and the parts are fastened together on a moving assembly line, Fig. 18-77.

Small parts need to be oriented and moved to a holding fixture, where the part is positioned or fastened, Fig. 18-78. Parts

Fig. 18-77. Assembly lines for large products will have the larger parts and subassemblies brought to the work stations by monorails and other conveyors. The assembly line and conveyors are being repaired.

Fig. 18-76. An in-line assembly machine making a bobbin assembly. The machine automatically transfers feeds and assembles the glass reed capsules. The group straps are welded to capsule leads, and a retainer is assembled. A conveyor ejects 20 parts per minute.

Fig. 18-78. Small part feeders orient and move parts into an assembly machine so they may be automatically fastened.

Fig. 18-79. The vibrating bowl moves the parts up the track and various orienting devices position the part or drop it back into the bowl.

Fig. 18-81. A multitrack part feeder. This machine feeds studs to a power-driving machine on an assembly machine.

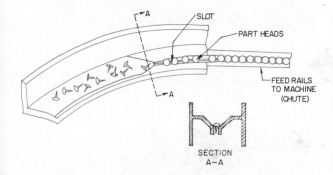

Fig. 18-80. Parts are positioned and transferred to a feed track. Feed for long-headed parts first hang through the bowl wall and are then turned into a vertical position.

may be oriented mechanically by the use of slides or rollers, but the most frequently used method of orientation is done with vibratory parts feeders. Vibratory feeders have a bowl into which the randomly positioned parts are placed. The bowl vibrates

about 3600 or 7200 times per minute. The parts move up an inclined track. The track has a number of operating devices such as wipers, spacers, cutouts, slots, and other shapes. The natural resting position of the part is used and vibrated onto one of the position devices, which either turns the part into the desired position or drops it back into the bowl and starts the process again, Figs. 18-79 and 18-80. In this manner, rivets, washers, clips, bolts, nuts, and various shaped parts may be oriented and started into a feed chute and elevator which lead to the assembly machine, Fig. 18-81. Parts which are positioned are placed into a track and transferred to another machine where the fastening will be done.

Fastening Automatically. The fastening in an assembly machine may be carried out by any of a number of processes, depending upon the product. Wire clips, staples, retainers, and bonding agents are special fasteners applied by assembly machines. Rivets, screws, and processes such as

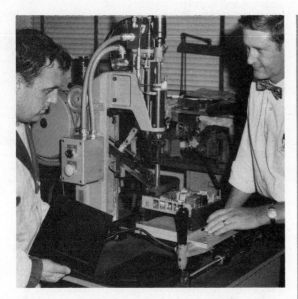

Fig. 18-82. Electronic parts and sheet metal chassis are rapidly and automatically fastened.

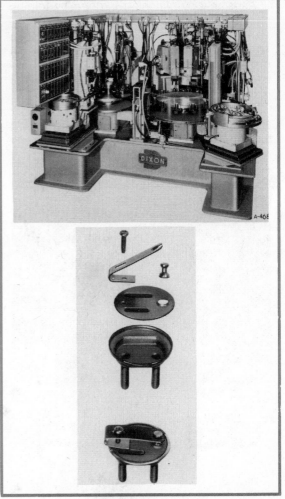

Fig. 18-83. Automatic assembly of a bi-metal sensor assembly. Dual synchronized dial machine assembles valve and plate, taps screw hole, assembles bi-metal, and drives screw. It transfers subassembly to second dial, inserts into sensor body, and receives a two-stage clinching operation, and final screw calibration at a rate of 20 per minute or more.

welding, pressing, staking, and casting may be used to fasten.

Fastening automatically with rivets has long been a satisfactory way to assemble small products, Fig. 18-82. Electronic and sheet metal parts can be inexpensively assembled with small rivets. Transformers and motor starters are examples of this type of assembly.

Power screwdrivers and nutrunners are automated to assemble small parts and subassemblies with screws, nuts, and bolts, Fig. 18-83. A screw assembly has more strength than is usually provided by rivets. It can also be disassembled or adjusted at a later time.

Welding is a method of assembling large products together quickly. Automated spot welders provide aligning and welding of the main body structures of automobiles. Spot welds are made in rapid sequence and are permanent.

Pressing and staking are used for fastening all manner of mating parts. In pressing, the parts are stamped or machined with tight tolerances and the two parts are pushed together with pressure to produce

a solidly fastened part. Staking is similar except that it has a burr which further tightens and locks the part onto the mating section. The staking is done by a punch which is driven into the metal next to the joint, expanding the metal to produce a friction joint.

Assembly by clinching is done by folding one metal stamping around another and crushing the parts closely together, Fig. 18-84. This assembly method is easily automated. It is efficient, and there are no fasteners to be purchased, since the fastener is made of the metal itself.

Another method of fastening is done by aligning the parts in a fixture and pouring the molten metal into a prepared cavity around the joint to be assembled. This process makes a permanent and efficient joint and may be automated.

A New Direction in Assembly

With the increasing problems of boredom and frustration experienced by the assembly line workers, manufacturing engineers are planning more automation of assembly. Achieving total assembly automation is slow and difficult. The final assembly of many products is still done by hand labor. Headlamps, water pumps, disk brakes, rear differentials, and numerous small light assemblies (holders, mounting brackets, and bearings) are assembled by automation. However, the aligning and fastening of parts, coupling of hydraulic lines, fastening of electrical wires, and the installation of trim and accessories are still done by hand labor. Soon many of these assembly operations will be redesigned and automated.

Machines that Manufacture and Assemble. New machines that can both machine parts and also assemble and inspect the subassembly are now being manufactured. These new machines combine a manufacturing machine and an assembly machine into one single unit. The advantage of this method of automation is that manufacturing operations may be performed in a functional sequence that saves storage and handling of parts which go into the assembly. The parts are all manufactured and assembled as needed.

Leaders in the field of manufacturing include a wider use of computers for manufacturing assembly, quality control, and production control, Fig. 18-85. Information is monitored and accomplished through a network of minicomputers, data collection terminals, and data transmission terminals. Production and assembly machines receive the information to perform the work through the same network of computer links.

Testing and inspection stations are located at manufacturing plants in the assembly machines so that constant and 100% effective surveillance provides feedback in-

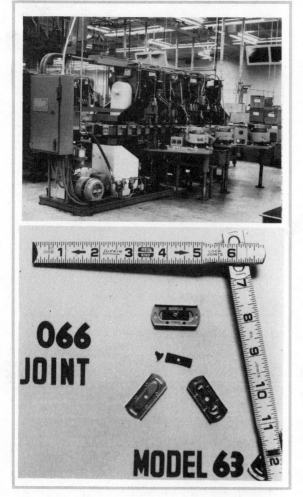

Fig. 18-84.   These carpenters' rules are made of brass and wood components. The components are assembled and fastened by clinching. The metal is folded over the part and crushed onto the wood rule.

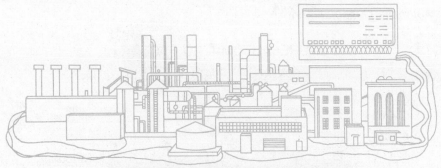

COMPUTER-CONTROLLED FACTORY

Fig. 18-85. Computers are being applied to the manufacturing process and are in the process of controlling the functions of a plant. Production lines, production reports, lab testing, quality control, inventory control, management reports, engineering calculations, and plant security become part of the computer's function.

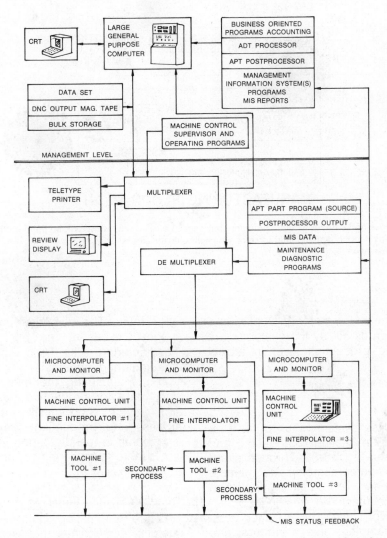

Fig. 18-86. A minicomputer provides computer power on the production floor and provides information and data for management.

FUTURE AUTOMATED FACTORY

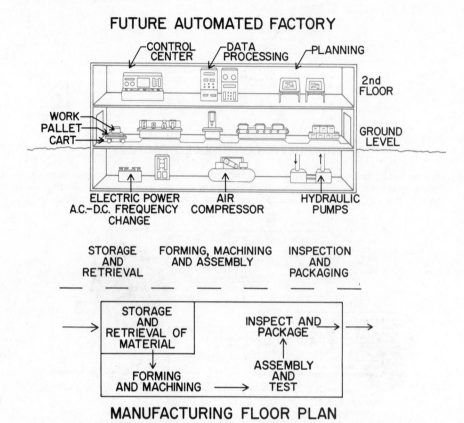

Fig. 18-87. Factories of the future will be computer-controlled. The data processing will be done on the top floor of the building. Production, shipping, and receiving will be on the ground floor. All power and energy sources will be in the basement.

formation for machine control. Quality and reliability standards set by the customer are assured at the production level in a real time frame.

Data Base System

The data base system is used for pre-production design, testing, scheduling, and method analysis, Fig. 18-86. Through simulation, the computer searches its memory for information and data on the configuration of the part manufactured, methods of assembly, conveyor speeds, work station positioning and clamping, time standards, and labor rates. The computer complex makes suggestions about the best methods for manufacturing with these variables to the manufacturing engineer.

In manufacturing the minicomputer provides data for production machine control and monitors shop activity, inventory status, process and production control, and part distribution in manufacturing. All of this data is available to management on a continuous or daily basis.

Trends in Manufacturing

Future automation factories will have a number of machines that will specialize in

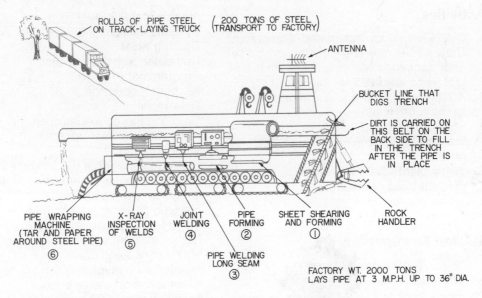

ROLLS OF PIPE STEEL
ON TRACK-LAYING TRUCK (200 TONS OF STEEL)
TRANSPORT TO FACTORY

ANTENNA

BUCKET LINE THAT
DIGS TRENCH

DIRT IS CARRIED ON
THIS BELT ON THE
BACK SIDE TO FILL
IN THE TRENCH
AFTER THE PIPE IS
IN PLACE

PIPE WRAPPING
MACHINE
(TAR AND PAPER
AROUND STEEL PIPE)
⑥

X-RAY
INSPECTION
OF WELDS
⑤

JOINT
WELDING
④

PIPE
FORMING
②

SHEET SHEARING
AND FORMING
①

ROCK
HANDLER

PIPE WELDING
LONG SEAM
③

FACTORY WT. 2000 TONS
LAYS PIPE AT 3 M.P.H. UP TO 36" DIA.

Fig. 18-88.   A mobile factory making and laying a natural gas line.

a size and family of operations such as a machine drilling on two axes, another machine boring on three axes, and another performing a family of milling operations. Complex work will be mounted on pallets that are coded for the part that they carry. The computer will read the pallet part code and direct it through the factory to the proper machine. When the machine is available for a part, the part code will be received by the computer. It, in turn, will select the proper program from the memory disc and send the proper machining program for that part to the machine control center and the machine.

Flexibility of Manufacturing. This will be gained by having direct computer-controlled machines performing standard operations located anywhere in the factory, Fig. 18-87. The computer will shuttle the parts to the proper machine at the proper time. When it has been shaped, formed, or machined, the part will be transported to assembly machines or to machines that are combination manufacturing-assembly machines.

Since adhesives are one of the simplest types of fasteners to automate, they will be further developed and used to fasten, seal, and to lock components together.

The Mobile Factory

The agricultural industries have explored the concept of mobile factories initially with the wheat combines; potato, beet, and tomato harvesters; and other like machines which are moved through the fields to pick, clean, and grade the product.

When the transportation of a product is difficult or expensive, mobile factories will be used, Fig. 18-88. An example of a mobile factory may be a large diameter pipeline machine. The machine forms strip steel and welds it into pipe while other parts of the machine dig the ditch and lower the completed pipe into the ditch. The factory slowly moves across the countryside, laying the pipeline. Long-distance power, telephone, and television cables can be produced on a ship or track-laying factory which turns the raw stock into the finished product as it is installed for service. The time from manufacturing to installation becomes a matter of minutes with the use of a mobile factory.

Activities

1. Disassemble an automated component. Draw a diagram showing the processes and sequence of processes that you feel were involved in the assembly.
2. Arrange a study trip to a computer center in your community.
3. Build a Binary counter.
4. Program your initials for a N/C cut plaque.
5. Visit a plant where numerically controlled machines are operating.

Related Occupations

These occupations are related to automation:

Systems engineer
Structural dynamics engineer
Mechanical engineer
Systems analyst
Drafting artist
Computer console operator
Programmer
Keypunch operator
Data typist
Card-to-tape converter operator
High-speed printer operator
Tape librarian
Programmer analyst
Digital computer systems technician
Graphics technician
Electronics service technician
Electromechanical technician
Numerical control service technician
Maintenance engineer
Maintenance mechanic
Assembly machine operator
Quality control test technician
Security officer

Appendix

TABLE A-1

Millimetre-Inch Equivalents for Drills, Reamers, and End Mills

MILLIMETRE-INCH EQUIVALENTS

Inch	Wire	Decimal	mm	Inch	Wire	Decimal	mm	Inch	Wire	Decimal	mm	Inch	Wire and Letter	Decimal	mm	Inch	Letter	Decimal	mm	Inch	Decimal	mm	Inch	Decimal	mm
	97	.0059	.15		61	.0390			35	.1100			5	.2055		21/64		.3281			.6693	17.	1 15/64	1.2344	
	96	.0063	.16			.0394	1.			.1102	2.8			.2067	5.25			.3307	8.4	43/64	.6719			1.2402	31.5
	95	.0067	.17		60	.0400			34	.1110				.2087	5.3		Q	.3320		11/16	.6875		1 1/4	1.2500	
	94	.0071	.18		59	.0410			33	.1130			4	.2090				.3346	8.5		.6890	17.5		1.2598	32.
	93	.0075	.19			.0413	1.05			.1142	2.9			.2126	5.4			.3386	8.6	45/64	.7031		1 17/64	1.2656	
	92	.0079	.20		58	.0420			32	.1160			3	.2130			R	.3390			.7087	18.		1.2795	32.5
	91	.0083	.21		57	.0430				.1181	3.			.2165	5.5			.3425	8.7	23/32	.7188		1 9/32	1.2812	
	90	.0087	.22			.0433	1.1		31	.1200		7/32		.2188		11/32		.3438			.7283	18.5	1 19/64	1.2969	
	89	.0091	.23			.0453	1.15			.1220	3.1			.2205	5.6			.3445	8.75	47/64	.7344			1.2992	33.
		.0094	.24		56	.0465		1/8		.1250			2	.2210				.3465	8.8		.7480	19.	1 5/16	1.3125	
	88	.0095		3/64		.0469				.1260	3.2			.2244	5.7		S	.3480		3/4	.7500			1.3189	33.5
		.0098	.25			.0472	1.2			.1280	3.25			.2264	5.75			.3504	8.9	49/64	.7656		1 21/64	1.3281	
	87	.0100				.0492	1.25		30	.1285			1	.2280				.3543	9.		.7677	19.5		1.3386	34.
		.0102	.26			.0512	1.3			.1299	3.3			.2283	5.8		T	.3580		25/32	.7812		1 11/32	1.3438	
	86	.0105			55	.0520				.1339	3.4			.2323	5.9			.3583	9.1		.7874	20.		1.3583	34.5
		.0106	.27			.0531	1.35		29	.1360			A	.2340		23/64		.3594		51/64	.7969		1 23/64	1.3594	
	85	.0110	.28		54	.0550				.1378	3.5	15/64		.2344				.3622	9.2		.8071	20.5	1 3/8	1.3750	
		.0114	.29			.0551	1.4		28	.1405				.2362	6.			.3642	9.25	13/16	.8125			1.3780	35.
	84	.0115				.0571	1.45	9/64		.1406			B	.2380				.3661	9.3		.8268	21.	1 25/64	1.3906	
		.0118	.30			.0591	1.5			.1417	3.6			.2402	6.1		U	.3680		53/64	.8281			1.3976	35.5
	83	.0120			53	.0595			27	.1440			C	.2420				.3701	9.4	27/32	.8438		1 13/32	1.4062	
	82	.0125				.0610	1.55			.1457	3.7			.2441	6.2			.3740	9.5		.8465	21.5		1.4173	36.
		.0126	.32	1/16		.0625			26	.1470			D	.2460		3/8		.3750		55/64	.8594		1 27/64	1.4219	
	81	.0130				.0630	1.6			.1476	3.75			.2461	6.25		V	.3770			.8661	22.		1.4370	36.5
		.0134	.34		52	.0635			25	.1495				.2480	6.3			.3780	9.6	7/8	.8750		1 7/16	1.4375	
	80	.0135				.0650	1.65			.1496	3.8	1/4	E	.2500				.3819	9.7		.8858	22.5	1 29/64	1.4531	
		.0138	.35			.0669	1.7		24	.1520				.2520	6.4			.3839	9.75	57/64	.8906			1.4567	37.
		.0142	.36		51	.0670				.1535	3.9			.2559	6.5			.3858	9.8		.9055	23.	1 15/32	1.4688	
	79	.0145				.0689	1.75		23	.1540			F	.2570			W	.3860		29/32	.9062			1.4764	37.5
		.0150	.38		50	.0700		5/32		.1562				.2598	6.6			.3898	9.9	59/64	.9219		1 31/64	1.4844	

TABLE A-1 (cont.)

Decimal Equivalents (continued)

mm	Decimal	Fraction
38.	1.4961	
	1.5000	1 1/2
	1.5156	1 33/64
38.5	1.5157	
	1.5312	1 17/32
39.	1.5354	
	1.5469	1 35/64
39.5	1.5551	
	1.5625	1 9/16
40.	1.5748	
	1.5781	1 37/64
	1.5938	1 19/32
40.5	1.5945	
	1.6094	1 39/64
41.	1.6142	
	1.6250	1 5/8
41.5	1.6339	
	1.6406	1 41/64
42.	1.6535	
	1.6562	1 21/32
	1.6719	1 43/64
42.5	1.6732	
	1.6875	1 11/16
43.	1.6929	
	1.7031	1 45/64
43.5	1.7126	
	1.7188	1 23/32
44.	1.7323	
	1.7344	1 47/64
	1.7500	1 3/4
44.5	1.7520	
	1.7656	1 49/64
45.	1.7717	
	1.7812	1 25/32

mm	Decimal	Fraction
23.5	.9252	
	.9375	15/16
24.	.9449	
	.9531	61/64
24.5	.9646	
	.9688	31/32
25.	.9843	
	.9844	63/64
	1.0000	1
25.5	1.0039	
	1.0156	1 1/64
26.	1.0236	
	1.0312	1 1/32
26.5	1.0433	
	1.0469	1 3/64
	1.0625	1 1/16
27.	1.0630	
	1.0781	1 5/64
27.5	1.0827	
	1.0938	1 3/32
28.	1.1024	
	1.1094	1 7/64
28.5	1.1220	
	1.1250	1 1/8
	1.1406	1 9/64
29.	1.1417	
	1.1562	1 5/32
29.5	1.1614	
	1.1719	1 11/64
30.	1.1811	
	1.1875	1 3/16
30.5	1.2008	
	1.2031	1 13/64
	1.2188	1 7/32
31.	1.2205	

mm	Drill	Decimal	Fraction
		.3906	25/64
10.		.3937	
	X	.3970	
	Y	.4040	
		.4062	13/32
	Z	.4130	
10.5		.4134	
		.4219	27/64
11.		.4331	
		.4375	7/16
11.5		.4528	
		.4531	29/64
		.4688	15/32
12.		.4724	
		.4844	31/64
12.5		.4921	
		.5000	1/2
13.		.5118	
		.5156	33/64
		.5312	17/32
13.5		.5315	
		.5469	35/64
14.		.5512	
		.5625	9/16
14.5		.5709	
		.5781	37/64
15.		.5906	
		.5938	19/32
		.6094	39/64
15.5		.6102	
		.6250	5/8
16.		.6299	
		.6406	41/64
16.5		.6496	
		.6562	21/32

mm	Drill	Decimal	Fraction
	G	.2610	
6.7		.2638	
		.2656	17/64
6.75		.2657	
	H	.2660	
6.8		.2677	
6.9		.2717	
	I	.2720	
7.		.2756	
	J	.2770	
7.1		.2795	
	K	.2810	
		.2812	9/32
7.2		.2835	
7.25		.2854	
7.3		.2874	
	L	.2900	
7.4		.2913	
	M	.2950	
7.5		.2953	
		.2969	19/64
7.6		.2992	
	N	.3020	
7.7		.3031	
7.75		.3051	
7.8		.3071	
7.9		.3110	
		.3125	5/16
8.		.3150	
	O	.3160	
8.1		.3189	
8.2		.3228	
	P	.3230	
8.25		.3248	
8.3		.3268	

mm	Drill	Decimal	Fraction
	22	.1570	
4.		.1575	
	21	.1590	
	20	.1610	
4.1		.1614	
4.2		.1654	
	19	.1660	
4.25		.1673	
4.3		.1693	
	18	.1695	
		.1719	11/64
	17	.1730	
4.4		.1732	
	16	.1770	
4.5		.1772	
	15	.1800	
4.6		.1811	
	14	.1820	
	13	.1850	
4.7		.1850	
4.75		.1870	
		.1875	3/16
	12	.1890	
4.8		.1890	
	11	.1910	
4.9		.1929	
	10	.1935	
	9	.1960	
5.		.1969	
	8	.1990	
5.1		.2008	
	7	.2010	
		.2031	13/64
	6	.2040	
5.2		.2047	

mm	Drill	Decimal	Fraction
1.8		.0709	
	49	.0728	
1.85		.0730	
1.9		.0748	
	48	.0760	
1.95		.0768	
		.0781	5/64
	47	.0785	
2.		.0787	
2.05		.0807	
	46	.0810	
	45	.0820	
2.1		.0827	
2.15		.0846	
	44	.0860	
2.2		.0866	
2.25		.0886	
	43	.0890	
2.3		.0906	
2.35		.0925	
	42	.0935	
		.0938	3/32
2.4		.0945	
	41	.0960	
2.45		.0965	
	40	.0980	
2.5		.0984	
	39	.0995	
	38	.1015	
2.6		.1024	
	37	.1040	
2.7		.1063	
	36	.1065	
2.75		.1083	
		.1094	7/64

mm	Drill	Decimal	Fraction
		.0156	1/64
.4		.0157	
	78	.0160	
.42		.0165	
.44		.0173	
.45		.0177	
	77	.0180	
.46		.0181	
.48		.0189	
.5		.0197	
	76	.0200	
	75	.0210	
.55		.0217	
	74	.0225	
.6		.0236	
	73	.0240	
	72	.0250	
.65		.0256	
	71	.0260	
.7		.0276	
	70	.0280	
	69	.0292	
.75		.0295	
	68	.0310	
		.0312	1/32
.8		.0315	
	67	.0320	
	66	.0330	
.85		.0335	
	65	.0350	
.9		.0354	
	64	.0360	
	63	.0370	
.95		.0374	
	62	.0380	

TABLE A-2
Recommended Cutting Fluids for Drilling

Material to Be Drilled	Cutting Fluid(s)
Aluminum and Alloys	Kerosene, Kerosene and Lard Oil, Soluble Oil
Brass and Bronze	Dry For Deep Holes: Kerosene and Mineral Oil, Lard Oil, Soluble Oil
Magnesium and Alloys	Mineral Lard Oil, Kerosene, Dry
Copper	Mineral Lard Oil and Kerosene, Soluble Oil, Dry
Monel Metal	Mineral Lard Oil, Soluble Oil
Mild Steels	Mineral Oil, Soluble Oil
Tough Alloy Steels	Sulfurized Oils, Mineral Lard Oil
Steel Forgings	Sulfurized Oil, Mineral Lard Oil
Cast Steel	Soluble Oil, Sulfurized Oil
Wrought Iron	Soluble Oil, Sulfurized Oil
High-Tensile Steels	Soluble Oil, Sulfurized Oil
Manganese Steel	Dry
Cast Iron	Dry
Malleable Iron	Dry, Soluble Oil
Stainless Steel	Soluble Oil, Sulfurized Oil
Titanium Alloys	Soluble Oil, Sulfurized Oil
Tool Steel	Mineral Lard Oil and Kerosene, Kerosene, Mineral Lard Oil
Abrasives, Plastics	Dry
Fibre, Asbestos, Wood	Dry
Hard Rubber	Dry

TABLE A-3
Sine Bar Constants (5" Bar)
(Multiply Constants by Two for a 10" Sine Bar)

Min.	0°	1°	2°	3°	4°	5°	6°	7°	8°	9°	10°	11°	12°	13°	14°	15°	16°	17°	18°	19°	Min.
0	.00000	.08725	.17450	.26170	.34880	.43580	.52265	.60935	.69585	.78215	.86825	.95405	1.0395	1.1247	1.2096	1.2941	1.3782	1.4618	1.5451	1.6278	0
2	.00290	.09015	.17740	.26460	.35170	.43870	.52555	.61225	.69875	.78505	.87110	.95690	.0424	.1276	.2124	.2969	.3810	.4646	.5478	.6306	2
4	.00580	.09310	.18030	.26750	.35460	.44155	.52845	.61510	.70165	.78790	.87395	.95975	.0452	.1304	.2152	.2997	.3838	.4674	.5506	.6333	4
6	.00875	.09600	.18320	.27040	.35750	.44445	.53130	.61800	.70450	.79080	.87685	.96260	.0481	.1332	.2181	.3025	.3865	.4702	.5534	.6361	6
8	.01165	.09890	.18615	.27330	.36040	.44735	.53420	.62090	.70740	.79365	.87970	.96545	.0509	.1361	.2209	.3053	.3893	.4730	.5561	.6388	8
10	.01455	.10180	.18905	.27620	.36330	.45025	.53710	.62380	.71025	.79655	.88255	.96830	1.0538	1.1389	1.2237	1.3081	1.3921	1.4757	1.5589	1.6416	10
12	.01745	.10470	.19195	.27910	.36620	.45315	.54000	.62665	.71315	.79940	.88540	.97115	.0566	.1417	.2265	.3109	.3949	.4785	.5616	.6443	12
14	.02035	.10760	.19485	.28200	.36910	.45605	.54290	.62955	.71600	.80230	.88830	.97405	.0594	.1446	.2293	.3137	.3977	.4813	.5644	.6471	14
16	.02325	.11055	.19775	.28490	.37200	.45895	.54580	.63245	.71890	.80515	.89115	.97690	.0623	.1474	.2322	.3165	.4005	.4841	.5672	.6498	16
18	.02620	.11345	.20065	.28780	.37490	.46185	.54865	.63530	.72180	.80800	.89400	.97975	.0651	.1502	.2350	.3193	.4033	.4868	.5699	.6525	18
20	.02910	.11635	.20355	.29070	.37780	.46475	.55155	.63820	.72465	.81090	.89685	.98260	1.0680	1.1531	1.2378	1.3221	1.4061	1.4896	1.5727	1.6553	20
22	.03200	.11925	.20645	.29365	.38070	.46765	.55445	.64110	.72755	.81375	.89975	.98545	.0708	.1559	.2406	.3250	.4089	.4924	.5755	.6580	22
24	.03490	.12215	.20940	.29655	.38360	.47055	.55735	.64400	.73040	.81665	.90260	.98830	.0737	.1587	.2434	.3278	.4117	.4952	.5782	.6608	24
26	.03780	.12505	.21230	.29945	.38650	.47345	.56025	.64685	.73330	.81950	.90545	.99115	.0765	.1615	.2462	.3306	.4145	.4980	.5810	.6635	26
28	.04070	.12800	.21520	.30235	.38940	.47635	.56315	.64975	.73615	.82235	.90830	.99400	.0793	.1644	.2491	.3334	.4173	.5007	.5837	.6663	28
30	.04365	.13090	.21810	.30525	.39230	.47925	.56600	.65265	.73905	.82525	.91120	.99685	1.0822	1.1672	1.2519	1.3362	1.4201	1.5035	1.5865	1.6690	30
32	.04655	.13380	.22100	.30815	.39520	.48210	.56890	.65550	.74190	.82810	.91405	.99970	.0850	.1700	.2547	.3390	.4228	.5063	.5893	.6718	32
34	.04945	.13670	.22390	.31105	.39810	.48500	.57180	.65840	.74480	.83100	.91690	1.0016	.0879	.1729	.2575	.3418	.4256	.5091	.5920	.6745	34
36	.05235	.13960	.22680	.31395	.40100	.48790	.57470	.66130	.74770	.83385	.91975	.0054	.0907	.1757	.2603	.3446	.4284	.5118	.5948	.6772	36
38	.05525	.14250	.22970	.31685	.40390	.49080	.57760	.66415	.75055	.83670	.92260	.0082	.0935	.1785	.2631	.3474	.4312	.5146	.5975	.6800	38
40	.05820	.14540	.23265	.31975	.40680	.49370	.58045	.66705	.75345	.83960	.92545	1.0110	1.0964	1.1813	1.2660	1.3502	1.4340	1.5174	1.6003	1.6827	40
42	.06110	.14835	.23555	.32265	.40970	.49660	.58335	.66995	.75630	.84245	.92835	.0139	.0992	.1842	.2688	.3530	.4368	.5201	.6030	.6855	42
44	.06400	.15125	.23845	.32555	.41260	.49950	.58625	.67280	.75920	.84530	.93120	.0168	.1020	.1870	.2716	.3558	.4396	.5229	.6058	.6882	44
46	.06690	.15415	.24135	.32845	.41550	.50240	.58915	.67570	.76205	.84820	.93405	.0196	.1049	.1898	.2744	.3586	.4423	.5257	.6085	.6909	46
48	.06980	.15705	.24425	.33135	.41840	.50530	.59200	.67860	.76495	.85105	.93690	.0225	.1077	.1926	.2772	.3614	.4451	.5285	.6113	.6937	48
50	.07270	.15995	.24715	.33425	.42130	.50820	.59490	.63145	.76780	.85390	.93975	1.0253	1.1106	1.1955	1.2800	1.3642	1.4479	1.5312	1.6141	1.6964	50
52	.07565	.16285	.25005	.33715	.42420	.51105	.59780	.68435	.77070	.85680	.94260	.0281	.1134	.1983	.2828	.3670	.4507	.5340	.6168	.6991	52
54	.07855	.16580	.25295	.34010	.42710	.51395	.60070	.68720	.77355	.85965	.94550	.0310	.1162	.2011	.2856	.3698	.4535	.5368	.6196	.7019	54
56	.08145	.16870	.25585	.34300	.43000	.51685	.60355	.69010	.77645	.86250	.94835	.0338	.1191	.2039	.2884	.3726	.4563	.5395	.6223	.7046	56
58	.08435	.17160	.25875	.34590	.43290	.51975	.60645	.69300	.77930	.86540	.95120	.0367	.1219	.2068	.2913	.3754	.4591	.5423	.6251	.7073	58
60	.08725	.17450	.26170	.34880	.43580	.52265	.60935	.69585	.78215	.86825	.95405	1.0395	1.1247	1.2096	1.2941	1.3782	1.4618	1.5451	1.6278	1.7101	60

Courtesy Brown & Sharpe Mfg. Co.

TABLE A-3 (cont.)

Sine Bar Constants (5" Bar)

(Multiply Constants by Two for a 10" Sine Bar)

Min.	39°	38°	37°	36°	35°	34°	33°	32°	31°	30°	29°	28°	27°	26°	25°	24°	23°	22°	21°	20°	Min.
0	3.1466	3.0783	3.0091	2.9389	2.8679	2.7959	2.7232	2.6496	2.5752	2.5000	2.4240	2.3473	2.2699	2.1918	2.1131	2.0337	1.9536	1.8730	1.7918	1.7101	0
2	.1488	.0806	.0114	.9413	.8702	.7984	.7256	.6520	.5777	.5025	.4266	.3499	.2725	.1944	.1157	.0363	.9563	.8757	.7945	.7128	2
4	.1511	.0829	.0137	.9436	.8726	.8008	.7280	.6545	.5802	.5050	.4291	.3525	.2751	.1971	.1183	.0390	.9590	.8784	.7972	.7155	4
6	.1534	.0852	.0160	.9460	.8750	.8032	.7305	.6570	.5826	.5075	.4317	.3550	.2777	.1997	.1210	.0416	.9617	.8811	.8000	.7183	6
8	.1556	.0874	.0183	.9483	.8774	.8056	.7329	.6594	.5851	.5100	.4342	.3576	.2803	.2023	.1236	.0443	.9643	.8838	.8027	.7210	8
10	3.1579	3.0897	3.0207	2.9507	2.8798	2.8080	2.7354	2.6619	2.5876	2.5126	2.4367	2.3602	2.2829	2.2049	2.1262	2.0469	1.9670	1.8865	1.8054	1.7237	10
12	.1601	.0920	.0230	.9530	.8821	.8104	.7378	.6644	.5901	.5151	.4393	.3627	.2855	.2075	.1289	.0496	.9697	.8892	.8081	.7265	12
14	.1624	.0943	.0253	.9554	.8845	.8128	.7402	.6668	.5926	.5176	.4418	.3653	.2881	.2101	.1315	.0522	.9724	.8919	.8108	.7292	14
16	.1646	.0966	.0276	.9577	.8869	.8152	.7427	.6693	.5951	.5201	.4444	.3679	.2906	.2127	.1341	.0549	.9750	.8946	.8135	.7319	16
18	.1669	.0989	.0299	.9600	.8893	.8176	.7451	.6717	.5976	.5226	.4469	.3704	.2932	.2153	.1368	.0575	.9777	.8973	.8162	.7347	18
20	3.1691	3.1012	3.0322	2.9624	2.8916	2.8200	2.7475	2.6742	2.6001	2.5251	2.4494	2.3730	2.2958	2.2179	2.1394	2.0602	1.9804	1.8999	1.8189	1.7374	20
22	.1714	.1034	.0345	.9647	.8940	.8224	.7499	.6767	.6025	.5276	.4520	.3755	.2984	.2205	.1420	.0628	.9830	.9026	.8217	.7401	22
24	.1736	.1057	.0369	.9671	.8964	.8248	.7524	.6791	.6050	.5301	.4545	.3781	.3010	.2232	.1447	.0655	.9857	.9053	.8244	.7428	24
26	.1759	.1080	.0392	.9694	.8988	.8272	.7548	.6816	.6075	.5327	.4570	.3807	.3036	.2258	.1473	.0681	.9884	.9080	.8271	.7456	26
28	.1781	.1103	.0415	.9718	.9011	.8296	.7572	.6840	.6100	.5352	.4596	.3832	.3061	.2284	.1499	.0708	.9911	.9107	.8298	.7483	28
30	3.1804	3.1125	3.0438	2.9741	2.9035	2.8320	2.7597	2.6865	2.6125	2.5377	2.4621	2.3858	2.3087	2.2310	2.1525	2.0734	1.9937	1.9134	1.8325	1.7510	30
32	.1826	.1148	.0461	.9764	.9059	.8344	.7621	.6889	.6149	.5402	.4646	.3883	.3113	.2336	.1552	.0761	.9964	.9161	.8352	.7537	32
34	.1849	.1171	.0484	.9788	.9082	.8368	.7645	.6914	.6174	.5427	.4672	.3909	.3139	.2362	.1578	.0787	.9991	.9188	.8379	.7565	34
36	.1871	.1194	.0507	.9811	.9106	.8392	.7669	.6938	.6199	.5452	.4697	.3934	.3165	.2388	.1604	.0814	2.0017	.9215	.8406	.7592	36
38	.1893	.1216	.0530	.9834	.9130	.8416	.7694	.6963	.6224	.5477	.4722	.3960	.3190	.2414	.1630	.0840	.0044	.9241	.8433	.7619	38
40	3.1916	3.1239	3.0553	2.9858	2.9153	2.8440	2.7718	2.6987	2.6249	2.5502	2.4747	2.3985	2.3216	2.2440	2.1656	2.0867	2.0070	1.9268	1.8460	1.7646	40
42	.1938	.1262	.0576	.9881	.9177	.8464	.7742	.7012	.6273	.5527	.4773	.4011	.3242	.2466	.1683	.0893	.0097	.9295	.8487	.7673	42
44	.1961	.1285	.0599	.9904	.9200	.8488	.7766	.7036	.6298	.5552	.4798	.4036	.3268	.2492	.1709	.0920	.0124	.9322	.8514	.7701	44
46	.1983	.1307	.0622	.9928	.9224	.8512	.7790	.7061	.6323	.5577	.4823	.4062	.3293	.2518	.1735	.0946	.0150	.9349	.8541	.7728	46
48	.2005	.1330	.0645	.9951	.9248	.8535	.7815	.7085	.6348	.5602	.4848	.4087	.3319	.2544	.1761	.0972	.0177	.9376	.8568	.7755	48
50	3.2028	3.1353	3.0668	2.9974	2.9271	2.8559	2.7839	2.7110	2.6372	2.5627	2.4874	2.4113	2.3345	2.2570	2.1787	2.0999	2.0204	1.9402	1.8595	1.7782	50
52	.2050	.1375	.0691	.9997	.9295	.8583	.7863	.7134	.6397	.5652	.4899	.4138	.3371	.2596	.1814	.1025	.0230	.9429	.8622	.7809	52
54	.2072	.1398	.0714	3.0021	.9318	.8607	.7887	.7158	.6422	.5677	.4924	.4164	.3396	.2621	.1840	.1052	.0257	.9456	.8649	.7837	54
56	.2095	.1421	.0737	3.0044	.9342	.8631	.7911	.7183	.6446	.5702	.4949	.4189	.3422	.2647	.1866	.1078	.0283	.9483	.8676	.7864	56
58	.2117	.1443	.0760	3.0067	.9365	.8655	.7935	.7207	.6471	.5727	.4975	.4215	.3448	.2673	.1892	.1104	.0310	.9510	.8703	.7891	58
60	3.2139	3.1466	3.0783	3.0091	2.9389	2.8679	2.7959	2.7232	2.6496	2.5752	2.5000	2.4240	2.3473	2.2699	2.1918	2.1131	2.0337	1.9536	1.8730	1.7918	60

Courtesy Brown & Sharpe Mfg. Co.

573

TABLE A-3 (cont.)

Sine Bar Constants (5" Bar)

(Multiply Constants by Two for a 10" Sine Bar)

Min.	59°	58°	57°	56°	55°	54°	53°	52°	51°	50°	49°	48°	47°	46°	45°	44°	43°	42°	41°	40°	Min.
0	4.2858	4.2402	4.1933	4.1452	4.0957	4.0451	3.9932	3.9400	3.8857	3.8302	3.7735	3.7157	3.6567	3.5967	3.5355	3.4733	3.4100	3.3456	3.2803	3.2139	0
2	.2873	.2418	.1949	.1468	.0974	.0468	.9949	.9418	.8875	.8321	.7754	.7176	.6587	.5987	.5376	.4754	.4121	.3478	.2825	.2161	2
4	.2888	.2433	.1965	.1484	.0991	.0485	.9967	.9436	.8894	.8339	.7773	.7196	.6607	.6007	.5396	.4774	.4142	.3499	.2847	.2184	4
6	.2903	.2448	.1981	.1500	.1007	.0502	.9984	.9454	.8912	.8358	.7792	.7215	.6627	.6027	.5417	.4795	.4163	.3521	.2869	.2206	6
8	.2918	.2464	.1997	.1517	.1024	.0519	4.0001	.9472	.8930	.8377	.7811	.7235	.6647	.6047	.5437	.4816	.4185	.3543	.2890	.2228	8
10	4.2933	4.2479	4.2012	4.1533	4.1041	4.0536	4.0019	3.9490	3.8948	3.8395	3.7830	3.7254	3.6666	3.6068	3.5458	3.4837	3.4206	3.3564	3.2912	3.2250	10
12	.2948	.2494	.2028	.1549	.1057	.0553	.0036	.9508	.8967	.8414	.7850	.7274	.6686	.6088	.5478	.4858	.4227	.3586	.2934	.2273	12
14	.2963	.2510	.2044	.1565	.1074	.0570	.0054	.9525	.8985	.8433	.7869	.7293	.6706	.6108	.5499	.4879	.4248	.3607	.2956	.2295	14
16	.2978	.2525	.2060	.1581	.1090	.0587	.0071	.9543	.9003	.8451	.7887	.7312	.6726	.6128	.5519	.4900	.4269	.3629	.2978	.2317	16
18	.2992	.2540	.2075	.1597	.1107	.0604	.0089	.9561	.9021	.8470	.7906	.7332	.6745	.6148	.5540	.4921	.4291	.3650	.3000	.2339	18
20	4.3007	4.2556	4.2091	4.1614	4.1124	4.0621	4.0106	3.9579	3.9039	3.8488	3.7925	3.7351	3.6765	3.6168	3.5560	3.4941	3.4312	3.3672	3.3022	3.2361	20
22	.3022	.2571	.2107	.1630	.1140	.0638	.0123	.9596	.9058	.8507	.7944	.7370	.6785	.6188	.5581	.4962	.4333	.3693	.3044	.2384	22
24	.3037	.2586	.2122	.1646	.1157	.0655	.0141	.9614	.9076	.8525	.7963	.7390	.6805	.6208	.5601	.4983	.4354	.3715	.3065	.2406	24
26	.3052	.2601	.2138	.1662	.1173	.0672	.0158	.9632	.9094	.8544	.7982	.7409	.6824	.6228	.5621	.5004	.4375	.3736	.3087	.2428	26
28	.3066	.2617	.2154	.1678	.1190	.0689	.0175	.9650	.9112	.8562	.8001	.7428	.6844	.6248	.5642	.5024	.4396	.3758	.3109	.2450	28
30	4.3081	4.2632	4.2169	4.1694	4.1206	4.0706	4.0193	3.9667	3.9130	3.8581	3.8020	3.7448	3.6864	3.6268	3.5662	3.5045	3.4417	3.3779	3.3131	3.2472	30
32	.3096	.2647	.2185	.1710	.1223	.0722	.0210	.9685	.9148	.8599	.8039	.7467	.6883	.6288	.5683	.5066	.4439	.3801	.3153	.2494	32
34	.3111	.2662	.2201	.1726	.1239	.0739	.0227	.9703	.9166	.8618	.8058	.7486	.6903	.6308	.5703	.5087	.4460	.3822	.3174	.2516	34
36	.3125	.2677	.2216	.1742	.1255	.0756	.0244	.9720	.9184	.8636	.8077	.7505	.6923	.6328	.5723	.5107	.4481	.3844	.3196	.2538	36
38	.3140	.2692	.2232	.1758	.1272	.0773	.0262	.9738	.9202	.8655	.8096	.7525	.6942	.6348	.5744	.5128	.4502	.3865	.3218	.2561	38
40	4.3155	4.2708	4.2247	4.1774	4.1288	4.0790	4.0279	3.9756	3.9221	3.8673	3.8114	3.7544	3.6962	3.6368	3.5764	3.5149	3.4523	3.3886	3.3240	3.2583	40
42	.3170	.2723	.2263	.1790	.1305	.0807	.0296	.9773	.9239	.8692	.8133	.7563	.6981	.6388	.5784	.5169	.4544	.3908	.3261	.2605	42
44	.3184	.2738	.2278	.1806	.1321	.0823	.0313	.9791	.9257	.8710	.8152	.7582	.7001	.6408	.5805	.5190	.4565	.3929	.3283	.2627	44
46	.3199	.2753	.2294	.1822	.1337	.0840	.0331	.9809	.9275	.8729	.8171	.7601	.7020	.6428	.5825	.5211	.4586	.3950	.3305	.2649	46
48	.3213	.2768	.2309	.1838	.1354	.0857	.0348	.9826	.9293	.8747	.8190	.7620	.7040	.6448	.5845	.5231	.4607	.3972	.3326	.2671	48
50	4.3228	4.2783	4.2325	4.1854	4.1370	4.0874	4.0365	3.9844	3.9311	3.8765	3.8208	3.7640	3.7060	3.6468	3.5866	3.5252	3.4628	3.3993	3.3348	3.2693	50
52	.3243	.2798	.2340	.1870	.1386	.0891	.0382	.9861	.9329	.8784	.8227	.7659	.7079	.6488	.5886	.5273	.4649	.4014	.3370	.2715	52
54	.3257	.2813	.2356	.1886	.1403	.0907	.0399	.9879	.9347	.8802	.8246	.7678	.7099	.6508	.5906	.5293	.4670	.4036	.3391	.2737	54
56	.3272	.2828	.2371	.1902	.1419	.0924	.0416	.9896	.9364	.8820	.8265	.7697	.7118	.6528	.5926	.5314	.4691	.4057	.3413	.2759	56
58	.3286	.2843	.2387	.1917	.1435	.0941	.0433	.9914	.9382	.8839	.8283	.7716	.7138	.6548	.5947	.5335	.4712	.4078	.3435	.2781	58
60	4.3301	4.2858	4.2402	4.1933	4.1452	4.0957	4.0451	3.9932	3.9400	3.8857	3.8302	3.7735	3.7157	3.6567	3.5967	3.5355	3.4733	3.4100	3.3456	3.2803	60

Courtesy Brown & Sharpe Mfg. Co.

Index